T0297661

TinyML for Edge Intelligence in IoT and LPWAN Networks

TinyML for Edge Intelligence in IoT and LPWAN Networks

Edited by

Bharat S. Chaudhari

Sheetal N. Ghorpade

Marco Zennaro

Rytis Paškauskas

Academic Press is an imprint of Elsevier
125 London Wall, London EC2Y 5AS, United Kingdom
525 B Street, Suite 1650, San Diego, CA 92101, United States
50 Hampshire Street, 5th Floor, Cambridge, MA 02139, United States

ISBN: 978-0-443-22202-3

For information on all Academic Press publications
visit our website at https://www.elsevier.com/books-and-journals

Publisher: Mara Conner
Acquisitions Editor: Tim Pitts
Editorial Project Manager: Namrata Lama
Production Project Manager: Anitha Sivaraj
Cover Designer: Matthew Limbert

Typeset by VTeX

Contents

List of contributors

Zeinab E. Ahmed
Department of Electrical and Computer Engineering, International Islamic University Malaysia, Kuala Lumpur, Malaysia
Department of Computer Engineering, University of Gezira, Madani, Sudan

Mounir Arioua
Laboratory of Innovative Systems Engineering (ISI), National School of Applied Sciences of Tétouan (ENSATe), Abdelmalek Essaadi University, Tétouan, Morocco

Anjali Askhedkar
Department of Electrical and Electronics Engineering, Dr. Vishwanath Karad MIT World Peace University, Pune, India

Diego Avellaneda
Department of Electronics Engineering, Pontificia Universidad Javeriana, Bogotá, Colombia

Cevat Balaban
Özdilek Ev Tekstil San. ve Tic. AŞ, Özveri Ar-Ge Merkezi, Bursa, Turkey

Aparna Bannore
SIES Graduate School of Technology, Navi Mumbai, India

Mamta Bhamare
Department of Computer Engineering and Technology, Dr. Vishwanath Karad MIT World Peace University, Pune, India

Sarika Bobde
Department of Computer Engineering and Technology, Dr. Vishwanath Karad MIT World Peace University, Pune, India

Irene Bru-Santa
Department of Information and Communication Engineering, University of Murcia, Murcia, Spain

Bharat S. Chaudhari
Department of Electrical and Electronics Engineering, Dr. Vishwanath Karad MIT World Peace University, Pune, India

Sachin Chougule

Department of Electrical and Electronics Engineering, Dr. Vishwanath Karad MIT World Peace University, Pune, India

Rubiscape Private Limited, Pune, India

Daniel Crovo

Department of Electronics Engineering, Pontificia Universidad Javeriana, Bogotá, Colombia

Ahmed El Oualkadi

Laboratory of Innovative Systems Engineering (ISI), National School of Applied Sciences of Tétouan (ENSATe), Abdelmalek Essaadi University, Tétouan, Morocco

Evangelia Fragkou

Department of Electrical and Computer Engineering, Volos, Greece

Jorge Gallego-Madrid

Department of Information and Communication Engineering, University of Murcia, Murcia, Spain

Sheetal N. Ghorpade

Rubiscape Private Limited, Pune, India

Antonio Guerrero Gonzales

Department of Automation, Electrical Engineering and Electronic Technology, Universidad Politécnica de Cartagena, Cartagena, Spain

Aisha A. Hashim

Department of Electrical and Computer Engineering, International Islamic University Malaysia, Kuala Lumpur, Malaysia

Savitri Jadhav

Department of Electrical and Electronics Engineering, Dr. Vishwanath Karad MIT World Peace University, Pune, India

Varsha Jujare

Department of Computer Engineering and Technology, Dr. Vishwanath Karad MIT World Peace University, Pune, India

Önder Karademir

Özdilek Ev Tekstil San. ve Tic. AŞ, Özveri Ar-Ge Merkezi, Bursa, Turkey

Aristeidis Karras

Computer Engineering and Informatics Department, University of Patras, Patras, Greece

Christos Karras
Computer Engineering and Informatics Department, University of Patras, Patras, Greece

Dimitrios Katsaros
Department of Electrical and Computer Engineering, Volos, Greece

Nadir Kocakır
Özdilek Ev Tekstil San. ve Tic. AŞ, Özveri Ar-Ge Merkezi, Bursa, Turkey

Pradnya V. Kulkarni
Department of Computer Engineering and Technology, Dr. Vishwanath Karad MIT World Peace University, Pune, India

Vrushali Kulkarni
Department of Computer Engineering and Technology, Dr. Vishwanath Karad MIT World Peace University, Pune, India

Diego Méndez
Department of Electronics Engineering, Pontificia Universidad Javeriana, Bogotá, Colombia

Hüseyin Özkan
Özdilek Ev Tekstil San. ve Tic. AŞ, Özveri Ar-Ge Merkezi, Bursa, Turkey

Rytis Paškauskas
Science, Technology and Innovation Unit, Abdus Salam International Centre for Theoretical Physics (ICTP), Trieste, Italy

Ruhi Patankar
Department of Computer Engineering and Technology, Dr. Vishwanath Karad MIT World Peace University, Pune, India

Rachana Yogesh Patil
Pimpri Chinchwad College of Engineering, Pune, India

Yogesh H. Patil
D. Y. Patil College of Engineering, Pune, India

Musa Peker
Özdilek Ev Tekstil San. ve Tic. AŞ, Özveri Ar-Ge Merkezi, Bursa, Turkey

Nilam Pradhan
Department of Electrical and Electronics Engineering, Dr. Vishwanath Karad MIT World Peace University, Pune, India

Rashmi Rane
Department of Computer Engineering and Technology, Dr. Vishwanath Karad MIT World Peace University, Pune, India

Mamoon M. Saeed
Department of Communications and Electronics Engineering, Faculty of Engineering, University of Modern Sciences (UMS), Sana'a, Yemen

Rashid A. Saeed
Department of Computer Engineering, College of Computers and Information Technology, Taif University, Taif, Saudi Arabia

Ramon Sanchez-Iborra
Department of Information and Communication Engineering, University of Murcia, Murcia, Spain

Antonio Skarmeta
Department of Information and Communication Engineering, University of Murcia, Murcia, Spain

Melek Turan
Özdilek Ev Tekstil San. ve Tic. AŞ, Özveri Ar-Ge Merkezi, Bursa, Turkey

Yassine Yazid
Department of Automation, Electrical Engineering and Electronic Technology, Universidad Politécnica de Cartagena, Cartagena, Spain
Laboratory of Innovative Systems Engineering (ISI), National School of Applied Sciences of Tétouan (ENSATe), Abdelmalek Essaadi University, Tétouan, Morocco

Mohamed Zbairi
Laboratory of Innovative Systems Engineering (ISI), National School of Applied Sciences of Tétouan (ENSATe), Abdelmalek Essaadi University, Tétouan, Morocco

Marco Zennaro
Science, Technology and Innovation Unit, Abdus Salam International Centre for Theoretical Physics (ICTP), Trieste, Italy

Preface

In the last two decades, the Internet of Things (IoT) has revolutionized society and businesses through its innovative applications. The challenges of wide-area IoT applications, such as energy efficiency, scalability, and coverage, have driven the development of Low-Power Wide-Area Networks (LPWAN). These networks are composed of resource-constrained edge nodes and offer low power consumption, extended battery life, scalability, and cost-effectiveness, making them promising candidates for IoT applications. Recent advancements in artificial intelligence, machine learning, and computing have transformed diverse fields, but IoT edge nodes face challenges with cloud-based processing due to connectivity requirements and network latency. Fog and edge computing offers a promising solution, with processing getting closer to where data is produced. Edge nodes include embedded systems based on microcontrollers with limited computing and storage capabilities, small batteries, and other constraints and cannot run conventional machine learning models. The new paradigm of TinyML enables the deployment of machine learning models on embedded system nodes, reducing latency, improving data privacy, and minimizing energy consumption with minimal connectivity requirements. This book is intended to provide a one-stop solution for the study of emerging TinyML for IoT and LPWAN applications. Primarily, the book focuses on the foundations of TinyML, embedded systems, cloud and edge intelligence, algorithms, software and hardware, energy efficiency, design aspects, applications, and future directions. The book is organized into 19 chapters:

Chapter 1, TinyML for Low-Power Internet of Things, presents a general introduction to IoT/LPWANs applications, edge AI/TinyML, and challenges.
Chapter 2, Embedded Systems for Low-Power Applications, focuses on key considerations and requirements, components selection, and hardware platforms.
Chapter 3, Cloud and Edge Intelligence, presents edge intelligence paradigms, advantages, and applications, as well as architectural layers in the roadmap for edge intelligence.
Chapter 4, TinyML: Principles and Algorithms, starts with an overview of TinyML, its benefits, workflow, edge computing, traditional and evolutionary algorithms, and optimization approaches.
Chapter 5, TinyML Using Neural Networks for Resource Constraint Devices, presents the background and use of neural networks for TinyML and its challenges.
Chapter 6, Reinforcement Learning for LoRaWANs, focuses on LoRa-based LPWANs, reinforcement learning algorithms, and multiarmed bandit-based algorithms for edge applications.
Chapter 7, Software Frameworks for TinyML, describes software and libraries for edge AI, and all essential frameworks and development environments are presented.
Chapter 8, Extensive Energy Modeling for LoRaWANs, covers LoRa and LoRaWAN, energy modeling: communication, computational, and others, along with a case study.

Chapter 9, TinyML for 5G Networks presents a detailed study of the 5G network and its requirements, virtualization, multiaccess edge computing, interfaces, machine learning modeling for 5G, and TinyML for 5G.
Chapter 10, Nonstatic TinyML for ad hoc networked Devices, covers the fundamentals of TinyML, neural network model reduction, transfer learning, purely distributed federated learning, and related details.
Chapter 11, Bayesian-Driven Optimizations of TinyML for Efficient Edge Intelligence in LPWAN Networks, covers Bayesian optimization and its practical applications in TinyML and LPWANs. It also presents a comprehensive examination of Bayesian optimization-driven algorithms designed to improve the performance of LPWANs.
Chapter 12, 6TiSCH Adaptive Scheduling for Industrial Internet of Things presents studies of Industrial IoT, the role of TinyML in IIoT, time-slotted channel hopping protocol scheduling, and a detailed case study of that.
Chapter 13, Securing TinyML in a Connected World, covers the study of security aspects, including issues and challenges in TinyML Security, threat models, and solutions.
Chapter 14, TinyML Applications and Use Cases for Healthcare, presents wearable devices, TinyML for wearables, and drug discovery, advantages, and use cases.
Chapter 15, Machine Learning Techniques for Indoor Localization on Edge Devices, offers an overview of the technical challenges for indoor positioning systems and presents different techniques and mechanisms, including TinyML, to deal with these issues.
Chapter 16, Embedded Intelligence in the Internet of Things Scenarios: TinyML Meets eBPF, covers the review of the extended Berkeley packet filter and related topics. It also explores the potential of ML and eBPF in enhancing the performance of IoT devices without augmenting their computational workload in excess.
Chapter 17, A Real-Time Price Recognition System using Lightweight Deep Neural Networks on Mobile Devices, presents a study on machine learning-based techniques for image processing. It also covers data collection techniques, detection and recognition algorithms, along with experimental study.
Chapter 18, TinyML Network Applications for Smart Cities, covers IoT and ICT applications in smart cities and also explains how TinyML can play a crucial role in such and new applications.
Chapter 19, Emerging Application Use Cases and Future Directions, presents various existing applications and future applications and challenges for TinyML.

Acknowledgments

The editors would like to acknowledge the interest and help of all the persons directly and indirectly involved during the preparation of this book. The editors are sincerely thankful to all the authors for their reader-friendly contributions to new and emerging technology. Our gratitude goes to all the reviewers for sparing their time and expertise in thoroughly reviewing the book chapters and helping with quality improvement. With the generous backing of both authors and reviewers, this book was brought to fruition. We are grateful to the entire Elsevier team, especially Mr. Tim Pitts, Senior Acquisitions Editor, Ms. Namrata Lama, Editorial Project Manager, and Ms. Anitha Sivaraj, Project Manager, for their relentless dedication in producing a high-quality publication. The editors are also grateful to leadership and colleagues at their respective serving organizations: Dr. Vishwanath Karad MIT World Peace University, Pune, India, Rubiscape Private Limited, Pune, and the Abdus Salam International Centre for Theoretical Physics, Trieste, Italy, for encouraging and providing all necessary support for this project. Finally, the editors extend their gratitude to their family members and supporters for their unwavering encouragement and understanding.

Bharat S. Chaudhari
Sheetal N. Ghorpade
Marco Zennaro
Rytis Paškauskas
Editors

CHAPTER

TinyML for low-power Internet of Things

1

Bharat S. Chaudhari[a], Sheetal N. Ghorpade[b], Marco Zennaro[c], and Rytis Paškauskas[c]

[a]*Department of Electrical and Electronics Engineering, Dr. Vishwanath Karad MIT World Peace University, Pune, India*

[b]*Rubiscape Private Limited, Pune, India*

[c]*Science, Technology and Innovation Unit, Abdus Salam International Centre for Theoretical Physics (ICTP), Trieste, Italy*

1.1 Introduction

The Internet of Things (IoT) has experienced exponential growth and disruptive trends, significantly influencing various aspects of human life. IoT applications permeate nearly all sectors, either directly or indirectly, with notable manifestations in smart cities, manufacturing, energy, environmental monitoring, smart grids, home automation, logistics, resource management, healthcare, education, and other settings where assets are interconnected via wired or wireless internet connections. Given the broad spectrum of applications and resource limitations, the characteristics and requirements of IoT vary across different applications. These applications may necessitate singular or heterogeneous technologies, combinations of proximity, personal, local neighborhood, or wide area network (WAN) technologies. In the context of wide area IoT applications, numerous challenges related to coverage, capacity, energy efficiency, cost, and others may arise. To address these challenges in WAN contexts, a new paradigm of low-power WANs (LPWANs), primarily focusing on long-range and low-power applications, has been developed. LPWANs are emerging as promising contenders in IoT applications due to their extremely low power consumption, extended battery life, low bandwidth requirements that support most IoT applications, large coverage area, scalability, license-free operation in some instances, and cost effectiveness. While several technologies fall under the umbrella of LPWANs, technology solutions such as LoRaWAN, NB-IoT, LTE-M, Sigfox, Ingenu, and others are gaining prominence and market acceptance.

Recent advancements in AI, machine learning, data science, and computing have revolutionized applications across various fields, opening up immense opportunities and presenting new challenges. Particularly in the realm of IoT, cloud-based data processing has been the norm for a considerable period. However, network-related issues such as latency, bandwidth availability, data privacy, and security pose significant concerns for cloud-based processing. Conversely, the resource constraints of IoT

TinyML for Edge Intelligence in IoT and LPWAN Networks. https://doi.org/10.1016/B978-0-44-322202-3.00006-3

nodes, such as lower processing speed, limited memory capacities, restricted battery power, and duty cycle restrictions, act as limiting factors. The advent of fog and edge computing has ushered in a promising direction for optimizing existing applications and developing emerging applications. In fog computing, data are transmitted to the so-called "fog nodes," which are located closer to the edge nodes instead of being sent to the cloud. These fog or edge nodes, equipped with AI and machine learning models, introduce a new paradigm known as edge AI. Fog nodes can comprise clusters of computers, routers, switches, gateways, coordinating/sink nodes, etc. The implementation of fog computing-based edge AI reduces the network latency and traffic, enhances data security and privacy, improves energy consumption, and boosts performance.

As the global technological landscape shifts towards more compact, intelligent, and miniaturized electronic hardware, the deployment of machine learning models on edge nodes has become feasible. These edge nodes typically comprise embedded systems, which include microprocessors or microcontroller units (MCUs) with limited computing and storage capabilities, battery power, and other constraints [1]. Due to these limitations, traditional machine learning models often fail to meet the performance requirements of various applications. However, the advent of edge AI has enabled the deployment of machine learning models on these embedded system nodes, leading to the emergence of a new paradigm known as tiny machine learning (TinyML) [2].

TinyML facilitates the operation of optimized machine learning models on these resource-constrained nodes, providing basic data analytics capabilities in a secure and efficient manner with the available resources. Recent years have seen a dramatic improvement in the accessibility to MPU computing, resulting in a significant reduction in the cost and time required to develop and deploy machine learning models. These advancements have spurred the development of innovative applications across various fields, including image detection and processing, event detection triggered by specific conditions, sound analysis, vibration monitoring, and others. As raw or primary data are not transmitted to the fog or cloud, this results in reduced latency, network traffic, and energy consumption while enhancing data security and privacy.

The edge AI market is projected to experience substantial growth, with forecasts predicting an increase from $15.60 billion in 2022 to $107.47 billion by 2029. This represents a compound annual growth rate of 31.7%. [3]. As per Gartner's estimation, by 2025, 55% of all data analyses by deep neural networks will take place at the edge [4].

1.2 IoT/LPWANs

The majority of contemporary IoT LPWAN technologies employ star and mesh-type network architectures. Utilizing a star or star-on-star network can enhance the communication range and cell size of the system. However, gateway nodes, positioned at the hub, are tasked with forwarding messages to the cloud and vice versa. This responsibility significantly impacts the battery life of the gateway unless it is powered by mains electricity. These gateways can engage in bidirectional communication with multiple edge (end) nodes. The end nodes, typically embedded systems based on microprocessors or microcontrollers, are often remotely located and are where sensing occurs. Packets transmitted by these end nodes are received and processed by all reachable gateway nodes for subsequent transmission to the network server on the cloud. The gateways and network servers are typically interconnected via a backhaul such as cellular, WiFi, Ethernet, or satellite link. Direct communication between end nodes is not possible.

The raw data sensed are transmitted from end nodes to the cloud for processing. Following processing, the results are returned to the originating end nodes for appropriate action. This process introduces latency and increases network traffic, which in turn escalates the energy consumption and raises concerns about data security and privacy. Therefore, the deployment of small and optimized machine learning models on the end nodes and using TinyML can alleviate these issues. A framework for the LoRaWAN-based LPWAN is depicted in Fig. 1.1. TinyML models can be deployed on IoT/LoRa embedded systems' resource-constrained nodes. These nodes wirelessly transmit LoRa packets to all reachable gateways. The operation of the network necessitates at least one reachable gateway. The gateways, in turn, forward the packets to the network server. The LPWAN server is further connected to application servers via the Internet. The network server performs packet redundancy, error checks, and security checks. The network server possesses all the intelligence and filters duplicate packets from different gateways, conducts security checks, and sends acknowledgments to the gateways. If a packet is intended for a specific application server, the network server forwards the packet to that particular application server. This type of network, where all reachable gateways can send the same packet to the network server, eliminates the need for hand-off or handover. This feature is beneficial in mobile applications such as asset tracking applications where assets transition from one location to another.

Owing to the availability of a diverse range of LPWAN technologies, their utilization across various applications, the evolution of network architectures, and the scalability of deployments, LPWANs are increasingly becoming heterogeneous. Consequently, these networks may encompass multiple LPWAN technologies. The integration of these technologies presents a significant challenge, as these technologies are predicated on distinct principles and specifications.

1.3 LPWAN applications/nodes constraints and requirements

A key requirement for LPWAN nodes is the need for energy-efficient operations to extend the lifespan of the nodes [5]. These devices employ energy-saving strategies such as the adaptation of extremely low-transmission duty cycles and the limitation of data rates. They possess a lightweight MAC, rendering LPWANs an appealing

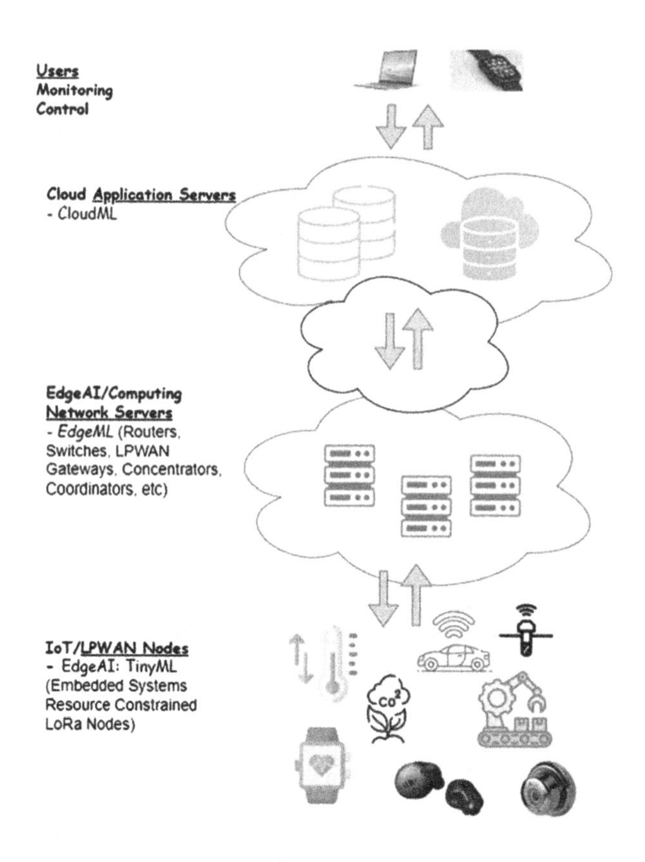

FIGURE 1.1

TinyML framework for LoRaWAN-based LPWAN.

choice for low-data-rate IoT applications across a broad range. To further enhance energy efficiency, reduce latency, and bolster security at a low cost, TinyML emerges as a promising approach. At the network's edge, diminutive nodes based on microprocessors or microcontrollers can execute rudimentary tasks without necessitating a connection to the cloud, thereby circumventing issues related to latency, cost, and security. With a forward-looking approach, TinyML holds significant potential in the era of the Internet of Everything (IoE), AR/VR/XR, where low-cost computation will be indispensable for achieving the desired performance. As the adoption of low-power IoT/LPWANs grows, innovative applications are being integrated, some of which are enumerated in Table 1.1. Many of these applications necessitate low latency, low power consumption, low cost, data security, and privacy.

Various IoT/LPWAN applications have different requirements, such as coverage, capacity, cost, and energy efficiency. However, balancing these factors can lead to trade-offs. Some applications require similar types of devices/nodes, e.g., utility metering, while others involve diverse devices with varied requirements. Certain

Table 1.1 Applications of IoT/LPWANs based on TinyML.

Field	Applications
Smart cities	Smart parking, structural health of the buildings, bridges, and historical monuments, air quality measurement, sound noise level measurement, traffic congestion and traffic light control, road toll control, smart lighting, trash collection optimization, waste management, street cleaning, utility meters, fire detection, elevator monitoring and control, manhole cover monitoring, flood management, construction equipment, labor health monitoring, environment, and public safety
Smart environment	Water quality, air pollution, temperature, forest fires, landslides, animal tracking, snow level monitoring, and early earthquake detection
Smart water	Water quality, water leakage, river flood monitoring, swimming pool management, and chemical leakage
Smart metering	Smart electricity meters, gas meters, and water flow meters, gas pipeline monitoring, and warehouse monitoring
Smart grid and energy	Network control, load balancing, remote monitoring and measurement, transformer health monitoring, and windmill/solar power installation monitoring
Security and emergencies	Perimeter access control, liquid presence detection, radiation levels, and explosive and hazardous gases
Retail	Supply chain control, intelligent shopping applications, smart shelves, and smart product management
Transport and logistics	Insurance, security and tracking, lease, rental, shared car management, detection of road surface hazards and obstacles, localization, vehicle routing, trip cost estimation, quality of shipment conditions, item location, storage incompatibility detection, fleet tracking, smart trains, and mobility-as-a-service
Industrial automation and smart manufacturing	Machine-to-machine applications, robotics, indoor air quality monitoring, temperature monitoring, production line monitoring, ozone measurement, indoor location, vehicle autodiagnosis, machine health monitoring, preventive maintenance, energy management, machine/equipment-as-a-service, and factory-as-a-service
Smart agriculture and farming	Temperature, humidity, and alkalinity measurement, wine quality enhancement, smart greenhouses, agricultural automation and robotics, meteorological station network, compost, fertilizer/spray control, hydroponics, offspring care, livestock monitoring and tracking, and measurement of toxic gas levels
Smart homes/buildings and real estate	Energy and water use, temperature, humidity, fire/smoke detection, CO/CO_2 level monitoring, flood, room occupancy (space-as-a-service), motion detection, remote control of appliances, intrusion detection systems, art and goods preservation

continued on next page

Table 1.1 *(continued)*

Field	Applications
eHealth, life sciences, and, wearables	Patient health and parameters, connected medical environments, healthcare wearables, patient surveillance, ultraviolet radiation monitoring, telemedicine, fall detection, assisted living, medical fridges, sportsmen care, tracking of chronic diseases, tracking populations of mosquitoes and other insects and their growth
Agriculture and food	Optimizing fish growth, automating fish farming processes, animal welfare, air quality monitoring in farms, water quality monitoring for irrigation, solar radiation detection, crop yield maximization, fire prevention, disease and pest prevention
Sustainability	Air quality monitoring, forest fire detection, monitoring of water quality in rivers and oceans, monitoring of water quality in protected natural areas, flood risk management

applications demand additional features, and hence, IoT/LPWAN solutions can be tailored to specific or broad sets of applications and characteristics, depending on the requirements [6] [7] [8]. Similarly, choosing the appropriate edge AI platform involves assessing multiple parameters, such as latency, processing power, energy efficiency, and physical characteristics like size and weight. For instance, applications requiring real-time processing prioritize platforms with high processing power and low latency. LPWAN and edge AI solutions can be tailored to specific or broad applications. Table 1.2 summarizes the mapping of selected applications with their requirements for LPWAN and edge AI. A mapping table assigns emphasis based on application requirements, including additional features like interoperability, categorized by relative importance: high (H), medium (M), or low (L). This aids in making architectural decisions tailored to the targeted applications.

1.4 Edge AI and TinyML

In recent years, the rapid advancements in AI and machine learning have precipitated transformative changes across society, industry, business, and everyday life. These developments have opened up vast opportunities while simultaneously presenting unprecedented challenges. Cloud AI emerges as a promising solution for complex problems that necessitate high computational power and can tolerate delays. Raw data from end or edge nodes is transmitted to cloud servers to execute machine learning models, and the results are subsequently relayed back to the end nodes for further action. However, this process introduces latency, consumes a relatively large bandwidth for uploads, increases power consumption during data transmission and reception, and raises concerns about data security and privacy.

With billions of IoT and LPWAN nodes already deployed and the world transitioning towards IoE, these nodes, typically based on microcontrollers, possess ex-

Table 1.2 Mapping applications with their requirements for LPWAN and edge AI.

Applications	LPWAN requirements					Edge AI requirements			
	Coverage	Capacity	Cost	Low power	Additional specific	Latency	Data processing power	Low power	Size and weight
Smart cities	H	H	H	M	H	H	M	M	H
Smart environment	M	H	H	H	M	M	M	H	M
Smart water	H	M	M	M	L	M	L	M	M
Smart metering	H	H	H	M	L	M	L	M	M
Smart grid and energy	H	H	M	M	M	H	H	M	M
Security and emergencies	H	L	M	H	H	H	H	H	H
Retail	H	H	H	L	M	M	M	L	H
Automotives and logistics	H	H	M	L	H	H	H	L	H
Industrial automation and smart manufacturing	L	H	H	L	L	H	H	L	M
Smart agriculture and farming, food	H	H	M	H	L	M	M	H	H
Smart homes/building and real estate	H	M	L	L	L	H	H	L	M
eHealth, life sciences, and wearables	H	H	M	H	H	H	H	H	H
Sustainable applications	H	H	H	H	H	H	M	H	H

tremely limited computational power and storage capacity and operate on battery power. Their primary function is to sense measurable parameters and perform basic onboard processing. These nodes may struggle to withstand the long delays and high energy consumption associated with cloud-based AI processing.

The emergence of edge AI has addressed these issues, revolutionizing AI deployment for resource-constrained nodes to facilitate real-time processing. By processing data on-device, edge AI enables real-time decision-making, reduces latency, enhances data security and privacy, and decreases reliance on network and bandwidth availability. The major advantages of edge AI and TinyML include reduced latency, real-time data processing at edge nodes, low bandwidth requirements, enhanced data and network security and privacy, low power consumption, low cost, on-device machine learning, and improved efficiency and flexibility.

In the near future, many applications are expected to gradually transition towards the edge, enhancing the AI landscape by bringing computation closer to data sources. However, due to resource constraints, edge AI/TinyML presents several challenges, some of which are discussed below.

1.5 Challenges for edge AI and TinyML

Edge AI revolutionizes modern computing by enabling local data processing on devices and reducing reliance on fog nodes and centralized servers. However, currently, the implementation of edge AI faces some challenges that affect its efficiency, performance, and overall success [1] [9–11]. Some of the critical challenges are discussed in this section.

- **Energy-efficient operations**

Energy-efficient operations necessitate hardware and software platforms of low complexity and optimized supporting resources. To prolong the lifespan of a node, it is crucial to utilize multiple low-power modes, such as sleep mode, particularly in LPWAN applications characterized by infrequent data transmission at ultralow speeds. Nodes should adhere to an extremely low-duty cycle to maximize time spent in sleep mode, with high-power components like data transceivers deactivated during idle periods, thereby minimizing energy consumption. Transceivers are activated solely for data transmission or reception, while access stations can either wait for sleep mode to conclude or transmit wake-up signals for data transfer. The machine learning algorithms and models intended for deployment on the nodes should be simple and optimized.

- **Scalability**

Edge AI encounters scalability issues when there is a significant increase in the number of nodes, impacting performance, reliability, and flexibility. Scalability involves managing an increased number of nodes and data without compromising efficiency. The available computational power of the nodes must be sufficient to process more data without exceeding the capabilities of the device. Data scalability deals with growing volumes of data hindered by limited transfer capacity and unreliable connectivity. System scalability efficiently expands edge devices and users, complicated by the needs of distributed processing. Techniques such as load balancing and parallel processing optimize scalability, while solutions such as edge orchestration and edge-to-cloud coordination enhance performance.

- **Limited memory**

Edge nodes, typically embedded system nodes, possess limited onboard memory and storage capabilities. In scenarios of scalability and large datasets or enhanced machine learning models, maintaining node performance may pose a challenge. The node may encounter traditional benchmarked databases, compromising performance. Therefore, benchmarking databases for edge AI/TinyML is a critical requirement.

- **Latency**

Although in edge AI the data are not sent to the fog or cloud every time for processing, it incorporates delay in the process from the sensor to electronic processing to the final action. The latency depends on the hardware configuration, sensor response time, available memory size, ML model complexity, etc. Techniques like edge caching and federated learning mitigate these challenges. Edge caching stores frequently accessed data near the network edge, enhancing real-time responsiveness. Federated learning enables distributed model training without compromising data privacy, enhancing prediction accuracy by leveraging diverse data sources.

- **Machine learning models**

The embedded system nodes face challenges in deploying and running complex machine learning models because of a lack of dedicated machine learning hardware. Moreover, ensuring the portability and compression of machine learning programs is complex, as they are developed using high-level languages [12].

- **Cost effectiveness**

With the exponential growth of the IoT and LPWANs, the sensor nodes at the edge experience massive growth. The high cost of the nodes can hinder the adoption and effectiveness of edge AI systems. The node hardware costs typically encompass specialized edge microcontroller/microprocessor-based hardware, sensors, and actuators. The software costs involve developing, deploying, and maintaining applications and algorithms for edge AI, requiring specialized skills and facing constraints due to limited processing power and storage capacity. Additionally, data storage, network connectivity, and computing costs pose significant concerns, with the need for robust infrastructure and energy-efficient architectures to optimize operations. Data privacy and security are critical, necessitating protection against breaches, while specialized network infrastructure adds to infrastructure costs. Mitigation strategies include leveraging existing networks, edge-to-cloud coordination, and open-source software adoption.

- **Network/data management and integration**

Edge AI/TinyML progress can be constrained due to improper network and data management. Computing at the edge faces challenges related to data movement, impacting power, efficiency, latency, and real-time decision-making. Volume and velocity of data influence data movement, with implications for security, power usage, computation, and real-time decision-making. Solutions like federated learning enhance data quality by training AI models on distributed edge data. Limited storage capacity in edge devices poses another data management challenge addressed through compression techniques. Secondly, integrating edge AI with other systems poses challenges, primarily due to compatibility issues between hardware, software, and communication protocols among edge devices. Each device has unique specifications, architecture, and interfaces, complicating integration efforts. Additionally, AI models often require specific software libraries and frameworks that may not align with other systems.

- **Data security and privacy**

In the context of edge AI, it is imperative to consider security measures to ensure the integrity and consistency of data and devices. Techniques such as secure boot and hardware root of trust can be utilized to maintain device integrity. These techniques can be complemented with secure software development practices, including threat modeling and code reviews. For network data security, potential security risks can be mitigated through the implementation of data encryption and secure authentication. In the realm of edge AI/TinyML, the emphasis is placed on minimizing the transmission of raw data to the cloud, thereby bolstering system security through on-device learning.

1.6 Hardware and software considerations

A large number of edge AI platforms are available to address wide requirements and needs of applications, facilitating the deployment of machine learning models on IoT/LPWAN edge devices. These platforms facilitate the execution of algorithms closer to the data source, reducing latency and bandwidth usage. They boast unique features pivotal for advanced AI applications. Prominent platform options for edge AI include TensorFlow Lite, PyTorch Mobile, OpenVINO, NVIDIA Jetson, Edge Impulse, Caffe2, MXNet, and others, each with its distinct strengths [1]. For TinyML, the options are TensorFlow Lite or TensorFlow Lite Micro. One can select the most fitting platform by carefully assessing application requirements.

While deploying AI on edge devices, it is crucial to consider the limitations of edge devices and their compatibility with other hardware devices and software tools. For example, microcontroller-based sensor IoT nodes have several limitations in terms of computation and storage capabilities, as well as battery life, potentially impacting performance for tasks like image recognition and natural language processing. Optimizing AI models for these constraints is vital, and specialized hardware like FPGAs, ASICs, and GPUs can enhance computation at the edge. The computing power and clock speed of the hardware are vital for achieving real-time processing, which is essential in edge AI applications. Additionally, the amount and type of memory available are critical considerations, as edge AI requires sufficient memory to store and process large volumes of data rapidly. Optimizing hardware for energy-efficient operations is essential for an extended device lifetime without the need for frequent battery recharging. Employing strategies like low-power chips, hardware accelerators, and intelligent power management systems, optimized duty cycle strategies can help in achieving low power consumption in edge AI.

Software tool selection for edge AI is equally important, necessitating compatibility with various devices and platforms, including different sensors, processors, operating systems, and programming languages, for seamless integration [1][2]. When selecting software, compatibility is to be ensured by efficient utilization of hardware resources for real-time data processing and seamless integration with other components like operating systems, libraries, and frameworks. Additionally, scalability is essential to handle and manage growing data volumes, processing demands, and user requests without compromising performance. The software should also demonstrate high precision and reliability in analyzing and processing data to have accurate insights and decision-making. The software should also be able to provide interpretable results clearly and logically.

References

[1] https://www.wevolver.com/article/2023-edge-ai-technology-report.

[2] https://www.tinyml.org.

[3] https://www.fortunebusinessinsights.com/edge-ai-market-107023.

[4] https://www.gartner.com/en/newsroom/press-releases/2023-08-01-gartner-identifies-top-trends-shaping-future-of-data-science-and-machine-learning.

[5] https://info.semtech.com/abi-research-white-paper.

[6] B. Chaudhari, S. Borkar, Design considerations and network architectures for low-power wide-area networks, in: LPWAN Technologies for IoT and M2M Applications, Elsevier, 2020, pp. 15–35, https://doi.org/10.1016/B978-0-12-818880-4.00002-8.

[7] B.S. Chaudhari, M. Zennaro (Eds.), LPWAN Technologies for IoT and M2M Applications, Academic Press, An Imprint of Elsevier, 2020.

[8] N.S. Chilamkurthy, O.J. Pandey, A. Ghosh, L.R. Cenkeramaddi, H.-N. Dai, Low-power wide-area networks: a broad overview of its different aspects, IEEE Access 10 (2022) 81926–81959, https://doi.org/10.1109/ACCESS.2022.3196182.

[9] N. Schizas, A. Karras, C. Karras, S. Sioutas, TinyML for ultra-low power AI and large scale IoT deployments: a systematic review, Future Internet 14 (12) (2022) 363, https://doi.org/10.3390/fi14120363.

[10] N.N. Alajlan, D.M. Ibrahim, TinyML: enabling of inference deep learning models on ultra-low-power IoT edge devices for AI applications, Micromachines 13 (6) (2022) 851, https://doi.org/10.3390/mi13060851.

[11] M. Xu, H. Du, D. Niyato, J. Kang, Z. Xiong, S. Mao, Z. Han, A. Jamalipour, D.I. Kim, X. Shen, V.C.M. Leung, H.V. Poor, Unleashing the power of edge-cloud generative AI in mobile networks: a survey of AIGC services, IEEE Communications Surveys and Tutorials (2024) 1–1, https://doi.org/10.1109/COMST.2024.3353265.

[12] Sachin B. Chougule, Bharat S. Chaudhari, Sheetal N. Ghorpade, Marco Zennaro, Exploring computing paradigms for electric vehicles: from cloud to edge intelligence, challenges and future directions, World Electric Vehicle Journal 15 (2) (2024) 39, https://doi.org/10.3390/wevj15020039.

CHAPTER

Embedded systems for low-power applications

2

Savitri Jadhav and Bharat S. Chaudhari

Department of Electrical and Electronics Engineering, Dr. Vishwanath Karad MIT World Peace University, Pune, India

2.1 Introduction

Due to the miniaturization of embedded hardware and the tremendous growth of energy-efficient applications, low-power technology has become one of the top trends in the technology landscape. The global market of low-power efficient microprocessors and controllers is rapidly growing. Due to the advent of new software technologies, the shift in focus from cloud-to-edge AI, and resource-constrained nodes, the design or selection of embedded devices has become a challenging task. The most critical requirements at the edge are energy-efficient and low-cost operations. Such low-power technology has numerous applications in the Internet of Things (IoT), low-power wide area networks (LPWANs) in for example the medical sector, industrial control, automation, and many more fields. For edge AI, sufficient processing power and memory at nodes help fast and accurate processing of the data for real-time applications [1]. For low power consumption, hardware optimization helps battery-operated nodes live longer.

The power budget and anticipated power consumption based on the high-level embedded design assist in achieving power efficiencies. The architectural and software techniques and the lower-level circuit techniques must play a major role in creating power-efficient embedded systems. This chapter offers a framework for low-power embedded system designs. The optimized power profile makes the component selection application-specific. It covers the analysis of the application's needs and constraints, followed by the selection and integration of suitable hardware and software components for low power consumption.

Several feature-level approaches can be employed when it comes to achieving low power consumption in embedded systems. These approaches focus on optimizing specific features or components of the system to minimize power consumption. By using them, embedded system designers can significantly reduce power consumption and extend the battery life of embedded systems. It is important to consider the specific requirements and constraints of the application to determine which approaches are most appropriate and effective for the given system. Fig. 2.1 shows three important approaches used for low-power embedded systems: component selection, microcontroller unit (MCU) architecture level, and software level considerations. The

TinyML for Edge Intelligence in IoT and LPWAN Networks. https://doi.org/10.1016/B978-0-44-322202-3.00007-5

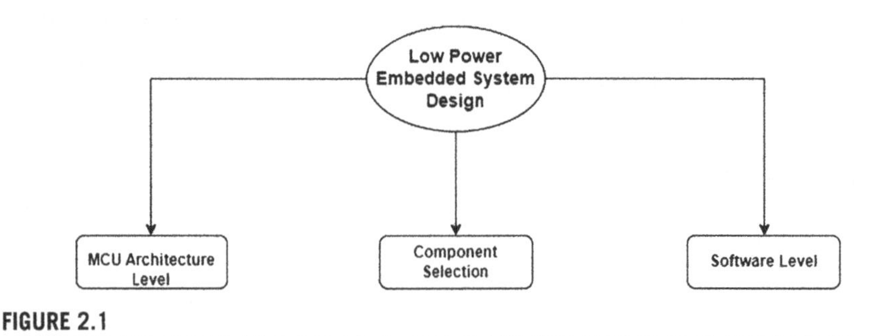

FIGURE 2.1

Different approaches to design low-power embedded systems.

typical node hardware includes an MCU and its peripherals such as analog-to-digital converters (ADCs), digital-to-analog converters (DACs), memories, modulators, demodulators, amplifiers, radio, sensors, clock circuitry, and other blocks.

2.2 Key considerations and requirements for microcontroller units

MCU architectures for low-power embedded systems vary in their specific features and capabilities [2]. Along with the advanced features, microcontrollers provide embedded systems with more power management flexibility. To reduce on-chip power consumption, there are different architecture-level approaches, viz. dynamic voltage and frequency scaling (DVFS), adaptive voltage scaling (AVS), clock gating, power gating, core pipelining, core design, cache management, etc. The architectural tradeoffs need to be considered to reserve every last microampere of electricity [3]. Table 2.1 shows the summary of these architectural approaches.

2.2.1 Dynamic voltage and frequency scaling

Dynamic voltage and frequency scaling (DVFS) is a technique used in embedded systems to adjust the operating voltage and clock frequency of the system based on its workload or processing requirements. It is a practical approach for achieving power efficiency and optimizing performance. DVFS works by dynamically adjusting the supply voltage and clock frequency of a microcontroller or processor to match the processing needs of the system. When the workload is light or the system is idle, the voltage and frequency can be lowered to reduce power consumption [4].

Conversely, during periods of high computational demand, the voltage and frequency can be increased to meet the performance requirements. They are managed independently by a power management block. The key considerations and benefits of DVFS in ultralow-power (ULP) embedded systems [5] are as follows:

Table 2.1 Architecture-level power management approaches of MCUs.

Approaches	Functional description	Outcome
Dynamic voltage and frequency scaling	Adjust the operating voltage and clock frequency of the system based on its workload or processing requirements	Achieve power efficiency and optimize performance
Adaptive voltage scaling	Adjust a supply voltage as per requirements	Optimize supply voltages while dropping energy usage
Clock gating	Disable clock signals to unused or idle peripherals and modules	Reducing dynamic power dissipation
Power gating	Cutting off the power supply to inactive components	Effectively reducing leakage currents and static power consumption
Core pipelining	Reduce the number of instructions per clock cycle	Reducing clock per instruction
Cache management	Balance between cache size, associativity, and cache policies	Minimizing cache-related power overhead
Power saving modes	Reducing the core voltage	Reduction in power

Power efficiency: DVFS allows the system to operate at the minimum voltage and frequency necessary for the given workload. By reducing the supply voltage and clock frequency during idle or low-demand periods, power consumption is significantly reduced, leading to longer battery life or reduced energy consumption.

Performance optimization: DVFS can dynamically scale up the voltage and frequency when there is a higher computational demand, ensuring the system meets performance requirements. By adapting to workload variations, DVFS enables optimal performance without sacrificing power efficiency.

Temperature management: Lowering the voltage and frequency during periods of lower activity can help in managing the system's temperature. By reducing power dissipation, DVFS can mitigate overheating issues and enhance the system's overall reliability.

System stability: DVFS techniques are designed to maintain the system's stability and ensure correct operation even when voltage and frequency levels change. Careful calibration and monitoring are required to ensure the system remains within acceptable operating limits.

Dynamic power management: DVFS can be integrated with other power management techniques, such as power gating and clock gating, to optimize power consumption further. By combining these approaches, embedded systems can achieve even greater power efficiency.

Implementing DVFS requires support from the microcontroller or processor architecture, which should include the ability to adjust voltage and frequency levels based on workload conditions dynamically. Additionally, software algorithms and monitoring mechanisms are necessary to assess the workload and make real-time adjustments to the voltage and frequency settings [6].

2.2.2 Adaptive voltage scaling and clock/power gating

Adaptive voltage scaling regulates the voltage of each component in the chip to handle the shifts in operating temperature, ensuring that each module of the chip supplies the required voltage. Clock gating disables clock signals to unused or idle peripherals and modules. It prevents unnecessary power consumption due to unused components. Power gating completely power downs the specific modules or peripheral blocks when not in use. It effectively reduces leakage currents and static power consumption.

2.2.3 Number of cores and core pipelining

A chip can have a large number of cores side by side, and the signal gets routed between the cores and memory. This approach keeps the complexity with routing logic rather than increasing the frequency of one core. Hence, a multicore system results in power saving, and furthermore, it ensures cooling by spreading power over a large piece of silicon [7]. For enhanced power-efficient operations, by activation of the mode option, the cores can slow down their clock and reduce the voltage.

The pipeline is a microarchitectural design that impacts performance and power consumption. It may contribute in both low-level and high-level performance and power optimization. Power saving solutions involving the core pipeline aim to optimize instruction execution while reducing superfluous power usage at various stages of instruction processing or executing several instructions simultaneously. These strategies boost performance while significantly increasing power consumption. The other power-saving measures may include reducing the usage of harsh instruction-level parallelism. In addition, processors frequently forecast the conclusion of branches and execute instructions on the fly. However, if the prediction is incorrect, this can require additional power [8].

2.2.4 Memory

Although the attention has been focused on MCUs, memory is typically the true power hog [9]. Flash memory is widely used for on-chip and off-chip temporary data storage. While selecting an MCU with on-chip flash, it is to be ensured that peripherals can access that memory without awakening the CPU in case of requirement. Static RAM is commonly employed in high-speed applications as cache memory, and it is quick and non-volatile and consumes little power. Asynchronous dynamic RAM is less expensive and faster. However, it must be refreshed regularly, which makes it unsuitable for use in situations where the processor would be in sleep mode most of the time. Synchronous dynamic RAM (SDRAM) is significantly faster than regular DRAM since it is synchronized with the system bus. One more option is low-power DRAM. Cache management is one of the controlling parameters for power consumption [10].

- **Cache size:** ULP microcontrollers may have smaller cache sizes and need to cautiously decide the right cache size that best meets the performance requirements of the given application with lower efficiency.
- **Cache associativity:** In low-power microcontrollers, as compared to fully associative, two-way set-associative or direct mapped caches are regularly chosen because they are simpler and have low power consumption.
- **Cache policies:** Picking suitable cache replacement and write policies is important. Furthermore, discerning and smart prefetching policies can be used to lessen needless data transfers. To reduce cache contention, separate data and instruction caches are preferable.
- **Compiler optimizations:** The compiler choice can be optimized using compiler flags/directives, loop unrolling, and loop tiling techniques, which reduce cache misses without consuming more energy.
- **Software profiling:** Profile the application to identify cache behavior and performance bottlenecks. Understanding cache usage patterns can lead to targeted optimizations that improve cache efficiency and overall power consumption.
- **Critical code placement:** Cache misses are reduced by profiling the application and placing critical code parts in cache-friendly regions.

Most cutting-edge two-level caches and prefetching approaches are developed for high-performance processors operating at high frequencies, where massive SRAM-based L1 caches work in tandem with complicated branch predictors to mask the substantial filling latency towards L2 memories, while SRAM-based caches and complicated prefetching algorithms require minor area overheads when integrated into high-performance processors. If not appropriately managed at the system level, they may become area- and power-dominant in a ULP scenario. In modern computer systems, direct memory access programmable hardware module is integrated.

2.2.5 Power down/saving modes

Higher device density has been made possible by shrinking line geometries but at the cost of increased leakage current. In low-power applications, most of the time, the MCU is in sleep mode. The leakage current may add more to overall consumption than a high active current [11].

Contemporary MCUs support as many as seven/eight power down modes. They are valued but complicated to program. During programming, it is necessary to verify the peripherals which are operating in each sleep mode, the wakeup time between each sleep mode and active mode, the peripherals in power down modes, the return at a faster rate from power down mode, and the capability of peripherals to poll and process without waking the CPU.

The MCU should be selected not only based on the availability of all the required peripherals but also on checking whether it is fast enough with low power requirements. So, it is necessary to look into the capability of all peripherals to be in each provided power saving mode. This MCU power profile helps to make better choices.

2.2.6 New trends

Almost every major MCU vendor provides various power optimization strategies for their MCUs [12]. Managing all these techniques has become so complex that enhancements are inevitable to tackle them. One of the developments is the inclusion of power management ICs to manage low-level power strategies and provide a complete power management solution. It usually contains several independent voltage regulators, such as linear low-dropout regulators, switching regulators, and controllers, and functionality such as voltage sequencing, current limiting, and surge protection. Next is dark silicon, which includes parts of the CPU that can operate most of the time in a deactivated state. RISC-V is a rapidly emerging technology. It is an open standard instruction set architecture based on RISC philosophies. Furthermore, wireless systems require frequency CPU intervention, resulting in power wastage in traditional MCU solutions. On the contrary, a dedicated and programmable data transfer manager working autonomously without depending on the MCU core can be used. A big.LITTLE contains clusters of rapid, high-power cores as well as slow, low-power cores. If a thread does not need to execute quickly, such as when waiting for input-output or memory, it can save power by executing it on the low-power core. When it can accelerate, the operating system can transfer it to a high-power core. The power wall is clearly impeding future scaling. Though only core scaling is saturated, the aid comes from a chip multiprocessor. An important question is how much more performance can be mined from the multicore path in the near future. Thus, there will again be the condition where fundamental physics and new replacements can play a vital role in the development of adaptive power saving mechanisms [13]. Incorporating energy harvesting techniques to capture and utilize energy from ambient sources such as solar, vibration, or thermal energy to power the system is another paradigm. It can supplement or even replace traditional power sources like batteries, thereby extending the system's operational time without solely relying on external power.

2.3 Component selection

For low-power embedded system design, one should consider component selection and its power dissipation rather than core-only power optimizations [14]. It is an important and practical exercise to estimate the power consumed by each component to have a clear picture of the power budget. The systematic and all-inclusive approach with all the applicable parameters leads to an effective low-power design. This section discusses the selection of components, viz. MCU, sensors, peripherals, interfaces, etc., for low-power embedded system design.

2.3.1 MCU selection

MCU selection is based on the evaluation of different MCU vendors, their product lines, and technical documentation based on specific requirements. It requires designers to traverse the numerous features claimed by vendors [15]. From a systems architecture viewpoint, it is challenging to identify which MCUs are low-power as different vendors use different metrics. To determine the low-power capabilities of MCUs, developers need to consider the MCU's supply current specifications and system-level optimizations. Each MCU vendor specifies operating conditions differently and, in some cases, a low-power value available in an unusable mode.

In addition to the specifications mentioned above, MCUs become more and more configurable and complex to reach the lowest power consumption. To cope with this, one should look at available evaluation platforms to decide the application specifications. Nonetheless, it will be difficult to configure the MCU to attain these on custom hardware. One solution is to get various code examples that can be used as a starting point. Another is graphical configuration tools to gain a profound understanding of where the total consumed power is going in MCUs. While developing low-power applications, it is helpful to know where the total consumed power is being utilized. It provides power tips and a power budget representation.

Generally, an MCU is considered ULP with an active mode in the range of 30 μA/MHz to 40 μA/MHz and a shutdown current of 50 nA to 70 nA. However, classifying a microcontroller as low-power is based on a complex combination of features, including architecture, system-on-chip (SoC) design, process technology, smart peripherals, and deep-sleep modes [19].

2.3.2 Peripheral selection

The discussion in the previous section focused on evaluating an MCU core's features aiming at low power [10]. It is important to note that on-chip and off-chip peripherals and interfaces also play an important role in the power budget of an embedded system. For power profiling of MCU peripherals, the active and sleep mode current consumptions should be compared as the application involves several peripheral transactions. The detailed discussion of these peripherals is categorized into analog peripherals and interfaces.

Although it is difficult for low-power and analog to go together, the developments in the circuits and architectures make analog peripherals appropriate for low-power applications. The selection of analog peripherals such as ADCs, DACs, etc., is considered for ULP applications.

ADC is a valuable resource in microcontrollers to read real-time signals and to digitize them for further processing. The first step towards it is signal conditioning, amplification, and filtering. First, to read the inputs of ADCs, it is advantageous to substitute Class A op-amps with Class AB op-amps. The reason is that Class AB op-amps are biased at small currents, resulting in less quiescent power dissipation. Secondly, digital filtering is suggested for very weak analog signals closer to the noise floor. After reading inputs, the sampling rate and resolution are important specifications of ADCs to be taken into account. There are many types of ADCs; the two foremost are successive approximation (SAR) and sigma-delta (or delta-sigma), and others are pipelined and flash type [16]. The SAR ADC principle of operation is based on comparisons between an input voltage and a known reference. Sigma-delta

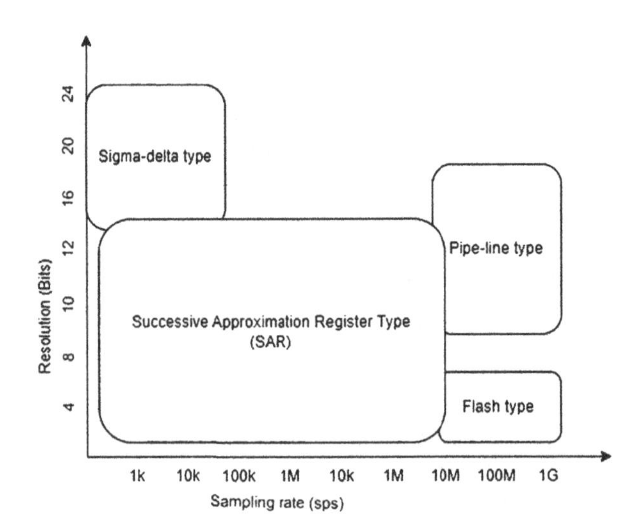

FIGURE 2.2

ADC selection.

ADCs use feedback and filtering to reduce noise and boost resolution by oversampling the input signal. As resolution and power are considered, SAR ADCs have a lower resolution and use less power than sigma-delta ADCs. The pipelined ADC is the preferred architecture for high sampling speeds, and the flash type is preferred for single-event acquisition. The newer applications demand ever-increasing ADC resolution and speed. Both manufacturers and developers are strongly apprehensive about ADC selection and its performance. Fig. 2.2 elaborates on the performance parameters of different types of ADCs and particular applications.

Different vendors provide numerous features of ADCs and flexibility for sampling rate, resolution, triggering options, periodic sampling, channels, modes of operations, and much more to minimize power consumption [17]. One way to reduce consumption is reducing the sampling rate. It also permits a decrease in the power consumption of the input amplifier, as it lowers the bandwidth necessity. Furthermore, reducing the supply voltages also diminishes power consumption. The selection can be made based on the lowest power in the power down mode rather than the lowest power in the active mode. To categorically diminish power consumption, it is essential to consider the time in each mode. Though a lower clock rate leads to lower power consumption, it is likely to reduce power consumption even more with burst-mode operation. Running the ADC at its maximum clock rate and powering down the device between conversions can help to reduce average power usage.

The DAC converts a digital signal to an analog voltage required to interface with the external world. Like ADCs, manufacturers provide several options/features aiming to achieve lower power consumption. One of the options is that the output can have a low-impedance buffer for driving external loads. Secondly, sample and hold mode can be designed in such a way that it reduces power consumption significantly.

Furthermore, DMA access allows to offload the CPU, and different triggering options allow to trigger the DAC. While operating DACs with very low-duty cycles results in extremely small power consumption. Hence, the duty cycle program should be not only flexible but autonomous too. DACs should also be allowed to be active in the low-power modes.

Interfaces

Interfaces are often dictated by the system into which a device will be connected. Different protocols are expected to be supported by the embedded devices for large application areas. They include serial interfaces such as UART, USART, I2C, SPI, USB to CAN, LIN, etc. Usually, interface protocols constantly poll the devices on the bus, resulting in drawing far more power than those that wait for an event. The slower interface consumes less power than a faster one. In the selection of buses for power application, buses that enable multiple idle states at different power levels must be chosen. Additionally, the clock scaling of the bus clock saves power. Furthermore, the DMA controller can also assist the interfaces during power saving modes. Such features help the design to be power-efficient.

One of the most recent applications of embedded systems going wireless and being portable requires integration with low-power wireless design. The important thing to consider for ULP is that less time should be spent for transmitting RF. The low-power nodes should wake up for very short time periods. During this short period, polling for data is followed by transmitting a short burst and going back to sleep mode. Different vendors have designed low-power wireless protocols which use small software stacks.

2.3.3 Sensor selection

The embedded system drives the sensor and actuator mechanism by performing different functions such as connect and activate; the MCU connects the sensor device and activates it by sending a start signal. During data collection and recording, data are collected through signal conditioning circuits and recorded in memory or displayed on the display devices. In data transmission, as per the requirements, the processed data are transmitted via data transmission networks to other devices or cloud platforms. For sensor selection, not only the sensor specification but also the application requirements should be considered. Furthermore, the sensor selection significantly impacts the hardware and software architecture of the embedded system. Therefore, in addition to the basic features of sensor selection, it is required to explore features like high precision, sensitivity, low power consumption, compactness, and easy connections.

Nowadays, higher levels of sensing requirements are emerging. Smart sensor technologies such as microelectromechanical systems (MEMS) have been effective in the physical sensing context. Additionally, with the technological progress, MEMS sensors have been powerfully moving together with communications technologies with low-power circuits. So, they enable compact and high-performance solutions with low power and low cost and bring machine learning capabilities to edge applications for a wide range of applications [18].

2.4 Embedded software tools for MCUs

Along with the hardware components of an embedded system, evaluation platforms and software tools play a critical role in overall performance. Software tools are important and required to be considered in the microcontroller selection process.

Firmware is nothing but the software on the hardware, and its development process is hardware/architecture-specific. MCU manufacturers are required to source firmware that can reconstruct the data sheet specifications so that software and hardware can work together. From a low power consumption perspective, firmware development should aim at battery life and power consumption optimization, delivering product requirements. The firmware optimization can be achieved with the following approaches:

- Keeping MCUs in power saving mode as long as possible.
- All peripherals and subsystems in the MCU are to be switched off if not in use.
- Though a simple notion, it is always beneficial to verify that everything is being shut down appropriately.
- The processor/other peripherals are put into their power down states at the appropriate time.

The application program flow should be determined while designing an embedded system. Two important types of flow can be considered, viz. polling and interrupt. The application program is usually the combination of both polling and interrupt. Essentially, for power saving the preferred flow is interrupt-driven. For example, sensor selection has a considerable impact on data sensing, timing control, and data gathering. These processes are mainly carried out by interrupt routines and a task priority mechanism. Moreover, it is also recommended for peripherals in order to avoid needless usage.

Typically, MCUs provide different options to configure port pins and it is also important to see that the GPIO configurations are made appropriately. Common options for the configuration are pins such as input or output, open-drain, push-pull, or enabling internal pull-ups/pull-downs. As discussed earlier it is always recommended to put MCU components to sleep or in low-power mode if not in use. This is true for the external components as well. Particularly, to optimize the code, it is to use minimum CPU blocks to have maximum power saving. More options include applying compression before transmission, using appropriate communication protocols, transferring tasks to the cloud, and minimizing the boottime, etc.

2.5 Important hardware platforms

The developments in IoT and computing will connect low-end embedded devices to the full hardware/software stack, including high-performance systems. Some of the important existing platforms to the power consumption problem are introduced.

2.5.1 Hardware platforms

- **Arduino**

The Arduino Nano 33 BLE Sense development board is a compact platform comprising a Cortex-M4 microprocessor and off-chip peripherals like a microphone, motion sensors, and Bluetooth Low Energy (BLE). The firmware is open source. The Nicla Sense ME is a small, low-power tool that redefines intelligent sensing technologies. The board combines four cutting-edge sensors from Bosch Sensortec with the ease of integration and scalability of the Arduino environment. With a Cortex-M7 and Cortex-M4 microcontroller, a BLE and WiFi radio, and an extension slot for attaching the Portenta vision shield, which contains a camera and two microphones, the Portenta H7 is a potent Arduino development board. Real-time tasks and high-level code are both run concurrently using Portenta H7. Two parallel-capable processors are part of the design. For instance, it is feasible to run MicroPython code alongside Arduino code and have both cores communicate with one another. Portenta has dual functionality; it can act as the central processing unit of an embedded computer or as any other embedded microcontroller board. With Portenta, it is simple to run programs made with TensorFlow Lite. For example, you could utilize one core to instantly calculate a computer vision algorithm and the other to perform low-level tasks like operating a motor or serving as a user interface [20].

- **Raspberry Pi**

The Raspberry Pi RP2040 SoC is a remarkably influential however completely low-cost microcontroller having dual Arm Cortex-M0+ processors with the most energy-efficient Arm processor available. It is designed to stretch maximum performance in the lowermost power region. It consumes significantly low power for different embedded applications. For data collection, the Raspberry Pi Pico makes it simple to connect external sensors. The Raspberry Pi 4 is also a beneficial Linux development board [21].

- **Jetson Nano**

The Jetson Nano is an embedded Linux development kit comprising a GPU accelerated processor NVIDIA Tegra for ULP applications. It does support a USB external device; for example, USB webcams work exceptionally well on the development board [22].

- **ST Microelectronics**

STM32 ULP microcontrollers provide engineers with energy-efficient embedded systems as well as applications having trade-off between performance, security, cost, and power. The range includes the STM32L0 with Cortex-M0+, the STM32L4 with Cortex-M4, and STM32L5 MCU with Cortex-M33. The STM32U5 series has the best-in-class power consumption. The ST IoT Discovery Kit is a development board with a Cortex-M4 microcontroller. It contains a microphone, MEMS motion sensors, and WiFi. The firmware for this development board is open source. ST's ULP MCU platform is established on a proprietary ultralow-leakage technology with optimized design [23].

- **Texas Instruments**

Support for the CC13xx, CC23xx, and CC26xx families of products is provided through the SimpleLink Low Power SDKs. The CC2651R3 has a 48 MHz Arm Cortex-M4 processor having support for BLE, Zigbee, and the 802.15.4 low-data-rate wireless personal area network (WPAN). It includes flash, ultralow-leakage SRAM, and cache SRAM. The MCU needs 2.9 mA in active mode, 61 μA/MHz running CoreMark, 0.8 μA in standby mode with RTC, and 0.1 μA in shutdown mode [24].

- **Analog Devices**

MAX32670 is a ULP microcontroller designed by Analog Devices with an Arm Cortex-M4. It supports industrial IoT applications with multifaceted sensor processing abilities; additionally, it has AES and CRC hardware acceleration. The chip has a low-dropout (LDO) regulator for the 1.5 V core. It offers low-power timers for PWM generation in the lowest-power sleep modes. It has support for I2C, SPIs, and UARTs [25].

- **Silicon Labs**

The Silicon Labs xG24 Dev Kit is a compact and feature-packed development platform. It is built for the EFR32MG24 Cortex-M33 microcontroller. It supports the development of prototype wireless IoT products. The platform besides has a varied variety sensor and BLE. The firmware for this development board is open source. Silicon Labs EFR32BG22 Wireless Gecko Bluetooth 5.2 SoC combines ultralow transmit and receive power (3.6 mA Tx at 0 dBm, 2.6 mA Rx) and a security-enhanced, single-core Arm Cortex-M33 CPU that draws 27 μA/MHz while active and 1.2 μA in sleep mode [26].

- **Infineon Technologies**

The CY8C4247LQQ-BL483 is a 32-bit MCU and it has an Arm Cortex-M0 core with BLE, a 12-bit, 1 MS/s SAR ADC, and a touch button interface. The supply current is 1.7 mA at 3 MHz in active mode and only 1.5 μA in sleep mode [27].

- **Microchip Technology**

The Microchip Technology ATSAML21E incorporates an ARM Cortex-M0. The chip possesses sophisticated power management technologies. It consumes as little as 35 μA/MHz in active mode and 200 nA in sleep mode. Standby current is as low as 1.5 μA. The MCU also features a backup state, powered by battery input [28].

- **NXP Semiconductors**

The LPC55S66 MCU has a 150 MHz Arm Cortex-M33, but it can manage little power consumption. The M33 core is crafted in Armv8-M architecture and has advanced security features. It has 256 MB of flash, 144 kB of SRAM, and nine flexible serial communication peripherals [29].

References

[1] 2023 Edge AI Technology Report, https://www.wevolver.com/article/2023-edge-ai-technology-report-chapter-iv-hardware-software-selection. (Accessed 20 December 2023).

[2] Pete Warden, Daniel Situnayake, TinyML: Machine Learning with TensorFlow Lite on Arduino and Ultra-Low-Power Microcontrollers, O'Reilly Media, 2019.

[3] Luca Benini, Giovanni de Micheli, System-level power optimization: techniques and tools, ACM Transactions on Design Automation of Electronic Systems (TODAES) 5 (2) (2000) 115–192.

[4] Jacob Murray, Teng Lu, Partha Pande, Behrooz Shirazi, Sustainable DVFS-enabled multicore architectures with on-chip wireless links, in: Advances in Computers, vol. 88, Elsevier, 2013, pp. 125–158.

[5] Wenheng Ma, Qiao Cheng, Yudi Gao, Lan Xu, Ningmei Yu, An ultra-low-power embedded processor with variable micro-architecture, Micromachines 12 (3) (2021) 292.

[6] David Brooks, Vivek Tiwari, Margaret Martonosi, Wattch: a framework for architectural-level power analysis and optimizations, ACM SIGARCH Computer Architecture News 28 (2) (2000) 83–94.

[7] D. Chinnery, K. Keutzer, Pipelining to reduce the power, in: Closing the Power Gap Between ASIC & Custom, Springer, Boston, MA, 2007.

[8] Ivan Ratković, Nikola Bežanić, Osman S. Ünsal, Adrian Cristal, Veljko Milutinović, An overview of architecture-level power-and energy-efficient design techniques, Advances in Computers 98 (2015) 1–57.

[9] Jie Chen, Igor Loi, Eric Flamand, Giuseppe Tagliavini, Luca Benini, Davide Rossi, Scalable hierarchical instruction cache for ultralow-power processors clusters, IEEE Transactions on Very Large Scale Integration (VLSI) Systems 31 (4) (2023) 456–469.

[10] Michele Magno, Luca Benini, Christian Spagnol, E. Popovici, Wearable low power dry surface wireless sensor node for healthcare monitoring application, in: 2013 IEEE 9th International Conference on Wireless and Mobile Computing, Networking and Communications (WiMob), IEEE, 2013, pp. 189–195.

[11] https://www.silabs.com/documents/public/white-papers/picking-the-right-microcontroller-based-on-low-power-specs.pdf.

[12] H. Han, J. Siebert, TinyML: a systematic review and synthesis of existing research, in: 2022 International Conference on Artificial Intelligence in Information and Communication (ICAIIC), Jeju Island, Republic of Korea, 2022, pp. 269–274, https://doi.org/10.1109/ICAIIC54071.2022.9722636.

[13] Maxime France-Pillois, Abdoulaye Gamatié, Gilles Sassatelli, A segmented adaptive router for near energy-proportional networks-on-chip, ACM Transactions on Embedded Computing Systems (TECS) 21 (4) (2022) 1–27.

[14] Amin Firoozshahian, Joel Coburn, Roman Levenstein, Rakesh Nattoji, Ashwin Kamath, Olivia Wu, Gurdeepak Grewal, et al., MTIA: first generation silicon targeting meta's recommendation systems, in: Proceedings of the 50th Annual International Symposium on Computer Architecture, 2023, pp. 1–13.

[15] Monica Redon, Strategies for choosing the appropriate microcontroller when developing ultra low power systems, Cortex 3 (2017) 10–920.

[16] Pieter Harpe, Low-power SAR ADCs: basic techniques and trends, IEEE Open Journal of the Solid-State Circuits Society 2 (2022) 73–81.

[17] Sergio Rapuano, Pasquale Daponte, Eulalia Balestrieri, Luca De Vito, Steven J. Tilden, Solomon Max, Jerome Blair, ADC parameters and characteristics, IEEE Instrumentation & Measurement Magazine 8 (5) (2005) 44–54.

[18] Masayuki Hirayama, Sensor selection method for IoT systems–focusing on embedded system requirements, in: MATEC Web of Conferences, vol. 59, EDP Sciences, 2016, p. 01002.

[19] Joakim Lindh, Christin Lee, Marie Hernes, Siri Johnsrud, Measuring CC13xx and CC26xx current consumption, Texas Instrument, Application Report, 2019.

[20] https://www.arduino.cc/en/hardware#boards.

[21] https://www.arm.com/blogs/blueprint/raspberry-pi-rp2040.

[22] https://developer.nvidia.com/embedded/jetson-nano-developer-kit.

[23] https://www.st.com/en/microcontrollers-microprocessors/stm32-ultra-low-power-mcus.html.

[24] https://www.ti.com/tool/SIMPLELINK-LOWPOWER-SDK.

[25] https://www.stg-maximintegrated.com/en/products/microcontrollers/MAX32670.html.

[26] https://www.silabs.com/development-tools/wireless/efr32xg24-dev-kit?tab=overview.

[27] https://www.infineon.com/cms/en/product/microcontroller/32-bit-psoc-arm-cortex-microcontroller/psoc-4-32-bit-arm-cortex-m0-mcu/cy8c4247lqq-bl483/.

[28] https://www.microchip.com/en-us/product/atsamd21g18.

[29] https://www.nxp.com/products/processors-and-microcontrollers/arm-microcontrollers/general-purpose-mcus/lpc5500-arm-cortex-m33/high-efficiency-arm-cortex-m33-based-microcontroller-family:LPC55S6x.

CHAPTER

Cloud and edge intelligence

3

Sachin Chougule[a,b], Bharat S. Chaudhari[a], Sheetal N. Ghorpade[b], and Marco Zennaro[c]

[a]*Department of Electrical and Electronics Engineering, Dr. Vishwanath Karad MIT World Peace University, Pune, India*

[b]*Rubiscape Private Limited, Pune, India*

[c]*Science, Technology and Innovation Unit, Abdus Salam International Centre for Theoretical Physics (ICTP), Trieste, Italy*

3.1 Introduction

Cloud computing has seen widespread adoption since its inception, significantly transforming the way people live. Major firms such as Google, Amazon, and Microsoft have introduced their respective cloud computing services. To put this in perspective, during the 1960s, the primary focus was on enhancing the computing power of individual devices. However, by the 1980s, the concept of distributed computing emerged as multiple "dumb terminals" were connected to a central mainframe. Over the past decade, the prevailing trend has leaned toward centralized data processing through cloud computing. The key advantages of cloud computing include seemingly limitless storage capacity, ample computing resources, reduced upfront costs, and a smaller environmental footprint. Despite cloud computing driving the era of ubiquitous computing, the centralized approach of capturing, storing, and processing data in the cloud falls short of meeting the demands of emerging requirements and workloads. The Internet of Things (IoT) has been buzzing for quite some time, but over the past decade, we have witnessed a significant increase in its adoption by businesses. This explosive growth of the IoT has pushed computing towards the edge of local networks and intelligent devices. These challenges have been further amplified by the ongoing proliferation of both mobile and fixed Internet-connected devices [1].

The issues like high latency and limited bandwidth, resulting in a reduced quality of experience (QoE) for users, prompted a reimagining of the cloud. Instead of viewing it as a uniform entity, the cloud was conceptualized as having a distinct "edge," separate from the core, where large-scale processing and storage occur [2]. Devices could communicate with local servers unless there were a specific need to connect with the core cloud infrastructure [3]. This concept challenged the growing reliance on massive data centers for hosting cloud computing [4]. Advocates argued that having multiple geographically diverse data centers would offer a superior model for

TinyML for Edge Intelligence in IoT and LPWAN Networks. https://doi.org/10.1016/B978-0-44-322202-3.00008-7

applications like email distribution. Local servers could filter out spam and block undesirable traffic closer to its source [5,6]. Soon after, the "cloudlet" concept emerged, catering to small groups of casual and transient mobile device users in places like coffee shops and restaurants [7]. Lastly, the concept of "fog computing" (FC) evolved from the growing use of fixed Internet-connected sensors, particularly in the context of the IoT. FC aims to provide swift responses to these sensors' data, effectively bridging the gap between local processing and cloud-based IoT platforms, which is essential for autonomous vehicles. Cloud computing encounters several challenges, as highlighted by existing research [1]. The surge in interconnected devices necessitates efficient data processing and robust decision-making within strict latency constraints. Despite the efficiency and speed of our networks, transporting the massive volume of information spawned by these devices to the cloud for investigation and storage is impractical. The transfer of such vast data over cloud networks introduces overheads that diminish throughput, escalate energy consumption, increase network traffic, and incur additional costs.

The cloud's complexity is exacerbated by the diverse and real-time data streams from numerous Autonomous Vehicles (AVs), significantly increasing the workload on the cloud infrastructure. In response to these challenges, edge computing has emerged as a solution to a distributed computation model deployed in nearby proximity to the data source. By deploying an edge intelligence model, the inference computing of AVs can experience substantial improvements in accuracy and latency. This shift toward edge computing addresses the limitations posed by centralized cloud processing and aligns with the demands of processing diverse, real-time data from interconnected devices in AVs. The increasing demands of autonomous driving have brought together machine learning, explicitly AI and mobile edge computing, giving rise to edge intelligence or edge AI. This convergence aims to enhance various routine activities [2–4] significantly. The core aim of edge intelligence is to orchestrate the collaboration among numerous edge devices and servers to handle data generated in close vicinity. Simultaneously, AI seeks to replicate intelligent human behavior in devices and machines by learning from data. The fusion of AI and edge intelligence is a logical progression due to the evident overlap between these two technologies, collectively referred to as edge intelligence.

The rest of the chapter is organized as follows. In Section 3.2, we discuss edge intelligence paradigms, which describe strategies for training and inference on the cloud or edge. The advantages and applications of edge intelligence are presented in Section 3.3. Edge intelligence presents more intriguing opportunities but encounters numerous challenges during its implementation. Section 3.4 elaborates on challenges and opportunities in edge intelligence. Section 3.5 presents an in-depth analysis of AI-based solutions for optimizing the computation of edge intelligence. Architectural layers in the roadmap for edge intelligence are discussed in detail in Section 3.6. Lastly, the chapter is concluded in Section 3.7. The chapter organization is shown in Fig. 3.1.

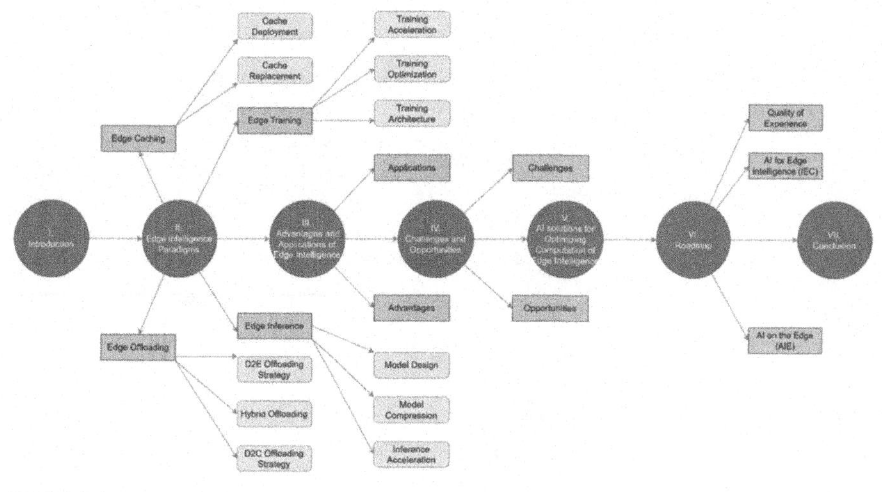

FIGURE 3.1

Organization of the chapter.

3.2 Edge intelligence paradigms

Edge intelligence is the execution of AI algorithms on edge devices using data generated on those devices and on sensor nodes [6,7]. This approach often involves high-performance AI chips but has limitations. It increases energy consumption as well as cost and is inappropriate for aged devices having restricted computation capabilities. However, it is essential to recognize that this narrow definition of edge intelligence does not fully leverage the potential of the technology. Recent studies have shown that for deep neural network (DNN) models, an amalgamation of edge and cloud computing can reduce latency and energy consumption compared to local implementations. Edge intelligence should encompass a broader concept, utilizing available data and resources across various levels, as shown in Fig. 3.2, from end nodes and edge devices to cloud data centers, for optimizing the training and inference of DNN models.

The levels of edge intelligence include:

- L0: Cloud intelligence. Complete DNN model training and interpretation in the cloud.
- L1: Cloud–edge cooperation and cloud training. Train the DNN-based model (DNNM) in the cloud, then perform inference in collaboration with the edge, partly offloading new additional data to the cloud.
- L2: In-edge cooperation and cloud training. Train the DNNM in the cloud but perform interpretation at the edge, potentially offloading data on edge devices or else on the adjacent devices.

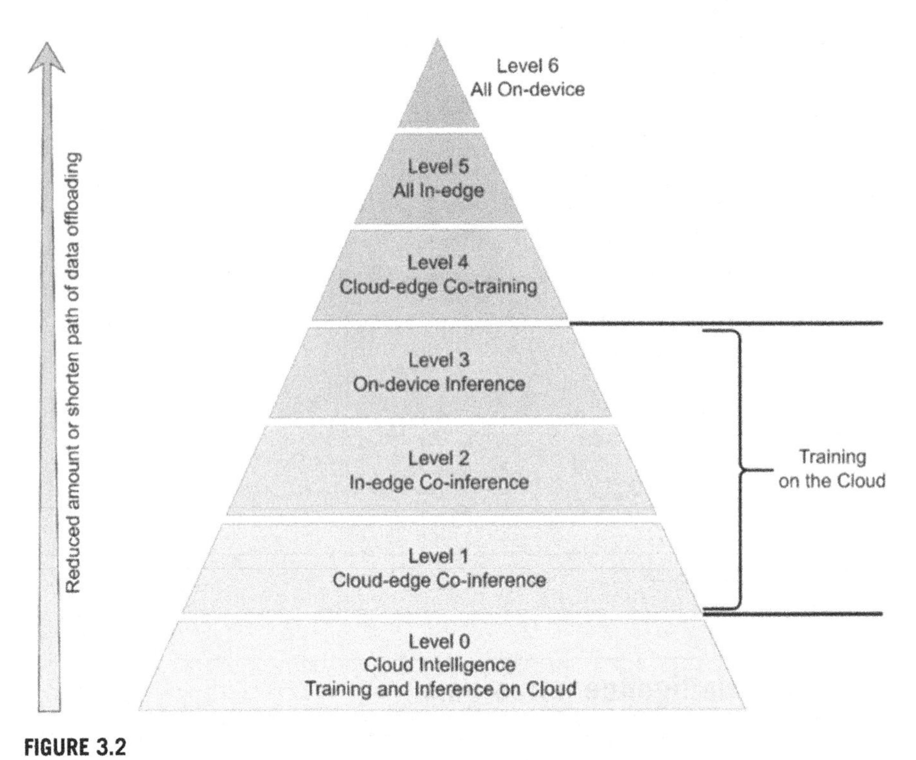

FIGURE 3.2

Six leveled ratings for edge intelligence.

- L3: On-device interpretation and cloud training. Train the DNNM in the cloud but perform on end nodes for interpretation with partial data offloading from cloud to end nodes.
- L4: Cloud–edge co-training and interpretation. Both training and interpretation of the DNNM model occur in cooperation between the cloud and edge.
- L5: All in-edge. Both training and interpretation of the DNNM take place in the edge environment.
- L6: All on-device. Both training and interpretation of the DNNM occur exclusively on the end node.

The choice of edge intelligence level depends on various factors, including latency, energy efficiency, privacy, and WAN bandwidth cost, making it application-dependent. Four crucial elements of edge intelligence are edge caching, edge training, edge interpretation, and edge offloading, and their subclasses are shown in Fig. 3.3. These elements are elaborated further in this section.

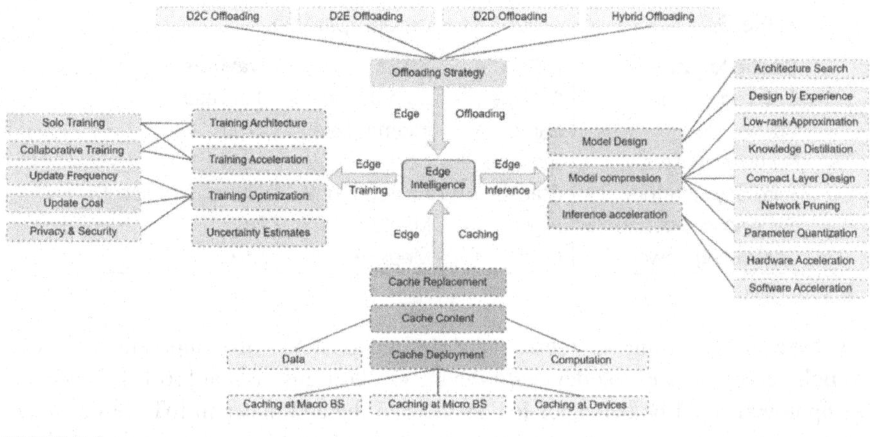

FIGURE 3.3

Elements of edge intelligence.

3.2.1 Edge caching

Edge caching involves storing information generated by edge devices, sensors, and IoT devices closer to users to enhance performance and reduce latency. This technique can reduce computational complexity and interpretation time, storing raw sensor data or previous computation results for reuse. Various caching methods have shown significant latency improvements [8–11].

3.2.1.1 Cache deployment

Cache elements are deployed at edge or end entities such as macro base stations, micro base stations, and end nodes.

- Caching at macro base stations:

Broader coverage and substantial cache size are the characteristics of macro base stations. A macro base station typically covers a radius of approximately 500 meters [12].

- Caching at micro base stations

Micro base stations refer to a group of low-energy access points with a coverage span ranging from 20 to 200 meters, including microcells, picocells, and femtocells [13]. Deploying small base stations or hotspots in strategic locations enhances the overall experience. This improvement is attributed to advantages like efficient spatial spectrum reuse, resulting in benefits such as higher end rates [14,15].

- Caching at devices

Device-level caching takes advantage of the storage capacity within end devices, which can optimize local transmission and computing redundancy. Additionally,

Table 3.1 Cache placement location comparison.

Cache location	MBSs	SBSs	Devices
Coverage range	500 meters	20–200 meters	10 meters
Cache size	Large	Intermediate	Small
Number of users	Massive	Low	Low
Topology structure	Stable	Altered slightly	Altered significantly
Redundancy capability	High	Medium	Low
Computing power	High	Medium	Low

they have the capability to retrieve the desired contents or computing outcomes through nearby devices via device-to-device (D2D) transmission [16,17]. However, this approach could be better for IoT as most of the end nodes in IoT are resource-constrained nodes having very low onboard memory and limited computational capability, and, most importantly, are battery-operated.

3.2.1.2 Cache replacement

In practical situations, the allocation of requests for cache access changes over time, and new content is continuously generated. Therefore, it is crucial to update caches periodically. This updating process is referred to as cache replacement. Various conventional strategies for cache replacement have been suggested, including first-in-first-out (FIFO), least frequently used (LFU), least recently used (LRU), and its modifications [18]. A comparison of edge cache placement locations is shown in Table 3.1.

3.2.2 Edge training

Edge training allows devices to learn patterns from cached edge data. It can occur on an edge server or device and includes independent training and collaborative training strategies. Collaborative training involves multiple devices and requires communication updates, posing challenges to data privacy and security. Various factors related to edge training are discussed in this section.

3.2.2.1 Training architecture

The training framework relies on the computational capabilities of both edge devices and servers. If a singular edge device or server possesses adequate power, it can employ the identical training structure as a centralized server, performing the training on a single device. Conversely, when the device or server lacks such capabilities, cooperation with other devices becomes imperative. This results in the emergence of two types of training frameworks: individual training, which entails executing training tasks on a lone edge device or server, and cooperative training, where limited devices and servers work together to execute training tasks.

A prevalent example of a cooperative training framework is the master–slave model, as illustrated by federated learning [19]. In federated learning, a server involves numerous devices and delegates training tasks individually. Another form of cooperative training architecture is peer-to-peer, where participants are regarded as equals in the training process.

3.2.2.2 Training acceleration

The emphasis is on expediting training at the edge, with some initiatives [20,21] exploring transfer learning to enhance training speed. Transfer learning involves utilizing features learned from previous models, resulting in a significant reduction in learning time. In a cooperative training approach, edge devices have the capability to acquire from one another, thereby enhancing overall knowledge proficiency. A framework known as Recycle ML employs cross-modal transmission to expedite the training of neural networks along mobile platforms throughout the diverse sensory system. Federated learning can also be employed to hasten model training on distributed edge devices, particularly in scenarios where labeled data are insufficient [22].

3.2.2.3 Training optimization

Training optimization involves streamlining the training procedure to attain specific goals, viz. minimizing energy consumption, enhancing precision, preserving secrecy, maintaining security, and more. The critical factors involved in optimization are communication frequency, communication cost, privacy, and security issues.

- Communication frequency

Communication frequency is a crucial aspect of federated learning, where the exchange of information among edge devices and the cloud server plays a vital role. This operation involves update uploading through edge devices to the cloud server and downloading the combined updates from the distributed model to local models. Given the potential for erratic network conditions in edge devices, it is essential to minimize update cycles.

- Communication cost

Besides the frequency of communication, another factor influencing the efficiency of communication among edge devices and the central server is the cost of communication. Minimizing communication costs has the potential to save bandwidth substantially and enhance overall communication efficiency.

- Privacy and security issues

Upon catching updates from edge devices, the central server is tasked with aggregating these updates to construct unified updates for the distributed universal model. The concern arises that malicious hackers may scrutinize the updates, posing a threat to the privacy of participating edge users. To address this, an aggregation process is implemented to combine updates from all edge devices, rendering individual updates indiscernible by the central server [23]. More precisely, every edge device transmits encrypted updates to the server. The server later combines these encrypted updates.

The counteraction of masks occurs when enough edge devices are involved. Consequently, the server gains the ability to unveil the aggregated update by unmasking it. Throughout this aggregation process, exclusive updates remain unscrutinized, and the server can solely approach the combined unmasked updates, thereby efficiently safeguarding the secrecy of participants.

The conventional deep learning approaches for classification and regression lack the ability to account for model uncertainty.

3.2.3 Edge interpretation

Edge interpretation occurs during the use of the trained model for computing output on edge devices and servers. However, many deep learning algorithms are designed for high-performance hardware and are not suitable for edge environments. Challenges include designing models for edge deployment and accelerating edge inference for real-time responses. These issues can be addressed through new model designs or model compression techniques.

3.2.3.1 Model design

The primary emphasis in model design revolves around creating neural network architectures that are lightweight and appropriate for execution on edge devices with lower hardware demands. This process involves either automated generation of the optimal architecture by machines or manual design by humans.

- Architecture search

Exploration in architecture search is a thriving research field with broad applications in the future. A recent notable advancement in this area is differentiable architecture search (DARTS) [24], which offers the potential to substantially decrease reliance on hardware. DARTS relies on the constant easing of architecture description and employs gradient descent for the process of hunting for architecture.

- Design by experience

The experience-driven design employs two distinct strategies. The initial approach involves the use of comprehensive detachable convolutions, which are utilized to construct a streamlined DNN known as MobileNets. This design specifically caters to the needs of mobile and embedded devices [25]. Another method employed is group convolution, which serves as an alternative means to diminish computation costs during the process of designing architecture.

3.2.3.2 Model compression

Model compression seeks to reduce the dimensions of a model, enhance energy efficiency, and accelerate inference on edge devices with a restricted number of resources, all without compromising precision. The five important methods to model compression are low-rank approximation, knowledge distillation, compact layer design, parameter quantization, and network pruning.

- Low-rank approximation

The fundamental concept behind low-rank approximation involves the replacement of high-dimensional kernels by low-rank convolutional kernel multiplication.

- Knowledge refinement

It relies on transfer learning, wherein a smaller neural network is trained employing distilled data by initiating a large model. The large, intricate model is known as the mentor model, whereas the more compacted model is called the student model. The student model gains advantages by assimilating knowledge from the teacher network.

- Designing condensed layers

In DNN, when weight values approach zero, computational resources are inefficiently utilized. A key strategy to address this issue involves creating a condensed layer in the neural network, excellently minimizing resource utilization such as memory and computation power. Christian and colleagues suggest addressing this by incorporating sparseness and substituting the entirely linked layers in GoogLeNet. In Residual-Net, an alternative approach is taken by replacing entirely linked layers through global regular merging to decrease resource demands. Substituting a large convolutional layer with several smaller and more compact layers can efficiently decrease the parameter count and subsequently lower computational requirements.

- Network pruning

The fundamental concept behind network pruning involves the removal of less significant parameters, recognizing that not all parameters play a crucial role in extremely accurate DNN. As a result, associations between lower weights are eliminated, transforming a heavy network into a sparser one.

- Parameter quantization

Achieving high performance in neural networks does not always require highly precise parameters, particularly when those parameters are unnecessary. Research has demonstrated that a relatively small quantity of parameters suffice for reconstructing a complete network.

3.2.3.3 Interpretation acceleration

The primary concept behind accelerating models in interpretation is to decrease the runtime of interpretation on edge devices and achieve instantaneous replies for precise applications based on neural networks, all deprived of modifying the architecture of the trained model. There are two main categories of acceleration: hardware and software acceleration. The software acceleration approach is centered on enhancing resource management, pipeline structure, and compiler optimization.

- Hardware acceleration

Methods for hardware acceleration concentrate on parallelizing inference tasks across accessible hardware, including CPU, GPU, and DSP. In recent times, the potency of mobile devices has seen a notable rise. A growing number of mobile platforms now feature GPUs. Given that mobile CPUs are less apt for DNN computations, leveraging embedded GPUs becomes a viable strategy to distribute computing tasks and expedite the inference process.

- Software acceleration

Software acceleration primarily centers on enhancing resource allocation, refining pipeline design, and optimizing compilers. Methods for software acceleration aim to optimize the utilization of limited resources to achieve faster performance, but this can sometimes result in a reduction in accuracy in specific scenarios.

3.2.4 Edge offloading

It is a distributed computation paradigm that furnishes computation services for edge caching, training, and interpretation. It allows tasks to be processed in cloud servers when edge hardware lacks capability. Four offloading strategies exist, including device-to-cloud (D2C), device-to-edge server (D2E), device-to-device (D2D), and hybrid offloading, each with its adaptiveness and resource utilization.

- D2C offloading strategy

In the D2C offloading strategy, devices transfer input data, such as audio data or images, to a cloud server. Powerful computers perform high-accuracy inference using a large neural model, and the outputs are sent backward via the identical network. However, it has some primary drawbacks. Mobile devices need to communicate large volumes of information to the cloud, creating a bottleneck in the overall process [26]. The execution is reliant on internet connectivity. The transmitted information from mobile devices might include users' personal information, such as personal photos, making it susceptible to attacks by mischievous hacks during the interpretation on the cloud server [27]. Various considerations, including energy efficiency, latency, and privacy, can guide the design of model partitioning and layer scheduling in this context.

- D2E offloading strategy

In contrast to D2C offloading, which involves transferring inferencing to a central server in the cloud, D2E offloading shifts inferencing to an Edge server. An Edge server, in this context, denotes robust servers that are physically close to mobile devices and possess greater processing power than typical edge devices.

- D2D offloading strategy

In the strategy of D2D offloading, devices like smartwatches are connected to smartphones or home gateways and have the capability to delegate model interpretation tasks to more influential connected devices. Binary decision-based offloading and partial offloading exist in this context. Binary decision offloading involves deciding upon the execution of the task locally or offloading it. On the other hand, partial offloading entails breaking down the interpretation task into various subtasks and offloading a handful of them to connected devices.

- Hybrid offloading

The hybrid computing framework efficiently leverages cloud services, edge computing, and mobile devices in a comprehensive approach. Distributed DNNs (DDNNs) derived from this holistic computing architecture represent a hybrid offloading technique that strategically allocates portions of a DNN across a distributed computing hierarchy [103]. The collective training of these segments occurs in the cloud and aims to reduce communication and resource usage on edge devices. During the interpretation phase, individual edge devices perform local computations, and the resultant outputs are combined to generate the final results.

3.3 Edge intelligence: advantages and applications

In recent times, there has been a noticeable trend towards enhanced intelligence in various aspects of life, ranging from smartwatches to automobiles and from agriculture to industrial processes and even urban environments, benefiting from the general benefits of edge intelligence, like reduced latency and bandwidth consumption. Edge intelligence can help enterprises make quicker data-driven decisions responding to the requirements of their clients. Reduced storage requirements in edge intelligence lead to improved operational cost savings for enterprises.

3.3.1 Advantages

- Enriching AI with richer data and application scenarios

Recent advancements in deep learning have been driven by four factors: algorithms, hardware, data, and application scenarios. Data play a pivotal role in enhancing AI performance. As IoT grows, vast quantities of information will be spawned at the edge, challenging cloud-based processing due to bandwidth constraints. Edge intelligence addresses this challenge by enabling low-latency data processing closer to the data source, potentially boosting AI performance. Edge intelligence and AI are each other's counterparts technically and also with regard to application and adoption [28].

- Key infrastructure for AI democratization

AI has made significant strides in digital products and services, from online shopping to self-driving cars. Major IT companies envision democratizing AI, making it accessible to everyone and everywhere. Edge intelligence is well suited to this goal and offers diverse application scenarios. Thus, edge intelligence serves as a crucial enabler for ubiquitous AI [29].

- Popularizing edge intelligence with AI applications

Edge intelligence is already bringing about significant transformations across various industries, such as manufacturing, energy, healthcare, agriculture, logistics, and transportation [30–32]. Real-time video analytics, built on computer vision, emerges as a killer application for edge intelligence due to its high computational demands, bandwidth requirements, privacy concerns, and low-latency needs [33]. Multiple benefits of edge intelligence have created a path for expanded progression in the near future [34].

3.3.2 Applications of edge intelligence

Gartner anticipates a significant surge in the adoption of edge intelligence use cases in the coming years. The projection is that by 2024, over 50% of the potential enterprises will have implemented a minimum of six edge intelligence use cases. This marks a remarkable expansion in comparison with the scenario in 2019, in which merely 1% of larger enterprises had very few edge intelligence deployments. Presently, some key applications of edge computing include the following.

3.3.2.1 Smart cities

With the rapid expansion of urban populations and the ongoing trend of urbanization, the concept of smart cities has emerged, garnering considerable attention. The fundamental idea behind smart cities involves utilizing intelligent methods to diminish energy consumption, enhance energy efficiency, alleviate traffic congestion, ensure urban and resident safety, and elevate residents' quality of life.

In the smart city framework, numerous hardware devices continually generate data. These devices encompass everyday smart tools like smartphones, smart bracelets, and portable medical devices, as well as surveillance cameras and various environmental detection sensors for urban security. AI stands out as a promising solution for smart cities to enhance the precision and effectiveness of data analysis, leveraging its capability to handle vast amounts of data.

Within densely populated urban areas, such as cities, smart cities impose more stringent demands on real-time responsiveness and network stability to guarantee the comfort and security of civilian life. However, the substantial computing tasks associated with AI training and reasoning present a significant challenge to meeting these requirements. In response to this challenge, some researchers have directed their focus toward edge intelligence.

3.3.2.2 Urban healthcare

With the rise in popularity of the IoT and cloud computing, an increasing number of personal medical devices are being integrated into daily life. These devices have the capability to gather users' physical data, which are then uploaded to a cloud server. The application of AI in analyzing these data holds significant potential for enhancing the accuracy of medical systems in terms of disease classification and diagnosis. However, the current cloud computing model faces challenges in meeting the specific requirements of telemedicine, particularly concerning time delay and data transmission.

In contrast to traditional cloud computing, the utilization of edge intelligence proves to be more aligned with the demands of medical systems for reliable data transmission, minimal transmission delay, and enhanced data security. In critical scenarios, such as emergencies, the occurrence of errors like extended response times or data loss could pose a direct threat to human life. Additionally, edge intelligence boasts robust location awareness characteristics, making it particularly well suited for location-sensitive medical systems where the accelerated processing speed of edge intelligence emerges as a crucial factor.

3.3.2.3 Autonomous vehicles

Autonomous vehicles rely on the swift exchange of vast amounts of data from diverse sources. The widespread adoption of these vehicles hinges on their capacity for rapid and high-capacity data transmission, given the critical importance of even milliseconds on the road. Edge intelligence plays a crucial role in enabling self-driving vehicles to effectively handle and transmit substantial data volumes. This is achieved by leveraging existing communication networks and cloud computing systems, all while prioritizing considerations of security, reliability, and scalability. Ultimately, the integration of edge intelligence enhances the long-term viability of autonomous vehicles.

3.3.2.4 Smart manufacturing

The information collected from sensors and interconnected devices in manufacturing processes, equipment, and finished products does not necessarily have to be managed on centralized servers. Instead, most manufacturers only require notification when the data indicate a problem. In contrast to cloud-based systems, edge intelligence enables manufacturers to swiftly extract and analyze necessary data for real-time interventions. In simpler terms, by speeding up the troubleshooting process, edge intelligence contributes to enhanced operational efficiency, increased cost savings, and the prevention of supply chain disruptions.

To enhance the automation and intelligence of real-time production control, the architecture can dynamically adjust the production line configuration with the assistance of edge intelligence. This involves collecting and processing various real-time data generated in the factory and utilizing AI to identify and judge, ultimately achieving more efficient feedback control. Collaboration among devices in the factory is crucial, emphasizing the need for coordinated group efforts rather than independent device operations. The utilization of edge intelligence technology in smart factories presents a new challenge.

Regarding industrial production site safety, monitoring the operational status of machinery is vital due to the inevitable quality issues that arise during prolonged use. To detect the machine's running status, an edge intelligent framework must incorporate a device layer, a local private edge cloud near the device layer, and a remote public cloud. A framework needs to be developed, utilizing a powerful public cloud to train the predictive model and then transferring the model to a private edge cloud for online diagnostic and prognosis tasks. This approach reduces delays to some extent and enhances the accuracy of diagnosis and prognosis.

3.3.2.5 Smart grids

Smart grids can undergo a significant transformation through edge intelligence, enabling enterprises to enhance energy efficiency. By connecting sensors and IoT devices to edge platforms within manufacturing plants and offices, real-time monitoring and analysis of energy consumption become possible. This immediate visibility allows companies to predict their energy generation or usage more accurately. Furthermore, enterprises can leverage this capability to sell surplus power back to the grid, thereby generating additional revenue.

Although edge intelligence presents more intriguing opportunities compared to cloud computing, organizations encounter numerous challenges during its implementation. Opportunities and challenges associated with edge intelligence are discussed in the next section.

3.4 Challenges and opportunities for edge intelligence

Edge intelligence is yet in its early stages, and currently, a framework to strengthen it is yet to be established. Such frameworks must meet specific requirements, including the ability to develop applications for instantaneous processing on edge nodes. While existing cloud computing frameworks can handle data for exhaustive purposes, enabling instantaneous data treatment at the network edge remains an area of ongoing research [35,36]. Moreover, we must gain a deep understanding of installing application capabilities on edge nodes, including strategies for workload placement, policies for connecting to edge nodes, and management of distinct node types when deploying applications at the edge.

3.4.1 Challenges

Five key research challenges spanning the hardware, middleware, and software layers are identified to create such a framework.

3.4.1.1 Enabling generic computing on edge nodes

In principle, the concept of edge intelligence involves utilizing various nodes linking the edge device and the cloud. For illustration, base stations are equipped with specialized digital signal processors (DSPs) designed to manage specific tasks. Nevertheless, in practical terms, base stations might not be ideal for handling critical assignments due to the fact that DSPs are not devised for versatile computation tasks. Furthermore, the situation remains uncertain since these nodes may not be able to execute additional computations along with the primary functions.

Research is underway to enhance the computing capabilities of edge nodes to assist generic tasks. For instance, it is possible to upgrade a wireless home router to handle added tasks [37]. Intel's Smart Cell Platform17 utilizes virtualization to accommodate supplementary tasks. An alternative solution involves exchanging specific DSPs with equivalent generic CPUs, although it demands substantial investment.

Cutting-edge AI techniques like neural networks have shown great potential in solving various challenges using remarkable precision. Nevertheless, this often arises at the expense of higher computation and memory demands. As a result, neural networks typically execute these algorithms on high-powered GPUs, which consume significant power. In contrast, embedded processors and DSPs propose a more power-efficient remedy and are capable of fixed-point processes [38]. To make neural networks functional for deployment on mobile devices, we need less complex CNN models that can execute on embedded processors without sacrificing precision. Additionally, it is essential to enhance both the efficacy of the inherent processes executed by neural networks and their overall structure to make them more suitable for resource-competent procedures.

3.4.1.2 Exploring edge node discovery

The exploration of resources and services within a distributed computing environment is an established field. It is accomplished in both tightly and loosely connected setups through various techniques integrated into monitoring tools [39,40] and service brokerages [41,42]. These methods, like benchmarking, create the foundation for yielding decisions about allocating tasks to highly suitable resources to enhance performance.

Nevertheless, the challenge arises when we aim to control the capability of the network's edge. In a decentralized cloud configuration, discovering appropriate nodes necessitates mechanisms that go beyond manual intervention due to the sheer number of available devices at this level. Additionally, such mechanisms should accommodate diverse devices from diverse generations and adapt to prevailing workloads like newly added exhaustive machine learning tasks. Benchmarking methods must rapidly communicate the attainability and capabilities of resources. It is also desirable for them to handle node failures consistently and enable independent recovery.

In this context, the conventional methods used in the cloud for discovering edge nodes, such as resource management and task scheduling, face the following limitations:

- Resource management

It entails guaranteeing an ample supply of resources within the edge network, as exemplified in smart parking setups in which sensor data are seamlessly and dependably transmitted to edge devices [43]. Essential components of resource management include dynamic load balancing [44] and the creation of platforms for resource allocation [45]. Nevertheless, addressing on-demand resource requirements, fluctuating workloads, and data streams originating from diverse devices across extensive geographical areas may require making trade-offs between computing power and communication speed [46].

- Resource management and task scheduling

Edge devices exhibit numerous models, diverse hardware architectures, various operating systems, and inconsistent creation environments. Established edge intelligence platforms struggle to effectively incorporate and administer such diverse edge devices, particularly when it comes to supporting AI workloads. Managing and orchestrating AI workloads, which have distinct characteristics compared to web loads, is a pressing challenge. Consequently, edge intelligence platforms must introduce novel resource perceptions tailored to the challenges of AI workloads, including GPU support, capability extension, and task dependence handling.

- Customized AI algorithms

While model compression preserves to foster AI execution on the edge, it usually leads to a deficit of model precision. Static model compression methods fail to amend the dynamical hardware configurations and loads of edge nodes. Thus, there is a growing need for dynamic compression methods tailored to the complex conditions of edge nodes. Additionally, current model-splitting techniques utilize the hierarchical constitution of deep learning models. Future research should focus on developing partitioning methods tailored to the specific characteristics of AI applications.

Furthermore, data availability presents another formidable challenge for edge devices attempting to process raw data for edge training. The usability of data is crucial, and raw data captured from edge devices often cannot be directly used for model training and inference due to potential bias. While federated learning offers a partial solution, the synchronization of training procedures across devices and communication remains a challenging aspect. In conclusion, discovering and effectively utilizing edge nodes in a decentralized cloud setup poses unique challenges that necessitate innovative approaches beyond traditional cloud-based methods.

3.4.1.3 Dividing and delegating discovery

The advancement of distributed computation environments leads to the creation of various methods for dividing tasks to be implemented in numerous geographical localities [47]. One instance is the partitioning of workflows to be executed in different places [48]. Task splitting is typically conveyed obviously using a semantic or administration tool. However, using edge nodes for offloading computations presents the challenge of efficiently dividing computational assignments inevitably, deprived of essentially compelling appropriate definitions of the competencies or locations of edge nodes.

With the increasing global demand for mobile applications, mobile devices face growing constraints such as limited resources and reduced battery life. Mobile fog architectures have been discussed in the context of mobile cloud computing and code offloading mechanisms. Existing investigations have predominantly depended on simulation techniques for examining task offloading. However, this methodology has limitations because it cannot accurately depict the authentic characteristics of AI workloads in industrial settings. AI algorithms utilized in diverse industrial sectors exhibit distinct model structures and processing steps.

Consequently, when devising a task offloading algorithm, it is crucial to customize the offloading strategy to align with the processing steps of the AI application and its model structure. Current research is transitioning towards a synergy between cloud and edge intelligence. In the future, the emphasis may pivot towards cooperative computing among edge nodes, where multiple edge nodes can collect information from various perspectives, contributing to heightened analysis and decision-making capabilities.

3.4.1.4 Unwavering quality of service and user experience

The value delivered by edge nodes can be measured using quality of Service (QoS), while QoE assesses the quality experienced by users. In the realm of edge intelligence, a critical principle to embrace is the avoidance of overburdening nodes with computationally demanding tasks [49,50]. The difficulties lie in ensuring that these nodes maintain high throughput and reliability while accommodating surplus workloads from data centers or other edge devices. Even though an edge node is fully utilized, users of edge devices and data centers rightfully expect a baseline stage of service. For instance, overloading a base station can negatively impact the service given to connected edge devices. It is imperative to have a comprehensive understanding of peak usage hours for edge nodes so that tasks can be effectively divided and organized in a versatile manner. While an administration framework could be beneficial, it also advances concerns associated with supervising, scheduling, and rescheduling at all levels.

Collaborative training is also an important task to be considered. Edge intelligence employs two AI training methods: distributed training and federated learning [51]. In distributed machine learning, data analysis tasks are performed on nodes that create data, with models and data exchanged among different nodes [52]. Google has introduced federated learning as a privacy-preserving technique, which has found applications in sensitive areas like healthcare and finance. Federated learning and distributed training are distinct. Generally, distributed training emphasizes utilizing data at the network's edge, whereas federated learning places a greater emphasis on safeguarding data privacy. When dealing with diverse edge devices that vary in computing power and communication protocols, adapting models and ensuring serviceability poses challenges. The same methods may yield different learning outcomes when applied to different device clusters. Establishing robust, flexible, and secure synchronization between edge devices, servers, and cloud resources at both hardware and software levels is of utmost importance. There are significant research opportunities in developing a standardized API/IO interface for edge learning across various ubiquitous edge devices.

3.4.1.5 Utilizing edge nodes safely and publicly

Hardware assets held by data centers, supercomputing facilities, and private entities using virtualization have the potential to be repurposed to provide computing services as a utility. This approach involves assessing the associated risks for both providers and users, ultimately enabling pay-as-you-go computing. Consequently, a competitive market has emerged, offering a multitude of options to cater to computing consumers while adhering to service level agreements (SLAs) [53].

Nevertheless, when contemplating the use of alternative devices like switches, routers, and base stations as publicly accessible edge nodes, several challenges must be confronted. Firstly, there is a necessity to clearly delineate and communicate the associated risks for both public and private organizations owning these devices and those planning their deployment. Secondly, it is imperative to ensure that the primary function of the device, such as a router managing internet traffic, remains unaltered when repurposed as an edge intelligence node. Thirdly, realizing multi-tenancy on edge nodes demands technology that prioritizes security; for instance, containers, which are potentially lightweight technology for edge nodes, must exhibit more robust security features [54]. Fourthly, a baseline level of service must be ensured for users of the edge node. Lastly, diverse factors like workloads, computation, data location and transfer, maintenance costs, and energy expenses need consideration when formulating appropriate pricing models for facilitating access to edge nodes.

The significance of data privacy and security cannot be overstated. AI serves as an effective tool in identifying malicious attacks and preventing privacy breaches. However, edge devices face constraints in computing resources, presenting a substantial challenge in designing lightweight and efficient AI algorithms suitable for edge computing.

3.4.2 Opportunities

Despite the difficulties that arise in the implementation of edge intelligence, several promising opportunities exist. We have identified five such opportunities.

3.4.2.1 Establishing standards, benchmarks, and a marketplace

The practical realization and public accessibility of edge intelligence hinge on the clear articulation of duties, associations, and consequences among all involved parties. Various efforts have been made to define cloud standards, including those by organizations such as the National Institutes of Standards and Technology (NIST) in 2021, the IEEE Standards Association, the International Standards Organization (ISO), the Cloud Standards Customer Council (CSCC), and the International

Telecommunication Union (ITU). However, these standards must now be revisited to account for added stakeholders, such as public and private organizations that acknowledge edge nodes, to address the communal, legitimate, and moral aspects of edge node utilization. This is undeniably a complex task that demands devotion and investment from both public and private organizations as well as academic institutions.

The implementation of standards relies on the ability to benchmark the performance of edge nodes beside established metrics. Benchmarking efforts for the cloud have been undertaken by organizations like the Standard Performance Evaluation Corporation (SPEC) and several academic scientists. In an environment as unpredictable as the cloud, benchmarking presents substantial difficulties. Benchmarking edge nodes will pose even greater challenges but will open up additional opportunities for research.

Utilizing edge nodes becomes an enticing prospect when duties, associations, and consequences are well defined. Much like a cloud marketplace, the creation of an edge intelligence marketplace offering a heterogeneity of edge nodes on a pay-as-you-go basis is reasonable. Research is needed to establish SLAs for edge nodes and develop pricing models to facilitate the creation of such a marketplace.

3.4.2.2 Frameworks and languages

There are numerous possibilities for running applications within the cloud paradigm. Besides widely used programming languages, there is a diverse range of offerings available for deploying cloud-based applications. In scenarios where resources beyond the cloud are utilized, such as running a bioinformatics workload on the public cloud with data sourced from a private database, a typical approach involves the use of workflows. Research has extensively explored software frameworks and toolkits for creating extensive workflows in a distributed environment [55]. However, as edge nodes capable of supporting general-purpose computing become more prevalent, the development of new frameworks and toolkits becomes necessary.

The potential applications of edge analytics are expected to vary significantly from established workflows, which have mainly been explored in scientific fields such as bioinformatics [56] or astronomy [57]. As edge analytics becomes relevant in user-driven scenarios, the current frameworks may not be well suited for representing edge analytics workflows. The programming model created to leverage the capabilities of edge nodes should be capable of handling task- and data-level parallelism while executing workloads across various hierarchical levels of hardware.

Additionally, the programming language that supports this model should take into account the diverse hardware landscape and resource capacities present in the workflow. In cases where edge nodes are highly specific to a particular vendor, the framework supporting the workflow must be adaptable. This level of complexity goes beyond that of existing models designed to make cloud computing accessible.

3.4.2.3 Utilizing lightweight libraries and algorithms

In contrast to large servers, edge nodes face limitations in supporting resource-intensive software due to hardware constraints. For instance, consider a small-cell base station (SBS) equipped with Intel's T3K Concurrent Dual-Mode system-on-chip. This device typically features a 4-core ARM-based CPU and limited memory, making it inadequate for executing complex data processing tools like Apache Spark. Apache Spark demands a minimum of eight CPU cores and 8 gigabytes of memory for optimal performance. In the context of edge analytics, there is a need for lightweight algorithms capable of performing reasonable machine learning or data processing tasks [58].

One example of a lightweight library is Apache Quarks, which can be utilized on compact edge devices such as smartphones to enable real-time data analytics. However, Quarks primarily supports basic data processing functions, such as filtering and windowed aggregations, which may not suffice for advanced analytical tasks like context-aware recommendations. There is a demand for machine learning libraries that consume less memory and disk space, benefiting data analytical tools designed for edge nodes.

TensorFlow is another framework to consider, supporting deep learning algorithms and heterogeneous distributed systems, although its potential for edge analytics remains to be explored.

AI algorithms play a pivotal role in extracting valuable insights from big data. Nevertheless, the information extracted by existing algorithms is somewhat limited. In the case of supervised learning, manual data labeling can introduce unknown errors. Furthermore, the future data acquisition systems for smart medical applications will predominantly rely on wearable devices. The rapid analysis of and response to data collected on these wearables presents a significant challenge in terms of energy supply. Balancing the accuracy and lightweight nature of AI models is an area that warrants further investigation.

3.4.2.4 Micro operating systems and virtualization

Research into micro operating systems or microkernels presents a potential avenue for addressing challenges associated with deploying applications on diverse edge nodes. These nodes, unlike traditional servers, typically have limited resources. Therefore, it is essential to optimize the general-purpose computing environment at the edge by conserving resources. Advantages such as rapid deployment, shorter boot-up times, and resource isolation are highly desirable [59]. Initial studies suggest that mobile containers, which distribute device hardware functions among multiple virtual devices, can offer performance comparable to native hardware [60]. Container technologies, such as docker, are advancing and enabling swift application deployment on various platforms. However, further research is needed to establish containers as a suitable method for deploying applications on edge nodes.

Virtualization plays a vital role in the evolution of IT technologies, enabling the simultaneous operation of multiple operating systems or numerous applications on a single server [61]. Its key function is to diminish the reliance on physical servers, leading to substantial reductions in power consumption and cooling costs. The growing prevalence of IoT, mobile devices, and sensors has amplified the requirement for remote data centers [62]. Consequently, there exists an opportunity to relocate applications and intelligence from the cloud to the edge network. This transition can usher in a new type of virtualization at the edge, wherein a physical server can provide adaptable and dedicated storage and cache resources [63].

3.4.2.5 Energy efficiency

The rapid proliferation of edge devices in urban areas has worsened the global energy crisis and the issue of global warming. One potential method to mitigate this problem involves harnessing renewable energy sources to power these edge devices. Given that these devices are dispersed throughout the city, adopting distributed renewable energy generators can significantly reduce the reliance on conventional energy sources. However, this approach is not without its challenges. It must address issues like minimizing the use of conventional energy while ensuring the uninterrupted operation of edge devices and establishing a complementary power system for various edge devices [64]. In the context of an Energy Internet system, the energy router, which serves as a control center, requires a certain level of computational capacity [65]. Hence, a plausible avenue for future research is to explore the integration of energy routers with edge intelligence.

3.5 AI solutions for optimizing computation of edge intelligence

As discussed, major problems in computing for edge intelligence are computing offload, resource allocation, privacy, and security. Enhancement in conventional approaches or hybridization can help improve and optimize computing, resource allocation, privacy, and security for edge intelligence. A detailed review of the techniques proposed by the researchers to address these objectives is presented in Table 3.2.

3.6 Architectural layers in the roadmap for edge intelligence

The architectural layers in the roadmap for edge intelligence distinguish between two main directions, viz. AI for the edge and AI on the edge, as shown in Fig. 3.4. Using a bottom-up strategy, our focus in Edge intelligence research is on dividing efforts into topology, content, and service segments, all of which stand to gain advantages from AI technologies. Conversely, a top-down method dissects AI research on the edge into model adaptation, framework design, and processor acceleration. Prior to exploring AI for the edge and AI on the edge as distinct entities, it is essential to establish a shared objective, termed QoE, which consistently takes precedence. The

Table 3.2 Summary of solutions for optimizing computing for edge intelligence.

Problem addressed	Objective	Technique/algorithm	Details of technique/algorithm
Computing offloading optimization	Reduction in energy consumption and latency	Deep reinforcement learning (DRL) based offloading scheme [66]	Lack of prior familiarity with transmission delay and energy consumption models reduces the complexity of the state space by employing DRL to augment the understanding speed. Additionally, the energy consumption (EC) scenario involving energy harvesting is considered.
		DRL-based computing offloading algorithm [67]	Utilizes a Markov decision process for the portrayal of computational offloading, employing DRL to acquire insights into network dynamics.
		A hybrid approach based on Q-function breakdown and double DQN [68]	Utilizes a double deep Q-network for the attainment of optimal computing offloading in the absence of prerequisites, employing a novel algorithm which is function approximator based DNN model which is designed to address high-dimensional state spaces.
		Reinforcement learning utilizing neural network architectures [69]	A continuous-time Markov decision process with an infinite horizon and average rewards is employed to model the optimization issue. Additionally, a novel value function approximator is introduced to address the challenges posed by high-dimensional state spaces.
	Optimization of the hardware structure for edge devices	Binary weight convolutional neural network-based algorithm [70]	Static random access memory (SRAM) is designed for binary weight convolutional neural networks (CNNs) with the aim of minimizing memory data output, facilitating parallel implementation of CNN operations.
		Approach based on DNN and FPGA [71]	Expeditor for weed species categorization utilizing a binarized DNN, which is employed on FPGA.
	Reduction in energy consumption	Distributed deep learning-based offloading technique [72]	A model was built by adding the cost of varying local implementation assignments in the cost function.

continued on next page

Table 3.2 (*continued*)

Problem addressed	Objective	Technique/algorithm	Details of technique/algorithm
	Reduction in latency	DL-based Smart-Edge-CoCaCo [73]	An approach based on joint optimization of wireless communication, combined filter caching, and computation offloading was developed to reduce the latency.
		A heuristic offloading technique [74]	Using electronic communication networks to estimate the distance between origin and destination, along with heuristic searching, to identify the most effective scheme for reducing the communication lag of deep learning tasks.
		Cooperative Q-learning [75]	Noticeable improvement in the searching pace of the conventional Q-learning approach.
		TD learning involves a method that incorporates postdecision states and utilizes a semi-gradient descent approach [76]	Utilizes approximate dynamical planning as a strategy to tackle the difficulties established by the curse of dimensionality.
		Online reinforcement learning [77]	Unique arrangements of state transitions are designed to address the difficulties raised by the curse of dimensionality. Moreover, it takes into account the energy-harvesting aspect of edge intelligence.
Security of edge intelligence		Hypergraph clustering [78]	Improves the identification rate by modeling the association among edge nodes and DDoS through hypergraph clustering.
		Extreme learning machine [79]	Demonstrates quicker convergence rates and enhanced generalization capabilities compared to the majority of traditional algorithms.
		Distributed deep learning [80]	Eases the load of training the model while enhancing its accuracy.
		An algorithm based on restricted Boltzmann machines [81]	Enhances the ability to identify unfamiliar attacks through the incorporation of active learning features.

continued on next page

Table 3.2 *(continued)*

Problem addressed	Objective	Technique/algorithm	Details of technique/algorithm
		Deep PDS-Learning [82]	Accelerates the training process by incorporating supplementary details, such as the energy consumption of edge devices.
Resource allocation optimization		Actor-critic RL [83]	Introduces an additional DNN for expressing a parameterized stochastic policy, aiming to enhance both performance and convergence speed. Additionally, a natural policy gradient approach to mitigate the risk of local convergence is incorporated.
		DRL-based resource allocation scheme [84]	Enhances the quality of service (QoS) through the integration of supplementary SDN
		Multi-task DRL [85]	Modifies the final layer of a DNN responsible for estimating the Q-function to accommodate action spaces with increased dimensions.
Privacy protection		Generative adversarial networks (GANs) [86]	An algorithm for objective perturbation and another for output perturbation, both ensuring adherence to the principles of differential privacy.
		Edge Sanitizer: A deep inference framework [87]	Proposes maximum utilization of data while ensuring privacy protection.
		Deep Q-learning [88]	Generate trust values through uncertain reasoning and prevent local convergence by regulating the learning rate.
Other ways to reduce energy consumption	Control device operational condition	DRL-based joint mode selection and resource management approach [89]	Minimizes energy consumption in the medium and long term by managing the communication mode of the operator apparatus and regulating the active state of processors.
	Merging into Energy Internet	Model-based DRL [90]	Solves the energy supply issue of the multi-access edge server.
		Reinforcement learning [91]	A fog computing device operating on energy generated from a renewable source.
		Minimax Q-learning [92]	Gradually learns the optimal strategy by raising the spectral efficiency throughput.

continued on next page

Table 3.2 *(continued)*

Problem addressed	Objective	Technique/algorithm	Details of technique/algorithm
		Online learning [93]	Minimizes bandwidth utilization by selecting the server with the highest reliability.
		Multiple AI-based algorithms [94]	A mechanism for selecting AI algorithms intelligently to choose the most suitable algorithm for a given task was developed.

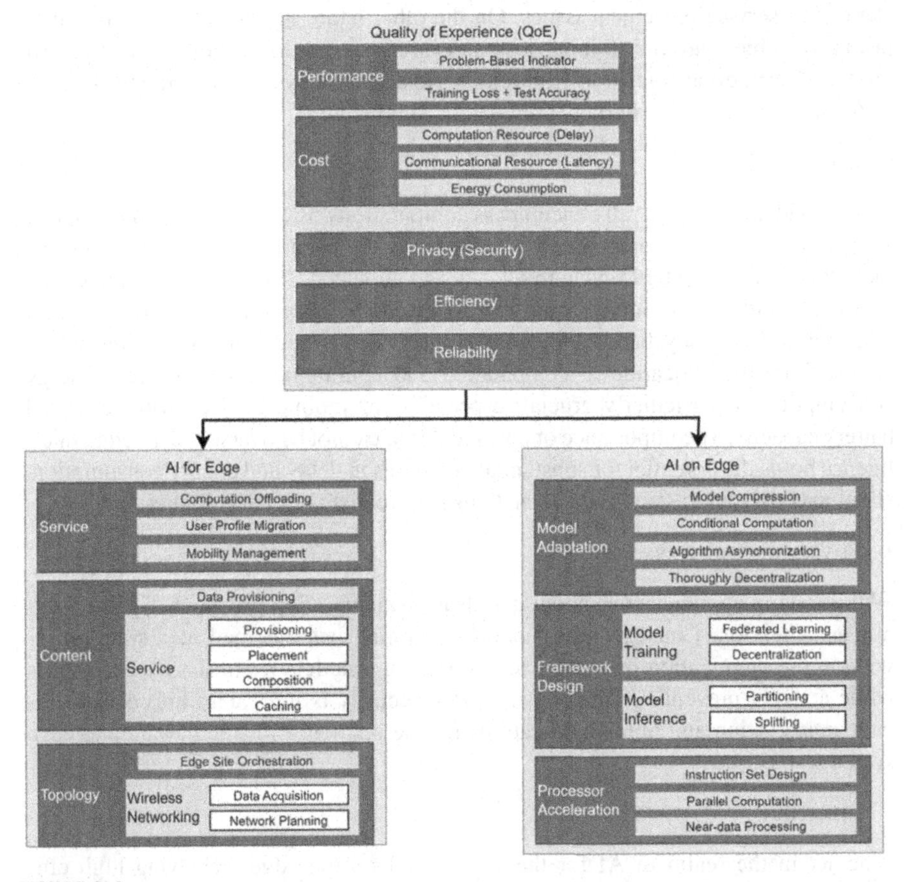

FIGURE 3.4

The architectural layers in the roadmap.

detailed discussion of QoE, AI for the edge, and AI on the edge is discussed further in this section.

3.6.1 Quality of experience

We believe that QoE should be tailored to specific applications and should be established by contemplating multiple criteria: performance, cost, privacy (security), efficiency, and reliability.

- Performance

Performance criteria differ between AI for the edge and AI on the edge. In the case of the former, performance metrics are tailored to specific problems. For instance, it might encompass metrics like the successful offloading ratio in computation offloading challenges or the efficient optimization of revenue and hiring costs for base stations in service placement issues. On the other hand, for the latter, performance primarily centers around training loss and inference accuracy, both critical for AI models. Despite the transition from cloud clusters to a system integrating devices, edge, and cloud, these criteria continue to be significant.

- Cost

Cost considerations generally encompass computation cost, communication cost, and energy consumption. Computation cost indicates the need for computing resources, including factors like CPU cycle frequency and allocated CPU time. Communication cost deals with the resource requirements for communication, considering aspects like power, frequency band, and access time, with a focus on minimizing delays arising from the allocation of computation and communication resources. Energy consumption is particularly crucial, especially for mobile devices with restricted battery capacity. The importance of cost reduction cannot be overstated, as edge intelligence holds the potential for substantial decreases in delay and energy consumption, simultaneously addressing critical challenges in realizing 5G capabilities.

- Privacy and security

With growing apprehensions about data leaks, safeguarding privacy has gained significant attention. Consequently, federated learning has emerged as a solution involving the aggregation of local machine learning models from distributed devices while actively preventing data leakage [95]. Security is intricately linked with privacy preservation and holds implications for the resilience of middleware and edge systems.

- Efficiency

Whether in the realm of AI for the edge or AI on the edge, achieving high efficiency is paramount to achieving outstanding performance with minimal overhead. The pursuit of efficiency is crucial for the improvement of existing algorithms and models, especially in the context of AI on the edge. Numerous strategies, including model compression, conditional computation, and asynchronous algorithms, have been suggested to enhance the efficiency of training and inference processes for deep AI models.

- Reliability

System reliability plays a pivotal role in ensuring the continuous operation of edge intelligence over specified durations, a critical element for user experience. In the domain of edge intelligence, system reliability holds particular importance for AI on the edge. This is especially true when considering that model training and inference frequently take place in a distributed and synchronized manner, and local users may encounter obstacles related to wireless network congestion when attempting to complete model uploads and downloads.

3.6.2 Edge intelligence/intelligent edge computing

The roadmap, as illustrated in Fig. 3.4, pertains to AI for edge intelligence, which we refer to as intelligent edge computing (IEC). AI offers potent tools for addressing intricate challenges in learning, planning, and decision-making. We adopt a bottom-up approach to categorize the primary concerns in edge intelligence into three layers: topology, content, and service.

- Topology

In terms of topology, our attention is directed towards orchestrating edge sites (OESs) and wireless networking (WN). Within our framework, an edge site is defined as a micro data center hosting deployed applications and connected to an SBS. OES focuses on the deployment and configuration of wireless telecom equipment and servers. Notably, recent years have seen a surge in interest surrounding the management and automation of unmanned aerial vehicles (UAVs). These UAVs, equipped with a small server and access point, can be viewed as mobile edge servers with exceptional maneuverability. Consequently, numerous studies explore scheduling and trajectory planning challenges with the aim of minimizing UAV energy consumption.

For instance, Chen et al. [96] investigated power consumption by caching popular content based on predictions, introducing a conceptor-based echo state network (ESN) algorithm to learn user mobility patterns. Leveraging this efficient machine learning technique, their algorithm significantly outperforms benchmarks in terms of transmission power and user satisfaction. On the other hand, WN encompasses data acquisition and network planning. The former focuses on swiftly acquiring data from widely distributed sources at edge devices, while the latter concentrates on network scheduling, operation, and management. Fast data acquisition involves elements such as multiple access, radio resource allocation, and signal encoding/decoding. Network planning explores efficient management through protocols and middleware.

Significantly, recent years have witnessed a rising trend in intelligent networking, utilizing AI technologies to construct intelligent wireless communication mechanisms. As an example, Zhu et al. [97] proposed learning-driven communication, exploiting the synergy between communication and learning in edge systems.

- Content

The focus lies on several vital aspects, including data provisioning, service provisioning, service placement, service composition, and service caching. In the realms of data and service provisioning, resources can be drawn from remote cloud data centers and edge servers. Recent initiatives have concentrated on developing lightweight QoS-aware service-based frameworks. Alternatively, shared resources may originate from mobile devices with suitable incentive mechanisms in place.

Service placement complements service provisioning and delves into the location and method of deploying complex services on potential edge sites. In recent times, numerous studies have approached service placement from the perspective of application service providers (ASPs). For instance, Chen et al. [98] endeavored to deploy services within a limited budget on fundamental communication and computation infrastructure. Subsequently, they applied the multi-armed bandit (MAB) theory, a branch of reinforcement learning, to optimize service placement decisions.

Service composition involves the selection of candidate services for composition, taking into account energy consumption and QoE for mobile end users. This domain presents opportunities for leveraging AI technologies to generate more effective service selection strategies. Service caching, akin to service provisioning, revolves around designing a caching pool to store frequently accessed data and services. It can also be explored in a cooperative manner, offering research prospects for applying multi-agent learning to enhance QoE in large-scale edge intelligence systems.

- Services

Regarding services, our focus is on computation offloading, user profile migration, and mobility management. Computation offloading addresses the load balancing of various computational and communication resources, involving edge server selection and frequency spectrum allocation. Recent research has concentrated on dynamically managing radio and computational resources for multi-user, multi-server edge intelligence systems, utilizing Lyapunov optimization techniques. Computation offloading decisions are also being optimized through Deep Q-Network (DQN), modeling the problem as a Markov decision process to maximize long-term utility performance, encompassing the aforementioned QoE indicators.

User profile migration involves adjusting the location of user profiles, encompassing configuration files, private data, and logs, as mobile users are constantly on the move. User profile migration is often intertwined with mobility management. For instance, the JCORM algorithm, proposed in [99], optimizes computation offloading and migration through cooperative networks, presenting research opportunities for the application of more advanced AI technologies to enhance optimality.

Mobility management, viewed through the lens of statistics and probability theory, is another area of interest. There is a notable inclination toward realizing mobility management with the assistance of AI.

3.6.3 AI on the edge

The right side of the roadmap focuses on AI on the edge involving the study of implementing AI model training and inference on the network edge. This area is categorized into four research endeavors: framework design, model training, inference, and adaptation, conditional computation, and processor acceleration. Given that model adaptation builds upon existing training and inference frameworks, the initial focus will be on introducing framework design.

- Framework design

Framework design aims to establish an improved training and inference architecture for the edge without altering existing AI models. Researchers strive to create new frameworks for both model training and model inference.

- Model training

Presently, the predominant frameworks for model training are largely distributed, with the exception of those based on knowledge distillation. Distributed training frameworks can be categorized based on data splitting and model splitting methods [100]. Data splitting involves master–device, helper–device, and device–device splitting, differing in how training samples are sourced and how the global model is aggregated. Model splitting involves separating neural network layers onto various devices, relying on complex pipelines. Knowledge distillation-based frameworks may or may not be decentralized, utilizing transfer learning technologies to enhance the accuracy of shallower student networks [97]. This process entails training a basic network on a standard dataset and transferring the learned features to student networks, which are then trained on their respective datasets by multiple mobile end devices. There is significant potential for exploration in knowledge distillation-based frameworks for model training at the edge.

The leading approach in model training is federated learning, which is designed to preserve privacy while training DNNs in a distributed manner [101]. It trains local models on multiple clients, optimizing a global model by averaging trained gradients. Due to limited resources in edge nodes, training a comprehensive model there is impractical, making distributed training more feasible. However, coordination between edge nodes becomes essential. The challenge lies in optimizing the global gradient from distributed local models. Regardless of the learning algorithms used, stochastic gradient descent (SGD) plays a crucial role in model training. Edge nodes utilize SGD to update local gradients based on their datasets, sending updates to a central node for global model enhancement. Balancing model performance and communication overhead is critical. Selective transmission of local gradients showing significant improvements can ensure global model performance while reducing communication overheads, preventing network congestion caused by simultaneous transmissions from all edge nodes.

- Model inference

While splitting a model during training poses challenges, it is a preferred method during model inference. Model splitting, or partitioning, serves as a framework for model inference, and various techniques, including model compression, input filtering, early exit, etc., are adaptations from existing frameworks. A well-described example of model inference on the edge can be found in reference [96], where a DNN is divided into two parts, each processed collaboratively. The computationally intense segment operates on the edge server, while the other part functions on the mobile device. The challenge lies in determining the optimal layer to split and when to exit the intricate DNN while maintaining inference accuracy.

- Model adaptation

Model adaptation is a process that fine-tunes existing training and inference frameworks, particularly in the context of federated learning, to suit edge intelligence better. While federated learning can potentially operate on the edge, its traditional version demands significant communication efficiency, as complete local models are transmitted back to the central server. Consequently, many researchers are concentrating on developing more efficient model updates and aggregation policies. Their endeavors aim to minimize costs, enhance robustness, and ensure system performance. Approaches to achieving model adaptation include techniques such as model compression, conditional computation, algorithm synchronization, and comprehensive decentralization.

Model compression exploits the inherent sparsity structure of gradients and weights. Potential strategies encompass quantization, dimensional reduction, pruning, precision downgrading, components sharing, and cutoff, among others. Implementation methods involve techniques like singular value decomposition (SVD), Huffman coding, principal component analysis (PCA), and similar approaches.

- Conditional computation

Conditional computation serves as an alternative approach to reduce computations by selectively disabling less crucial calculations in DNNs. Feasible methods include component shutdown, input filtering, early exit, and results caching. This concept can be likened to block-wise dropout [102]. Additionally, random gossip communication can help reduce unnecessary calculations and model updates. Asynchronization in algorithms aims to aggregate local models asynchronously, mitigating the inefficiency and lengthy synchronous steps of model updates in federated learning. Thorough decentralization involves eliminating the central aggregator to prevent potential data leaks and address issues stemming from the central server's malfunction. Strategies for achieving complete decentralization encompass blockchain technologies, game-theoretical approaches, and similar methods.

- Process acceleration

Processor acceleration aims to optimize the structure of DNNs, specifically targeting the frequently used computation-heavy multiply-and-accumulate operations for improvement. Strategies for enhancing DNN computations on hardware involve various methods, such as creating specialized instruction sets for DNN training and inference, developing highly parallel computing paradigms, and implementing near-data processing to bring computation closer to memory. Highly parallelized computing paradigms can be categorized into temporal and spatial architectures. Temporal architectures like CPUs and GPUs can be accelerated by reducing the number of multiplications and increasing throughput. Spatial architectures, on the other hand, can be accelerated by enhancing data reuse with data flows.

3.7 Conclusions

The critical components of edge intelligence, including edge caching, edge training, edge inference, and edge offloading, are discussed. The efficient deployment of AI to the network's edge hinges on improving the efficacy of AI algorithms with limited computing and energy resources, necessitating the design of lightweight AI models. Beyond the significance of individual technologies, this study also delves into the challenges that must be overcome and the areas that require reinforcement. Investigating the incorporation of essential technologies, issues, opportunities, and roadmap in this study will be a valuable direction for the community engaged in research on edge intelligence.

References

[1] H. Khayyam, B. Javadi, M. Jalili, R.N. Jazar, Artificial intelligence and Internet of Things for autonomous vehicles, in: Nonlinear Approaches in Engineering Applications, Springer, Berlin/Heidelberg, Germany, 2020, pp. 39–68.

[2] K. Zhang, Y. Zhu, S. Leng, Y. He, S. Maharjan, Y. Zhang, Deep learning empowered task offloading for mobile edge computing in urban informatics, IEEE Internet of Things Journal 6 (2019) 7635–7647.

[3] U. Arul, R. Gnanajeyaraman, A. Selvakumar, S. Ramesh, T. Manikandan, G. Michael, Integration of IoT and edge cloud computing for smart microgrid energy management in VANET using machine learning, Computers & Electrical Engineering 110 (2023) 108905.

[4] Y. Dai, D. Xu, S. Maharjan, G. Qiao, Y. Zhang, Artificial intelligence empowered edge computing and caching for the Internet of vehicles, IEEE Wireless Communications 26 (2019) 12–18.

[5] J. Mendez, K. Bierzynski, M. Cuéllar, D.P. Morales, Edge intelligence: concepts, architectures, applications and future directions, ACM Transactions on Embedded Computing Systems (TECS) 21 (2022) 1–41.

[6] S. Ghorpade, M. Zennaro, B. Chaudhari, Survey of localization for Internet of things nodes: approaches, challenges and open issues, Future Internet 13 (8) (Aug. 2021) 210, https://doi.org/10.3390/fi13080210.

[7] S.N. Ghorpade, M. Zennaro, B.S. Chaudhari, R.A. Saeed, H. Alhumyani, S. Abdel-Khalek, A novel enhanced quantum PSO for optimal network configuration in heterogeneous industrial IoT, IEEE Access 9 (2021) 134022–134036, https://doi.org/10.1109/ACCESS.2021.3115026.

[8] U. Drolia, K. Guo, J. Tan, R. Gandhi, P. Narasimhan, Cachier: edge-caching for recognition applications, in: Proc. IEEE 37th Int. Conf. Distrib. Comput. Syst. (ICDCS), Jun. 2017, pp. 276–286.

[9] U. Drolia, K. Guo, P. Narasimhan, Precog: prefetching for image recognition applications at the edge, in: Proc. Symp. Edge Comput., 2017, p. 17.

[10] P. Guo, B. Hu, R. Li, W. Hu, Foggy cache: cross-device approximate computation reuse, in: Proc. 24th Annu. Int. Conf. Mobile Comput. Netw., Oct. 2018, pp. 19–34.

[11] S.N. Ghorpade, M. Zennaro, B.S. Chaudhari, Binary grey wolf optimization-based topology control for WSNs, IET Wireless Sensor Systems 9 (6) (2019) 333–339, https://doi.org/10.1049/iet-wss.2018.5169.

[12] T. Li, Z. Xiao, H.M. Georges, Z. Luo, D. Wang, Performance analysis of co-and cross-tier de-vice-to-device communication underlaying macro-small cell wireless networks, KSII Transactions on Internet and Information Systems 10 (4) (2016).

[13] S.N. Ghorpade, M. Zennaro, B.S. Chaudhari, GWO model for optimal localization of IoT-enabled sensor nodes in smart parking systems, IEEE Transactions on Intelligent Transportation Systems 22 (2021) 1217–1224.

[14] S.N. Ghorpade, M. Zennaro, B.S. Chaudhari, IoT-based hybrid optimized fuzzy threshold ELM model for localization of elderly persons, Expert Systems with Applications 184 (Dec. 2021) 115500.

[15] S.N. Ghorpade, M. Zennaro, B.S. Chaudhari, R.A. Saeed, H. Alhumyani, S. Abdel-Khalek, Enhanced differential crossover and quantum particle swarm optimization for IoT applications, IEEE Access 9 (2021) 93831–93846.

[16] M. Ji, G. Caire, A.F. Molisch, Wireless device-to-device caching networks: basic principles and system performance, IEEE Journal on Selected Areas in Communications 34 (1) (2016) 176–189.

[17] W. Chen, T. Li, Z. Xiao, D. Wang, On mitigating interference under device-to-device communication in macro-small cell networks, in: 2016 International Conference on Computer, Information and Telecommunication Systems (CITS), IEEE, 2016, pp. 1–5.

[18] A. Ioannou, S. Weber, A survey of caching policies and forwarding mechanisms in information-centric networking, IEEE Communications Surveys and Tutorials 18 (4) (2016) 2847–2886.

[19] B. McMahan, D. Ramage, Federated learning: collaborative machine learning without centralized training data, Google Research Blog 3 (2017).

[20] O. Valery, P. Liu, J.-J. Wu, CPU/GPU collaboration techniques for transfer learning on mobile devices, in: 2017 IEEE 23rd International Conference on Parallel and Distributed Systems (ICPADS), IEEE, 2017, pp. 477–484.

[21] O. Valery, P. Liu, J.-J. Wu, Low precision deep learning training on mobile heterogeneous platform, in: 2018 26th Euromicro International Conference on Parallel, Distributed and Network-Based Processing (PDP), Cambridge, UK, 2018, pp. 109–117, https://doi.org/10.1109/PDP2018.2018.00023.

[22] T. Xing, S.S. Sandha, B. Balaji, S. Chakraborty, M. Srivastava, Enabling edge devices that learn from each other: cross modal training for activity recognition, in: Proceedings of the 1st International Workshop on Edge Systems, Analytics and Networking, ACM, 2018, pp. 37–42.
[23] K. Bonawitz, V. Ivanov, B. Kreuter, A. Marcedone, H.B. McMahan, S. Patel, D. Ramage, A. Segal, K. Seth, Practical secure aggregation for privacy-preserving machine learning, in: Proceedings of the 2017 ACM SIGSAC Conference on Computer and Communications Security, ACM, 2017, pp. 1175–1191.
[24] H. Liu, K. Simonyan, Y. Yang, DARTS: differentiable architecture search, arXiv preprint, arXiv:1806.09055, 2018.
[25] A.G. Howard, M. Zhu, B. Chen, D. Kalenichenko, W. Wang, T. Weyand, M. Andreetto, H. Adam, MobileNets: efficient convolutional neural networks for mobile vision applications, arXiv preprint, arXiv:1704.04861, 2017.
[26] Y. Kang, J. Hauswald, C. Gao, A. Rovinski, T. Mudge, J. Mars, L. Tang, Neurosurgeon: collaborative intelligence between the cloud and mobile edge, ACM SIGARCH Computer Architecture News 45 (1) (2017) 615–629.
[27] N. Raval, A. Srivastava, A. Razeen, K. Lebeck, A. Machanavajjhala, L.P. Cox, What you mark is what apps see, in: Proceedings of the 14th Annual International Conference on Mobile Systems, Applications, and Services, ACM, 2016, pp. 249–261.
[28] S. Wendelken, C. MacGillivray, Worldwide and U.S. IoT cellular connections forecast, 2021–2025, https://www.idc.com/getdoc.jsp?containerId=US47296121. (Accessed 17 February 2022).
[29] T.-T. Wong, Performance evaluation of classification algorithms by k-fold and leave-one-out cross validation, Pattern Recognition 48 (9) (2015) 2839–2846.
[30] S. Mittal, A survey of FPGA-based accelerators for convolutional neural networks, Neural Computing & Applications 32 (Oct. 2018) 1–31.
[31] J. Manokaran, G. Vairavel, An empirical comparison of machine learning algorithms for attack detection in Internet of things edge, ECS Transactions 107 (2022) 2403.
[32] D.S. Watson, On the philosophy of unsupervised learning, Philosophy & Technology 36 (2023) 28, https://doi.org/10.1007/s13347-023-00635-6.
[33] N. Thomos, T. Maugey, L. Toni, Machine learning for multimedia communications, Sensors 22 (3) (2022) 819.
[34] S.N. Ghorpade, M. Zennaro, B.S. Chaudhari, Towards green computing: intelligent bio-inspired agent for IoT-enabled wireless sensor networks, International Journal of Sensor Networks 35 (2) (Feb 2021) 121–131, https://doi.org/10.1504/IJSNET.2021.113632.
[35] W. Li, Y. Zhao, S. Lu, D. Chen, Mechanisms and challenges on mobility-augmented service provisioning for mobile cloud computing, IEEE Communications Magazine 53 (3) (2015) 89–97.
[36] H. Hromic, D. Le Phuoc, M. Serrano, A. Antonic, I.P. Zarko, C. Hayes, S. Decker, Real time analysis of sensor data for the Internet of things by means of clustering and event processing, in: Proceedings of the IEEE International Conference on Communications, 2015, pp. 685–691.
[37] C. Meurisch, A. Seeliger, B. Schmidt, I. Schweizer, F. Kaup, M. Mühlhäuser, Upgrading wireless home routers for enabling large-scale deployment of cloudlets, in: Mobile Computing, Applications, and Services, 2015, pp. 12–29.
[38] M. Shafique, T. Theocharides, C.S. Bouganis, M.A. Hanif, F. Khalid, R. Hafz, S. Rehman, An overview of next-generation architectures for machine learning: roadmap, opportunities and challenges in the IoT era, in: 2018 Design, Automation Test in Europe Conference Exhibition (DATE), March 2018, pp. 827–832.

[39] J. Povedano-Molina, J.M. Lopez-Vega, J.M. Lopez-Soler, A. Corradi, L. Foschini, DARGOS: a highly adaptable and scalable monitoring architecture for multi-tenant clouds, Future Generation Computer Systems 29 (8) (2013) 2041–2056.
[40] J.A. Perez-Espinoza, V.J. Sosa-Sosa, J.L. Gonzalez, E. Tello-Leal, A distributed architecture for monitoring private clouds, in: 2015 26th International Workshop on Database and Expert Systems Applications (DEXA), Valencia, Spain, 2015, pp. 186–190, https://doi.org/10.1109/DEXA.2015.51.
[41] N. Grozev, R. Buyya, Inter-cloud architectures and application brokering: taxonomy and survey, Software, Practice & Experience 44 (3) (2014) 369–390.
[42] S.K. Garg, S. Versteeg, A framework for ranking of cloud computing services, Future Generation Computer Systems 29 (4) (2013) 1012–1023.
[43] P. Singh, et al., Machine learning for cloud, fog, edge and serverless computing environments: comparisons, performance evaluation benchmark and future directions, International Journal of Grid and Utility Computing 13 (4) (2022) 447–457.
[44] L. Stacker, J. Fei, P. Heidenreich, F. Bonarens, J. Rambach, D. Stricker, C. Stiller, Deployment of deep neural networks for object detection on edge AI devices with runtime optimization, in: Proceedings of the IEEE/CVF International Conference on Computer Vision, 2021, pp. 1015–1022.
[45] S. Iftikhar, M.M.M. Ahmad, et al., HunterPlus: AI based energy-efficient task scheduling for cloud–fog computing environments, IEEE Internet of Things Journal 21 (2023) 100667.
[46] S. Mousavi, S.E. Mood, A. Souri, M.M. Javidi, Directed search: a new operator in NSGA-II for task scheduling in IoT based on cloud-fog computing, IEEE Transactions on Cloud Computing 11 (2) (2023) 2144–2157, https://doi.org/10.1109/TCC.2022.3188926.
[47] T. Ghafariana, B. Javadi, Cloud-aware data intensive workflow scheduling on volunteer computing systems, Future Generation Computer Systems 51 (2015) 87–97.
[48] W. Tang, J. Jenkins, F. Meyer, R. Ross, R. Kettimuthu, L. Winkler, X. Yang, T. Lehman, N. Desai, Data-aware resource scheduling for multicloud workflows: a fine-grained simulation approach, in: Proceedings of the IEEE International Conference on Cloud Computing Technology and Science, 2014, pp. 887–892.
[49] M.T. Beck, M. Maier, Mobile edge computing: challenges for future virtual network embedding algorithms, in: Proceedings of the International Conference on Advanced Engineering Computing and Applications in Sciences, 2014, pp. 65–70.
[50] P. Simoens, L. Van Herzeele, F. Vandeputte, L. Vermoesen, Challenges for orchestration and instance selection of composite services in distributed edge clouds, in: Proceedings of the IFIP/IEEE International Symposium on Integrated Network Management, 2015, pp. 1196–1201.
[51] https://timestech.in/federated-learning-collaborative-ml-without-centralized-data/.
[52] L. Valerio, A. Passarella, M. Conti, A communication efficient distributed learning framework for smart environments, Pervasive and Mobile Computing 41 (2017) 46–68.
[53] S.A. Baset, Cloud SLAs: present and future, ACM SIGOPS Operating Systems Review 46 (2) (2012) 57–66.
[54] T. Bui, Analysis of Docker security, CoRR, arXiv:1501.02967 [abs], 2015.
[55] E. Deelman, K. Vahi, G. Juve, M. Rynge, S. Callaghan, P.J. Maechling, R. Mayani, W. Chen, R. Ferreira da Silva, M. Livny, K. Wenger, Pegasus: a workflow management system for science automation, Future Generation Computer Systems 46 (2015) 17–35.
[56] B. Serrano-Solano, et al., Galaxy: a decade of realizing CWFR concepts, Data Intelligence 4 (2) (2022) 358–371.

[57] J. Ruiz, J. Garrido, J. Santander-Vela, S. Snchez-Expsito, L. Verdes Montenegro, AstroTaverna—building workflows with virtual observatory services, Astronomy and Computing 78 (2014) 3–11.

[58] S. Kartakis, J.A. McCann, Real-time edge analytics for cyber physical systems using compression rates, in: Proceedings of the International Conference on Autonomic Computing, 2014, pp. 153–159.

[59] L. Xu, Z. Wang, W. Chen, The study and evaluation of ARM-based mobile virtualization, International Journal of Distributed Sensor Networks 11 (7) (2015), https://doi.org/10.1155/2015/310308.

[60] J. Andrus, C. Dall, A.V. Hof, O. Laadan, J. Nieh, Cells: a virtual mobile smartphone architecture, in: Proceedings of the ACM Symposium on Operating Systems Principles, 2011, pp. 173–187.

[61] K. Bansal, et al., DeepBus: machine learning based real time pothole detection system for smart transportation using IoT, Internet Technology Letters 3 (3) (2020) e156.

[62] R. Morabito, N. Beijar, Enabling data processing at the network edge through lightweight virtualization technologies, in: Sensing, Communication and Networking (SECON Workshops), 2016 IEEE International Conference on, IEEE, 2016, pp. 1–6.

[63] A. Barker, B. Varghese, J.S. Ward, I. Sommerville, Academic cloud computing research: five pitfalls and five opportunities, in: Proceedings of the USENIX Conference on Hot Topics in Cloud Computing, 2014.

[64] Y. Liu, C. Yang, L. Jiang, S. Xie, Y. Zhang, Intelligent edge computing for IoT-based energy management in smart cities, IEEE Network 33 (2) (2019) 111–117.

[65] H. Liang, H. Hua, Y. Qin, M. Ye, S. Zhang, J. Cao, Stochastic optimal energy storage management for energy routers via compressive sensing, IEEE Transactions on Industrial Informatics 18 (4) (2022) 2192–2202.

[66] M. Min, L. Xiao, Y. Chen, P. Cheng, D. Wu, W. Zhuang, Learning-based computation offloading for IoT devices with energy harvesting, IEEE Transactions on Vehicular Technology 68 (2) (2019) 1930–1941.

[67] X. Cheng, L. Feng, W. Quan, C. Zhou, H. He, W. Shi, X. Shen, Space/aerial-assisted computing offloading for IoT applications: a learning-based approach, IEEE Journal on Selected Areas in Communications 37 (5) (2019) 1117–1129.

[68] X. Chen, H. Zhang, C. Wu, S. Mao, Y. Ji, M. Bennis, Optimized computation offloading performance in virtual edge computing systems via deep reinforcement learning, IEEE Internet of Things Journal 6 (3) (2019) 4005–4018.

[69] L. Lei, H. Xu, X. Xiong, K. Zheng, W. Xiang, X. Wang, Multiuser resource control with deep reinforcement learning in IoT edge computing, IEEE Internet of Things Journal 6 (6) (2019) 10119–10133.

[70] B.S. Chaudhari, M. Zennaro, Introduction to low-power wide-area networks, in: LPWAN Technologies for IoT and M2M Applications, Elsevier, 2020, pp. 1–13, https://doi.org/10.1016/B978-0-12-818880-4.00001-6.

[71] C. Lammie, A. Olsen, T. Carrick, M. Rahimi Azghadi, Low-power and high-speed deep FPGA inference engines for weed classification at the edge, IEEE Access 7 (2019) 51171–51184.

[72] L. Huang, X. Feng, A. Feng, Y. Huang, L. Qian, Distributed deep learning-based offloading for mobile edge computing networks, Mobile Networks and Applications 66 (12) (2018) 6353–6367.

[73] Y. Hao, Y. Mian, L. Hu, M.S. Hossain, G. Muhammad, S.U. Amin, Smart-edge-CoCaCo: AI-enabled smart edge with joint computation, caching, and communication in heterogeneous IoT, IEEE Network 33 (2) (2019) 58–64.

[74] X. Xu, D. Li, Z. Dai, S. Li, X. Chen, A heuristic offloading method for deep learning edge services in 5G networks, IEEE Access 7 (2019) 67734–67744.
[75] N. Kiran, C. Pan, S. Wang, C. Yin, Joint resource allocation and computation offloading in mobile edge computing for SDN based wireless networks, Journal of Communications and Networks 22 (1) (2020) 1–11, https://doi.org/10.1109/JCN.2019.000046.
[76] L. Lei, H. Xu, X. Xiong, K. Zheng, W. Xiang, Joint computation offloading and multiuser scheduling using approximate dynamic programming in NB-IoT edge computing system, IEEE Internet of Things Journal 6 (3) (2019) 5345–5362.
[77] J. Xu, L. Chen, S. Ren, Online learning for offloading and autoscaling in energy harvesting mobile edge computing, IEEE Transactions on Cognitive Communications and Networking 3 (3) (2017) 361–373.
[78] X. An, J. Su, X. Lu, F. Lin, Hypergraph clustering model-based association analysis of DDOS attacks in fog computing intrusion detection system, EURASIP Journal on Wireless Communications and Networking 1 (2018) 249–258.
[79] R. Kozik, M. Ficco, M. Choraś, F. Palmieri, A scalable distributed machine learning approach for attack detection in edge computing environments, Journal of Parallel and Distributed Computing 119 (2018) 18–26.
[80] A. Abeshu, N. Chilamkurti, Deep learning: the frontier for distributed attack detection in fog-to-things computing, IEEE Communications Magazine 56 (2) (2018) 169–175.
[81] Y. Chen, Y. Zhang, S. Maharjan, M. Alam, T. Wu, Deep learning for secure mobile edge computing in cyber-physical transportation systems, IEEE Network 33 (4) (2019) 36–41.
[82] X. He, R. Jin, H. Dai, Deep PDS-learning for privacy-aware offloading in MEC-enabled IoT, IEEE Internet of Things Journal 6 (3) (2019) 4547–4555.
[83] Y. Wei, F. Yu, M. Song, Z. Han, Joint optimization of caching, computing, and radio resources for fog-enabled IoT using natural actor-critic deep reinforcement learning, IEEE Internet of Things Journal 6 (2) (2019) 2061–2073.
[84] J. Wang, L. Zhao, J. Liu, N. Kato, Smart resource allocation for mobile edge computing: a deep reinforcement learning approach, IEEE Transactions on Emerging Topics in Computing 9 (3) (2021) 1529–1541, https://doi.org/10.1109/TETC.2019.2902661.
[85] J. Chen, S. Chen, Q. Wang, B. Cao, G. Feng, J. Hu, iRAF: a deep reinforcement learning approach for collaborative mobile edge computing IoT networks, IEEE Internet of Things Journal 6 (4) (2019) 7011–7024.
[86] M. Du, K. Wang, Z. Xia, Y. Zhang, Differential privacy preserving of training model in wireless big data with edge computing, IEEE Transactions on Big Data 6 (2) (2020) 283–295.
[87] C. Xu, J. Ren, L. She, Y. Zhang, Z. Qin, K. Ren, EdgeSanitizer: locally differentially private deep inference at the edge for mobile data analytics, IEEE Internet of Things Journal 6 (2019) 5140–5151.
[88] Y. He, F. Yu, Y. He, S. Maharjan, Y. Zhang, Secure social networks in 5G systems with mobile edge computing, caching, and device-to-device communications, IEEE Wireless Communications 25 (3) (2019) 103–109.
[89] Y. Sun, M. Peng, S. Mao, Deep reinforcement learning-based mode selection and resource management for green fog radio access networks, IEEE Internet of Things Journal 6 (2) (2019) 1960–1971.
[90] M.S. Munir, S.F. Abedin, N.H. Tran, C.S. Hong, When edge computing meets microgrid: a deep reinforcement learning approach, IEEE Internet of Things Journal 6 (5) (2019) 7360–7374.

[91] S. Conti, G. Faraci, R. Nicolosi, S.A. Rizzo, G. Schembra, Battery management in a green fog-computing node: a reinforcement-learning approach, IEEE Access 5 (2017) 21126–21138.

[92] B. Wang, Y. Wu, K.R. Liu, T.C. Clancy, An anti-jamming stochastic game for cognitive radio networks, IEEE Journal on Selected Areas in Communications 29 (4) (2011) 877–889.

[93] B. Li, T. Chen, G.B. Giannakis, Secure mobile edge computing in IoT via collaborative online learning, IEEE Transactions on Signal Processing 67 (23) (2019) 5922–5935.

[94] Y. Wang, W. Meng, W. Li, Z. Liu, Y. Liu, H. Xue, Adaptive machine learning-based alarm reduction via edge computing for distributed intrusion detection systems, Concurrency and Computation: Practice and Experience 31 (19) (2019) 1–12.

[95] H.B. McMahan, E. Moore, D. Ramage, S. Hampson, B.A.Y. Arcas, Communication-efficient learning of deep networks from decentralized data, in: Proc. Int. Conf. Artif. Intell. Stat. (AISTATS), 2016, pp. 1273–1282.

[96] E. Li, Z. Zhou, X. Chen, Edge intelligence: on-demand deep learning model co-inference with de-vice-edge synergy, in: Proc. Workshop Mobile Edge Commun. (MECOMM@SIGCOMM), Budapest, Hungary, Aug. 2018, pp. 31–36.

[97] J. Wang, J. Zhang, W. Bao, X. Zhu, B. Cao, P.S. Yu, Not just privacy: improving performance of private deep learning in mobile cloud, in: Proc. 24th ACM SIGKDD Int. Conf. Knowl. Disc. Data Mining, 2018, pp. 2407–2416, https://doi.org/10.1145/3219819.3220106.

[98] M. Chen, U. Challita, W. Saad, C. Yin, M. Debbah, Artificial neural networks-based machine learning for wireless networks: a tutorial, IEEE Communications Surveys and Tutorials 21 (4) (Fourthquarter 2019) 3039–3071.

[99] C. Zhang, H. Zhao, S. Deng, A density-based offloading strategy for IoT devices in edge computing systems, IEEE Access 6 (2018) 73520–73530.

[100] J. Park, S. Samarakoon, M. Bennis, M. Debbah, Wireless network intelligence at the edge, arXiv:1812.02858, 2018.

[101] K. Arulkumaran, M.P. Deisenroth, M. Brundage, A.A. Bharath, Deep reinforcement learning: a brief survey, IEEE Signal Processing Magazine 34 (6) (Nov. 2017) 26–38.

[102] N. Srivastava, G. Hinton, A. Krizhevsky, I. Sutskever, R. Salakhutdinov, Dropout: a simple way to prevent neural networks from overfitting, Journal of Machine Learning Research 15 (56) (2014) 1929–1958.

[103] J. Lee, J.K. Eshraghian, K. Cho, K. Eshraghian, Adaptive precision CNN accelerator using radix-X parallel connected memristor crossbars, arXiv:1906.09395 [abs], 2019.

CHAPTER

TinyML: principles and algorithms

4

Sheetal N. Ghorpade[a], Sachin Chougule[a,b], Bharat S. Chaudhari[b], and Marco Zennaro[c]

[a]*Rubiscape Private Limited, Pune, India*

[b]*Department of Electrical and Electronics Engineering, Dr. Vishwanath Karad MIT World Peace University, Pune, India*

[c]*Science, Technology and Innovation Unit, Abdus Salam International Centre for Theoretical Physics (ICTP), Trieste, Italy*

4.1 Introduction

Over the past decade, remarkable advancements have been made in machine learning algorithms, particularly in conjunction with the Internet of Things (IoT) in microelectronics, communication, and information technology. This emerging area has fueled the proliferation of ubiquitous and unified devices across diverse areas, including wearables, smart cities, infrastructure, and many more. The Internet now connects billions of IoT devices, leading to the development of computing platforms and the generation of a huge amount of heterogeneous data [1–3].

Machine learning algorithms have become crucial for managing such heterogeneous data efficiently. A large number of smart devices will get connected to the Internet in the near future. Typically, most IoT nodes have limited resources and processing capabilities, so their integration with the cloud is an attractive solution. Cloud computing provides significant resources that could empower resource-constrained devices with improved processing, storage, and memory [4]. However, the large geographical distance between sensor nodes and cloud resources adds latency and bandwidth constraints, impeding real-time vital applications and impacting the overall quality of service (QoS). As data scale and complexities have grown, the cloud has challenges to keep up with and concerns over data breaches [5].

Nevertheless, increasing data from cloud-dependent machines can overwhelm cloud resources. To address the issues, one potential solution is that the processing mechanisms available in the cloud must be integrated at the edge. This measure could reduce the frequency of accessing the cloud, alleviating the burden on its resources. Now, data processing is being moved near the end, giving rise to edge computing, also called edge AI. Edge computing can be an intercessor between the data sources and the cloud servers, in which traffic is communicated to the cloud as per the requirements [6]. Furthermore, this approach can significantly enhance privacy and security [7]. Transferring data processing from the cloud to IoT edge devices min-

TinyML for Edge Intelligence in IoT and LPWAN Networks. https://doi.org/10.1016/B978-0-44-322202-3.00009-9

imizes reliance on long backhaul data transceiving. However, it is essential to note that edge hardware's resource-constrained nature restricts it from supporting high-end and complex services. Despite such constraints, it is expected to evolve in the near future and cater to a broader range of applications [8].

One of the challenges lies in the heterogeneous behavior between hardware and web-based technologies. Machine learning applications, for instance, demand resourceful infrastructures for training, weight updates, and model deployment [9]. The current significance of IoT edge lies in its deployability for smart use case development involving machine learning. However, the platform's dependence on embedded system architectures poses difficulties in establishing a standard machine learning framework for all IoT edge systems. The lack of software to complement dedicated machine learning-embedded hardware further widens the gap [10]. Many of the existing embedded processors are capable of handling general-purpose sensor data processing and web-based applications. At the same time, machine learning toolsets trust GPUs and ASICs, which necessitate substantial power and memory capability to execute deep neural network (DNN) models. This challenges the envisioned "cloud-to-embedded" aspect of edge computing. One of the crucial aspects of embedded intelligence is signal processing, which aims to shift cloud intelligence to tiny embedded devices for machine learning, also known as Tiny Machine Learning (TinyML) [11–13]. The TinyML paradigm is yet in its early stages and expects appropriate alignment with existing edge IoT frameworks.

4.2 Overview of TinyML

TinyML enables the deployment of small machine learning models onto tiny edge devices, which face severe resource limitations such as constrained computation, minuscule memory, and low power consumption. Using TinyML, data analysis and interpretation can be performed locally on these devices, allowing real-time actions [14]. TinyML is not an alternative to the fog or cloud paradigms, but it is an auxiliary system that works as an accelerator to the existing paradigms. Fig. 4.1 illustrates TinyML's capability to process sensor data locally on microcontroller-based tiny edge devices. TinyML offers several benefits, including significant cost savings, energy efficiency, and improved privacy protection [15].

The preliminary objective of TinyML is to enhance the efficiency of learning models by reducing computing complexity and data requirements. This advancement may assist the rapid growth of edge AI and the IoT market. In this context, TinyML refers to incorporating structures, frameworks, methodologies, tools, and strategies crafted for conducting on-device analysis across diverse sensing modes, including vision, audio, speech, motion, chemical, physical, textual, and cognitive functions. These functionalities operate within a mW or lower power range, tailored explicitly for battery-powered embedded edge devices. The primary emphasis is deployment in extensive use cases, preferably within IoT or wireless sensor networks [12]. Consequently, TinyML encompasses three fundamental elements: software, hardware,

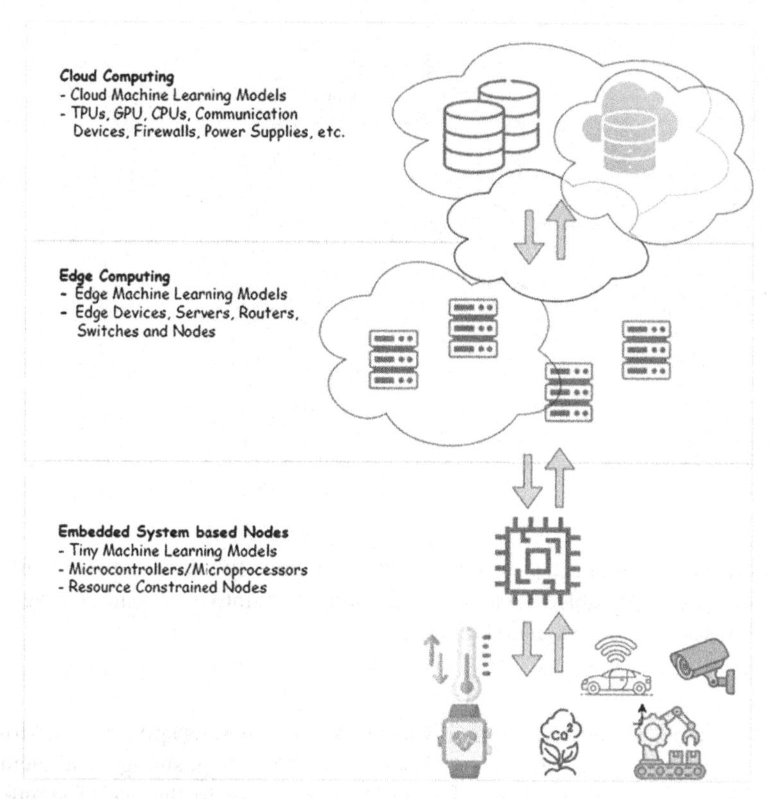

FIGURE 4.1

Cloud computing, edge computing, and TinyML architecture.

and algorithms. Fig. 4.2 depicts the essential components of TinyML, underscoring the importance of an efficient hardware–software co-design. These systems should conform to optimized machine learning principles, ensuring high data quality and a well-organized software architecture.

4.2.1 Benefits of TinyML

The major advantages of TinyML when integrated with IoT devices, especially for edge or daily-use devices, include the following.

- **Energy efficiency**

One of the significant benefits of adopting TinyML on microcontroller units (MCUs) for IoT applications is its energy efficiency. TinyML on MCUs consumes much less energy than powerful processors and GPUs. This low power consumption is crucial for IoT devices that rely on batteries or energy harvesting. As a result, IoT devices can be deployed everywhere, without requiring continual connection to the power

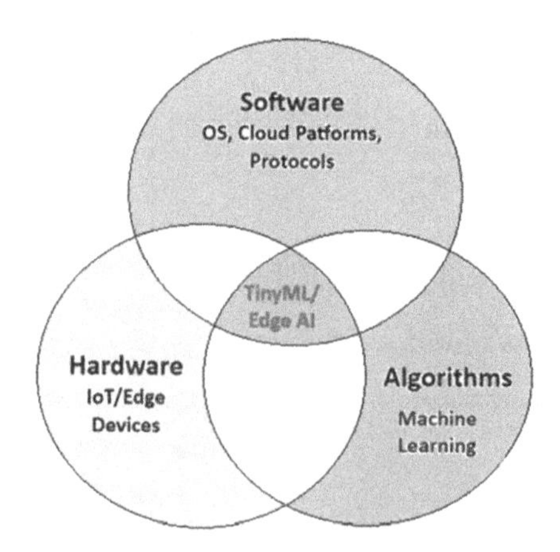

FIGURE 4.2

TinyML components.

grid. Moreover, the reduced power consumption permits IoT devices to be combined with sizable battery-enabled devices, transforming them into linked smart things, like scooters or seaways, for extended lifetimes [16].

- **Low cost**

IoT devices are designed to perform diverse tasks in numerous applications. Most of the tasks demand higher computational power for processing, storage, and memory. These extensive requirements translate to high costs due to the use of significant resources and reliance on cloud-based processing and storage. In contrast, TinyML processes data locally on microcontroller devices, leading to a lower cost and high-performance AI processing on the device itself. Microcontrollers are cost-effective solutions compared to other alternatives, with processors typically operating in the range of 1 MHz to 400 MHz, memory ranging from 2 kB to 512 kB, and storage capacities of 32 KB to 2 MB. This affordability makes microcontrollers attractive for local data processing [16].

- **Latency**

With TinyML, data handling occurs locally on the device, reducing the latency associated with transmitting data to remote servers or the cloud. By performing computations on the device, IoT devices benefit from quicker response times and speedy analysis, particularly in emergency and time-critical scenarios. Additionally, local data processing reduces the burden on the cloud, enhancing overall system efficiency [17,18].

- **System reliability and data security**

Traditional IoT devices often rely on communication channels to transmit raw data from the device to the cloud for processing. However, this approach exposes the data to potential transmission errors, cyberattacks, and compromises in data security. Data breaches can result in significant financial losses. TinyML performs data processing locally on the device to address these security concerns, limiting unnecessary cloud traffic. This approach reduces the risk of data interception and ensures that only aggregated or less sensitive data are transmitted, mitigating potential attacks [16–18].

4.3 Workflow of TinyML

The standard procedure for implementing TinyML, as shown in Fig. 4.3, comprises three distinct phases: training, optimization, and deployment. This architecture can be abstracted into two main components: traditional machine learning and TinyML. The former encompasses data collection, algorithm selection, model training, and optimizing phases, aligning with conventional practices in traditional machine learning approaches. On the other hand, the latter focuses on TinyML models and involves the model porting and deployment phases [19].

The first step for developing a TinyML model is selecting a relevant algorithm and gathering the necessary data. These data can be sourced from precompiled repositories or collected in real time from sensors. After the required data are collected, the model training process can be done on a server or desktop. Similar to conventional machine learning, modern machine learning frameworks like TensorFlow, PyTorch, and scikit-learn can be utilized for training purposes.

After the model training, the next step is optimization. Techniques such as pruning, knowledge distillation, quantization, and encoding are employed to meet specific requirements [20,21]. Pruning involves iteratively eliminating redundant/unnecessary parameters from a neural network [22], while knowledge distillation entails training a small, well-optimized model to mimic the behavior of a pretrained model [23].

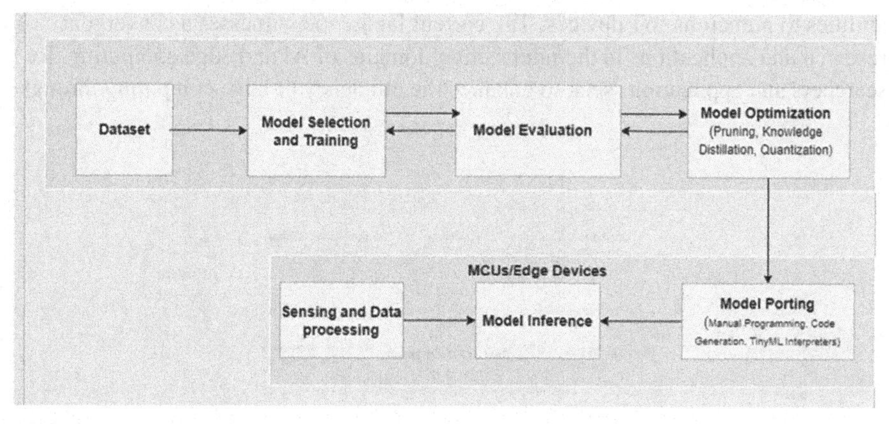

FIGURE 4.3

TinyML at MCUs/edge devices.

Quantization refers to reducing the numerical precision of a network, often achieved by representing weights and activations as integers instead of floating points (32/64 bit) [24]. However, caution is needed, as quantizing a network from single to low precision may impact performance.

The final stages in a TinyML solution are porting and deployment. Before deployment, the model must be converted to a language understandable by MCUs. TinyML interpreters, like TensorFlow Lite Micro [25], are employed to convert a model built in a standard programming language into a frozen graph, ideally represented as a C array [26,27]. Lastly, the ported model is embedded into an MCU, where inference is conducted on sensor data with the assistance of the underlying machine learning mechanism.

4.4 AI and edge computing

This section analyzes the individual progress of AI and edge computing, exploring the reasons behind their integration. Subsequently, a summary of relevant AI algorithms is discussed. Lastly, a brief discussion on AI-driven algorithms pertaining to various areas, including optimization for computing offloading, methods for reducing energy consumption without computing offloading, security in edge computing, safeguarding data privacy, and optimizing resource allocation, is presented.

Fig. 4.4 presents a mutually beneficial association between traditional AI and edge computing. It shows that the optimization and advancement of edge computing rely on AI algorithms' support, particularly in areas such as computation offloading optimization [28]. On the other hand, edge computing must be deployed near terminal devices. This proximity is essential to fulfill the demands of specific AI applications sensitive to latency, such as those found in smart city scenarios. In the era of big data, AI stands out as a crucial technology, imparting intelligence and interpretation abilities to numerous IoT devices. The current landscape witnesses a convergence of research and applications in the intersecting domains of AI and edge computing. Researchers and applications seek to enhance the efficiency of edge computing through

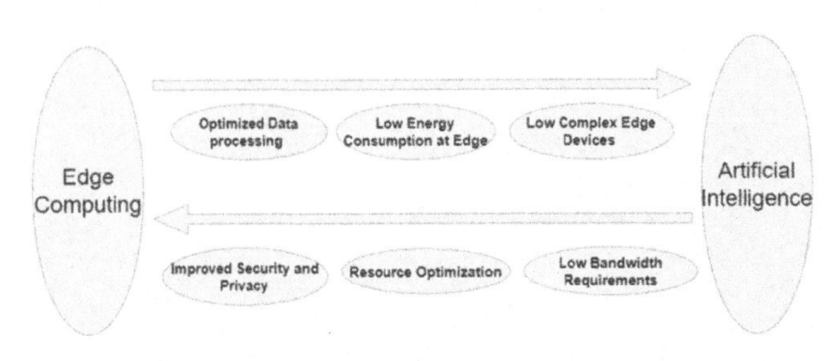

FIGURE 4.4

Association between AI and edge computing.

the incorporation of AI algorithms. This synergy aids in optimizing and deploying edge computing resources effectively.

4.4.1 Edge computing for AI applications

Edge computing plays a pivotal role in supporting AI applications that require proximity to terminal devices for reduced latency and enhanced network stability. This collaboration ensures that necessary computing functions are provided in close proximity to facilitate optimal AI application performance [29]. The computational power and energy support required for AI's reasoning and training often exceed the capabilities of terminal devices. In the past few years, cloud computing has addressed this gap by offloading AI model training and reasoning tasks that exceed the capacities of terminal devices to cloud servers. However, relying solely on cloud computing presents challenges like inadequate bandwidth and high latency, mainly when numerous terminal devices simultaneously utilize many AI models [30]. The emergence of edge computing provides a solution by enabling the deployment of AI near terminal devices and users at the edge, equipped with specific computing and storage resources. This approach caters to the demands for low latency and high network permanence [31]. Additionally, edge computing introduces a few additional concepts to enhance the application of AI in various domains.

- **Preprocessing huge data locally and cloud-based AI training/reasoning**

In this approach, massive amounts of data undergo local preprocessing before being uploaded to the cloud for AI training and reasoning [32]. While this strategy alleviates the pressure on bandwidth and transmission costs, it fails to meet latency requirements for specific applications, such as those in the Internet of Vehicles (IoV) and augmented reality/virtual reality domains.

- **Edge- or end-based AI reasoning for reduced latency**

To minimize latency, AI reasoning tasks are executed at the edge or the end, while model training tasks continue to be executed in the cloud [32]. This configuration optimizes latency-sensitive applications without compromising the advantages of cloud-based model training.

4.4.1.1 Delegating AI training and reasoning tasks to the edge

This concept involves assigning either a portion or the entirety of AI training and reasoning tasks to the edge. This approach enhances the location awareness of AI models while concurrently alleviating latency and bandwidth pressures [33]. It is important to note that the energy consumption and computing power requirements of edge devices will rise with an increasing number of tasks shifted to the edge. We may have three approaches: preprocessing data locally and cloud-based AI training/reasoning, edge- or end-based AI reasoning for reduced latency, or delegating tasks to the edge, all with their advantages and disadvantages. Consequently, existing studies tend to select the most suitable approach based on the specific circumstances.

4.4.2 AI algorithms in edge computing

AI is pivotal in enhancing the optimization of edge computing [34]. Given the distributed nature of edge computing and the dynamic fluctuations in workload across edge devices concerning both time and location, the resulting uncertainty and unpredictability pose substantial challenges to edge computing applications. Consequently, there is an ongoing need for optimization and improvement in various facets of edge computing, including computing offloading, resource allocation, latency reduction, energy consumption, and user experience enhancement.

Most of the optimization issues within edge computing are intricate and fall into complex non-convex problems. The scale of these problems rapidly escalates with the increasing number of devices and users [35]. Machine learning proves to be particularly well suited for addressing these optimization challenges in edge computing, offering superior outcomes compared to traditional methods [36]. Furthermore, AI algorithms effectively extract hidden information and discerning patterns from data in the intricate and noisy environments typical of edge computing. This capability addresses longstanding challenges that have plagued traditional optimization methods.

Various AI algorithms are employed in edge computing, primarily emphasizing machine learning. The important traditional machine learning algorithms, deep learning, reinforcement learning, and deep reinforcement learning (DRL), are discussed here. While the focus lies within the realm of machine learning, it is worth noting that other algorithms, such as evolutionary algorithms, play an important role in machine learning and hence are briefly touched upon in this section.

We will delve into each category of AI algorithms, providing insights into their applications within the context of edge computing. Examples will be highlighted to illustrate how these algorithms contribute to optimizing various aspects of edge computing, aligning with the overarching theme of our discussion on machine learning in the AI algorithm landscape for edge computing.

4.4.2.1 Traditional machine learning

Traditional machine learning algorithms specifically encompass those machine learning techniques distinct from deep learning and reinforcement learning. Based on label information in the data, conventional machine learning algorithms can be categorized into supervised learning, unsupervised learning, and semi-supervised learning. In supervised learning, labeled data are necessary for training the model. Common supervised learning methods include support vector machine (SVM), boosting, random forest, and others. Unsupervised learning autonomously discovers underlying patterns in data. Common techniques include clustering algorithms, which are distance-based, centroid, and hierarchical, and dimensionality reduction approaches, such as principal component analysis (PCA). Semi-supervised learning, a fusion of supervised and unsupervised learning, utilizes labeled as well as unlabeled data. It includes label propagation and graphical models.

Traditional machine learning algorithms exhibit certain drawbacks, such as sensitivity to datasets, diminished effectiveness with large datasets, and the need for intricate artificial feature engineering. Despite these limitations, traditional machine learning possesses advantages, such as low energy consumption, minimal computing power costs, and ease of deployment, compared to deep learning and reinforcement learning. The distributed nature of edge computing enables the well-judged selection of AI algorithms based on the objective and task requirements of each edge and terminal device. Leveraging these advantages, traditional machine learning can find a fitting role within edge computing [28,37].

4.4.2.2 Deep learning

Deep learning mirrors the functionalities of the human brain, showcasing the ability to autonomously learn high-level features from raw data, making it adept at tasks like classification and prediction [38,39]. Deep learning architectures typically consist of multiple layers, including fully connected, convolutional, pooling, normalization, and activation layers. The arrangement of these layers forms a deep learning algorithm, with the depth of the algorithm determined by the number of layers it incorporates. Neurons in each layer receive as input the weighted sum of outputs from neurons in the preceding layer. Following activation by a specified function, the resulting number serves as the neuron's output [40]. Deep learning excels in extracting high-level features from extensive data compared to traditional machine learning algorithms, courtesy of its multilayer structure [41]. Regular deep learning models comprise the following neural networks:

- **Deep neural networks**

DNNs, also known as multiple linear perceptron (MLP), comprise multiple hidden layers, with layers categorized as input, hidden, and output. The addition of hidden layers enhances the learning capacity of the DNN model.

- **Convolutional neural networks**

Convolutional neural networks (CNNs) comprise various convolution layers, utilizing convolution operations to extract high-level features from input data. They demonstrate strong representation and image recognition capabilities, with fault detection and video surveillance applications within edge computing [42,43].

- **Recurrent neural networks**

Recurrent neural networks are adept at modeling and processing sequential data. However, they suffer from the issue of forgetting, where the impact of initial inputs diminishes over time. To address this, an improved version called long short-term memory (LSTM) was introduced [44]. Some edge computing studies have leveraged LSTM to address challenges in anomaly detection [45], task scheduling, resource allocation [46], privacy protection [47], and others.

In scenarios with abundant labeled data, deep learning outperforms traditional machine learning algorithms in fields like natural language processing and computer vision [39]. The characteristics of edge computing, where data can be processed locally in the physical environment, align with deep learning requirements. Consequently, deep learning finds applications in edge computing for anomaly detection, task scheduling, resource allocation, and privacy protection [45–47].

4.4.2.3 Reinforcement learning and deep reinforcement learning

Reinforcement learning stands apart from supervised and unsupervised learning by engaging in dynamic interaction with the environment rather than relying on static data. The fundamental concept involves training models through agents that receive the state of the environment and take actions to maximize rewards based on historical experiences. Due to its aptitude for addressing decision-making problems, some studies have integrated reinforcement learning algorithms into the decision-making processes of edge computing resource managing, allocation, and scheduling [48,49].

A prominent algorithm in reinforcement learning is the model-free and value-based Q-learning algorithm [50]. In every iteration, this algorithm calculates an anticipated cumulative reward, known as the Q-value, based on the existing condition and a given action. Nevertheless, as the environment becomes more complex, the state and action spaces grow rapidly, reducing convergence speed and increasing memory usage [51]. To address these challenges, the deep Q-network (DQN) [52] utilizes a DNN to approximate Q-values. Compared to classical reinforcement learning algorithms, DQN offers multiple benefits in managing edge computing with higher complexity [53]. It can handle high-dimensional and complex systems, learn the regularities of the system environment, and make optimum decisions by considering the present and historical long-term rewards.

Few of the techniques proposed in the literature leverage DQN algorithms to enhance control decision-making problems in edge computing, yielding favorable outcomes [54]. However, DQN has its drawbacks, particularly when using nonlinear functions, such as neural networks, to approximate the Q-function results in uneven or divergent learning. To mitigate such issues, an experience repetition method that incorporates previous experiences into DQN is proposed [55]. Multi-armed bandit-based algorithms are promising candidates for low-power reinforcement learning applications [56,57].

4.4.2.4 Federated learning

Federated learning stands as a distributed machine learning framework designed to effectively facilitate the training of models across multiple organizations while adhering to user privacy protection, data security, and government regulations [58]. In this framework, it is not required to transmit all the raw data to a central server for training. Instead, they train local models using secrecy-sensitive data, and subsequently, these local models are combined into a global model on the central server [59]. As outlined earlier, edge computing aims to deploy computation tasks on the network's edge, closer to the client. However, the data from an individual edge node might not suffice for comprehensive model training. Consequently, collaborative model training among different nodes, focusing on data privacy protection, has become a prominent area of research [60].

4.4.2.5 Evolutionary algorithms

Evolutionary algorithms represent a class of optimization methods inspired by the mechanisms and behaviors observed in biological evolution [61]. This category encompasses various techniques like particle swarm optimization (PSO), genetic algorithm (GA), differential evolution (DE), gray wolf optimization (GWO), whale optimization algorithm (WOA), and particle swarm gray wolf optimization (PSGWO), among others [62–65]. The procedural steps of evolutionary algorithms involve initializing variables, followed by continuous iteration through three key phases: fitness evaluation and selection, population reproduction and variation, and population updating [61]. This iterative process continues until a termination condition is met, marking the completion of the second step. Currently, evolutionary algorithms find applications in numerous problems within evolutionary computation, including resource scheduling optimization [66], load balancing [67], and task scheduling [68]. However, in this discussion focusing on machine learning, an emerging subset of AI, we briefly introduce evolutionary algorithms.

4.5 Edge computing optimization using AI solutions

A thorough overview of research endeavors employing AI methodologies to enhance edge computing across diverse contexts, encompassing computing offloading, energy consumption reduction, bolstering edge computing security, ensuring data privacy, and optimizing resource allocation.

4.5.1 Computing offloading optimization

This section provides an overview of AI-based computing offloading schemes designed to address the objectives of reducing energy consumption, minimizing latency, and achieving a balance between the two. Ali et al. [69] introduced a partial computing offloading scheme relying on deep learning decision-making. This novel decision-making process intelligently selects the most suitable computing offloading strategy, effectively lowering the overall energy consumption during task execution. This approach, compared to their earlier work [70], notably incorporates user equipment's energy consumption in the cost function, resulting in a 3% reduction in energy consumption.

In the context of latency reduction, Smart-Edge-CoCaCo [71] is proposed to minimize latency by optimizing the wireless communication model, collaborative filter caching model, and computing offloading model collectively. Despite the inherent low latency advantage of edge computing compared to cloud computing, there is still room for improvement. Meanwhile, Xu et al. [72] introduced a deep learning-based heuristic offloading method that leverages origin–destination electronic communications network distance estimation and heuristic searching to identify the optimal computing offloading strategy, considering the limited computing power of edge devices.

Some approaches leverage reinforcement learning to reduce energy consumption and latency. Kiran et al. [73] presented a scheme employing Q-learning for optimal control decisions, aiming to reduce delay in edge computing. This scheme also incorporates constraints in the cost function to curtail energy consumption in edge computing. Despite its effectiveness in achieving both objectives, it is important to note that this approach does not explicitly address the curse of dimensionality problem associated with edge computing.

The curse of dimensionality poses a challenge where problem complexity increases exponentially with growing dimensionality [74,75]. To tackle this issue, Xu et al. [75] introduced an algorithm that leverages the unique structure of state transitions in the considered edge computing system to mitigate the curse of dimensionality problem. Notably, the authors incorporated energy harvesting [76] to diminish reliance on traditional energy sources by maximizing the use of renewable energy. However, this approach requires knowledge of both the transmission delay model and the energy consumption model, a requirement alleviated by the method proposed in reference [77].

In contrast to reinforcement learning algorithms, DRL algorithms demonstrate enhanced capabilities in handling high-dimensional state spaces. Cheng et al. [78] presented a model-free DRL-based computing offloading method within a space–air–ground integrated network, utilizing a Markov decision process to represent the computing offloading decision process and DRL to learn network dynamics. However, the effectiveness of DRL algorithms in coping with high-dimensional state spaces is not flawless. Chen et al. [79] introduced a new DNN model based on a function approximator, incorporating a double deep Q-network to discover optimal offloading strategies without prior knowledge. Similarly, Lei et al. [80] proposed a novel value function approximator to address high-dimensional state equations. They employed an infinite-horizon average reward continuous-time Markov decision process to represent the optimal problem. DRL is applied to optimize computing offloading decisions for reduced energy consumption and latency in edge computing.

The DRL-based methods mentioned above primarily adopt a centralized style for model learning, assuming that edge devices in edge computing possess sufficient computing power. Acknowledging the limitations of this assumption, Ren et al. proposed a distributed computing offloading strategy, integrating federated learning and multiple DRLs [81]. Experimental evidence indicates that this method outperforms centralized learning in reducing transmission costs in edge computing. Furthermore, distributed learning demonstrates the advantage of rapid convergence, as confirmed in reference [82], by optimizing computing offloading using distributed machine learning.

4.5.2 Energy consumption reduction using non-computation offloading methods

Edge computing offers computing capabilities near data sources, providing high response speed for various applications. However, this model raises concerns about increased energy consumption on the edge, especially when AI algorithms are required for real-time decisions in applications like intelligent driving and monitoring systems. The computational intensity of AI algorithms poses a significant challenge for devices with limited power. Managing global overall energy consumption and enhancing energy efficiency are crucial considerations in the context of the growing prevalence of AI.

Beyond computation offloading, several factors impact the energy consumption of edge devices, including the choice of AI algorithms and hardware structures. This discussion explores AI-driven solutions to reduce edge computing energy consumption by optimizing hardware structures, controlling operating statuses, and integrating with the Energy Internet.

4.5.3 Hardware structure optimization

Utilizing static random access memory (SRAM) with parallel CNNs for simultaneous access to different memory blocks significantly reduces energy consumption compared to traditional digital accelerators.

By designing a binarized DNN accelerator for weed species classification on an FPGA, a sevenfold reduction in energy consumption compared to GPU-based accelerators was achieved.

4.5.3.1 Controlling operating status

A method based on DRL controls the communication modes of user devices and the light-on state of processors to reduce medium- and long-term energy consumption in edge computing. It uses a Markov process to model energy consumption and DRL for decision-making, optimizing user device precoding based on constraints.

4.5.3.2 Combining Energy Internet

Integrating the Energy Internet (smart grid and microgrid) with edge computing addresses edge devices' dynamic workload and unpredictable energy consumption. The combination leverages renewable energy sources to achieve local energy self-sufficiency, reducing reliance on non-renewable energy.

DRL-based control strategies aim to balance energy supply and demand in edge computing, considering the uncertainties in renewable energy production.

The distributed nature of edge computing, coupled with the integration of the Energy Internet, provides a framework for optimizing energy consumption and enhancing sustainability. Energy management complexities arising from deploying edge computing devices into the Energy Internet are addressed through DRL, often combined with curriculum learning, to implement effective bottom-up energy management schemes.

4.5.4 Security of edge computing

Transferring computational and storage responsibilities from centralized cloud systems to edge computing can mitigate security issues arising from network congestion and centralization. Nevertheless, the decentralized nature of edge computing introduces new security challenges, including distributed denial-of-service (DDoS) attacks and jamming attacks leading to the illicit distribution of distributed system resources [33,83]. Approaches effective in centralized environments like cloud computing may not suffice to address these emerging security concerns. This section explores research endeavors focusing on enhancing edge computing security through the application of AI algorithms, encompassing traditional machine learning and deep learning methods.

Conventional machine learning methods play a role in identifying and classifying various attacks. Wang et al. [84] proposed a stochastic game framework utilizing minimax Q-learning to counter jamming attacks, enhancing spectral efficiency throughput. Despite requiring additional bandwidth, an alternative approach involves selecting the most reliable server through online learning to mitigate security risks associated with jamming attacks [85]. To enhance the performance of traditional intrusion detection systems, an algorithm selection mechanism on the edge side helps intelligently choose the optimal machine learning algorithm for edge devices, reducing false alarms and data transmission delays [86]. Experimental results underscore the efficacy of AI algorithms in enhancing edge computing security compared to non-AI alternatives.

Among common network attacks, DDoS is addressed using hypergraph clustering [87] to model the relationships between edge nodes, improving recognition rates [88]. Kozik et al. utilized a single-layer neural network for extreme learning machine classification, where the attack detection model is trained in the cloud and then offloaded to edge devices for deployment. This approach exhibits faster convergence and more robust generalization performance than traditional algorithms like SVM or the single-layer perceptron.

Transitioning to deep learning methods, researchers recognized the limitations of traditional machine learning algorithms in automatic feature extraction, making them less sensitive to slightly modified known attacks and ineffective against zero-day attacks [89,90]. With its ability to automatically extract hidden features from extensive data, deep learning becomes a focus for cybersecurity in edge computing. Abeshu et al. [38] proposed a deep learning-based attack detection method, leveraging a pretrained stacked autoencoder for feature selection and softmax for classification, showcasing advantages in availability, scalability, and effectiveness compared to traditional machine learning algorithms. However, this method neglects to enhance the detection rate for new attacks, a gap addressed by unsupervised learning.

A deep learning-based algorithm [91] incorporates unsupervised learning-restricted Boltzmann machines, an active learning stochastic artificial neural network, to improve the recognition rate for previously unseen attacks. Utilizing a deep belief network and softmax function, this model effectively learns to attack characteristics in edge computing, demonstrating the potential of unsupervised learning to enhance security against novel threats.

4.5.5 Data privacy

To certain degrees, edge computing mitigates the risk of privacy breaches associated with uploading data to uncontrollable cloud servers. Nevertheless, the challenge of data privacy leakage also persists on the edge. The distributed nature of edge computing introduces new obstacles to privacy protection, and the utilization of AI on edge demands substantial data for training models, inevitably entailing a significant amount of user privacy. Certain models may inadvertently store portions of the training set containing private data during training, enabling attackers to access user privacy through model analysis [92] illicitly. Subsequently, it is crucial to guarantee data privacy and security for edge users without compromising edge computing performance.

A postdecision state (PDS) learning method, proposed in reference [93], addresses the edge computing offloading problem by factoring the state transition function into known and unknown components. This approach employs the Markov decision process to describe edge computing's offloading challenge. It combines PDS-learning techniques with the traditional deep Q-network algorithm to achieve a well-balanced solution for task scheduling and privacy protection. Notably, this new algorithm expedites model training by incorporating additional information, such as edge device energy utilization, compared to the traditional deep Q-network.

Federated learning introduces a privacy-preserving asynchronous federated learning mechanism (PAFLM) for edge computing, enabling multiple edge nodes to enhance federated learning efficiency without sharing private data and compromising inference accuracy [60]. Local model training in each node often leads to local optima. However, federated learning helps optimize the local model using parameters from other nodes, addressing the local optimum problem and improving model accuracy.

Differential privacy is frequently integrated with AI algorithms to safeguard user privacy in edge computing training datasets. Differential privacy ensures that the inclusion or exclusion of any data piece does not significantly alter the results of related data analysis. Du et al. [94] proposed two AI-based algorithms satisfying differential privacy: the objective and output perturbation algorithms. The former adds Laplace noise to objective functions, while the latter adds noise to outputs. This injection of Laplace noise enhances the efficiency and accuracy of machine learning algorithms in prediction and proves more effective in safeguarding training data privacy in edge computing. A differential privacy-based deep reasoning framework called Edge Sanitizer has also been introduced [95]. This framework maximizes the utilization of helpful information through a deep learning-based data minimization method. It removes sensitive private information by introducing random noise via local differential privacy, ensuring maximal data usage while preserving privacy in edge computing.

4.5.6 Resource allocation optimization

DRL has demonstrated its ability to address dynamic decision challenges involving high-dimensional states and action spaces [96]. There is a focus on applying DRL to tackle resource allocation issues in edge computing. In reference [56], a method is introduced to account for the constantly changing state of the edge computing environment. Information such as wireless channel conditions, node trust values, cache contents, and available computational capacity is fed into a DNN to estimate the Q-function. While this technique yields positive results, there is still room for enhancing convergence and performance.

Despite the effectiveness of DQN in optimizing dynamic decision problems in high-dimensional state space, limitations persist when addressing problems rooted in high-dimensional action space. Therefore, Chen et al. [96] proposed a novel DRL-based resource allocation decision framework, contributing in the following ways.

The framework employs DNN training with a self-supervised process to predict resource allocation actions, utilizing training data generated by the Monte Carlo tree search (MCTS) algorithm [97].

The authors modified the last layer of the conventional DNN used for Q-function estimation, enabling support for a higher-dimensional action space. Experimental results demonstrate that, compared to the direct use of DQN, this method reduces delay by 51.71%.

References

[1] S. Hamdan, M. Ayyash, S. Almajali, Edge-computing architectures for Internet of Things applications: a survey, Sensors 20 (2020) 6441.

[2] Z. Wu, K. Qiu, J. Zhang, A smart microcontroller architecture for the Internet of Things, Sensors 20 (2020) 1821.

[3] G. Signoretti, M. Silva, P. Andrade, I. Silva, E. Sisinni, P. Ferrari, An evolving TinyML compression algorithm for IoT environments based on data eccentricity, Sensors 21 (2021) 4153.

[4] Bharat S. Chaudhari, Marco Zennaro, Introduction to low-power wide-area networks, in: Bharat S. Chaudhari, Marco Zennaro (Eds.), LPWAN Technologies for IoT and M2M Applications, Academic Press, ISBN 9780128188804, 2020, pp. 1–13, https://doi.org/10.1016/B978-0-12-818880-4.00001-6.

[5] Marcelo Antonio Marotta, Leonardo Roveda Faganello, Matias Artur Klafke Schimuneck, Lisandro Zambenedetti Granville, Juergen Rochol, Cristiano Bonato Both, Managing mobile cloud computing considering objective and subjective perspectives, Computer Networks 93 (2015) 531–542.

[6] Michael Armbrust, Armando Fox, Rean Griffith, Anthony D. Joseph, Randy Katz, Andy Konwinski, Gunho Lee, David Patterson, Ariel Rabkin, Ion Stoica, et al., A view of cloud computing, Communications of the ACM 53 (4) (2010) 50–58.

[7] J. Singh, Y. Bello, A.R. Hussein, A. Erbad, A. Mohamed, Hierarchical security paradigm for IoT multiaccess edge computing, IEEE Internet of Things Journal 8 (7) (2021) 5794–5805, https://doi.org/10.1109/JIOT.2020.3033265.

[8] D. Wu, X. Huang, X. Xie, X. Nie, L. Bao, Z. Qin, LEDGE: leveraging edge computing for resilient access management of mobile IoT, IEEE Transactions on Mobile Computing 20 (3) (2021) 1110–1125, https://doi.org/10.1109/TMC.2019.2954872.

[9] C. Guleria, K. Das, A. Sahu, A survey on mobile edge computing: efficient energy management system, in: 2021 Innovations in Energy Management and Renewable Resources(52042), 2021, pp. 1–4, https://doi.org/10.1109/IEMRE52042.2021.9386951.
[10] W. Ren, Y. Sun, H. Luo, M. Guizani, A demand-driven incremental deployment strategy for edge computing in IoT network, IEEE Transactions on Network Science and Engineering 9 (2) (2022) 416–430, https://doi.org/10.1109/TNSE.2021.3120270.
[11] T. Ogino, Simplified multi-objective optimization for flexible IoT edge computing, in: 2021 4th International Conference on Information and Computer Technologies (ICICT), 2021, pp. 168–173, https://doi.org/10.1109/ICICT52872.2021.00035.
[12] P. Warden, D. Situnayake, TinyML: machine learning with TensorFlow lite on Arduino and ultra-low-power microcontrollers, O'Reilly Media. 6G White Paper on Edge Intelligence, https://arxiv.org/abs/2004.14850, 2019.
[13] TinyML, https://cms.tinyml.org/wp-content/uploads/emea2021/tinyML_Talks_Felix_Johnny_Thomasmathibalan_and_Fredrik_Knutsson_210208.pdf. (Accessed November 2021).
[14] TinyML, https://www.tinyml.org/. (Accessed November 2021).
[15] S. Vadera, S. Ameen, Methods for pruning deep neural networks, IEEE Access 10 (2022) 63280–63300, https://doi.org/10.1109/ACCESS.2022.3182659.
[16] Norah N. Alajlan, Dina M. Ibrahim, TinyML: enabling of inference deep learning models on ultra-low-power IoT edge devices for AI applications, Micromachines 13 (6) (2022) 851, https://doi.org/10.3390/mi13060851.
[17] Simone Disabato, Manuel Roveri, Incremental on-device tiny machine learning, in: AIChallengeIoT'20, Association for Computing Machinery, New York, NY, USA, 2020, pp. 7–13, https://doi.org/10.1145/3417313.3429378.
[18] Hiroshi Doyu, Roberto Morabito, Martina Brachmann, A TinyMLaaS ecosystem for machine learning in IoT: overview and research challenges, in: 2021 International Symposium on VLSI Design, Automation and Test (VLSI-DAT), 2021, pp. 1–5, https://doi.org/10.1109/VLSI-DAT52063.2021.9427352.
[19] Lachit Dutta, Swapna Bharali, TinyML meets IoT: a comprehensive survey, IEEE Internet of Things Journal 16 (2021) 100461.
[20] Visal Rajapakse, Ishan Karunanayake, Nadeem Ahmed, Intelligence at the extreme edge: a survey on reformable TinyML, ACM Computing Surveys 55 (13s) (December 2023) 282, https://doi.org/10.1145/3583683.
[21] Sachin B. Chougule, Bharat S. Chaudhari, Sheetal N. Ghorpade, Marco Zennaro, Exploring computing paradigms for electric vehicles: from cloud to edge intelligence, challenges and future directions, World Electric Vehicle Journal 15 (2) (2024) 39, https://doi.org/10.3390/wevj15020039.
[22] Miguel de Prado, Manuele Rusci, Alessandro Capotondi, Romain Donze, Luca Benini, Nuria Pazos, Robustifying the deployment of tinyML models for autonomous mini-vehicles, Sensors 21 (4) (2021) 1339.
[23] Davis Blalock, Jose Javier Gonzalez Ortiz, Jonathan Frankle, John Guttag, What is the state of neural network pruning?, arXiv preprint, arXiv:2003.03033, 2020.
[24] Wonpyo Park, Dongju Kim, Yan Lu, Minsu Cho, Relational knowledge distillation, in: Proceedings of the IEEE/CVF Conference on Computer Vision and Pattern Recognition, 2019, pp. 3967–3976.
[25] Yaohui Cai, Zhewei Yao, Zhen Dong, Amir Gholami, Michael W. Mahoney, Kurt Keutzer, ZeroQ: a novel zero shot quantization framework, in: Proceedings of the IEEE/CVF Conference on Computer Vision and Pattern Recognition, 2020, pp. 13169–13178.

[26] Robert David, Jared Duke, Advait Jain, Vijay Janapa Reddi, Nat Jeffries, Jian Li, Nick Kreeger, Ian Nappier, Meghna Natraj, Tiezhen Wang, et al., TensorFlow Lite Micro: embedded machine learning for TinyML systems, Proceedings of Machine Learning and Systems 3 (2021).
[27] S.N. Ghorpade, M. Zennaro, B.S. Chaudhari, Towards green computing: intelligent bio-inspired agent for IoT-enabled wireless sensor networks, International Journal of Sensor Networks 35 (2) (Feb 2021) 121–131, https://doi.org/10.1504/IJSNET.2021.113632.
[28] Igor Fedorov, Ryan P. Adams, Matthew Mattina, Paul N. Whatmough, SpArSe: sparse architecture search for CNNs on resource-constrained microcontrollers, in: Proceedings of the 33rd International Conference on Neural Information Processing Systems, 2019, pp. 4977–4989.
[29] Haochen Hua, Yutong Li, Tonghe Wang, Nanqing Dong, Wei Li, Junwei Cao, Edge computing with artificial intelligence: a machine learning perspective, ACM Computing Surveys 55 (9) (September 2023) 184, https://doi.org/10.1145/3555802.
[30] M.S. Hossain, G. Muhammad, S.U. Amin, Improving consumer satisfaction in smart cities using edge computing and caching: a case study of date fruits classification, Future Generations Computer Systems 88 (2018) 333–341.
[31] F. Samie, L. Bauer, J. Henkel, From cloud down to things: an overview of machine learning in Internet of things, IEEE Internet of Things Journal 6 (3) (2019) 4921–4934.
[32] Z. Zhou, X. Chen, E. Li, L. Zeng, K. Luo, J. Zhang, Edge intelligence: paving the last mile of artificial intelligence with edge computing, Proceedings of the IEEE 107 (8) (2019) 1738–1762.
[33] C. Liu, Y. Cao, L. Yan, G. Chen, H. Peng, A new deep learning-based food recognition system for dietary assessment on an edge computing service infrastructure, IEEE Transactions on Services Computing 11 (2018) 249–261.
[34] D. Liu, Z. Yan, W. Ding, M. Atiquzzaman, A survey on secure data analytics in edge computing, IEEE Internet of Things Journal 6 (3) (2019) 4946–4967.
[35] S. Deng, H. Zhao, W. Fang, J. Yin, S. Dustdar, A.Y. Zomaya, Edge intelligence: the confluence of edge computing and artificial intelligence, IEEE Internet of Things Journal 7 (8) (2020) 7457–7469.
[36] Z. Yang, C. Pan, K. Wang, M. Shikh-Bahaei, Energy efficient resource allocation in UAV-enabled mobile edge computing networks, IEEE Transactions on Wireless Communications 18 (9) (2019) 4576–4589.
[37] N. Kiran, C. Pan, S. Wang, C. Yin, Joint resource allocation and computation offloading in mobile edge computing for SDN based wireless networks, Journal of Communications and Networks 22 (1) (2020) 1–11, https://doi.org/10.1109/JCN.2019.000046.
[38] Y. Guo, S. Wang, A. Zhou, J. Xu, J. Yuan, C. Hsu, User allocation-aware edge cloud placement in mobile edge computing, Software, Practice & Experience 50 (10) (2019) 489–502.
[39] A. Abeshu, N. Chilamkurti, Deep learning: the frontier for distributed attack detection in fog-to-things computing, IEEE Communications Magazine 56 (2) (2018) 169–175.
[40] Y. LeCun, Y. Bengio, Deep learning, Nature 521 (7553) (2015) 436–444.
[41] D.L. Elliot, A Better Activation Function for Artificial Neural Networks, University of Maryland, Systems Research Center, 1993.
[42] H. Li, K. Ota, M. Dong, Learning IoT in edge: deep learning for the Internet of things with edge computing, IEEE Network 32 (1) (2018) 96–101.
[43] P. Monkam, S. Qi, H. Ma, W. Gao, Y. Yao, W. Qian, Detection and classification of pulmonary nodules using convolutional neural networks: a survey, IEEE Access 7 (2019) 78075–78091.

[44] X. Zhang, J. Lin, et al., An efficient neural-network-based micro seismic monitoring platform for hydraulic fracture on an edge computing architecture, Sensors 18 (6) (2018) 1828.
[45] S. Hochreiter, J. Schmidhuber, Long short-term memory, Neural Computation 9 (8) (1997) 1735–1780.
[46] B. Hussain, Q. Du, S. Zhang, A. Imran, M.A. Imran, Mobile edge computing-based data-driven deep learning framework for anomaly detection, IEEE Access 7 (2019) 137656–137667.
[47] R. Dong, C. She, W. Hardjawana, Y. Li, B. Vucetic, Deep learning for hybrid 5G services in mobile edge computing systems: learn from a digital twin, IEEE Transactions on Wireless Communications 18 (10) (2019) 4692–4707.
[48] S.A. Osia, A.S. Shamsabadi, A. Taheri, H.R. Rabiee, H. Haddadi, Private and scalable personal data analytics using hybrid edge-to-cloud deep learning, Computer 51 (5) (2018) 42–49.
[49] Y. Wang, K. Wang, H. Huang, T. Miyazaki, S. Guo, Traffic and computation co-offloading with reinforcement learning in fog computing for industrial applications, IEEE Transactions on Industrial Informatics 15 (2) (2019) 976–986.
[50] S. Conti, G. Faraci, R. Nicolosi, S.A. Rizzo, G. Schembra, Battery management in a green fog-computing node: a reinforcement-learning approach, IEEE Access 5 (2017) 21126–21138.
[51] X. Zhao, G. Huang, L. Gao, M. Li, Low load DIDS task scheduling based on Q-learning in edge computing environment, Journal of Network and Computer Applications 188 (1) (2021) 103095.
[52] B. Guo, X. Zhang, Y. Wang, H. Yang, Deep-Q-network-based multimedia multi-service QoS optimization for mobile edge computing systems, IEEE Access 7 (2019) 160961–160972.
[53] V. Mnih, K. Kavukcuoglu, et al., Human-level control through deep reinforcement learning, Nature 518 (7540) (2015) 529–533.
[54] F. Xu, F. Yang, S. Bao, C. Zhao, DQN inspired joint computing and caching resource allocation approach for software defined information-centric Internet of things network, IEEE Access 7 (2019) 61987–61996.
[55] J. Wang, L. Zhao, J. Liu, N. Kato, Smart resource allocation for mobile edge computing: a deep reinforcement learning approach, IEEE Transactions on Emerging Topics in Computing 9 (3) (2021) 1529–1541, https://doi.org/10.1109/TETC.2019.2902661.
[56] Z. Qin, D. Liu, H. Hua, J. Cao, Privacy-preserving load control of residential microgrid via deep reinforcement learning, IEEE Transactions on Smart Grid 12 (5) (2021) 4079–4089.
[57] Anjali R. Askhedkar, Bharat S. Chaudhari, Multi-armed bandit algorithm policy for LoRa network performance enhancement, Journal of Sensor and Actuator Networks 12 (3) (2023) 38, https://doi.org/10.3390/jsan12030038.
[58] Sheetal N. Ghorpade, Marco Zennaro, Bharat S. Chaudhari, IoT-based hybrid optimized fuzzy threshold ELM model for localization of elderly persons, Expert Systems with Applications (ISSN 0957-4174) 184 (2021) 115500, https://doi.org/10.1016/j.eswa.2021.115500.
[59] S. Yu, X. Chen, Z. Zhou, X. Gong, D. Wu, When deep reinforcement learning meets federated learning: intelligent multitimescale resource management for multiaccess edge computing in 5G ultradense network, IEEE Internet of Things Journal 8 (4) (2021) 2238–2251.

[60] X. Lu, Y. Liao, P. Lio, P. Hui, Privacy-preserving asynchronous federated learning mechanism for edge network computing, IEEE Access 8 (2020) 48970–48981.

[61] J. Zhang, Z. Zhan, Y. Lin, N. Chen, Y. Gong, Evolutionary computation meets machine learning: a survey, IEEE Computational Intelligence Magazine 6 (4) (2020) 68–75.

[62] S.N. Ghorpade, M. Zennaro, B.S. Chaudhari, GWO model for optimal localization of IoT-enabled sensor nodes in smart parking systems, IEEE Transactions on Intelligent Transportation Systems 22 (2021) 1217–1224.

[63] F. Durao, F. Carvalho, A. Fonseka, V.C. Garcia, A systematic review on cloud computing, Journal of Supercomputing 68 (3) (2014) 1321–1346.

[64] W. Shi, S. Dustdar, The promise of edge computing, Computer 49 (5) (2016) 78–81.

[65] M. Qin, L. Chen, N. Zhao, Y. Chen, F.R. Yu, G. Wei, Power-constrained edge computing with maximum processing capacity for IoT networks, IEEE Internet of Things Journal 6 (3) (2018) 4330–4343.

[66] Y. Li, S. Wang, An energy-aware edge server placement algorithm in mobile edge computing, in: Proceedings of the IEEE International Conference on Edge Computing (EDGE'18), IEEE, San Francisco, 2018, pp. 66–73.

[67] H. Gao, W. Li, R. Banez, Z. Han, H. Poor, Mean field evolutionary dynamics in ultra dense mobile edge computing systems, in: Proceedings of the IEEE Global Communications Conference (GLOBECOM), IEEE, 2019, pp. 1–6.

[68] C. Dong, W. Wen, Joint optimization for task offloading in edge computing: an evolutionary game approach, Sensors 19 (3) (2019) 740–764.

[69] Z. Ali, L. Jiao, T. Baker, G. Abbas, Z.H. Abbas, S. Khaf, A deep learning approach for energy efficient computational offloading in mobile edge computing, IEEE Access 7 (2019) 149623–149633.

[70] S. Yu, X. Wang, R. Langar, Computation offloading for mobile edge computing: a deep learning approach, in: Proceedings of the IEEE Annual International Symposium on Personal, Indoor, and Mobile Radio Communications (PIMRC'17), IEEE, 2017, pp. 1–6.

[71] Y. Hao, Y. Mian, L. Hu, M.S. Hossain, G. Muhammad, S.U. Amin, Smart-edge-CoCaCo: AI-enabled smart edge with joint computation, caching, and communication in heterogeneous IoT, IEEE Network 33 (2) (2019) 58–64.

[72] X. Xu, D. Li, Z. Dai, S. Li, X. Chen, A heuristic offloading method for deep learning edge services in 5G networks, IEEE Access 7 (2019) 67734–67744.

[73] S.N. Ghorpade, M. Zennaro, B.S. Chaudhari, R.A. Saeed, H. Alhumyani, S. Abdel-Khalek, Enhanced differential crossover and quantum particle swarm optimization for IoT applications, IEEE Access 9 (2021) 93831–93846.

[74] L. Lei, H. Xu, X. Xiong, K. Zheng, W. Xiang, Joint computation offloading and multiuser scheduling using approximate dynamic programming in NB-IoT edge computing system, IEEE Internet of Things Journal 6 (3) (2019) 5345–5362.

[75] J. Xu, L. Chen, S. Ren, Online learning for offloading and autoscaling in energy harvesting mobile edge computing, IEEE Transactions on Cognitive Communications and Networking 3 (3) (2017) 361–373.

[76] D. Mishra, S. De, S. Jana, S. Basagni, K. Chowdhury, W. Heinzelman, Smart RF energy harvesting communications: challenges and opportunities, IEEE Communications Magazine 53 (4) (2015) 70–78.

[77] M. Min, L. Xiao, Y. Chen, P. Cheng, D. Wu, W. Zhuang, Learning-based computation offloading for IoT devices with energy harvesting, IEEE Transactions on Vehicular Technology 68 (2) (2019) 1930–1941.

[78] X. Cheng, L. Feng, W. Quan, C. Zhou, H. He, W. Shi, X. Shen, Space/aerial-assisted computing offloading for IoT applications: a learning-based approach, IEEE Journal on Selected Areas in Communications 37 (5) (2019) 1117–1129.

[79] X. Chen, H. Zhang, C. Wu, S. Mao, Y. Ji, M. Bennis, Optimized computation offloading performance in virtual edge computing systems via deep reinforcement learning, IEEE Internet of Things Journal 6 (3) (2019) 4005–4018.

[80] L. Lei, H. Xu, X. Xiong, K. Zheng, W. Xiang, X. Wang, Multiuser resource control with deep reinforcement learning in IoT edge computing, IEEE Internet of Things Journal 6 (6) (2019) 10119–10133.

[81] J. Ren, H. Wang, T. Hou, S. Zheng, C. Tang, Federated learning-based computation offloading optimization in edge computing-supported Internet of things, IEEE Access 7 (2019) 69194–69201.

[82] L. Huang, X. Feng, A. Feng, Y. Huang, L. Qian, Distributed deep learning-based offloading for mobile edge computing networks, Mobile Networks and Applications 66 (12) (2018) 6353–6367.

[83] R. Kozik, M. Ficco, M. Choraś, F. Palmieri, A scalable distributed machine learning approach for attack detection in edge computing environments, Journal of Parallel and Distributed Computing 119 (2018) 18–26.

[84] B. Wang, Y. Wu, K.R. Liu, T.C. Clancy, An anti-jamming stochastic game for cognitive radio networks, IEEE Journal on Selected Areas in Communications 29 (4) (2011) 877–889.

[85] B. Li, T. Chen, G.B. Giannakis, Secure mobile edge computing in IoT via collaborative online learning, IEEE Transactions on Signal Processing 67 (23) (2019) 5922–5935.

[86] Y. Wang, W. Meng, W. Li, Z. Liu, Y. Liu, H. Xue, Adaptive machine learning-based alarm reduction via edge computing for distributed intrusion detection systems, Concurrency and Computation: Practice and Experience 31 (19) (2019) 1–12.

[87] L. Yu, X. Shen, J. Yang, K. Wei, R. Xiang, Hypergraph clustering based on game-theory for mining microbial high-order interaction module, Evolutionary Bioinformatics 16 (2020) 117693432097057.

[88] X. An, J. Su, X. Lu, F. Lin, Hypergraph clustering model-based association analysis of DDOS attacks in fog computing intrusion detection system, Journal on Wireless Communications and Networking 1 (2018) 249–258.

[89] L. Fernández Maimó, A. Huertas Celdrán, M. Gil Pérez, F. García Clemente, G. Martínez Pérez, Dynamic management of a deep learning-based anomaly detection system for 5G networks, Journal of Ambient Intelligence and Humanized Computing 10 (8) (2019) 3083–3097.

[90] M. Zhang, L. Wang, S. Jajodia, A. Singhal, A. Massimiliano, Network diversity: a security metric for evaluating the resilience of networks against zero-day attacks, IEEE Transactions on Information Forensics and Security 11 (5) (2016) 1071–1086.

[91] Y. Chen, Y. Zhang, S. Maharjan, M. Alam, T. Wu, Deep learning for secure mobile edge computing in cyber-physical transportation systems, IEEE Network 33 (4) (2019) 36–41.

[92] M. Du, K. Wang, Y. Chen, X. Wang, Y. Sun, Big data privacy preserving in multi-access edge computing for heterogeneous Internet of things, IEEE Communications Magazine 56 (8) (2018) 62–67.

[93] X. He, R. Jin, H. Dai, Deep PDS-learning for privacy-aware offloading in MEC-enabled IoT, IEEE Internet of Things Journal 6 (3) (2019) 4547–4555.

[94] M. Du, K. Wang, Z. Xia, Y. Zhang, Differential privacy preserving of training model in wireless big data with edge computing, IEEE Transactions on Big Data 6 (2) (2020) 283–295.

[95] C. Xu, J. Ren, L. She, Y. Zhang, Z. Qin, K. Ren, EdgeSanitizer: locally differentially private deep inference at the edge for mobile data analytics, IEEE Internet of Things Journal 6 (2019) 5140–5151.

[96] J. Chen, S. Chen, Q. Wang, B. Cao, G. Feng, J. Hu, iRAF: a deep reinforcement learning approach for collaborative mobile edge computing IoT networks, IEEE Internet of Things Journal 6 (4) (2019) 7011–7024.

[97] G. Chaslot, S. Bakkes, I. Szita, P. Spronck, Monte-Carlo tree search: a new framework for game AI, in: Proceedings of the 4th AAAI Conference on Artificial Intelligence and Interactive Digital Entertainment, 2008, pp. 216–217.

CHAPTER

TinyML using neural networks for resource-constrained devices

5

Vrushali Kulkarni and Varsha Jujare
Department of Computer Engineering and Technology, Dr. Vishwanath Karad MIT World Peace University, Pune, India

5.1 Introduction

The term "machine learning" was first used in 1959 by Arthur Samuel, an early American pioneer in the fields of AI and computer gaming. Samuel was working for IBM at that time. "The field of study that gives computers the ability to learn without being explicitly programmed," was how he defined machine learning [1]. This chapter gives an overall idea about TinyML. Section 5.1 gives an introduction into machine learning. Section 5.2 explains the need for TinyML. Section 5.3 gives details about how different neural network models can be fitted into tiny devices. Section 5.4 focuses on some issues or challenges imposed by TinyML. Various applications are introduced in Section 5.5. Finally, Section 5.6 provides the conclusion of the chapter.

The creation of algorithms and statistical models that enable computers to learn from experience and previous data is referred to as machine learning. It gives conclusions or inferences based on the patterns that are otherwise hidden inside the data. Without being given explicit instructions, these models as well as algorithms are created to learn from data and make predictions or decisions. Machine learning comes in a variety of forms, such as supervised learning, unsupervised learning, semi-supervised learning, and reinforcement learning. Models are trained using labeled data in supervised learning, while models are trained using unlabeled data in unsupervised learning. A reinforcement learning agent can observe and understand its surroundings, act, and learn from mistakes. The basic task of machine learning is to extract important features from the dataset and build a model using these features [2]. A wide range of applications is based on machine learning, including natural language processing, recommender systems, object recognition, and weather forecasting, to name a few.

There is a subfield of machine learning named deep learning which specifically uses image data. Deep learning does feature extraction by itself and uses extensive computing power and memory for its task. Since AI and machine learning as well

TinyML for Edge Intelligence in IoT and LPWAN Networks. https://doi.org/10.1016/B978-0-44-322202-3.00010-5

as deep learning have a lot of advantages, there are some prominent disadvantages that we cannot neglect. The first point is that for a machine learning application, a common consumer CPU uses 65 to 85 W of power, while a common consumer GPU uses 200 to 500 W. Another consideration is that for machine learning to become effective, the algorithms need the necessary time to develop and learn to the point where they can accomplish their objectives with a high level of relevance and accuracy. Furthermore, they use a lot of resources to function. This needs extra processing power from the computer as a result [3].

5.2 Background

To overcome the drawbacks listed above, TinyML is an additional area of machine learning and embedded systems that investigate various model types that can operate on low-power gadgets like microcontrollers. When machine learning technology faced difficulties with high power consumption, TinyML evolved as an efficient solution [4]. As a result, TinyML has fixed the issue and made it possible to run with low power consumption. It gives access to edge devices with little power, low bandwidth, and low latency. A typical TinyML-powered microcontroller uses milliwatts or microwatts of power. TinyML sits squarely at the nexus of algorithms, hardware, software, and embedded machine learning applications. It is the meeting point of conventional machine learning and embedded systems. It requires knowledge of embedded systems as well as software. TinyML is a relatively new area of research with the goal of lowering the computational resources needed for machine learning solutions and enabling their deployment to embedded devices with limited resources. Data are moved to more powerful computing infrastructures, like workstations or the cloud, to generate models and solutions for machine learning. However, it is well known that energy requirements for data transmission are orders of magnitude higher than for information processing or memory access. In addition to energy economy, data transfer has other limits such as bandwidth and reliability. It might not be possible to communicate all the data the device has gathered in a timely manner to provide acceptable latency [5]. For current machine learning to be successful, high-performance computing is required. However, edge and embedded devices are a new category of computer infrastructures that are getting more attention nowadays. Devices that are resource-constrained and specialized are inexpensive, widely dispersed, and energy-efficient. These are perfect for an AI solution that is data-centric.

Monitoring audio input for speech recognition is one application that has propelled development in this area. Modern machine learning techniques can be used to tackle the speech recognition issue. On the other hand, a vocal user interface necessitates constant input processing. That suggests a significant energy investment. The short battery life of cell phones makes this a challenge for implementation. Additionally, the majority of the input that needs to be processed lacks relevant data. As a result, the user-speaking portions of the input only require high-power processing to be applied. This issue has been resolved by Google researchers by creating a cascade architecture made up of one module intended to identify the input segments to which the system must respond. This first module is always on and has very low power consumption and good energy efficiency [6]. Only when necessary does the first module wake up the second module, which is a high-power system that can respond to input. Designing and training a machine learning model in frameworks like TensorFlow constitute the first step of development. TensorFlow Lite is a customized variation of the software created by Google that is independent of any libraries, with a binary footprint of less than 16 kB.

Integrated development environments (IDEs) like the Arduino IDE, Keil MDK, and Mbed studio are used to combine the created model from TensorFlow Lite with the application code. The software then prepares the code for deployment on the proper hardware by compiling it into a binary file. TinyML applications can be created using a variety of development boards, such as the Arduino Nano 33 BLE, Sparkfun Edge, STM32F746NG MCU discovery kit, ESP32-DevKitC, Sony Spresense, etc.

Machine learning models that have already been trained and are ready to use are offered by TensorFlow Lite. These include:

- Object detection: Recognizes up to 80 different objects in an image.
- Smart response: Generates responses that are clever and resemble those that can be received from chatbots/conversational AI.
- Recommendation systems: offer specially created recommendation algorithms on the basis of consumer behavior.

TinyML can also be used for edge processing. Edge is about processing data more quickly and in larger volume near the point of generation, providing action-driven solutions in real time. Compared to conventional models, where processing power is centralized at an on-premise data center, it has some distinctive features. In order to be processed, data collected on edge nodes must be sent to a central node or the cloud. Much more energy is required for data transmission than for data processing. Because of the local elaboration it has significant energy savings and a decrease in the time between input gathering and response creation. The capacity, network accessibility, and transmission reliability of data transfer are additional constraints. TinyML is a data-centric strategy that opens the door to advancements in dependability, security, and data efficiency [7].

When it comes to hardware, TinyML is imposing because it plans to work with some really inferior technology. In some aspects, TinyML's real goal is to perform machine learning inference with the least amount of processing resources. Pete Warden, who is widely recognized as the creator of TinyML, asserts in his original work on the issue that TinyML should strive to operate at energy usage of less than 1 mW. This arbitrary number was chosen because a gadget can operate for months to a year on a typical coin battery with 1 mW consumption. When examining TinyML's power sources, one should take into account small Li-Po batteries, coin batteries, and energy-collecting gadgets. Additional parts one could find on a TinyML device include sensors (such as a camera or microphone), connectivity for Bluetooth Low

Energy (BLE), and possibly additional parts. Building machine learning models is frequently done using Python [8]. However, machine learning models may be created using TensorFlow Lite by using C, C++, or Java [18]. The process of connecting to networks requires energy. One can build machine learning models with TensorFlow Lite without explicitly connecting with the Internet. Additionally, this allays security worries because embedded systems are very less susceptible to hacking.

5.3 Neural networks for TinyML

The development of on-device learning is the key objective of this chapter. Inference on the device is necessary for many applications. For instance, in some circumstances, such as drone rescue operations, internet access could not always be available. Moving data to the cloud for processing is highly difficult in several businesses due to regulations and privacy norms, such as healthcare. Additionally, the roundtrip delay to the cloud is too long for applications that require real-time machine learning inference. These requirements have made on-device machine learning appealing from a scientific and business standpoint. Face and speech recognition are now available on the iPhone. People can use on-device translation on their Android phones. Machine learning is used by Apple Watch to recognize ECG patterns and movements. These on-device machine learning models have made it possible by improving the methods used to compress and make the neural networks more memory- and computing-efficient. However, they were also made possible by improvements in the hardware. Today's wearables and smartphones are much more powerful than servers were in the 1990s. Some devices even have co-processors with machine learning inference specificity.

5.3.1 Architecture of neural networks

Algorithms called artificial neural networks are designed to resemble how the brain operates. A representation of a neural network can be seen in Fig. 5.1. A neural network is made up of layers, and each layer is made up of a number of nodes that represent neurons. Edges show how neurons are connected to one another. Each neuron processes the input data using a mathematical function (such as linear regression) and then outputs a value that is propagated to the connected neuron [16]. A "bias" value is contained in each neuron, and a "weight" value (W1–W16) is contained in each connection. In order to minimize the loss function and produce better outcomes, the model searches for the optimal weight and bias values throughout the training phase. However, the neuron first applies a function known as the activation function (such as the sigmoid) before providing the output value. The system is frequently made non-linear by using the activation function. The neural network can recognize far more intricate patterns in data than a non-linear system.

Some optimization algorithms are used for the purpose of training these models and neural networks. An optimization method focused in this chapter to train neural

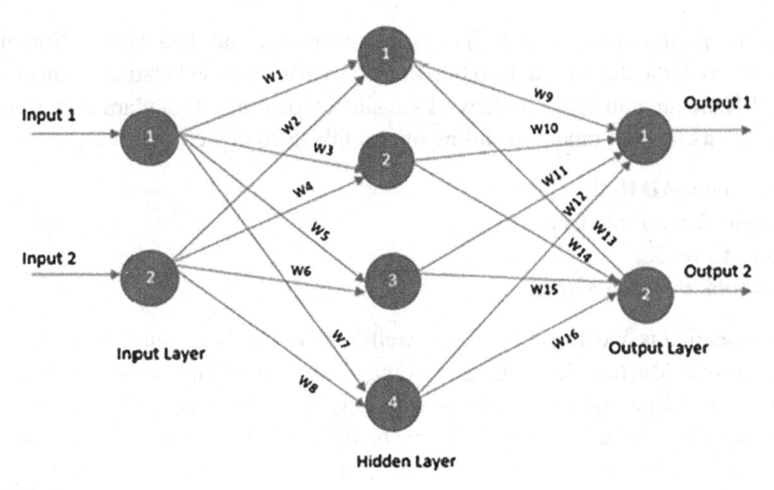

FIGURE 5.1

Representation of a neural network.

networks is stochastic gradient descent. The technique starts by attempting to find the right pattern in a labeled dataset (the training dataset). The error is then returned to the input layers via backpropagation [16]. The settings of each neuron (the model's parameters) are altered during backpropagation to decrease the error. The neural network can find patterns from the training dataset (and presumably from undiscovered data) with a very small error after numerous training rounds. TinyML advances edge AI by enabling the execution of deep learning models on microcontroller units (MCUs), which have significantly fewer resources than the tiny computers we wear on our wrists and carry in our pockets. An embedded system's microcontroller is a small integrated circuit that controls a single process. Microcontrollers are widely used in both consumer and industrial products. At the same time, they lack the resources found in standard computing devices. The majority of them lack an operating system. They lack networking hardware, have a modest CPU, and only contain very few kilobytes of low-powered memory (SRAM) and very few megabytes of storage [9]. They must run for years on cell and coin batteries because they typically lack a main electrical source. Fitting deep learning models to MCUs can therefore enable a wide range of applications [10].

5.3.2 Compression of models to fit into the tiny microprocessor

Applications of deep learning are increasingly being used in situations with fewer resources, such as cell phones, agricultural sensors, and medical equipment. The transition to environments with limited resources prompted initiatives for more compact and effective model architectures and also a greater focus on model optimization approaches [21]. The models which are trained using deep learning are quite challenging and time-consuming because they require processing tens of terabytes of data over a period of hours or even days in numerous cloud data centers. Therefore, how can we enable these models to function on such a small and resource-constrained device? There are numerous frameworks available to reduce a standard deep learning model into a compact model for fitting on an embedded device:

- Qualcomm AIMET,
- Google TensorFlow Lite,
- Apple CoreML,
- Facebook PyTorch Mobile.

To extend this we begin with the well-known TinyML TensorFlow Lite microframework. Machine learning models may be executed on microcontrollers such as TensorFlow Lite and other devices with only a few kilobytes of RAM. An Arm Cortex M3 can execute a variety of simple models [11]. It does not need any conventional C or C++ libraries, dynamic memory allocation, or operating system support. This model does not initially support training. The restricted available processing power (soft restriction) and, most significantly, the constrained memory (hard constraint) prevent training machine learning models on small devices [12]. To fit these models onto resource-constrained devices, model optimization and reduction of the size of the neural network are the ways. This is achieved by reducing the number of weights or the number of bits/weight. Since not all neurons are created equal and some contribute more than others, it makes sense to eliminate all of the unnecessary neurons in order to free up space; this technique is known as shrinking [13]. Shrinking techniques comprise four steps: compression, quantization, encoding, and compilation.

5.3.2.1 Compression

After establishing a neural network model, training is to be done for achieving minimum loss. During the training process, one needs to pass the training examples through the network using a forward propagation step. This involves feeding the input data into the network and calculating the output predictions. Each layer performs a linear transformation followed by an activation function. After obtaining the predictions, it calculates the loss by comparing them to the true labels using a suitable loss function. Common loss functions include the mean squared error (MSE) for regression tasks and cross-entropy loss for classification tasks [14]. The main step in training a neural network is backpropagation. It involves computing the gradients of the loss function with respect to the network's weights and biases. This is done by propagating the error from the output layer back to the input layer. This process is repeated iteratively on batches of training data until a predefined stopping criterion is reached. By performing training on the model, arrays get generated which represent the weights of the connections between each layer across the neural network for making up a significant portion of the model size. Once the models are trained, the model can be compressed with a very minimal accuracy loss. Pruning is the most commonly used compression method [20].

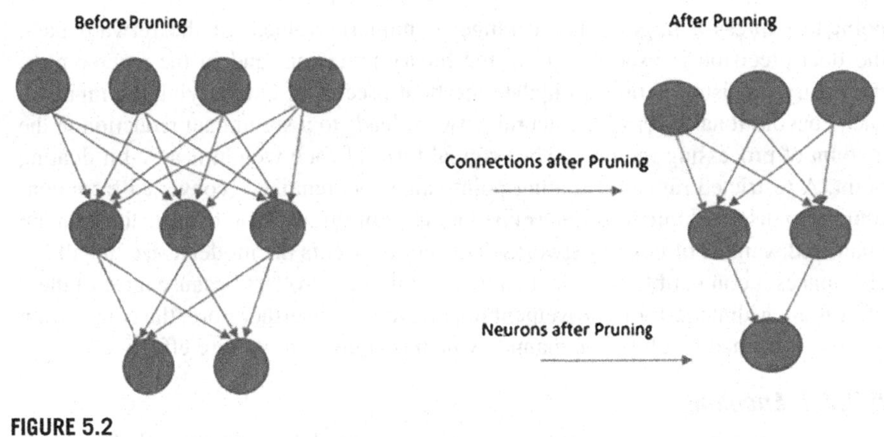

FIGURE 5.2

Pruning.

5.3.2.2 Pruning

The success of deep learning has been largely attributed to the construction of increasingly bigger neural networks. This makes it possible for these models to execute certain jobs more effectively, but it also increases the cost of using them. Larger models are more difficult to distribute since they require more storage space. They can also take more time to run. If the model is developed for a real-world application, this is extremely important to consider.

Model compression seeks to minimize performance loss while reducing model size. A trained model's weights are removed as part of the compression technique known as neural network pruning [16]. Pruning in agriculture refers to the removal of unneeded plant branches or stems. In the same way, pruning in machine learning is the removal of pointless neurons or weights. Pruning is a method for reducing the size of the network. Fig. 5.2 illustrates the pruning technique. It entails removing weights and neurons from the network, which has the dual benefits of requiring less computing to evaluate the model and less memory to store it. Performance is also affected because some neurons or filters lose information when some of their weight is removed [19]. This trade-off is desirable since it is possible to achieve large compression at a little cost to precision performance.

5.3.2.3 Quantization

After the training is complete, the neural network can be quantized. However, quantization during training is by far the most efficient strategy for maintaining high accuracy. Choudry et al. [14] described the core concept of the quantization of neural network parameters during training. There are actually two networks operating during training: float-precision and binary-precision. The total number of binary or decimal digits by omitting the sign determines the precision of binary integers and decimal values. The single-precision floating-point format is a type of computer number that typically takes up 32 bits in computer memory and uses a floating radix point to express a large dynamic range of numeric values. In the forward pass, the float-precision is used to update the binary-precision, and in the reverse pass, the binary-precision is used to update the float-precision. Quantifying the inputs to each convolutional layer of the neural network leads to a significant reduction in the amount of processing required. The original form of each weight is a 32-bit floating point. A restricted range of floating-point values is quantized (converted from continuous to discrete) into a predetermined number of information buckets that contain quantized weights of neural networks. This not only cuts the model's size in half but also makes it compatible with the majority of microcontrollers because most of them offer 8-bit arithmetic for improvement in performance. Furthermore, the convolution can be performed in a bit-wise manner, which is significantly more effective.

5.3.2.4 Encoding

Utilizing the principles of Huffman encoding, the model is compressed. This algorithm is a lossless data compression technique. Input characters are given variable-length codes, the lengths of which are determined by the frequency of the matching characters. This phase encrypts the weights by assigning the most common weights along with codes in the minimum number of bits. Consider Table 5.1 containing already quantized weights.

Table 5.1 Quantized weights.

Quantized weights	
5	16
9	45
12	13

Every weight carries 5 bits (ranging from 0 to 31). Fig. 5.3 illustrates the tree after applying Huffman encoding for encoding the weights based on the frequency.

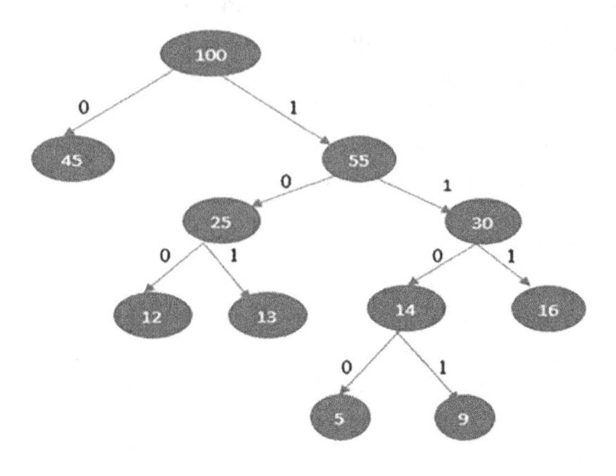

FIGURE 5.3

Weight tree after applying Huffman encoding.

Table 5.2 Compressed weights.

Weight	Size	Frequency	Code	Code size
5		1	1100	4 bits
9		1	1101	4 bits
12	5 bits	1	100	3 bits
13		1	101	3 bits
16		1	111	3 bits
45		1	0	1 bit

The original weights were compressed from 5 bits to 3 or 4 bits. The model will then decode the codes into the original weights.

Every circle represents the "weights" and these weights will be encoded with the "binary path." For example, 16 is encoded as "111," 45 is encoded as "0," etc. (See Table 5.2.)

5.3.2.5 Compilation

Since C and C++ are quicker and more memory-efficient than Python, they are typically used in microcontrollers. In this phase, the model will be converted into a format that interpreters like the TF Lite Micro Interpreter found in some microcontrollers and the neural network interpreter found in some Android devices can read and execute. Comparing neural networks to conventional algorithms, there are many benefits. They are able to learn from data and be applied to the resolution of challenging issues. Additionally, capable of generalization, neural networks can identify patterns in data that conventional algorithms might not be able to. It is not practical to design a universal neural network model for MCUs. The model expects to be instantly updated with field data due to the fact that different users have various patterns. TinyML's promotion and applicability on devices with limited resources are hampered by the training model's resource-intensive consumption.

5.4 Concerns/issues in TinyML

For object detection and classification, the ultrawideband (UWB) Doppler radar dataset is used along with a traditional neural network algorithm. This dataset has 17,000 samples of drones, cars, and people. Results of accuracy in the range of 0.9 are obtained. The role and place of the radar have a unique effect on the model's performance, and data are also sensitive to the changes [2]. The study found that as there are varying forms and sizes of different vehicles available, there are observable changes in the data collected. Furthermore, gathering a large enough dataset to generalize over all these differences is prohibitively expensive and time-consuming in this particular scenario. As a result, the process cannot be easily automated. So, on-device learning for TinyML is developed to deal with this issue which has been deployed via the TensorFlow Lite model [2]. Real-time data are produced when on-device learning is developed.

Traditional ways of assessing accuracy will not be able to handle these data well. Therefore, more complicated issues like concept drift, incremental learning, and transfer learning should be the focus of functionality testing in the case of on-device learning.

5.4.1 Incremental learning

The learning paradigm of incremental learning contrasts with batch learning. When new labeled data become available, a model is gradually trained. There are two categories in which incremental learning has proven itself. The first is the rehearsal technique, which entails remembering prior events in order to make up for lost performance on earlier tasks [15]. The second category is using the regularization technique to prevent undesirable interference by limiting changes to the weights. A large model or an ensemble is typically condensed into a single smaller model through quantization. This technique can be used in the context of incremental learning between iterations of a model to preserve comparable activations on prior tasks when learning the new one. Accidental forgetting is a major issue for incremental learning. The response seen in trained neural networks when they are adjusted to a new task, a changing environment, or changes in the data distribution is known as catastrophic forgetting, sometimes known as interference. The network's performance on the previous job may drastically suffer while adjusting to the new context. The network forgets the previous task as it gains expertise in solving the new one [4]. This problem can also be seen as a challenge to generalize, to develop a solution that offers acceptable performance on both tasks. This is problematic since it restricts the difficulty of the tasks that the network can handle. One logical response to this tendency is to train the network using data that are reflective.

5.4.2 Concept drift

When solving problems in the actual world, the models may be dependent on unmeasurable attributes or hidden concepts. Concept drift is the term used to describe how changes in this concealed concept can lead to changes in the problem targets. The situation when the problem description stays constant but the data distributions or the pertinent features vary is a different form of drift. Both of these occur primarily as a result of environmental evolution. If such a situation arises, give the model more latitude to use the attributes and guidelines it has learned to solve the current problem in a novel setting; it may even be preferable in this case to ignore the past data. Concept drift is not always easy to identify, and forgetting is not always a bad thing. The challenge is separating noise from genuine concept drift [2].

5.4.3 Transfer learning

This type of learning is a method for transferring a network that needs to train on a particular source domain to a target domain. The lack of specific data in the target domain is one of the key justifications for adopting transfer learning. The method entails fine-tuning the network from the source domain to the target domain in order to adapt it. A pretrained model that has been trained on a sizable dataset for a particular task is used as a starting point for solving a separate but related task using the transfer learning technique in deep neural networks. Transfer learning utilizes the information gained by the previously trained model on a source task and applies it to a target task as opposed to building a new model from scratch. Deep neural networks learn hierarchical representations of data in which earlier layers capture low-level features and later layers catch high-level information. This is the fundamental notion underpinning transfer learning. These learned representations can be applied to various tasks and datasets.

5.5 Applications

Deep neural networks can be utilized in TinyML applications, although they may require some modifications and optimizations to fit within the limited resources. Here are a few examples of TinyML being used in various domains.

5.5.1 Agriculture

Deep neural networks can be utilized within TinyML applications in agriculture to address a range of challenges. Here are some examples:

Crop disease detection: Deep neural networks can be trained to classify and detect diseases in crops by analyzing images of leaves or plants. With TinyML, these models can be deployed on low-power devices, such as edge devices or drones, to perform real-time disease identification in the field. This allows farmers to quickly detect and respond to crop diseases, preventing their spread and reducing yield losses.
Pest detection: Similarly, TinyML models can be trained to identify pests or insects that affect crops. By analyzing images or sounds captured by sensors or drones, deep neural networks can provide early warnings about the presence of pests, enabling targeted interventions and minimizing the use of pesticides.
Soil analysis: Deep neural networks can be used to analyze soil composition and quality. By processing sensor data, such as temperature, humidity, and pH levels, TinyML models can provide insights into soil health, nutrient deficiencies, and optimal irrigation strategies. This information can guide farmers in making data-driven decisions regarding fertilization, irrigation, and crop selection.
Weed identification and control: Deep neural networks can be trained to distinguish between crops and weeds based on visual cues. By deploying these models on small, energy-efficient devices, farmers can use them in smart weeding robots or

handheld devices to precisely identify and target weeds for effective weed control, reducing the need for herbicides.

Yield prediction: TinyML models can analyze various environmental and crop-specific parameters, such as weather data, soil conditions, and growth stages, to predict crop yield. By providing accurate yield forecasts, farmers can optimize harvesting schedules, plan logistics, and make informed decisions regarding crop management practices.

Livestock monitoring: Deep neural networks can be used in TinyML applications to monitor the health and behavior of livestock. For example, models can analyze sensor data from wearable devices on animals to detect anomalies in activity patterns, identify signs of distress or illness, or predict the onset of diseases.

5.5.2 Healthcare

TinyML has the potential to revolutionize various applications by enabling real-time analysis, monitoring, and decision-making at the edge. Here are a few examples of TinyML applications in healthcare:

Remote patient monitoring: TinyML can be used to develop wearable devices or implantable sensors that continuously monitor vital signs such as heart rate, blood pressure, or glucose levels. By processing and analyzing these data on the device itself, healthcare providers can receive real-time updates and alerts regarding the patient's condition, enabling early intervention and reducing the need for frequent hospital visits.

Fall detection and prevention: Falls are a major concern for the elderly, leading to serious injuries and hospitalizations. TinyML algorithms can be deployed on wearable devices or ambient sensors to detect falls and issue alerts to caregivers or emergency services. Additionally, these algorithms can analyze gait patterns and identify individuals at risk of falling, allowing for proactive interventions and personalized care.

Epilepsy monitoring: TinyML can be used in implantable or wearable devices to detect and predict epileptic seizures. By analyzing data from various sensors such as electroencephalograms (EEGs) or heart rate monitors, the system can identify patterns or anomalies associated with seizures. This information can be used to trigger timely interventions, administer medications, or notify caregivers [17].

Asthma management: Asthma is a chronic respiratory condition that requires continuous monitoring and management. TinyML algorithms can be employed in portable devices or inhalers to analyze environmental factors, inhaler usage patterns, and physiological data to provide personalized feedback, medication reminders, and early warnings of exacerbations.

Point-of-care diagnostics: TinyML models can be deployed on handheld or portable devices to perform rapid diagnostics at the point of care. For example, image classification algorithms can analyze medical images such as X-ray, ultrasound, or dermatological images to detect abnormalities or diseases. This enables faster diagnosis, reduces dependence on centralized laboratories, and improves accessibility to healthcare in remote areas.

Medication adherence: Non-adherence to medication regimens is a common problem, leading to suboptimal treatment outcomes. TinyML can be utilized in smart pill dispensers or wearable devices to monitor medication usage patterns and provide reminders to patients. These systems can also detect and alert healthcare providers if a patient misses a dose, enabling timely interventions and improved medication adherence.

5.5.3 Defense

When it comes to defense applications, TinyML with deep neural networks can have various use cases. Here are a few examples:

Intrusion detection: TinyML can be used to deploy deep neural network-based models on embedded systems or IoT devices to detect potential cyberthreats or cyberattacks. These models can analyze network traffic patterns, identify anomalies, and trigger appropriate defense mechanisms.

Object detection: Deep neural network-based models can be trained on larger systems and then deployed on resource-constrained devices for real-time object detection. This can be useful in defense scenarios for applications like surveillance, target tracking, or situational awareness.

Speech recognition: TinyML can be leveraged to run deep neural network-based speech recognition models on small devices, enabling voice-based command and control in defense systems. This can be particularly beneficial for hands-free operations or in situations where other input methods are not feasible.

Health monitoring: In defense applications, monitoring the health and well-being of soldiers or personnel is crucial. TinyML with deep neural networks can be utilized to analyze sensor data, such as heart rate, respiration, or movement patterns, on wearable devices to provide real-time health insights and alert for potential issues.

Autonomous systems: Autonomous systems, such as unmanned aerial vehicles or ground robots, can benefit from TinyML with deep neural networks. Onboard neural networks can process sensor data and make real-time decisions for navigation, target identification, or obstacle avoidance [21].

In all these scenarios, deploying neural network models on resource-constrained devices requires model compression techniques, such as quantization and pruning, to reduce memory footprint and computational requirements. Additionally, efficient inference engines and hardware accelerators designed for TinyML can be utilized to improve performance and energy efficiency.

5.6 Conclusion

The purpose of this chapter was to review TinyML and deep neural networks for diversified resource constraints, the limitations imposed by using low-powered devices, and the potential use of microcontrollers to implement complex machine learning algorithms. The integration of TinyML with deep neural networks brings numerous advantages; for example, deep neural networks can provide high accuracy and robustness in various tasks, including image recognition, audio processing, and sensor data analysis. This allows for sophisticated inference capabilities on small devices without relying on cloud- or server-based processing. Additionally, deep neural networks can be optimized for deployment on resource-limited devices, leveraging techniques such as model quantization, pruning, and compression. These methods reduce the model size and computational requirements while maintaining a reasonable level of accuracy. As a result, TinyML models can run efficiently on microcontrollers with limited memory, processing power, and energy resources. Moreover, TinyML in deep neural networks enables real-time and low-latency inference, facilitating rapid decision-making at the edge without requiring constant internet connectivity. This is particularly beneficial in applications where latency, privacy, or bandwidth constraints are critical, such as in autonomous systems, smart sensors, and wearable devices. Furthermore, by enabling on-device machine learning, TinyML in deep neural networks addresses privacy and security concerns associated with transmitting sensitive data to the cloud. Instead, data can be processed locally, reducing the risk of data breaches and ensuring user privacy. However, there are also challenges and limitations to consider. Developing TinyML models for deep neural networks requires specialized knowledge and expertise in model optimization techniques, efficient algorithm design, and hardware constraints. Balancing model complexity and accuracy with the limitations of resource-constrained devices remains a significant challenge. TinyML in deep neural networks holds great promise for enabling intelligent and autonomous applications at the edge. It empowers a wide range of industries, including healthcare, agriculture, smart homes, industrial automation, and defense. As technology advances and more tools and frameworks become available, we can expect to see further progress and innovation in the field of TinyML, enabling a new era of intelligent edge devices.

References

[1] A.L. Samuel, Some studies in machine learning using the game of checkers, IBM Journal of Research and Development 3 (3) (1959) 210–229, https://doi.org/10.1147/rd.33.0210.

[2] N. Jouppi, C. Young, N. Patil, D. Patterson, Motivation for and evaluation of the first tensor processing unit, IEEE MICRO 38 (3) (2018) 10–19, https://doi.org/10.1109/mm.2018.032271057.

[3] Vitaly Bushaev, Medium, https://towardsdatascience.com/how-do-we-train-neural-networks-edd985562b73, 2017.

[4] M.F. Bajestani, M. Ghasemi, S. Vrudhula, Y. Yang, Enabling incremental knowledge transfer for object detection at the edge, arXiv [cs.CV], http://arxiv.org/abs/2004.05746, 2020.
[5] R. Immonen, T. Hämäläinen, Tiny machine learning for resource-constrained microcontrollers, Journal of Sensors 2022 (2022) 1–11, https://doi.org/10.1155/2022/7437023.
[6] R. Budjac, M. Barton, P. Schreiber, M. Skovajsa, Analyzing embedded AIoT devices for deep learning purposes, in: Proceedings of the Computer Science On-Line Conference, Springer, 2022, pp. 434–448.
[7] H. Ren, D. Anicic, T. Runkler, TinyOL: TinyML with online-learning on microcontrollers, arXiv [cs.LG], http://arxiv.org/abs/2103.08295, 2021.
[8] Eugeniu Ostrovan, TinyML on-device neural network training, POLITECNICO, MILANO,1863, https://www.politesi.polimi.it/bitstream/10589/187690/6/TinyML_on_device_neural_network_training%20fin.pdf, 2022.
[9] R. Kolcun, D.A. Popescu, V. Safronov, P. Yadav, A.M. Mandalari, Y. Xie, R. Mortier, H. Haddadi, The case for retraining of ML models for IoT device identification at the edge, arXiv [cs.NI], http://arxiv.org/abs/2011.08605, 2020.
[10] C. Benzmuller, C. Lisetti, M. Theobald, Implementation of incremental learning in artificial neural networks, in: GCAI 2017. 3rd Global Conference on Artificial Intelligence, 2017.
[11] Subir Maity, Medium, https://medium.com/@subirmaity/a-simple-neural-network-implementation-approach-in-micropython-for-deep-learning-application-760ab35cb538, 2022.
[12] MCUNetV2: Memory-Efficient Patch-based Inference for Tiny Deep Learning, https://tinyml.mit.edu/mcunet/#mcunetv2.
[13] Aditya Sharma, Nikolas Wolfe, Bhiksha Raj, The incredible shrinking neural network: new perspectives on learning representations through the lens of pruning, https://arxiv.org/pdf/1701.04465.pdf, 2017.
[14] E. Fiesler, A. Choudry, H.J. Caulfield, Weight discretization paradigm for optical neural networks, in: H. Bartelt (Ed.), Optical Interconnections and Networks, SPIE, Aug 1990, https://doi.org/10.1117/12.20700.
[15] N. Schizas, A. Karras, C. Karras, S. Sioutas, TinyML for ultra-low power AI and large scale IoT deployments: a systematic review, Future Internet 14 (12) (2022) 363, https://doi.org/10.3390/fi14120363.
[16] Francisco Costa, Medium, https://medium.com/marionete/tinyml-models-whats-happening-behind-the-scenes-5e61d1555be9, 2021.
[17] T. Pimentel, M. Monteiro, A. Veloso, N. Ziviani, Deep active learning for anomaly detection, in: 2020 International Joint Conference on Neural Networks (IJCNN), 2020.
[18] R. David, J. Duke, A. Jain, V.J. Reddi, N. Jeffries, J. Li, N. Kreeger, I. Nappier, M. Natraj, S. Regev, R. Rhodes, T. Wang, P. Warden, TensorFlow Lite Micro: embedded machine learning on TinyML systems, arXiv [cs.LG], http://arxiv.org/abs/2010.08678, 2020.
[19] J.O. Neill, An overview of neural network compression, https://doi.org/10.48550/ARXIV.2006.03669, 2020.
[20] B.L. Deng, G. Li, S. Han, L. Shi, Y. Xie, Model compression and hardware acceleration for neural networks: a comprehensive survey, Proceedings of the IEEE 108 (4) (2020) 485–532, https://doi.org/10.1109/jproc.2020.2976475, Institute of Electrical and Electronics Engineers.
[21] W. Yu, C. Zhao, Broad convolutional neural network based industrial process fault diagnosis with incremental learning capability, IEEE Transactions on Industrial Electronics 67 (6) (2020) 5081–5091, https://doi.org/10.1109/tie.2019.2931255.

CHAPTER

Reinforcement learning for LoRaWANs

Anjali Askhedkar[a], Bharat S. Chaudhari[a], and Marco Zennaro[b]

[a]*Department of Electrical and Electronics Engineering, Dr. Vishwanath Karad MIT World Peace University, Pune, India*

[b]*Science, Technology and Innovation Unit, Abdus Salam International Centre for Theoretical Physics (ICTP), Trieste, Italy*

6.1 Introduction

The primary requirement for LoRa or any other LPWAN technology is low power consumption for wide area coverage. With the advent of low-power embedded devices and lightweight machine learning frameworks such as TinyML, there are prospects of using machine learning algorithms such as reinforcement learning for LoRa. This chapter discusses the different reinforcement learning approaches used for LoRa for performance enhancement. A novel multi-armed bandit (MAB) policy developed for LoRa parameter selection is discussed in detail with the simulation results. At the end of the chapter, a few applications that employ reinforcement learning algorithms for LoRa are presented.

6.1.1 LoRa technology overview

LoRa is a promising LPWAN technology suitable for a variety of low-power Internet of Things (IoT) applications. LoRa, proposed by Semtech and the LoRa Alliance, uses chirp spread spectrum (CSS) modulation at the physical layer. LoRaWAN is the medium access layer protocol that specifies a star-of-stars topology including several end devices connected to the gateways, network servers, and application servers. LoRa operates in the unlicensed industrial, scientific, and medical (ISM) band with the duty cycle regulations governed by the region of operation. CSS modulation employs different spreading factors (*SFs*), SF being the ratio of symbol rate to chirp rate in a LoRa transmission. Different *SFs* are orthogonal to each other and thus CSS offers resilience to interference. LoRa transmission depends on the following parameters with the typical values as mentioned. The SF varies as {7, 8, 9, 10, 11, 12}, with higher *SF* leading to a longer range and more time on air (ToA). Transmit power values can vary as {2, 5, 8, 11, 14} dBm and in turn determine the energy consumption. Bandwidth governs the data rate, and the options are {125, 250, 500} kHz. LoRa uses forward error correction with a coding rate of {4/5, 4/6, 4/7, 4/8}. A higher coding rate increases ToA, but provides a better guard against errors. Channel frequency is

TinyML for Edge Intelligence in IoT and LPWAN Networks. https://doi.org/10.1016/B978-0-44-322202-3.00011-7

decided by the ISM band governed by the region of operation. The choice of transmit parameters determines the LoRa network performance. LoRa data rates vary from approximately 0.3 kbps to about 50 kbps. The adaptive data rate (ADR) method is used by LoRa to manage the end device data rate centrally so as to enhance the battery life and the network capacity [1].

6.1.2 LoRaWAN challenges and machine learning

LoRaWAN network consists of a large number of nodes connected together that lead to several challenges such as coverage, energy consumption, resource allocation, collisions, interference, scalability, reliability, and security [2]. Machine learning techniques can be used to address these issues and improve the LoRa network performance as LoRa-based IoT applications continue to grow [3]. LoRa devices being low-power and battery-operated, learning algorithms with low complexity are required.

The LoRa transmission parameters such as *SF*, transmit power, coding rate, bandwidth, and channel frequency determine the energy consumption, throughput, and the network performance as a whole. Reinforcement learning algorithms such as MAB can be employed for optimal selection of these parameters. LoRa uses a simple ALOHA access mechanism and as a result if the node density increases, the probability of packet collisions also increases. As the coverage is large, the same gateway is shared by multiple nodes, further increasing the packet collision probability. Collisions result in packet losses, leading to retransmissions, hence increasing energy consumption and latency. MAB algorithms present a promising solution to handle these problems [4]. This chapter focuses on the use of MAB-based reinforcement learning algorithms for LoRaWAN.

6.2 Reinforcement learning algorithms

AI technology empowers machines with intelligence and the capacity to learn from experience instead of being explicitly programmed. It is based on the concept that machines learn from previous understanding and make decisions with algorithms. The main drawbacks of machine learning include the fact that training of models requires huge amounts of data and the complexity of the model. Sometimes, the data may be insufficient, unreliable, or unavailable. Machines need to become independent and perform actions on their own. These problems are addressed by reinforcement learning. Reinforcement learning is a type of machine learning in which a trained AI agent takes action and gets a reward as depicted in Fig. 6.1. The important terms in the context of reinforcement learning [5] are summarized below:

- An agent is a model that the reinforcement learning trains.
- Actions are the steps taken by the model.
- The state is the present condition of the model.

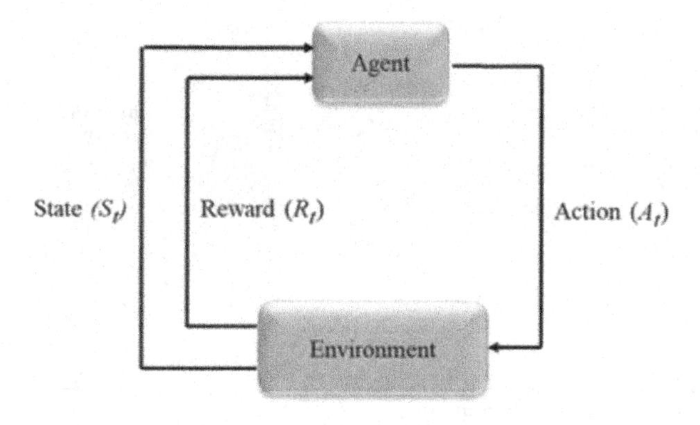

FIGURE 6.1

Reinforcement learning process.

- Environment is the training situation that the agent must optimize.
- Rewards are received if the model takes the right action.
- Policy decides which action an agent will take at a given time.

The agent gets a positive reward when it takes a correct action and a negative reward when a wrong action is taken. The agent executes commands as per the policy to maximize rewards and on its own might even learn a new action. The agent may also explore its environment during the exploration process. Its trend to maximize rewards is called exploitation. Exploration may lead to lesser rewards; hence, there is always a trade-off between exploration and exploitation. The agent is not explicitly told which action to take, but instead discovers which action produces more rewards through trial and error. The main objective of the agent is to maximize rewards; the agent learns from its environment and using the feedback decides its action, thus improving its performance. The difference between the cumulative mean rewards and the reward that may have been obtained using an optimal policy is termed regret, the aim of the policy being to lower the regret.

MAB problems fit into this framework of reinforcement learning and can be categorized as stochastic and adversarial problems based on the reward model. Stochastic MAB models have rewards with a stochastic distribution. Depending on the distribution, the stochastic MAB problems can be categorized as stationary and non-stationary. Stochastic stationary MAB problems have a stationary and stochastic reward distribution. It is assumed that rewards are independently generated and every action is associated with a constant and unknown distribution. The commonly used stochastic stationary MAB algorithms are Thompson sampling (TS) and upper confidence bound (UCB) [6,7]. In a stochastic non-stationary MAB problem, rewards follow a non-stationary stochastic distribution. This means that rewards from the same action may vary at different times which makes these algorithms more suitable to practical scenarios. Epsilon-greedy (ε-greedy), exponential-weight algorithm for exploration and exploitation (EXP3), and discounted UCB (DUCB) are some of

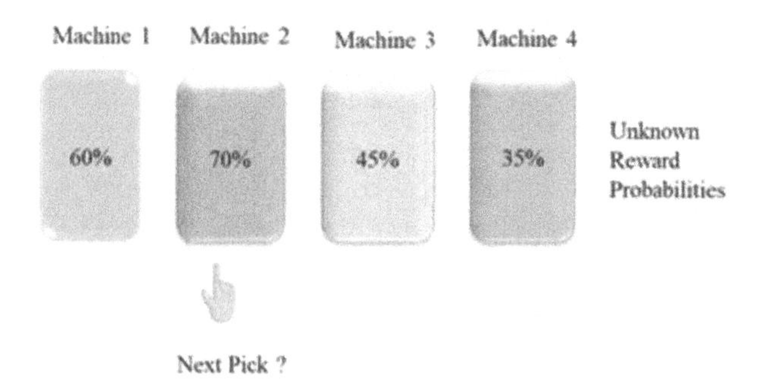

FIGURE 6.2

Multi-armed bandit problem illustration.

the MAB algorithms that address the stochastic non-stationary problem [7]. Fig. 6.2 illustrates an MAB problem where the optimal selection is done based on the rewards obtained [5].

- **Thompson sampling**

TS is a reinforcement learning algorithm that makes use of a probability distribution and Bayes rules to forecast the success rate of each action. At the start, it is assumed that all actions have a uniform probability distribution of getting a reward or success. For each trial obtained from an action, based on rewards obtained, a new probability distribution is generated for every action and based on these prior probabilities, the success distribution is updated. After sufficient observations, each action will have a success distribution associated with it which can help the agent to choose wisely so as to get maximum rewards. TS balances between exploring new actions and exploiting known ones and solves the MAB problem proficiently. TS has the benefit of its tendency to be more search-oriented when data are scarce and less search-oriented when data are more abundant [8].

- **Upper confidence bound algorithm**

UCB makes use of the uncertainty in action value estimation to solve the exploration–exploitation dilemma. UCB follows the principle of optimism in the face of uncertainty, which implies that if there is uncertainty about an action, it can be optimistically assumed as the correct action.

Initially, UCB explores more to methodically reduce uncertainty, and gradually the exploration reduces. Thus, UCB obtains greater rewards on average than most of the other reinforcement learning algorithms.

- **Discounted upper confidence bound algorithm**

For solving a non-stationary problem, more recent plays will be important. So the UCB algorithm for stationary problem is modified for the non-stationary problem by using a discounting factor. The DUCB policy averages past rewards with a discount factor which gives more weight to recent plays [9].

Studies show that UCB and TS perform better, give better rewards, and are faster than the ϵ-greedy algorithm. UCB and TS algorithms use exploration more effectively. TS is more stable for higher numbers of trials. UCB is more tolerant of noisy data and converges more rapidly. DUCB addresses the practical non-stationary wireless scenario more effectively. MAB is an important approach in reinforcement learning for taking decisions in case of uncertainty. MAB problems can be solved using different algorithms, and as every algorithm has its own set of benefits and drawbacks, the selection of the algorithm is based on the problem to be addressed and the adjustment of exploration and exploitation [5].

6.3 Related work

Since the choice of LoRa transmission parameters affects the capacity and overall network performance, if multiple end devices on the same channel use the same *SF* concurrently, collisions occur. As node density increases, collision probability increases and performance drops [10,11]. Selection of LoRa parameters or resources can be done in a centralized or distributed manner. Either the network controls the transmission parameters of the devices or the device can decide its own data rate and transmission power. LoRa Alliance recommends ADR, which is a centralized approach wherein the network server decides the end device transmit parameters. It adjusts the transmit power of the node by adjusting the data rate. The network estimates the transmit power of the node for the next transmission by changing the data rate and taking into consideration the node's transmit power for at least 20 prior transmissions. This is then communicated to the node by the server and the node accordingly adapts its parameters. The ADR approach is more suitable in stable RF conditions where the end devices are not mobile [12] which is a drawback in some practical scenarios wherein nodes may be mobile. A simple LoRa configuration assumes a Poisson point process distribution with nodes uniformly distributed around the gateway, transmitting at the same power level in a single channel and with no interference from non-LoRa nodes. Selection of optimal transmit parameters is a complex task even for the simplest case. The ADR algorithm also has some disadvantages. Depending on the uplink signal-to-noise ratio (SNR), ADR allots *SF* to the node. Lower *SFs* are allotted to nodes near the gateway and higher *SFs* are assigned to farther nodes. If all the nodes in the network are closer, then the same *SF* may be allocated to them by ADR, resulting in collisions due to overusage of the same *SF* and underusage of other *SFs* [13]. Although ADR reduces energy consumption, it suffers huge packet losses. Also, the experiments indicate that almost all MAB algorithms investigated in the study, such as UCB, TS, sliding window UCB (SWUCB), and switching TS (STS), outperform the ADR algorithm [10]. In such settings, distributed learning algorithms where the end devices work intelligently to choose a particular parameter from a given set at a given time are suitable. The selection of the SF and other parameters impacts interference avoidance and energy efficiency of the end devices in a LoRa network. Machine learning approaches can be useful in this perspective for appropriate parameter selection so as to reduce interference and energy consumption [14]. MAB algorithms, which form a special class of reinforcement learning, adhere to such a setting and can be effectively used for optimal parameter selection. The UCB algorithm executed by the end device for LoRa channel selection is demonstrated to reduce collisions with other devices in the ISM band [15]. Experimentation shows extended battery life with minimum processing and memory along with improved performance compared to the conventional random selection method. These inexpensive algorithms can be also used to reduce interference from other neighboring gateways. Work using less complex MAB algorithms such as TS and UCB1 along with GNU radio implementation illustrates that the intelligent end nodes improvise network access and performance [16]. Both TS and UCB1 are efficient in stationary environments, with TS giving slightly better performance and UCB1 learning faster. The use of learning algorithms apparently accommodates more devices when all the end devices are intelligent in a network. The use of UCB1 and TS algorithms with a time-frequency slotted ALOHA-based protocol validates increased successful transmissions even in non-stationary settings [7]. An adversarial MAB algorithm, an EXP3-based simulator that considers inter-*SF* collision and capture effects, is designed for resource allocation in IoT and LoRa networks [17]. Although EXP3 has long convergence times, the success rate, node lifetime, and energy efficiency are improved. EXP3.S, a modified alternative, needs less convergence time but the convergence time increases with increasing number of parameters to be selected. EXP3.S is also suitable for non-uniform device distribution in IoT and LoRa networks [18]. Stochastic and adversarial-based distributed learning algorithms, such as Updated UCB (UUCB), Updated UCB1 (UUCB1), and Updated EXP3 (UEXP3), are proposed to select the device communication parameters. Simulation results display improved reliability and energy efficiency for low-power IoT networks [14]. Strategies based on UCB algorithms for channel selection and retransmissions have been explored, and the results indicate an improved successful transmission rate even when the number of devices is large [19]. Reinforcement learning methods to reduce power consumption by using appropriate transmit parameters for LoRa networks have also been studied. A method that is based on a combination of centralized reinforcement learning and distributed reinforcement learning has been proposed [20], and the results demonstrate a reduction in energy consumption; however, the learning rate is higher. Further work to improve the learning rate and throughput can be carried out. Optimal parameter selection at the node level using different MAB algorithms has been investigated, confirming their advantage over ADR with respect to energy requirements and packet loss. In the context of cognitive radio, the use of MAB algorithms has been examined for spectrum sensing and spectrum access. A DUCB algorithm has been proposed for spectrum sensing in a dynamic scenario, and it gives better detection efficiency [21]. For a non-stationary cognitive radio scenario, the DUCB algorithm is used for frequency band selection with improved performance [9].

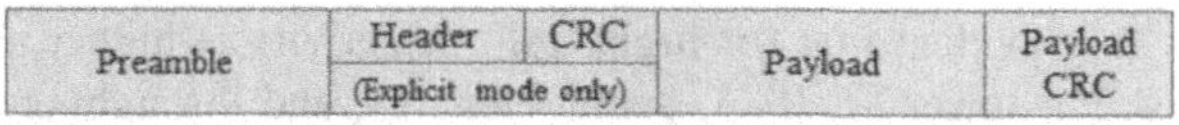

Preamble	Header (Explicit mode only)	CRC	Payload	Payload CRC

FIGURE 6.3

LoRa packet format.

6.4 Interference avoidance and energy efficiency for LoRa

For the transmission of a LoRa packet, the average energy required is given by

$$E_{avg} = P_t T_{pkt} N_p, \tag{6.1}$$

where E_{avg} denotes the average energy, P_t is transmission power, T_{pkt} is the time required to transmit a packet, and N_p is the number of transmissions required to send a packet successfully. The time for packet transmission, the number of transmissions needed for successful packet transmission, and the transmission power are significant factors determining the network performance. If interference and collisions are reduced, retransmissions are decreased. As SF increases, sensitivity improves, the need for retransmissions decreases, and consequently the average energy required for packet transmission also decreases. The LoRa packet format [22] is illustrated in Fig. 6.3.

The time required to transmit a packet or ToA can be calculated as

$$T_{pkt} = \left(n_p + 4.25\right) \frac{2^{SF}}{BW} + \left(8 + max\left(ceil\left(\frac{8PL - 4SF + (28 + 16C) - 20H}{4\left(SF - 2DE\right)}\right)\left(CR + 4\right), 0\right)\right) \frac{2^{SF}}{BW}, \tag{6.2}$$

where n_p is the number of programmed preamble symbols, PL is the packet payload, H is 0 when the header is present and 1 when the header is absent, and DE is 1 when low data rate optimization is enabled and 0 when low data rate optimization is disabled. Cyclic redundancy checksum (CRC) is by default enabled (C = 1), resulting in the term (28 + 16C) to be 44, whereas if it is disabled (C = 0), the term becomes 28. The CRC field is present only in uplink packet transmissions.

An LPWAN network can include LoRa and non-LoRa nodes with a single gateway or multiple gateways. LoRa uses CSS with quasi-orthogonal spreading factors (SF) having typical values of 7, 8, 9, 10, 11, or 12. The network can be affected by interference from other LoRa nodes with the same SF (Co-SF), other LoRa nodes using different SFs (Inter-SF), and other nodes using a different technology but the same carrier frequency. A signal is detected at the gateway when the received signal-to-interference-plus-noise ratio (SINR) exceeds the receiver sensitivity at the desired LoRa node for a particular SF [23,24]. For SINR to be higher so as to ensure signal detection, signal power needs to be higher and interference power should be lower. However, signal power must be less for lower energy consumption. From

(6.1) and (6.2), it is evident that for higher *SF*, sensitivity is higher and the SINR required reduces. For lower *SF*, ToA, average energy, and throughput are also low. The probability of successful transmission reduces with increasing network size and consequently an increasing number of devices using the same *SF*. Thus, there is always a trade-off between interference avoidance and energy efficiency. The selection of *SF* and other transmit parameters has a significant impact on the same [25]. Hence, optimal *SF* selection is primarily studied and investigated in this work.

6.4.1 Developed multi-armed bandit algorithm for LoRa

Studies suggest that machine learning algorithms can be applied for selecting parameters optimally to reduce interference and improve energy efficiency, thus enhancing the performance of the network [14]. For enhancing the energy efficiency in LPWANs, adaptive transmission and efficient resource utilization are the studied approaches [15]. Different centralized and decentralized approaches for selection of transmit parameters have been investigated. It is observed that the centralized optimization problem for parameter selection is highly complex even for the simplest configuration with assumptions such as a Poisson point process distribution where nodes are uniformly distributed around the gateway and have a constant transmit power, a single channel, and no interference from non-LoRa nodes. The distributed learning algorithms prove useful in such settings. In decentralized learning resource allocation, the end device can select *SF*, the subchannel for each packet transmission, and the transmission power to optimize performance with respect to energy efficiency, interference avoidance, and reliability. The optimal parameter selection problem fits into the framework of MAB algorithms, and this approach of MAB algorithms can be explored and examined further.

Consider SF selection in LoRa using an MAB algorithm as shown in Fig. 6.4. This is a distributed learning method that assumes an intelligent end device. The end device selects an *SF* or a strategy $s(t) = \{SFs\}$ from a given set of *SFs*. The device is unaware of its position or channel state. Therefore, the device may choose any *SF* from the set $s \in S$. Every end device chooses a strategy $s(t)$ at every packet arrival time t, depending on a particular distribution over S, that results in a reward $r_{s(t)} \in \{0, 1\}$. As the device transmits a packet by selecting an *SF*, the resulting trans-

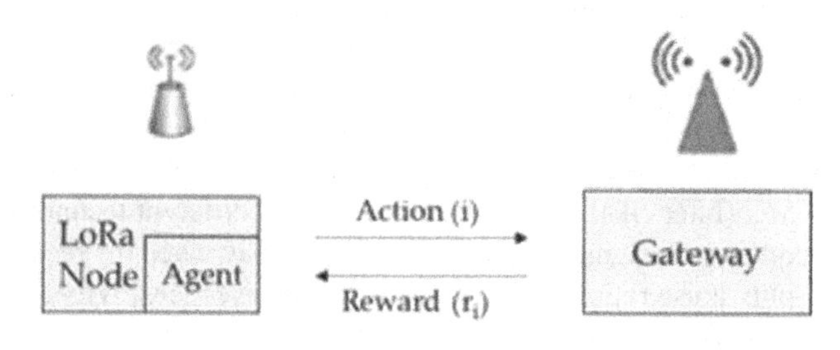

FIGURE 6.4

MAB model for LoRa.

mission can be successful or may fail. If the packet is successfully received by the gateway, it sends an acknowledgment to the gateway. This is similar to an MAB problem. The *SF* value and the state of *SF* can be modeled as an action and reward, respectively. The state of *SF* means whether choosing that *SF* results in successful transmission and acknowledgment receipt or otherwise. The end device receives an acknowledgment indicating reward = 1 or reward = 0 when it does not receive an acknowledgement. The end device uses the locally available information that acknowledgment is received or not and accordingly chooses an optimal *SF* that suffers the fewest collisions. The end devices are dynamic and hence modeling the *SF* selection problem as a non-stationary MAB problem is more appropriate.

- **Discounted upper confidence bound algorithm for LoRa**

Stochastic MAB algorithms such as TS and UCB are applicable for stationary distribution scenarios, whereas the advanced DUCB algorithm is suitable for non-stationary problems. By using an appropriate discount factor, the UCB algorithm can be modified to suit a non-stationary problem [9]. This is the idea behind the DUCB algorithm. The discount factor gives more weight to the most recent plays and averages past rewards in the DUCB policy. This approach fits the time-varying wireless environment. Therefore, the DUCB policy can also be optimized by modifying the discount factor and exploration bonus to adapt to the varying and complex LoRa network environment [9,26]. The DUCB algorithm core index is given as

$$U_k(t) = X_k(t) + B_k(t), \tag{6.3}$$

where $X_k(t)$ is the discounted average for exploitation and $B_k(t)$ is the exploration bonus [17]. If the discount function is a power function which is defined as $f(x) = \gamma^x$, then the term $X_k(t)$ can be written as

$$X_k(t) = \frac{\sum_{s=1}^{t} \gamma^{t-s} X_t^k 1_{i_s=i}}{\sum_{s=1}^{t} \gamma^{t-s} 1_{i_s=i}}, \tag{6.4}$$

where $X_k(t)$ is the average reward of action k at time step t, s is the sample, γ^{t-s} denotes the discount function, and 1_{i_s} is the indicator function with a value of 1 if true and 0 if false.

The exploration bonus $B_k(t)$ is given as

$$B_k(t) = 2B\sqrt{\frac{\xi\left(\log \sum_{i=1}^{k} Ni(t)\right)}{N_k(t)}}, \tag{6.5}$$

where N is the maximum number of trials, i is the index of the actions, X_k is the average reward for action k, N_k is the number of times action k is chosen, and ξ is the bias parameter. $N_k(t)$ is given as

$$N_k(t) = \sum_{s=1}^{t} \gamma^{t-s} 1_{i_s=i}. \tag{6.6}$$

A different exploration bonus depending on statistical variance is explored [9], given as

$$B_k(t) = \xi \sqrt{\frac{X_k(t) - X_k(t)^2}{N_k(t)}}, \tag{6.7}$$

where ξ is the bias parameter and the term $X_k(t) - X_k(t)^2$ denotes the statistical variance of the reward of each action. In the DUCB algorithm, the recent plays are given more weight by use of a suitable discount function. A monotonically decreasing function is generally chosen as the discount function, designed according to the application environment. Exponential functions, power functions, and window functions are some of the commonly used discount functions. An appropriate exploration bonus function decides the exploration of an action which has not been recently tried. Usually a monotonically decreasing function is used as the exploration bonus, which needs to be attuned to ensure that an optimal action is explored when rewards change.

- **Developed optimal policy UCB-P-1/2+O for LoRa**

LoRa devices have low data rates and low power consumption, and as they use the ISM band, devices need to comply with the duty cycle limitations. The number of transmissions per device per day depends on the duty cycle, which may be 1% or 0.1% as per the operation region. The parameter selection method for LoRa, therefore, needs to be fast and non-complex [18]. We propose a new discount function and exploration bonus especially suitable for LoRa requirements to develop a new and modified DUCB algorithm that aims to reduce complexity and improve the learning rate. It is the discount function in the core index of the DUCB policy that decides the weights assigned to the samples. Hence the application scenario and design requirements govern the discount function. For a setting with frequent changes such as cognitive radio systems, the discount function $y = \left[\frac{N-x}{N}\right]^a$, where N is the number of trials or time steps, is appropriate [9]. Since LoRa devices have low data rates with duty cycle restrictions, we have designed a new discount function:

$$y = \left[\frac{\mathbf{N} - \mathbf{x}}{\mathbf{N}}\right]^{1/a}, \tag{6.8}$$

where N is the number of trials or time steps and a is a constant that primarily decides the discount function variation. The proposed discount function is also a monotonically decreasing function, but it is more apt for scenarios where changes are less frequent as in LoRa. The exploration bonus decides the method of exploration of the actions as per the algorithm. A modified DUCB policy, UCB-P-1/2+O, is developed for transmit parameter selection in LoRa. The core index of the proposed policy is given as

$$U_k(t) = X_k(t) + 0.5\sqrt{\frac{(X_k(t) - X_k(t)^2)}{N_k(t)}}, \tag{6.9}$$

where $X_k(t)$ is the exploitation or discounted average with a discount function of $\left[\frac{N-x}{N}\right]^{1/2}$ and an exploration bonus of $0.5\sqrt{\frac{(X_k(t)-X_k(t)^2)}{N_k(t)}}$. The proposed algorithm is as follows.

Algorithm 6.1: Proposed UCB-P-1/2+O algorithm.

Inputs: Discount function $f(x)$, exploration bonus $B(x)$ as per the policy
Output: Received rewards
1: Initially select each action once
2: For every trial $t = k+1, k+2, k+3, \ldots$:
3: Set $U_k(t) = X_k(t) + B_k(t)$
4: Select action i_t = argmax $U_k(t)$
5: Receive the reward $X_k(t) \in [0, 1]$
6: For all the actions, $k = 1, 2, 3, \ldots K$ set:
7: $X_k(t) = \frac{\sum_{s=1}^{t} f(s) X_t^k 1_{i_s=i}}{\sum_{s=1}^{t} f(s) 1_{i_s=i}}$
8: $B_k(t) = B(t)$

The flowchart of the developed UCB-P-1/2+O policy is illustrated in Fig. 6.5. An intelligent end node can utilize this policy for selection of *SF* as one of the transmission parameters. At the start, the node randomly selects an *SF* from the set $\{SF\}$ and transmits a data packet using this *SF* parameter. The gateway sends an acknowledgment if it receives the packet successfully and the reward equals one. If the packet is not received successfully, no acknowledgment is received by the node and the reward is equal to zero. Accordingly, the policy is updated and for the next trial, the selection of *SF* is done as per the updated policy. Gradually, the node explores and exploits its choices and improves its learning, consequently leading to optimal *SF* selection.

6.4.2 Performance analysis

In this study, the *SF* selection problem is primarily considered as it significantly impacts LoRa network performance. A homogeneous LoRaWAN situation with a single gateway and multiple nodes using the same transmit power and channel frequency are assumed. Each node transmits packets with a particular *SF* and is not aware of the *SF* used by other nodes. If the packet is received by the gateway, an acknowledgment is sent to the node (reward = 1). Otherwise, if no acknowledgment is received (reward = 0), the node needs to retransmit. A node selecting one *SF* out of the available set is analogous to the selection of an action in an MAB algorithm. Simulations with a dataset of different *SF* values and mean rewards for varying numbers of trials are carried out. As per the strategy defined by the developed MAB policy, the *SF* value is selected in every trial. The developed UCB-P-1/2+O policy is evaluated and compared with other policies used in the literature to handle the *SF* selection problem. The SF is chosen from the set $SF = \{7, 8, 9, 10, 11, 12\}$ as per the LoRa specifications, so there are six actions. For the simulations, *SF* selection is consid-

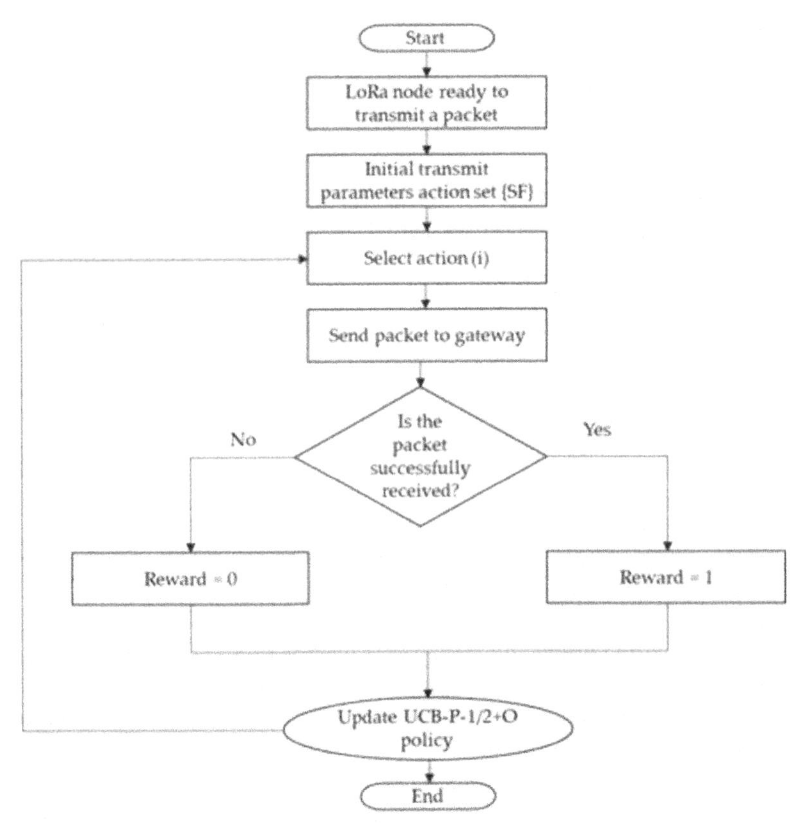

FIGURE 6.5

Flowchart of the developed DUCB-P-1/2+O policy for LoRa.

ered assuming a constant transmit power and the same channel frequency, resulting in six actions according to the six defined *SF* values. It is assumed that the policy action does not change the reward of any action. Also, actions are independent of each other, so the distribution or state of any action does not affect that of the other actions. LoRa transmissions are less frequent as they follow duty cycle restrictions [27]; therefore, the algorithms are evaluated over a lower number of trials, however with adequate iterations to support the observations. Increasing the number of trials further yields findings that are similar.

- **Different scenarios**

Three different datasets depicting three different network scenarios are designed for analysis of the proposed policies [9]. The mean rewards distribution or variation of different actions for a few practical scenarios is represented by the datasets. Scenario A, as elucidated in Fig. 6.6, represents a stable or stationary situation wherein

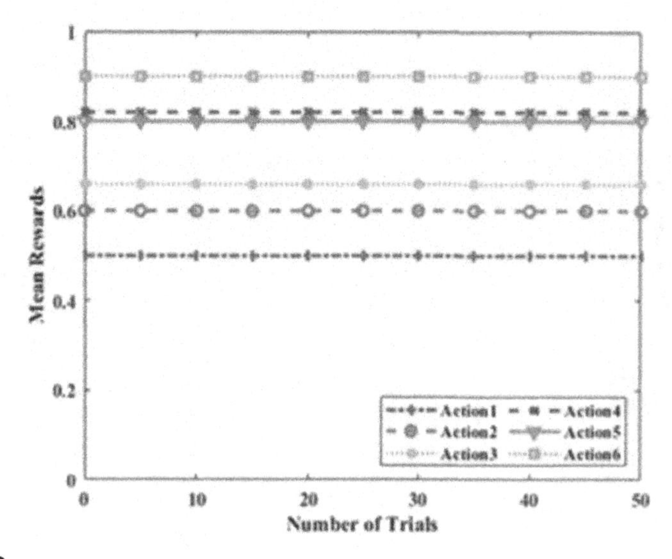

FIGURE 6.6

Dataset for scenario A with constant mean rewards for all actions.

the mean rewards of all actions remain constant and do not change and there is only a single optimal action. This portrays a comparatively simple situation but maps the case of distant, sparse LoRa nodes in a network and hence the algorithm evaluation over this is significant.

Scenario B exemplifies a more practical LoRa network wherein mean rewards of multiple actions vary simultaneously. This dataset wherein the mean rewards change four times and the optimal action changes five times for a set of six actions is shown in Fig. 6.7.

Scenario C characterizes the situation when the mean rewards for all actions remain unchanged except for one action, for which it varies. This is a representation of the case when the mean rewards of an action decrease when it is already busy and increase when it becomes free. This is analogous to the case of a small network with a low number of end devices. Here, the optimal action also varies with time. As shown in Fig. 6.8, the mean reward of action 6 changes two times and the optimal action changes once, from action 6 to action 5.

For the scenarios as previously mentioned, the proposed algorithm is investigated and compared with other algorithms used in the literature for a LoRa network consisting of multiple intelligent nodes using the MAB algorithm simultaneously. The simulation results for different algorithms such as TS, random sampling (RS), UCB, DUCB, and the designed DUCB-P-1/2+O algorithm are presented further [28]. The number of actions considered is six in accordance with the six *SF* values and the MAB algorithms are evaluated for scenarios A, B, and C, respectively, for different numbers of trials. Mean rewards and execution time are the parameters considered for investigation.

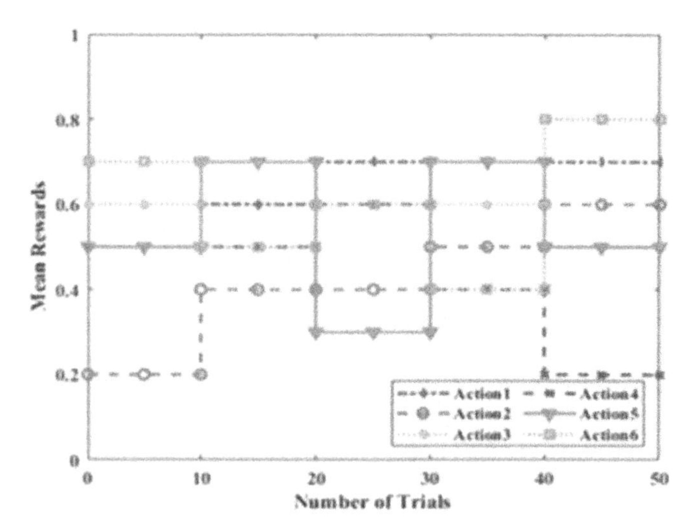

FIGURE 6.7

Dataset for scenario B with mean rewards of multiple actions that change simultaneously.

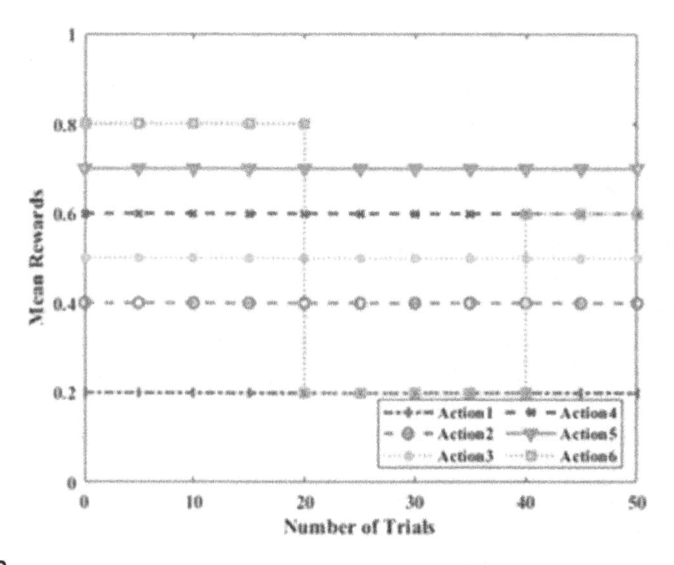

FIGURE 6.8

Dataset for scenario C with constant mean rewards for all actions, except for one.

For scenario A with six actions and five intelligent nodes, the variation in mean rewards with changing number of trials is presented in Fig. 6.9. For all the policies, as the number of trials increases, the mean rewards are also seen to increase, as expected.

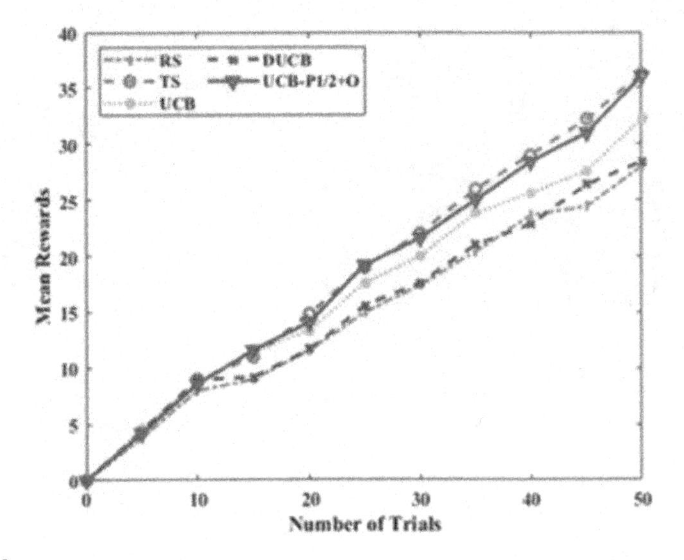

FIGURE 6.9

Mean rewards for scenario A.

The proposed UCB-P-1/2+O policy consistently yields better mean rewards with TS also faring well, while other policies result in lesser rewards.

The change in execution time with changing number of trials for different algorithms is shown in Fig. 6.10. The designed UCB-P-1/2+O algorithm needs similar execution times as DCUB but is faster than TS with an increasing number of trials. The conclusion is that the UCB-P-1/2+O algorithm provides a better reward and has a short execution time.

Fig. 6.11 illustrates that the UCB-P-1/2+O algorithm results in marginally better mean rewards in comparison with other studied ones for scenario B. DUCB follows the proposed UCB-P-1/2+O algorithm closely in this case.

For scenario B, although the execution time for UCB-P-1/2+O algorithm is similar to that of DUCB, also lesser for a few cases, as shown in Fig. 6.12, the UCB-P-1/2+O policy offers better mean rewards. It is also observed that RS requires less execution time; however depicts inconsistent nature with respect to the rewards obtained during different trials. Fig. 6.11 and Fig. 6.12 also lead to the similar conclusion that UCB-P-1/2+O policy executes faster and performs better.

Fig. 6.13 displays the mean rewards for scenario C with six actions and multiple intelligent nodes. It is observed that the UCB-P-1/2+O policy gives better mean rewards than the other policies with TS performing better too.

As observed in Fig. 6.14, for most of the trials, UCB-P-1/2+O involves less execution time than TS, UCB, and DUCB. As anticipated, RS needs the least execution time. Similar conclusions can be drawn from Fig. 6.13 and Fig. 6.14: UCB-P-1/2+O exhibits both fast execution and better rewards.

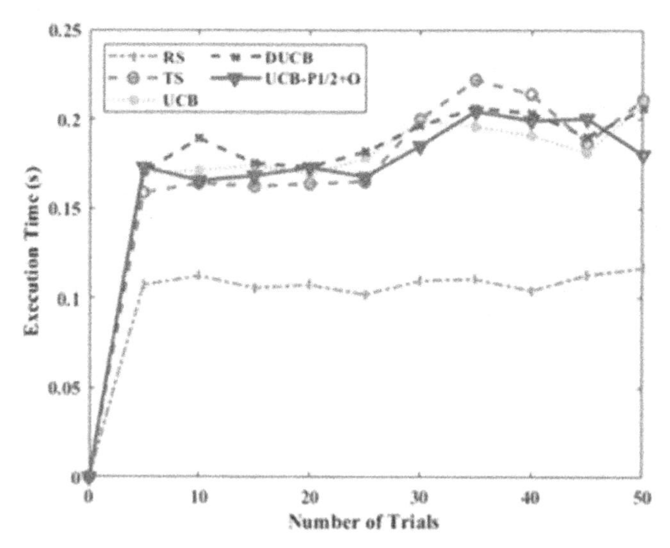

FIGURE 6.10

Execution time for scenario A.

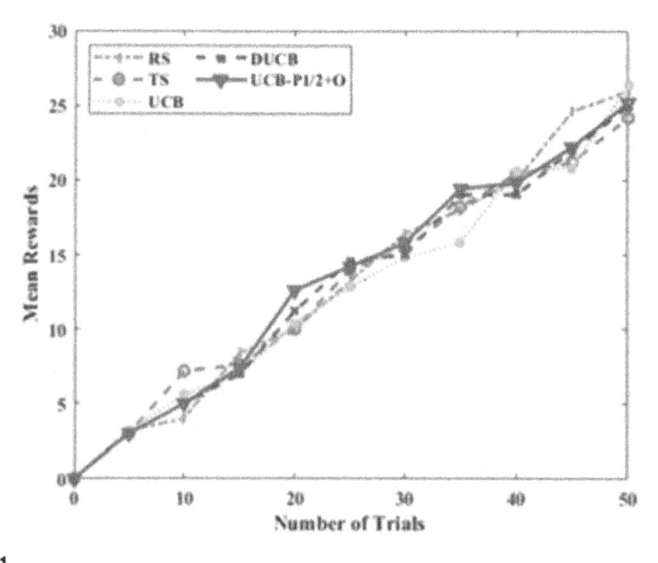

FIGURE 6.11

Mean rewards for scenario B.

The simulation results suggest that for multiple intelligent nodes too, in case of scenarios A and B, the UCB-P-1/2+O policy yields about 15% more mean rewards than DUCB, say for 50 trials, and a similar improvement is also observed for lower numbers of trials. For scenario C, it performs fairly similar to DUCB. For all the

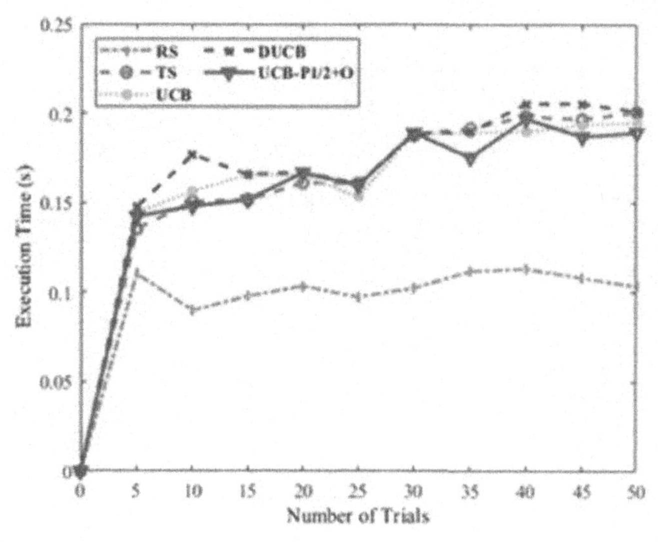

FIGURE 6.12

Execution time for scenario B.

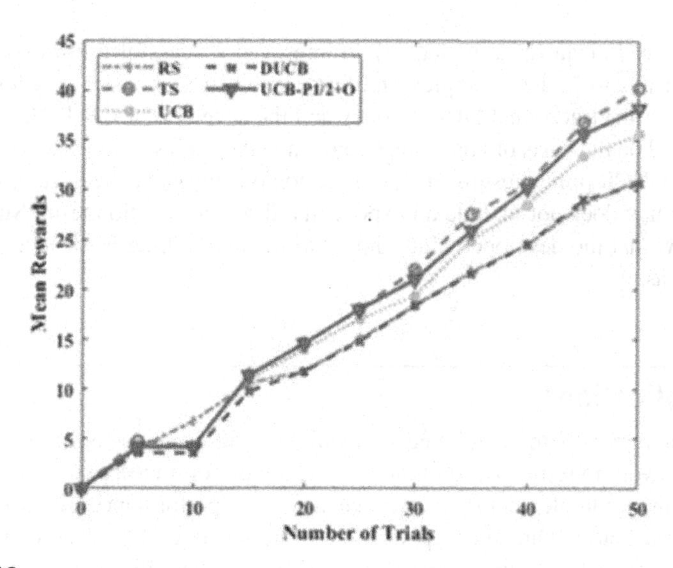

FIGURE 6.13

Mean rewards for scenario C.

various cases, scenarios, and numbers of trials, simulation results obtained lead to the conclusion that for multiple intelligent nodes, the designed UCB-P-1/2+O algorithm outperforms, gives better mean rewards than, and is faster than other policies. The developed policy with the designed discount function and exploration bonus

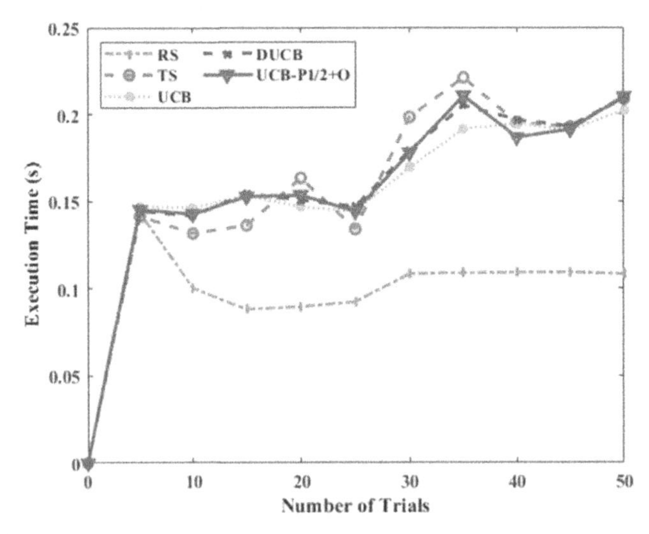

FIGURE 6.14

Execution time for scenario C.

yields improved mean rewards, leading to improved successful transmissions. The algorithm needs to be less complex and fast, since LoRa is basically a low-power technology. If the traditional DUCB policy and the developed UCB-P-1/2+O policy are compared on the basis of algorithm complexity, the exponential discount function term in the DUCB policy results in more time complexity, since the developed UCB-P-1/2+O policy does not include an exponential discount function term. Simulation results show that the developed algorithm is faster, underlining its benefit for LoRa transmissions.

6.5 Applications

LoRa technology is being employed for many IoT and machine-to-machine (M2M) applications with long-range requirements, and the use of reinforcement learning algorithms can be complementary to enhance the overall performance. A few examples are discussed below. One such application is the smart grid, for the communication of the advanced metering infrastructure (AMI) backhaul. Reinforcement learning algorithms can be applied to select the channel frequency and reduce the AMI communication latency. Low-complexity MAB algorithms such as UCB and TS that considerably reduce the collision probability and latency are used [29]. A novel MAB algorithm, LP-MAB, has been designed to adjust the end device parameters centrally by the network server, leading to less energy consumption. The proposed algorithm maintains a high packet delivery rate with reduced energy requirements. This is an important requirement in applications with mobility and varying speed requirements such as traffic monitoring, animal tracking, smart bicycles, and smart metering [30]. The radio propagation environment in an LPWAN is dynamically varying, and maintaining the desired battery lifetime for massive machine-type communication (mMTC) is a critical problem. Using reinforcement learning algorithms to select the optimal LPWAN technology at the end device dynamically, the battery lifetime can be extended. It is observed that TS performs better than ϵ-greedy and UCB algorithms, extending the battery lifetime in stationary as well as mobile end devices [31].

6.6 Conclusion

The performance of LoRa networks is significantly influenced by the choice of LoRa parameters. Resource allocation is a complicated process in LoRa as it requires selecting from a wide range of transmit parameter combinations. In this chapter, a developed UCB-P-1/2+O policy is presented with a new discount function and exploration bonus for optimal parameter selection in LoRa. Simulations indicate that the developed strategy performs better than the other approaches examined in the literature in terms of execution time as well as mean rewards, indicating that it is a viable lightweight optimized solution that is energy-efficient as the number of retransmissions is reduced. LoRa parameter selection using the developed algorithm can thus help in interference avoidance, enhancing the energy efficiency and the overall network performance. It is also evident that the MAB-based reinforcement learning method used by this algorithm is beneficial over the conventional random access techniques. This algorithm can also be used for the selection of multiple transmission parameters simultaneously, further enhancing LoRa network performance.

References

[1] LoRa Alliance, LoRaWAN. What is it? A technical overview of LoRa and LoRaWAN Technical Marketing Workgroup 1.0, https://lora-alliance.org/resource/hub/what-is-lorawan/, 2015.

[2] B. Chaudhari, S. Borkar, Design considerations and network architectures for low-power wide-area networks, in: B. Chaudhari, M. Zennaro (Eds.), LPWAN Technologies for IoT and M2M Applications, Elsevier/Academic Press, Amsterdam, The Netherlands/Cambridge, MA, USA, 2020, pp. 15–35.

[3] Sheetal N. Ghorpade, Marco Zennaro, Bharat S. Chaudhari, IoT-based hybrid optimized fuzzy threshold ELM model for localization of elderly persons, Expert Systems with Applications 184 (5) (December 2021) 115500, https://doi.org/10.1016/j.eswa.2021.115500, Pergamon.

[4] Seham Ibrahem Abd Elkarim, Basem M. ElHalawany, Ola Mohammed Ali, M.M. Elsherbini, Machine Learning Approaches for LoRa Networks: A Survey, Springer Nature, 2021.

[5] Richard S. Sutton, Andrew G. Barto, Reinforcement Learning: An Introduction, second edition, MIT Press, Cambridge, MA, 2018.

[6] R. Bonnefoi, L. Besson, C. Moy, E. Kaufmann, J. Palicot, Multi-armed bandit learning in IoT networks: learning helps even in non-stationary settings, in: Proceedings of the Cognitive Radio Oriented Wireless Networks, Lisbon, Portugal, 20–21 September 2017.
[7] A. Garivier, E. Moulines, On upper-confidence bound policies for switching bandit problems, in: J. Kivinen, C. Szepesvári, E. Ukkonen, T. Zeugmann (Eds.), Proceedings of the Algorithmic Learning Theory, Springer, Berlin/Heidelberg, Germany, 2011, pp. 174–188, https://doi.org/10.1007/978-3-642-24412-4_16.
[8] Daniel J. Russo, Benjamin Van Roy, Abbas Kazerouni, Ian Osband, Zheng Wen, A tutorial on Thompson sampling, Foundations and Trends® in Machine Learning 11 (1) (2018) 1–96, https://doi.org/10.1561/2200000070.
[9] Y. Chen, S. Su, J.A. Wei, Policy for optimizing sub-band selection sequences in wideband spectrum sensing, Sensors 19 (2019) 4090, https://doi.org/10.3390/s19194090.
[10] R. Kerkouche, R. Alami, R. Féraud, N. Varsier, P. Maillé, Node-based optimization of LoRa transmissions with multi-armed bandit algorithms, in: Proceedings of the 2018 25th International Conference on Telecommunications (ICT), Saint-Malo, France, 2018, pp. 521–526, https://doi.org/10.1109/ICT.2018.8464949.
[11] S. Gupta, B.S. Chaudhari, B. Chakrabarty, Vulnerable network analysis using war driving and security intelligence, in: Proceedings of the International Conference on Inventive Computation Technologies (ICICT), Coimbatore, India, 26–27 August, 2016, pp. 1–5, https://doi.org/10.1007/978-981-10-3812-9_49.
[12] LoRa and LoRaWAN Technical Overview, Semtech, December 2019.
[13] M.N. Ochoa, A. Guizar, M. Maman, A. Duda, Toward a self-deployment of LoRa networks: link and topology adaptation, in: Proceedings of the 2019 International Conference on Wireless and Mobile Computing, Networking and Communications (WiMob), Barcelona, Spain, 21–23 October, 2019, pp. 1–7, https://doi.org/10.1109/WiMOB.2019.8923427.
[14] A. Azari, C. Cavdar, Self-organized low-power IoT networks: a distributed learning approach, in: 2018 IEEE Global Communications Conference (GLOBECOM), Abu Dhabi, United Arab Emirates, 2018, pp. 1–7, https://doi.org/10.1109/GLOCOM.2018.8647894.
[15] C. Moy, IoTligent: first world-wide implementation of decentralized spectrum learning for IoT wireless networks, in: Proceedings of the 2019 URSI Asia-Pacific Radio Science Conference (AP-RASC), New Delhi, India, 9–15 March, 2019, pp. 1–4, https://doi.org/10.48550/arXiv.1906.00614.
[16] L. Besson, R. Bonnefoi, C. Moy, GNU radio implementation of MALIN: multi-armed bandits learning for Internet-of-things networks, in: Proceedings of the 2019 IEEE Wireless Communications and Networking Conference (WCNC), Marrakesh, Morocco, April 2019, pp. 15–18.
[17] Duc-Tuyen Ta, Kinda Khawam, Samer Lahoud, Cédric Adjih, Steven Martin, LoRa-MAB: a flexible simulator for decentralized learning resource allocation in IoT networks, in: 12th IFIP Wireless and Mobile Networking Conference, Paris, France, Sep 2019, pp. 55–62, hal-02431653, https://doi.org/10.23919/WMNC.2019.8881393.
[18] B.S. Chaudhari, M. Zennaro, Introduction to low-power wide-area networks, in: LPWAN Technologies for IoT and M2M Applications, Elsevier, Amsterdam, The Netherlands, 2020, pp. 1–13, https://doi.org/10.1016/B978-0-12-818880-4.00001-6.
[19] R. Bonnefoi, L. Besson, J. Manco-Vasquez, C. Moy, Upper-confidence bound for channel selection in LPWA networks with retransmissions, in: Proceedings of the 2019 IEEE Wireless Communications and Networking Conference, Workshop (WCNCW), Marrakech, Morocco, 15–18 April 2019, Hal-02049824v2f.

[20] G. Park, W. Lee, I. Joe, Network resource optimization with reinforcement learning for low power wide area networks, EURASIP Journal on Wireless Communications and Networking 2020 (2020) 176, https://doi.org/10.1186/s13638-020-01783-5.
[21] W. Ning, X. Huang, K. Yang, F. Wu, S. Leng, Reinforcement learning enabled cooperative spectrum sensing in cognitive radio networks, Journal of Communications and Networks 22 (2020) 12–22, https://doi.org/10.1109/JCN.2019.000052.
[22] LoRa Modem Design Guide. Semtech Wireless & Sensing, https://www.openhacks.com/uploadsproductos/loradesignguide_std.pdf. (Accessed 15 November 2023), July 2013.
[23] Zhiyu Nie, Lingyun Jiang, Interference modelling and analysis of LoRa network, in: 14th International Conference on Wireless Communications, Networking and Mobile Computing (WiCOM), 2018, https://doi.org/10.12783/dtcse/wicom2018/26271.
[24] D. Mankar, B.S. Chaudhari, Dynamic performance analysis of IEEE 802.15.4 devices under various RF interferences, in: 2016 International Conference on Inventive Computation Technologies (ICICT), Coimbatore, India, 2016, pp. 1–4, https://doi.org/10.1109/INVENTIVE.2016.7823212.
[25] A.R. Askhedkar, B.S. Chaudhari, Energy efficient LoRa transmission over TV white spaces, International Journal of Information Technology 15 (2023) 4337–4347, https://doi.org/10.1007/s41870-023-01453-x.
[26] P. Auer, N. Cesa-Bianchi, Y. Freund, R.E. Schapire, The nonstochastic multiarmed bandit problem, SIAM Journal on Computing 32 (2002) 48–77, https://doi.org/10.1137/S009753970139837.
[27] A. Askhedkar, B. Chaudhari, M. Zennaro, E. Pietrosemoli, TV white spaces for low-power wide-area networks, in: B. Chaudhari, M. Zennaro (Eds.), LPWAN Technologies for IoT and M2M Applications, Elsevier/Academic Press, Amsterdam, The Netherlands/Cambridge, MA, USA, ISBN 9780128188804, 2020, pp. 167–179, https://doi.org/10.1016/B978-0-12-818880-4.00009-0.
[28] A.R. Askhedkar, B.S. Chaudhari, Multi-armed bandit algorithm policy for LoRa network performance enhancement, Journal of Sensor and Actuator Networks 12 (2023) 38, https://doi.org/10.3390/jsan12030038.
[29] Remi Bonnefoi, Christophe Moy, Jacques Palicot, Improvement of the LPWAN AMI backhaul's latency thanks to reinforcement learning algorithms, EURASIP Journal on Wireless Communications and Networking 2018 (2018) 34, https://doi.org/10.1186/s13638-018-1044-2.
[30] B. Teymuri, R. Serati, N.A. Anagnostopoulos, M. Rasti, LP-MAB: improving the energy efficiency of LoRaWAN using a reinforcement-learning-based adaptive configuration algorithm, Sensors 23 (2023) 2363, https://doi.org/10.3390/s23042363.
[31] Martin Stusek, Pavel Masek, Dmitri Moltchanov, Nikita Stepanov, Jiri Hosek, Yevgeni Koucheryavy, Performance assessment of reinforcement learning policies for battery lifetime extension in mobile multi-RAT LPWAN scenarios, IEEE Internet of Things Journal 9 (24) (2022) 25581–25595, https://doi.org/10.1109/JIOT.2022.3197834.

CHAPTER

Software frameworks for TinyML

7

Sachin Chougule[a,b], Sheetal N. Ghorpade[a], Bharat S. Chaudhari[b], and Marco Zennaro[c]

[a]*Rubiscape Private Limited, Pune, India*

[b]*Department of Electrical and Electronics Engineering, Dr. Vishwanath Karad MIT World Peace University, Pune, India*

[c]*Science, Technology and Innovation Unit, Abdus Salam International Centre for Theoretical Physics (ICTP), Trieste, Italy*

7.1 Introduction

Modern applications are evolving from mere tools to central companions, with users increasingly expecting intuitive, almost human-like interactions with them. Autonomous vehicles mounted with cameras, radars, and various sensors generate gigabytes of data per second [1]; uploading these data to the cloud in real time to make decisions is a challenging task. Mobile phone applications like speech translation and face recognition require high computing skills to run online or offline. Driven by edge computing techniques and propelled by AI applications, edge intelligence has been pushed to new limits. Employing the collective strength of Internet of Things (IoT) and machine learning provides solutions for the unique problem of scaling for processing data where they are generated [2]. Lots of previously complex problems that require human expertise and complex hardware–software systems for decision-making are currently being automated by implementing cutting-edge machine learning models on lean embedded IoT. Low latency and high availability for several network-aware services are the main benefits of edge computing, and it also improves privacy, security, and reliability for the network end users [3].

In the current technological landscape, there is a gap between embedded hardware and the required software for executing machine learning workload [4]. The edge is always resource-constrained in comparison with the cloud, which has a high capability for executing AI algorithms using deep neural networks (DNN) as DNNs require both large storage (as big as 500 MB for the VGG-16 model [5]) and much computing power (as high as 15,300 MMA for executing the VGG-16 model [6]). Generally, existing embedded processors are used for sensor data processing and web-based applications. The current situation presents a significant obstacle to the realization of the "cloud-to-embedded" vision. Signal processing also played a vital role in the breakthrough move from cloud intelligence to edge devices – tiny embedded devices for performing machine learning, i.e., tiny machine learning (TinyML) [7,8].

TinyML for Edge Intelligence in IoT and LPWAN Networks. https://doi.org/10.1016/B978-0-44-322202-3.00012-9

7.2 Overview of TinyML

TinyML is a promising paradigm that has brought many innovations and contributed to the sharp upswing in IoT applications such as autonomous driving, smart manufacturing, and smart healthcare. TinyML is a framework designed to enable the implementation of machine learning on embedded edge devices with limited processor and memory resources [9,10]. The main objective of TinyML is to maximize the opulence of deep learning systems by means of reduced computation and data requirements, which opens an enormous market of edge intelligence and IoT [11]. TinyML empowers edge devices with machine learning, boosting privacy and responsiveness while reducing energy efficiency and cost for wireless transmission; at scale, this cost is significantly higher than the cost of computation. TinyML can be visualized as the combination of three essential components: software, hardware, and algorithms. The earliest TinyML applications can be housed in Linux, embedded Linux, and cloud-based software. Hardware used in TinyML consists of low-complexity IoT nodes with or without hardware accelerators. A diverse toolbox of computing options – in-memory, analog, and neuromorphic – empowers TinyML devices to deliver richer and more tailored learning experiences. The requirement is for kB-sized models to be implemented in resource-constrained edge devices, which requires innovative algorithms. Effective quantization and compression techniques are essential for TinyML [12].

TinyML models should be adequately condensed to be accommodated in confined microcontroller unit (MCU)-class devices. This means that the model should limit input size and the number of layers [13] or consider lightweight methods other than neural networks [14], the TensorFlow framework, effective inference tools [15,16], aggressive quantum methods [17], and memory-conscious neural architecture search [18].

Now, big corporations have turned their focus towards TinyML. The frameworks released by these corporations include:

- Google TensorFlow Lite [19],
- Microsoft EdgeML [14],
- ARM Cortex Microcontroller Software Interface Standard Neural Network,
- STMicroelectronics X-Cube-AI [16].

7.3 Challenges

This section outlines the difficulties encountered while deploying TinyML.

7.3.1 Low power

TinyML systems are known for their low energy consumption, so a good calibration should hypothetically explain the energy efficiency of every device. The exact measurement of energy usage by TinyML systems poses a number of constraints. For early users, the energy consumption of TinyML devices is dramatically changing, thus making it hard to ensure precision over a broad spectrum of devices. Disparate data flows and preprocessing steps between devices create a significant hurdle in precisely defining the scope of power measurements. Beyond the core functionalities, factors like chip peripherals and underlying firmware can significantly influence the outcome of power measurement. Like standard high-power machine learning systems, TinyML systems lack redundant cores, making it simpler to load systems for testing.

7.3.2 Limited memory

Memory issues are common in TinyML systems because of their small size. Smartphones, like normal machine learning systems, usually have a few GBs, whereas TinyML systems frequently have memories that are multifold smaller than those of normal machine learning systems. Memory is a critical parameter that impacts how TinyML systems perform. On traditional machine learning platforms, model inference requires substantially more memory than available on TinyML systems. Ensuring a creative optimization strategy suitable for most algorithms is a challenge [20].

7.3.3 Processor power

MCUs (like the ARM Cortex-M series) have comparatively poor processing power when compared with cloud-based systems. Quality of service may not be maintained when data are transferred to the cloud for analysis. A range of software frameworks attempt to resolve the problem of reducing machine learning algorithm density, making these frameworks feature-rich and more versatile.

7.3.4 Extensive machine learning models and resource requirements

Dedicated machine learning hardware does not exist in embedded systems compared to consistent and seamless cloud systems. Machine learning algorithms and software runtimes required by both of them are constructed using complex high-level languages, which contribute significantly to compressing or porting machine learning models to embedded systems, which is challenging. Laborious and slow machine learning activities cannot be executed on embedded systems, so in some cases, it is difficult to move from cloud to embedded.

7.3.5 Heterogeneous hardware

Regardless of being in the early stages, there is a big diversity in TinyML systems with respect to their abilities, performance, and power. A few examples of these devices are general-purpose MCUs, memory computing, and event-based brain chips

[21]. Differences in system architecture and software systems make it challenging to port machine learning algorithms on TinyML systems [22].

7.3.6 An absence of suitable datasets

Low memory and processing capacity puts limitations on the use of existing datasets to be utilized in TinyML systems. Datasets for sensor data analysis need to be meticulously calibrated in terms of both temporal and spatial resolution to perfectly reflect the real-world data captured by diverse sensors. Additionally, these low-power edge systems need to be exposed to diverse datasets with different noise levels. This requires the availability of reliable datasets for training TinyML systems.

7.3.7 Network and data administration

TinyML systems require development to control data and networks efficiently. It has been found that the present edge networks pose limitations in recognizing various data types, impacting the accuracy of machine learning models. Irregularities in sensor data and their structures add one more difficulty to the challenges for data management at the edge. Data and network-blending techniques can help to regulate the network at the edge. Augmentative learning algorithms, together with knowledge extraction strategies, can improve knowledge sharing. Lightweight machine learning algorithms might be studied to encourage independence and self-adaptivity for network management in the edge.

7.3.8 Advanced machine learning models

For TinyML, advanced machine learning models and algorithms are essential. Taking quick action should be a critical characteristic of these models. They may employ online learning, federated learning, reinforcement learning, transfer learning, and training using knowledge distillation techniques. Real-time solutions require machine learning models to encompass control and communication components. Model development processing should include model trimming and quantization steps. It should also achieve cost savings while using edge devices.

7.3.9 Benchmarking

TinyML is considered a significant tool in the development of modern systems and is used in various applications in diverse industry sectors. This has increased the demand for functional design and testing, which in turn propelled the large-scale implementation of TinyML systems in a number of applications. TinyML systems and applications can achieve all-round performance through real-time performance, power efficiency, and system efficiency. A significant breakthrough in TinyML systems was made by academia and the industry, which increased the need for benchmarking these machine learning systems [23,24].

7.4 Software and libraries

Research studies concerning the design environment are elaborated here, focusing on frameworks such as TensorFlow Lite Micro and various libraries. The primary goal is to provide solutions for developers dealing with resource-constrained devices, ensuring optimal performance, including high accuracy and low inference time. Given the widespread interest in TinyML and its immense potential for transforming various industries, there is a continuous effort to develop and deploy numerous libraries and tools which aim to streamline the implementation of machine learning algorithms on platforms with limited resources.

7.4.1 Frameworks and libraries

The absence of a unified TinyML framework has resulted in the adoption of custom frameworks. However, these custom frameworks, often with limited availability, pose challenges as they necessitate intricate manual optimization when applied to different hardware configurations. Nonetheless, recent years have witnessed advancements in the development of the TinyML framework. One of the initial frameworks to emerge was Arm uTensor, an open-source machine learning framework tailored for microcontrollers. Subsequently, in 2019, uTensor collaborated with Google's TensorFlow to create the TensorFlow Lite for the microcontroller framework jointly [25].

Arm has advanced this achievement by introducing a comprehensive range of network kernels within the Cortex Microcontroller Software Interface Standard-NN (CMSIS-NN) software library [26]. Apache has also broadened its open-source machine learning framework TVM to cover microcontrollers with MicroTVM [27]. Another noteworthy edge machine learning framework is PyTorch Mobile, an expansion of the PyTorch ecosystem [28]. In addition to these adaptable frameworks, the emlearn library is noteworthy as an open-source machine learning inference engine crafted for microcontrollers, beginning from an 8-bit architecture [29]. All these frameworks are described in detail below.

7.4.1.1 TensorFlow Lite

TensorFlow Lite is an open-source deep learning framework equipped with tools designed for the deployment and execution of machine learning models on a diverse range of platforms, including Android, iOS, embedded Linux devices, and microcontrollers [30]. This framework effectively addresses on-device machine learning at the edge, taking into account crucial constraints such as latency, privacy, connectivity, size, and power consumption. TensorFlow Lite supports multiple programming languages, such as Java, Swift, Objective-C, C++, and Python. Notably, TensorFlow Lite facilitates model optimization through hardware acceleration.

In scenarios where highly constrained microcontrollers possess only a few hundred or dozen kilobytes of RAM, TensorFlow Lite for Microcontrollers (TFLM) emerges as a valuable tool, complementing TensorFlow Lite. TFLM efficiently supports machine learning inference on devices, although it currently lacks on-device training capabilities. Its core runtime demands a mere 16 kB of memory, making it compatible with numerous Arm Cortex-M architecture microcontrollers. Extensive testing has been conducted with Espressif ESP32 and various digital signal processors [31]. Moreover, TFLM works without the need for an operating system and can be easily obtained as an Arduino library.

The TensorFlow Model Optimization Toolkit plays a crucial role in decreasing a model's latency, memory footprint, and power consumption. Different techniques within this toolkit, like posttraining quantization, quantization-aware training, pruning, and clustering, contribute to achieving these improvements [32]. Additionally, TensorFlow Lite incorporates the TensorFlow Lite converter, enabling the postquantization of pretrained models and their conversion into the TensorFlow Lite format optimized for devices [33].

Posttraining integer quantization is particularly well suited for constrained microcontrollers.

7.4.1.2 uTensor

uTensor, a free embedded learning environment introduced in 2021, serves as a valuable tool for prototyping and swiftly deploying IoT edge devices. This environment comprises an inference engine, a graph processing tool, and an upcoming data collection architecture. uTensor streamlines the process by taking a neural network model trained using Keras and converting it into C++. The resulting model is then aptly deployed on boards such as Mbed, ST, and K64. Boasting a compact size, uTensor occupies only 2 kB on disk. Customization of uTensor is facilitated through a Python software development kit (SDK), allowing users to tailor it to their specific needs. The toolkit relies on essential toolsets, including Python, uTensor-CLI, Jupyter, Mbed-CLI, and ST-link (for ST boards). The workflow involves initial model creation, definition with quantization effects, and subsequent code generation tailored for deployment on suitable edge devices.

7.4.1.3 Cortex Microcontroller Software Interface Standard-NN

The CMSIS-NN library is specifically designed for neural network development on Arm Cortex-M processors, delivering notable improvements in throughput and energy efficiency. Inference based on its functions yields a substantial 4.6-fold increase in throughput and a significant 4.9-fold reduction in energy consumption [34]. This library encompasses a dedicated set of neural network functions and supports all functionalities [26].

The functions within the CMSIS-NN library utilize either 8-bit or 16-bit integers as parameters, with a predominant use of 16-bit multiply and accumulate (MAC) instructions for operations like matrix multiplications [34]. While these 16-bit SIMD instructions require an Arm processor equipped with a SIMD unit, it is noteworthy that the CMSIS-NN library can still be utilized with older Arm processors, such as Arm Cortex-M0, even in the absence of a SIMD unit [35].

7.4.1.4 Apache Tensor Virtual Machines

Micro Tensor Virtual Machines (MicroTVM) extends prevailing tensor virtual machines (TVM) to enable the implementation of tensor programs on microcontroller boards. This extension facilitates program optimization through the AutoTVM platform, which is designed to optimize tensor programs [36]. In practical terms, the process begins with connecting a microcontroller to a desktop or high-end machine running TVM in the background via a USB-JTAG port. The desktop employs OpenOCD to establish the link among the microcontroller and itself. OpenOCD, in turn, supports lTVM to control the microcontroller through a device-agnostic TCP port.

Various aspects of the lTVM association with TinyML can be encountered, including lazy execution, tensor loading, function calling, and module loading. These aspects collectively contribute to the integration of MicroTVM with microcontrollers for efficient execution of tensor programs.

7.4.1.5 PyTorch Mobile

PyTorch Mobile belongs to the PyTorch ecosystem, focusing on supporting all stages of machine learning model development, from training to deployment on smartphones. This versatile tool is compatible with robust mobile operating systems such as iOS, Android, and Linux. To facilitate machine learning integration into mobile applications, PyTorch Mobile offers several APIs [37].

This platform supports both the scripting and tracing of TorchScript Intermediate Representation (IR). PyTorch Mobile incorporates the XNNPACK floating-point and QNNPACK 8-bit quantized kernel libraries, enhancing its capabilities for mobile-optimized neural network inference. The mobile interpreter within PyTorch Mobile provides optimization features crucial for deploying models on mobile phones.

While PyTorch Mobile currently cannot be directly employed with the most resource-constrained microcontrollers, it is possible to utilize PyTorch models on such microcontrollers through Open Neural Network Exchange (ONNX) format conversion. This conversion can be achieved with various software tools, including TensorFlow, STM32Cube.AI, and Cainvas [38–40]. Presently, PyTorch Mobile supports a range of tasks, including image segmentation, object detection, video processing, speech recognition, and question-answering.

7.4.1.6 emlearn

The emlearn library offers a Python-C model converter and inference engine tailored for microcontrollers and devices utilizing C-code [29]. This library proves versatile as it can convert various classic machine learning and neural network models, including random forest (RF), decision trees (DT), naive Bayes (NB), multilayer perceptron (MLP), and sequential models built with Keras and scikit-learn frameworks. A notable feature is its support for fixed-point math, and it operates without dynamic memory allocations.

Unlike many other frameworks predominantly designed for 32-bit computer architectures, emlearn stands out by being compatible with 8-bit AVR processors [41].

In terms of functionality, emlearn shares similarities with other libraries such as MicroMLgen [42], FogML [43], and sklearn-porter [44]. These libraries collectively contribute to the ecosystem of tools available for deploying machine learning models on microcontrollers and C-code-based devices.

7.4.1.7 Embedded Learning Library

Microsoft has introduced the Embedded Learning Library (ELL) to support the TinyML ecosystem [45]. It is compatible with Raspberry Pi, Arduino, and the micro:bit platform. Notably, models deployed on these devices operate independently of the internet, making them internet-agnostic and eliminating the need for cloud access.

Currently, ELL supports tasks related to image and audio classification, showcasing its applicability in diverse scenarios within the embedded learning domain. This framework aligns with the broader goal of enabling machine learning capabilities on resource-constrained embedded devices.

Based on all these advances, TinyML frameworks can be categorized into three distinct groups:

- The initial approach involves the conversion of preexisting trained models to overcome MCU limitations. These tools typically utilize inference tools derived from well-known machine learning libraries like TensorFlow [46], scikit-learn [47], or PyTorch [48]. The code is then adapted to run on devices with limited resources.
- The second category revolves around the development of machine learning libraries specifically tailored for MCUs, offering offline training and inference capabilities. This approach enables models to generate predictions from data collected on the device, thereby enhancing accuracy and facilitating the implementation of unsupervised learning algorithms. However, it is not a widely adopted approach due to its potential to significantly raise the cost and complexity of processing platforms.
- Lastly, the third technique involves integrating a fully dedicated co-processor to support the primary computing unit in machine learning-specific tasks. While this strategy enhances computing power, it is the least commonly employed approach due to its propensity to increase the cost and complexity of processing platforms substantially.

7.4.2 Development environments

Various development environments for TinyML are discussed in this section.

7.4.2.1 Edge Impulse Studio

The Edge Impulse (EI) SDK enables the implementation of neural networks on embedded devices and incorporates functionalities such as real sensor data collection, live signal processing, testing, and code deployment to the target device [49]. This SDK facilitates AutoML processing for edge platforms [50] and is compatible with various devices, including smartphones, for deploying learning models. The training process takes place on a cloud platform, and the resulting trained model can be transferred to an edge device via a data forwarder-enabled pathway. The impulse can be executed on a local machine using the built-in C++, Node.js, Python, and Go SDKs. Impulses are also deployable as a Web Assembly library. Additionally, the SDK allows for the collection of actual data from sensors in IoT devices and mobile phones. An existing dataset can be uploaded to the EI SDK using an uploader tool that supports JSON, CBOR, JPG, and WAV formats [51].

7.4.2.2 Rubiscape

AI/machine learning low-code platforms like Rubiscape are unified services that provide tools for building, training, and deploying machine learning models. It supports most of the open AI/machine learning libraries and services, provides AutoML, and is suitable for both data scientists and non-experts. Rubiscape has RubiStudio (AI/machine learning), RubiFlow (orchestration), RubiSight (data visualization) and RubiThings (IoT) modules.

RubiThings provides connected intelligence for internetworking physical devices, vehicles, buildings, machines, electronics, software, and sensors with IoT machine-to-machine applications. It connects, acquires data right from the source, monitors, processes for elimination of errors and noise, lastly using processed data it executes machine learning at the edge in real time [52].

7.4.2.3 Imagimob

Imagimob offers two software products designed for constructing edge AI applications. The first is the Imagimob AI software suite, presenting a comprehensive end-to-end development solution for creating edge AI and TinyML applications [53]. This suite is applicable to various types of time-series data, with a particular emphasis on deep learning. The Imagimob AI development process comprises five key steps:

- data capture and labeling,
- centralized data management,
- automated model construction using an AI training service,
- model verification with visualization of all models and predictions,
- edge optimization and application packaging.

Notably, Imagimob supports the quantization of long short-term memory (LSTM) layers, a challenging yet essential aspect when working with time-series data [54].

The second offering is Imagimob Edge, a user-friendly software-as-a-service (SaaS) solution designed to simplify the complexities associated with edge AI and TinyML development [55]. This tool can automatically convert TensorFlow and Keras h5 file formats into highly efficient C code used in edge devices. The conversion process, which could be a challenging task even for proficient programmers, can be accomplished within seconds when utilizing Imagimob Edge. This suite is well suited for running deep learning models on highly constrained embedded devices, including Arm Cortex-M0 microcontrollers with a RAM memory size as small as 10 kB [55,56].

7.4.2.4 Qeexo AutoML

Qeexo AutoML presents an automated machine learning platform designed for Arm Cortex processors, including the challenging M0 and M0+ processors [57]. Deploying machine learning on the M0+ can pose challenges due to its limitations, such as the ability to perform only 32-bit fixed-point mathematics, low memory capacity, low CPU speed, and the absence of support for saturation arithmetic and digital signal processing. To address these constraints, Qeexo AutoML has developed a highly optimized Arm Cortex-M0+ fixed-point machine learning pipeline that encompasses sensor data handling, feature computation, and inference, all performed with fixed-point data. For the M0+, this pipeline leverages tree-based machine learning algorithms like gradient boosting machine (GBM), RF, and eXtreme Gradient Boosting (XGBoost). Qeexo AutoML offers a diverse machine learning algorithm portfolio, including NB, DT, isolation forest (IF), support vector machine (SVM), local outlier factor (LOF), logistic regression (LR), convolutional neural network (CNN), convolutional recurrent neural network (CRNN), recurrent neural network (RNN), and artificial neural network (ANN) [57]. Intelligent pruning and posttraining quantization methods employed by Qeexo AutoML result in an impressive 90% compression of the model size. Furthermore, the application of 8-bit quantization can further reduce the model size by up to 75% when compared to models utilizing 32-bit precision [58].

7.4.2.5 STM32Cube.AI

STMicroelectronics STM32Cube.AI serves as a neural network and machine learning toolkit tailored for STM32 developers, enabling optimized inferences on microcontrollers [59]. The toolkit encompasses prevalent deep learning libraries and decision-making processes, featuring resource-optimized algorithms like a DT classifier. To enhance its capabilities, STM32Cube.AI can be extended with the X-CUBE-AI package, offering automatic conversion of pretrained neural network and classical machine learning models. X-CUBE-AI is compatible with various frameworks utilizing the ONNX format, including PyTorch, Microsoft Cognitive Toolkit, and MATLAB®. It also supports popular deep learning and machine learning frameworks like TensorFlow Lite, Keras, Caffe, Lasagne, ConvnetJS, scikit-learn (isolation forest, SVM, k-means clustering, etc.), and the XGBoost package [38,60,61]. Additionally, X-CUBE-AI optimizes networks through 8-bit quantization and enables the storage of weight and activation parameters in external flash and RAM memories, which is particularly beneficial for more extensive networks.

7.4.2.6 NanoEdge AI Studio

Cartesiam NanoEdge AI Studio is a software package coupled with a set of AI libraries designed for embedded developers, functioning as a search engine to assist in selecting the most suitable machine learning algorithm [62]. The studio encompasses signal preprocessing, hyperparameter tuning, anomaly detection, and various classification models, including k-nearest neighbor (kNN), SVM, and neural networks [63]. NanoEdge AI Studio facilitates the development of application-specific ma-

Table 7.1 Summary of TinyML development tool features.

Development tool	Algorithms and support
Edge Impulse Studio	Proprietary neural networks
Rubiscape	Neural network
Imagimob	Proprietary neural networks
Qeexo AutoML	DT, RF, XGBoost, GBM, NB, LR, LOF, SVM, ANN, CNN, CRNN, RNN
STM32Cube.AI	Scikit-learn, XGBoost, Keras, TensorFlow Lite, Caffe, Lasagne, ConvetJS
NanoEdge AI Studio	Neural network, KNN, SVM, proprietary machine learning

chine learning libraries, supporting unsupervised learning, inference, and prediction within a microcontroller environment [64]. The program automates testing, optimization, and the determination of the best algorithmic combination as a C library using an emulator before the final deployment on the edge. Once the NanoEdge AI Studio identifies the optimal library for the project, it learns normal behaviors and identifies anomalies [65]. The software offers several key features, such as restricting maximum flash memory usage during project creation, frequency filtering, flash memory optimization, serial data plotting, real-time search, and library selection based on benchmarks. Remarkably, it can perform iterative learning in just 30 ms on an Arm Cortex-M4 at 80 MHz and consumes a mere 4 kB of RAM in a typical configuration [66]. Notably, Cartesiam AI has been successfully employed in one of the pioneering commercial TinyML products, namely, a sensor named Bob Assistant. This sensor utilizes automated on-device learning techniques for online machine monitoring, generating predictive maintenance reports automatically once the machine's normal behavior learning period is concluded [67].

Features of TinyML development tools are summarized in Table 7.1, which incorporates offered machine learning algorithms, compatible frameworks, and architecture.

7.4.3 TinyML benchmarking

When addressing the design of a TinyML performance benchmarking test, challenges like varying power consumption, limited and varying memory resources, lack of hardware heterogeneity, lack of software heterogeneity, and coverage of model deployment have to be overcome.

Currently, benchmarking tests are typically designed to assess either machine learning inference or microcontroller performance independently rather than focusing on the intersection of these technologies. Notably, the MLPerf Inference Benchmark [68], while robust, is more tailored for powerful computers and may not suit the specific requirements of TinyML. In response to this gap, the authors of [69] introduced the MLPerf Tiny Benchmark Suite, specifically designed to address TinyML needs. This open-source suite [70] is instrumental in evaluating the accuracy, latency, and energy consumption of TinyML inference.

MLPerf Tiny v0.5, within this suite, includes tasks like Visual Wake Words, keyword spotting, anomaly detection, and image classification for benchmarking. Reference implementations for these tasks are provided using TensorFlow Lite and TFLM [69]. This benchmark suite is designed to assess embedded devices with clock speeds ranging from 10 MHz to 250 MHz, typically consuming less than 50 mW per inference [70].

7.4.4 In-depth overview of TinyML framework: TensorFlow Lite

In this section, a detailed explanation of TFLM, the most widely accepted and commonly used framework, is presented.

In the preceding sections, it was emphasized that there is a requirement for a suitable framework to facilitate the acceptance and deployment of TinyML across various products. The envisioned framework should not only facilitate the installation of a model on embedded devices but also support model training on more advanced computing platforms. Furthermore, the TinyML framework should leverage a diverse set of machine learning capabilities, incorporating tools for model organization and debugging that are instrumental for deployment on production devices [71]. To address the challenges we discussed earlier, TFLM was put forth [72]. TFLM prioritizes portability and flexibility while mitigating the slowdown and increased cost associated with training and deploying models on embedded hardware. TFLM facilitates the execution of TinyML applications on various architectures and empowers hardware suppliers to optimize cores for their specific devices gradually. The performance enhancement benefits of TFLM include:

- a versatile and portable compiler-based solution that is easily adaptable for incorporating new applications and features,
- reduction of library requests and external dependencies to achieve hardware independence,
- empowering hardware suppliers to deliver kernel-specific optimization platforms without the necessity for developing hardware-specific compilers,
- provision of benchmarks recognized by leading organizations such as MLPerf,
- compatibility with popular, well-maintained Google apps currently in development,
- facilitation for hardware suppliers to seamlessly integrate optimizations into their kernels, ensuring production performance and hardware benchmarking,
- compatibility of the model architecture framework with TensorFlow Lite.

7.4.4.1 Implementation of TensorFlow Lite

This section elaborates on the implementation of TFLM; a schematic overview is shown in Fig. 7.1.

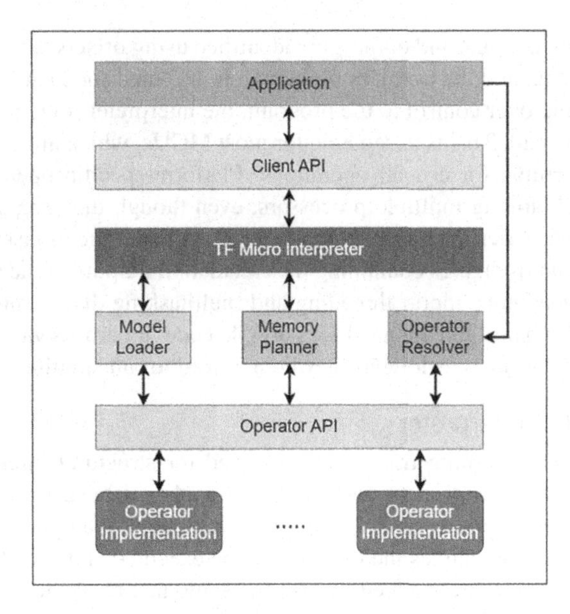

FIGURE 7.1

Implementation overview TFLM [71].

7.4.4.1.1 System overview

The initial phase of developing a TFLM application involves memorizing an active neural network model. Using the client API, the application developer creates an "operator resolver" object. To optimize file size, the "OpResolver" API governs the operators associated with the final binary. The subsequent step is to provide the interpreter with a continuous "region" of memory containing intermediate results and other variables. Since dynamic memory allocation is presumed to be unavailable, this step is crucial. The final step entails creating an interpreter instance and loading the model onto it, along with a resolution operator and a specified region as parameters. During initialization, the interpreter allocates all necessary memory from the arena. Any additional allocation can be managed to prevent heap fragmentation and potential issues in long-running applications.

Operator programs can allocate memory for evaluation, allowing the interpreter to express memory requirements. The OpResolver, provided by the implementation, maps operator types of the serialized model to their corresponding implementation functions. This ensures that operator implementations remain independent of interpreter details and modular, with all communication between the interpreter and operators handled via C API calls.

The implementation phase involves obtaining pointers to memory spaces representing model inputs and populating them with data, often derived from sensors or other user-supplied sources. Once inputs are prepared, the application invokes the interpreter to compute the model. During this process, topologically ordered operations are iterated through, inputs and outputs are identified using offsets determined during memory scheduling, and the evaluation function is executed for each operation.

Before handing over control to the program, the interpreter reviews all actions in a simple blocking call. This is acceptable for most MCUs, which are single-threaded and rely on interrupts for crucial operations. Platform-specific operators can distribute their work among multiple processors, even though the program itself runs on a single thread. After the call is complete, the application can instruct the interpreter to locate and use tables containing the calculation outputs of the model. While the framework does not support threading and multitasking due to minimal operating system requirements and the need for portable code, it permits various use cases, executing multiple models as long as they do not need to run simultaneously [71].

7.4.4.1.2 TFLM interpreter

TFLM is a machine learning framework designed for structural inference, relying on an interpreter-based approach. The interpreter loads a data structure that directly represents a machine learning model, and during runtime, it manipulates the model's data. This manipulation defines the operators to be executed and specifies where the arguments of the model are sourced despite the static nature of the execution code. The choice of an interpreter is based on expertise in designing models for embedded hardware in production use. Recognizing the need for easy model updates in the field, an interpreter is preferred over production code. The interpreter facilitates greater code sharing between models and applications, expedites code maintenance by allowing changes without reexporting the model, and reduces the long-term complexity of the kernel compared to traditional interpreters with multiple branches related to function calls. The interpreter overhead is minimized since each kernel runtime is lengthy.

Alternatively, a model can generate C or C++ source code by rendering operator function calls in fixed machine code at export, replacing an interpreter-based inference engine. While recompiling is necessary for each model, this approach may improve performance at the expense of portability. Code generation, which incorporates parameters such as weights, model architecture, and layer dimensions into the binary, completely replaces an executable when modifying a model. Since an interpreter method keeps all these data in a distinct file or memory region, updating a model involves substituting it with a single file or continuous memory area.

7.4.4.1.3 Loading of model

The interpreter loads a data structure that explicitly represents a model, adhering to the TensorFlow Lite portable data schema. This schema facilitates the importation of a wide range of models with minimal technical effort by reusing the extraction methods from TensorFlow Lite.

7.4.4.1.4 Model serialization

TensorFlow Lite for mobile devices utilizes the Flat Buffer serialization standard to store models. This approach typically uses less than 2 kB of binary space for the access code. A header-only library further reduces space and eases compilation by eliminating the need to decompress the serialization protocol into another format. However, a downside of this format is that the C++ header requires the platform compiler to implement the C++11 specification. Another challenge is that many embedded devices lack file storage systems. Nevertheless, due to its use of a memory-mapped representation, it is simple to transform the files into C source code files containing data tables. All these files are then consolidated into binary files accessible to the application.

7.4.4.1.5 Model representation

The model is represented by stored data, the value schema, and the TensorFlow Lite representation. This schema is designed with storage efficiency and quick access for mobile systems in mind. Unlike a directed acyclic graph, the functions are kept in a topologically ordered list. While a comprehensive representation of the graph would require preprocessing to match input dependencies, computations are straightforward by iterating over the list of operations. However, this representation is designed for portability between systems, requiring runtime processing to provide the necessary data for inference. For example, it extracts operator parameters from inputs before passing them to the functions executing those operations.

Consequently, each action necessitates a small piece of runtime code to convert from the serial representation to the underlying structure of the implementation. Although there is no additional code cost, the readability and compactness of operator implementations may degrade. Programming in memory poses a similar challenge, as TensorFlow Lite for mobile devices allows variable-size inputs, leading to potential size variations in independent operations. After determining all buffer dimensions, the optimal arrangement of intermediate buffers for computations must be decided at runtime.

7.4.4.1.6 Memory management

The framework operates under the assumption that the operating system may not support dynamic memory allocation. Consequently, it utilizes a predetermined memory space for allocation and maintenance. During model initialization, the interpreter calculates the total size and lifespan of all necessary buffers required for model execution. These buffers encompass runtime tensors, providing persistent memory for data storage, and scratch memory, utilized for temporary storage during model execution (as further discussed below). The framework generates a memory plan, utilizing non-persistent buffers whenever possible and ensuring their validity at the end of the memory plan's lifespan.

7.4.4.1.7 Persistent memory and scratchpads

Applications must provide a memory field with a specified size during interpreter formation, maintaining this memory field constant throughout the interpreter's lifespan. This region can be treated as a stack for assignments with the same lifespan, and an application-level error is generated if an assignment consumes too much space. Allocations only occur during the interpreter's setup phase, preventing memory faults during prolonged program execution. This straightforward approach is effective for initial development but may lead to memory wastage when assignments overlap in time, such as with data structures required only during startup.

The distribution system is modified to store initialization and evaluate lifetime distributions independently from the interpreter's lifetime objects. Stacks rising from the lowest direction and decreasing from the highest direction are employed for interpreter lifetime assignments and function lifetime objects, respectively. A limitation of capacity is indicated when the two stack indices cross.

The two-stack allocation technique benefits both shared and permanent buffers. However, model initialization retains allocation data that are no longer needed for model inference. Consequently, during memory planning, the space between the two stacks can be utilized for temporary allocation. The permanent assignment stack, part of the permanent stack, maintains any temporary data necessary for model inference.

7.4.4.1.8 Memory management

Due to limitations in embedded systems, developers of application models might find themselves compelled to create multiple specialized models instead of a single large model. Consequently, there may be a need for support for numerous models within the same embedded system. If an application involves various models that do not need to run simultaneously, it is possible to have two separate instances that operate independently. However, this approach is inefficient due to the inability to reuse the cache.

On the contrary, TFLM provides support for multi-tenancy along with specific, programmer-transparent changes to memory scheduling. TFLM enables the reuse of memory regions, allowing multiple model interpreters to allocate memory from a single area. This facilitates the reuse of the lifetime section of functions for both model evaluation and the stacking of interpreter lifetime regions in the same area. The reusable (non-permanent) component is set to the strictest standard based on all models assigned to the arena. Model-adjusted allocations involve raising the non-reusable (permanent) allocations for each model. This concept is referred to as model-adjusted allocation [66].

7.4.4.1.9 Multi-threading

TFLM ensures thread safety as long as the memory allocation of the model remains confined within the designated area and no model-related information is stored outside the interpreter. Within the region, only interpreter variables are retained, and each interpreter instance is linked to a specific model. As a distinct task or thread manages each interpreter instance, TFLM can effectively handle a large number of interpreter instances, even across multiple MCU cores. This is made possible by the fact that only the necessary interpreter variables are stored within the domain. The regions guarantee the absence of threading issues, even when the executable code is shared.

7.4.4.1.10 Operator support

Operators involve extensive computation, sometimes requiring hundreds or even billions of sequential arithmetic operations, such as additions or multiplications. These operators function seamlessly, encompassing inputs/outputs and specified state variables, with no apparent additional side effects. The customization of these processes for specific platforms is common to leverage hardware characteristics, as considerations like power consumption, code size, and model execution latency significantly influence the implementation of these techniques.

To address the need for platform-specific optimization, an API can be developed to connect inputs and outputs while concealing the execution details behind an abstraction, given that the boundaries of the operators are clearly defined. Some chip manufacturers have developed libraries of neural network kernels, aiming to enhance the performance of neural networks when executed on their CPUs. Arm's CMSIS-NN libraries provide optimization for various operations like fully connected layers, softmax, convolution, aggregation, activation, and optimized fundamental math. TFLM utilizes CMSIS-NN to achieve improved performance [66].

7.4.4.1.11 Building an all-in-one system

In order to tackle the challenges posed by the fragmentation of the embedded systems market, code must be capable of being compiled across a diverse range of platforms. Consequently, the code needs to be developed with a high degree of portability, minimizing dependencies to ensure it can run on various devices without sacrificing performance. The majority of embedded system programmers commonly utilize platform-specific integrated development environments or toolchains. These tools abstract away many complexities associated with building individual components and provide libraries as interface modules.

Despite developers being presented with a hierarchy of folders containing source code files, the process of creating and compiling code into a functional library involves several intermediate steps. This ensures that the code can effectively traverse the intricacies of different platforms and deliver a consistent and satisfactory experience on various devices.

7.5 Advantages of TinyML

By crunching numbers directly on the device, TinyML unlocks the door to new data-processing techniques that were earlier inaccessible for resource-constrained environments. Distinctive characteristics of TinyML systems, like simplicity, efficiency, and flexibility, are likely to change the whole IoT ecosystem and need to be considered in modern applications.

The following are primary performance metrics that illustrate the efficiency of TinyML as a crucial tool [73].

7.5.1 Transition from basic to smart IoT devices

The capability of sensor systems to create huge volumes of raw data has put limitations on the use of cloud computing for processing these raw data. Limited transmission of data from the edge to the cloud has led to the waste of plenty of data generated at the edge.

TinyML facilitates resource-constrained systems to perform data analysis and embed machine learning models on IoT devices, making them smart. Smart systems like smart cars generate 1 GB of data per second [74]. A Boeing-787 aircraft creates 5 GB of data per second [75]. It is essential to exclude worthless data at the edge by conducting an initial analysis of raw data, deploying machine learning models at IoT devices, and passing only useful data to the cloud.

7.5.2 Network bandwidth

Traditional IoT systems use a sensor network to gateways to the cloud for sending data so that machine learning models can run on the cloud and evaluate [76]. Unlike traditional approaches, TinyML's innovative techniques can profoundly redesign performance benchmarks in resource-limited environments. This is achieved by scaling back reliance on pervasive cloud services and offering optional deployment on other IoT layers. Reduced data transmission in limited-bandwidth environments makes TinyML systems more independent. Also, systems with dense IoT devices need high bandwidth for transmitting raw data. Therefore, it is essential to exclude worthless data at the edge by conducting an initial analysis of raw data and deploying machine learning models at IoT devices.

7.5.3 Security and privacy

With loads of sensitive data migrating to the cloud, apprehensions over data security have arisen as a critical obstacle to broader IoT adoption. In the case of third-party suppliers hired for IoT services, end users do not have transparency about ownership of data and the location where their personal data are upheld. Data transfer also leads to the risk of suspicious individuals spying on data. Processing data on devices locally, TinyML can minimize security issues [77,78]. That is why TinyML has data privacy and data security by default.

7.5.4 Latency

IoT devices send sensor data to the cloud, and cloud systems analyze data and generate decisions. Then, this info reaches the IoT device in the end. These series of events have more latency and require critical surveillance. In applications like driverless cars and healthcare systems, delays in communication from the cloud can have catastrophic effects. In such cases where systems cannot rely on external systems and connectivity, TinyML systems can fill in the infrastructure gap and minimize latency for machine learning decisions. Quicker reactions in case of emergency circumstances are enabled by on-device data processing in real time. This also reduces the load on cloud systems [79–81].

7.5.5 Energy efficiency

MCUs provide a significant advantage when using TinyML systems. IoT devices utilizing batteries like CR2032 or energy harvesting systems require much less energy; in contrast, powerful CPUs and GPUs require much power. These IoT devices can be installed everywhere without connecting to the power grid. This provides an opportunity for innovative, intelligent, and portable applications. Minimal energy use makes it possible to integrate tiny IoT sensors with present battery-powered devices, transforming them into intelligent and connected machines [77].

7.5.6 Reliability

When internet access becomes scarce, TinyML shines. Its ability to process data right on the device unlocks reliable IoT services even in remote locations like offshore platforms and rural communities.

7.5.7 Low cost

IoT systems execute many tasks in a wide range of applications; therefore, they have high requirements such as large memory and high computing power. They also need vast storage and processing capabilities in the cloud, increasing costs. TinyML executes data locally using MCU devices with high performance and enables AI on devices at a low cost [77]. Microcontrollers have small resources, such as computing power (processor) with a capacity of 1 MHz to 400 MHz and memory with a capacity of 2 kB to 512 kB, leading to low costs.

7.5.8 Data filtration

The cognitive capabilities of IoT devices give designers the ability to inspect data and filter useless data. In high-traffic situations, edge-based intelligent support systems leave traditional approaches inefficient. Their proximity to data and fast reaction times make them champions of efficiency and performance.

Take an anomaly detection system with a network of cameras: local processing shines in such scenarios. Filtering out redundant footage on the device itself, instead of overloading the cloud with every frame, significantly boosts efficiency and minimizes unnecessary bandwidth consumption.

References

[1] B.L. Mearian, Self-driving cars could create 1GB of data a second, https://www.computerworld.com/article/2707396/self-driving-cars-could-create-1gb-of-data-a-second.html, 2013.

[2] W. Shi, J. Cao, Q. Zhang, Y. Li, L. Xu, Edge computing: vision and challenges, IEEE Int. Things J. 3 (5) (2016) 637–646.

[3] W. Bao, C. Wu, S. Guleng, J. Zhang, K.-L.A. Yau, Y. Ji, Edge computing-based joint client selection and networking scheme for federated learning in vehicular IoT, China Commun. 18 (6) (2021) 39–52, https://doi.org/10.23919/JCC.2021.06.004.

[4] T. Ogino, Simplified multi-objective optimization for flexible IoT edge computing, in: 2021 4th International Conference on Information and Computer Technologies (ICICT), 2021, pp. 168–173, https://doi.org/10.1109/ICICT52872.2021.00035.

[5] K. Simonyan, A. Zisserman, Very deep convolutional networks for large-scale image recognition, arXiv preprint, arXiv:1409.1556, 2014.

[6] A.G. Howard, M. Zhu, B. Chen, D. Kalenichenko, W. Wang, T. Weyand, M. Andreetto, H. Adam, MobileNets: efficient convolutional neural networks for mobile vision applications, arXiv preprint, arXiv:1704.04861, 2017.

[7] P. Warden, D. Situnayake, TinyML: machine learning with TensorFlow Lite on Arduino and ultra-low-power icrocontrollers, O'Reilly Media, https://www.oreilly.com/library/view/tinyml/9781492052036/.

[8] TinyML, https://cms.tinyml.org/wp-content/uploads/emea2021/tinyML_Talks_Felix_Johnny_Thomasmathibalan_and_Fredrik_Knutsson_210208.pdf. (Accessed November 2021).

[9] ARM-TinyL, https://www.arm.com/blogs/blueprint/tinyml. (Accessed November 2021).

[10] Forbes-TinyML, https://www.forbes.com/sites/janakirammsv/2020/11/03/how-tinyml-makes-artificial-intelligence-ubiquitous/?sh=26e83c8f7622. (Accessed November 2021).

[11] J. Lin, W.M. Chen, Y. Lin, J. Cohn, C. Gan, S. Han, MCUNet: tiny deep learning on IoT devices, arXiv:2007.10319v2, 2020.

[12] N. Schizas, A. Karras, C. Karras, S. Sioutas, TinyML for ultra-low power AI and large scale IoT deployments: a systematic review, Future Internet 14 (2022) 363, https://doi.org/10.3390/fi14120363.

[13] Y. Zhang, N. Suda, L. Lai, V. Chandra, Hello edge: keyword spotting on microcontrollers, arXiv:1711.07128, 2017.

[14] A. Kumar, S. Goyal, M. Varma, Resource-efficient machine learning in 2 KB RAM for the Internet of Things, in: D. Precup, Y.W. Teh (Eds.), Proceedings of the 34th International Conference on Machine Learning, vol. 70, Sydney, Australia, 6–11 August, 2017, pp. 1935–1944.

[15] L. Lai, N. Suda, V. Chandra, CMSIS-NN: efficient neural network kernels for ARM Cortex-M CPUs, arXiv:1801.06601, 2018.

[16] A. Garofalo, G. Tagliavini, F. Conti, D. Rossi, L. Benini, XpulpNN, Accelerating quantized neural networks on RISCV processors through ISA extensions, in: Proceedings of the 23rd Conference on Design, Automation and Test in Europe, DATE'20, Grenoble, France, 9–13 March 2020, EDA Consortium, San Jose, CA, USA, 2020, pp. 186–191.

[17] K. Wang, Z. Liu, Y. Lin, J. Lin, S. Han, HAQ: hardware-aware automated quantization, arXiv:1811.08886, 2018.

[18] I. Fedorov, R.P. Adams, M. Mattina, P.N. Whatmough, SpArSe: sparse architecture search for CNNs on resource-constrained microcontrollers, arXiv:1905.12107, 2019.

[19] H. Doyu, R. Morabito, J. Höller, Bringing machine learning to the deepest IoT edge with TinyML as-a-service, IEEE IoT Newsl. (2020) 11.

[20] Z. Wu, M. Jiang, H. Li, X. Zhang, Mapping the knowledge domain of smart city development to urban sustainability: a scientometric study, J. Urban Technol. 28 (2021) 29–53.

[21] H. Kim, Q. Chen, T. Yoo, T.T.H. Kim, B. Kim, A 1-16b precision reconfigurable digital in-memory computing macro featuring column-MAC architecture and bit-serial computation, in: Proceedings of the ESSCIRC 2019-IEEE 45th European Solid State Circuits Conference (ESSCIRC), Cracow, Poland, 23–26 September 2019, pp. 345–348.

[22] M.S. Mahdavinejad, M. Rezvan, M. Barekatain, P. Adibi, P. Barnaghi, A.P. Sheth, Machine learning for Internet of Things data analysis: a survey, Digit. Commun. Netw. 4 (2018) 161–175.

[23] C.R. Banbury, V.J. Reddi, M. Lam, W. Fu, A. Fazel, J. Holleman, X. Huang, R. Hurtado, D. Kanter, A. Lokhmotov, Benchmarking TinyML systems: challenges and direction, arXiv:2003.04821, 2020.

[24] A. Osman, U. Abid, L. Gemma, M. Perotto, D. Brunelli, TinyML platforms benchmarking, in: Proceedings of the International Conference on Applications in Electronics Pervading Industry, Genova, Italy, 26–27 September 2022, Environment and Society, pp. 139–148.

[25] N. Tan, P. Warden, Z. Shelby, uTensor and TensorFlow announcement, https://os.mbed.com/blog/entry/uTensor-and-Tensor-Flow-Announcement/, 2019.

[26] Arm, CMSIS NN software library, https://arm-software.github.io/CMSIS_5/NN/html/index.html, 2021.

[27] L. Weber, A. Reusch, TinyML – how TVM is taming tiny, https://tvm.apache.org/2020/06/04/tinyml-how-tvm-is-taming-tiny.html, 2021.

[28] PyTorch, PyTorch mobile, https://pytorch.org/mobile/home/, 2021.

[29] J. Nordby, emlearn: Machine Learning Inference Engine for Microcontrollers and Embedded Devices, GitHub, 2019, https://github.com/emlearn/emlearn.

[30] Google, TensorFlow Lite: deploy machine learning models on mobile and IoT devices, https://www.tensorflow.org/lite.

[31] R. David, J. Duke, A. Jain, et al., TensorFlow Lite Micro: embedded machine learning for TinyML systems, in: Proceedings of Machine Learning and Systems, vol. 3, 2021, pp. 800–811.

[32] Google, TensorFlow model optimization, https://www.tensorflow.org/model_optimization, 2020.

[33] Google, Model optimization, https://www.tensorflow.org/lite/performance/model_optimization, 2020.

[34] https://github.com/ARM-software/CMSIS_5.

[35] Riku Immonen, Timo Hämäläinen, Tiny machine learning for resource-constrained microcontrollers, J. Sens. 2022 (2022) 7437023, https://doi.org/10.1155/2022/7437023.

[36] uTVM, https://octoml.ai/blog/tinyml-tvm-taming-the-final-ml-frontier.

[37] PyTorch, https://pytorch.org/.

[38] STMicroelectronics, Artificial intelligence (AI) software expansion for STM32Cube, https://www.stmicroelectronics.com.cn/resource/en/data_brief/x-cube-ai.pdf, 2020.

[39] Partha Pratim Ray, A review on TinyML: state-of-the-art and prospects, J. King Saud Univ, Comput. Inf. Sci. (ISSN 1319-1578) 34 (4) (2022) 1595–1623, https://doi.org/10.1016/j.jksuci.2021.11.019.

[40] A. Singh, Converting a model from PyTorch to TensorFlow: guide to ONNX, https://analyticsindiamag.com/converting-a-model-from-pytorch-to-tensorflow-guide-to-onnx/, 2021.

[41] R. Sanchez-Iborra, A.F. Skarmeta, TinyML-enabled frugal smart objects: challenges and opportunities, IEEE Circuits Syst. Mag. 20 (3) (2020) 4–18.

[42] MicroMLgen, https://github.com/eloquentarduino/micromlgen, 2021.

[43] T. Szydlo, J. Sendorek, R. Brzoza-Woch, Enabling machine learning on resource constrained devices by source code generation of the learned models, in: International Conference on Computational Science, in: Lecture Notes in Computer Science, vol. 10861, Springer, 2018, pp. 682–694.
[44] D. Morawiec, sklearn-porter, https://github.com/nok/sklearn-porter, 2021.
[45] ELL, https://microsoft.github.io/ELL/.
[46] Martín Abadi, et al., TensorFlow: a system for large-scale machine learning, in: 12th USENIX Symposium on Operating Systems Design and Implementation (OSDI 16), 2016.
[47] Fabian Pedregosa, et al., Scikit-learn: machine learning in Python, J. Mach. Learn. Res. 12 (2011) 2825–2830.
[48] Adam Paszke, et al., PyTorch: an imperative style, high-performance deep learning library, arXiv preprint, arXiv:1912.01703, 2019.
[49] Edge Impulse, TinyML for all developers with Edge Impulse, https://www.hackster.io/news/tinyml-for-alldeveloperswith-edge-impulse-2cfbbcc14b90, 2020.
[50] Edge Impulse, https://www.edgeimpulse.com/.
[51] Edge Impulse, Documentation, https://docs.edgeimpulse.com/docs, 2021.
[52] Rubiscape, https://www.rubiscape.com/.
[53] Imagimob AB, Imagimob AI, https://developer.imagimob.com/#/./imagimob-ai, 2021.
[54] J. Malm, Quantization of LSTM layers – a technical white paper, https://www.imagimob.com/blog/quantization-of-lstm-layers-a-technical-white-paper, 2022.
[55] Imagimob AB, Introducing Imagimob Edge: making TensorFlow AI models edge device ready at the click of a button, https://www.imagimob.com/news/introducingimagimob-edge-making-tensorflow-aimodelsedge-device-ready-at-the-click-of-a-button, 2020.
[56] EDGE Computing World, Edge startup of the year CXO interviews: Anders Hardebring, CEO and co-founder Imagimob AB, https://www.edgecomputingworld.com/2020/09/01/startup-of-the-year-finalist-anders-hardebring/, 2020.
[57] Qeexo, Enabling the new era of machine learning at the Edge, https://qeexo.com/, 2021.
[58] R. Bhatt, T. Shyuan, Building effective IoT applications with TinyML and automated machine learning, https://www.embedded.com/building-effective-iot-applications-with-tinyml-and-automated-machine-learning/, 2021.
[59] STMicroelectronics, STM32Cube.AI: convert neural networks into optimized code for STM32, https://blog.st.com/stm32cubeai-neural-networks/, 2020.
[60] STMicroelectronics, AI expansion pack for STM32CubeMX, https://www.st.com/en/embedded-software/x-cube-ai.html, 2020.
[61] STMicroelectronics, X-CUBE-AI documentation, https://wiki.st.com/stm32mcu/wiki/AI:X-CUBE-AI_documentation, 2022.
[62] Cartesiam, Cartesiam: leader in the edge AI market, with proven industrial reference, https://cartesiam.ai, 2020.
[63] STMicroelectronics, NanoEdge AI Studio, https://wiki.st.com/stm32mcu/wiki/AI:NanoEdge_AI_Studio, 2022.
[64] M. Vetrano, Cartesiam AI development environment brings artificial intelligence, learning and inference to everyday objects, https://www.prweb.com/releases/cartesiam-ai-development-environment-brings-artificial-intelligence-learning-and-inference-to-everyday-objects-893639520.html, 2020.
[65] NanoEdge AI Library for anomaly detection (AD), https://wiki.st.com/stm32mcu/wiki/AI:NanoEdge_AI_Library_for_anomaly_detection_(AD), 2023.

[66] Design and Reuse, Cartesiam transforms edge AI development for industrial IoT, http://www.design-reuse.com/news/49170/cartesiamnanoedge-ai-studio-ide-armcortex-m-mcu.html, 2020.

[67] nkeWATTECO, BoB Assistant, https://bobassistant.-com/en/offer, 2021.

[68] V.J. Reddi, B. Plancher, S. Kennedy, L. Moroney, P. Warden, A. Agarwal, C.R. Banbury, M. Banzi, M. Bennett, B. Brown, et al., Widening access to applied machine learning with TinyML, arXiv:2106.04008, 2021.

[69] H. Han, J. Siebert, TinyML: a systematic review and synthesis of existing research, in: Proceedings of the 2022 International Conference on Artificial Intelligence in Information and Communication (ICAIIC), Jeju Island, Republic of Korea, 21–24 February, 2022, pp. 269–274.

[70] TensorFlow Lite (TFL), https://www.tensorflow.org/lite. (Accessed 15 October 2022).

[71] R. David, J. Duke, A. Jain, V.J. Reddi, N. Jeffries, J. Li, N. Kreeger, I. Nappier, M. Natraj, S. Regev, et al., TensorFlow Lite Micro: embedded machine learning on TinyML systems, arXiv:2010.08678, 2020.

[72] TensorFlow Lite Guide, https://www.tensorflow.org/lite/guide.

[73] D.L. Dutta, S. Bharali, TinyML meets IoT: a comprehensive survey, IEEE Int. Things J. 16 (2021) 100461.

[74] Self-driving Cars Will Create 2 Petabytes of Data, What Are the Big Data Opportunities for the Car Industry, https://datafloq.com/read/self-driving-cars-create-2-petabytes-data-annually/. (Accessed 7 December 2016).

[75] S. Wang, Edge computing: applications, state-of-the-art and challenges, Adv. Netw. 7 (2019) 8–15.

[76] P. Mishra, D. Puthal, M. Tiwary, S.P. Mohanty, Software defined IoT systems: properties, state of the art, and future research, IEEE Wirel. Commun. 26 (2019) 64–71.

[77] B. Sudharsan, P. Patel, J.G. Breslin, M.I. Ali, Ultra-fast machine learning classifier execution on IoT devices without SRAM consumption, in: 2021 IEEE International Conference on Pervasive Computing and Communications Workshops and other Affiliated Events (PerCom Workshops), Kassel, Germany, 2021, pp. 316–319, https://doi.org/10.1109/PerComWorkshops51409.2021.9431061.

[78] MNIST Handwritten Digit Database, Yann LeCun, Corinna Cortes and Chris Burges. Available online: http://yann.lecun.com/exdb/mnist/.

[79] D. Puthal, S.P. Mohanty, S. Wilson, U. Choppali, Collaborative edge computing for smart villages [energy and security], IEEE Consum. Electron. Mag. 10 (2021) 68–71.

[80] M. Merenda, C. Porcaro, D. Iero, Edge machine learning for AI-enabled IoT devices: a review, Sensors 20 (2020) 2533.

[81] W. Niu, X. Ma, S. Lin, S. Wang, X. Qian, X. Lin, Y. Wang, B. Ren, PatDNN: achieving real-time DNN execution on mobile devices with pattern-based weight pruning, in: Proceedings of the Twenty-Fifth International Conference on Architectural Support for Programming Languages and Operating Systems, Lausanne, Switzerland, 16–20 March, 2020, pp. 907–922.

CHAPTER

Extensive energy modeling for LoRaWANs

Yassine Yazid[a,b], Mohamed Zbairi[b], Antonio Guerrero Gonzales[a], Mounir Arioua[b], and Ahmed El Oualkadi[b]

[a]*Department of Automation, Electrical Engineering and Electronic Technology, Universidad Politécnica de Cartagena, Cartagena, Spain*

[b]*Laboratory of Innovative Systems Engineering (ISI), National School of Applied Sciences of Tétouan (ENSATe), Abdelmalek Essaadi University, Tétouan, Morocco*

8.1 Introduction

Internet of Things (IoT) devices hold the potential to simplify daily life and enhance people's well-being both at home and work, as well as in various environments. With the considerable advancement in communication technology, the IoT concept is gradually being embraced across numerous applications in diverse sectors. Various research efforts have been undertaken to better understand and address the communication challenges associated with these technologies.

A rising segment within the IoT landscape revolves around low-power wide area networks (LPWANs) [1]. LPWAN technologies are categorized into cellular and non-cellular networks, aiming to facilitate extensive communication while ensuring scalable coverage for a broad spectrum of IoT applications across different settings and application categories. In IoT applications, several long-range communication paradigms have been employed, including Sigfox, INGENU, Weightless SIG, DASH7, and LoRa [2]. Among these, LoRa has garnered significant attention in the IoT domain. This technology operates within unlicensed industrial–scientific–medical (ISM) bands. Given that these end devices (EDs) are small and often rely solely on battery power, accurately estimating the energy model of the IoT-based LoRa system is crucial. Consequently, determining the optimal communication settings for LoRa in IoT networks has proven challenging.

LoRa networks involve multiple processing stages for transmitted data, encompassing whitening, channel encoding, interleaving, and modulation. Evaluating the energy efficiency of various parameter selections necessitates careful consideration of the trade-offs between transmission energy and computational energy. This chapter delineates a comprehensive communication system and energy model for LoRaWAN, an LPWAN technology. Moreover, it presents a detailed mathematical model calculating energy usage for transmission, processing, and sensing functions in LoRa's compact sensors. It encompasses transceiver modeling utilizing LoRa's default algorithm function procedure. Several state-of-the-art works have addressed the subject of

TinyML for Edge Intelligence in IoT and LPWAN Networks. https://doi.org/10.1016/B978-0-44-322202-3.00013-0

energy modeling and transmission parameter selection in LoRa networks [3][4][5]. However, those works have not considered the overall key parameters influencing the energy consumption of LoRa devices, which potentially leads to overestimating the lifespan of terminal devices and consequently the overall network longevity. For instance, the energy model presented in [6] disregards energy usage variations considering accessible parameters like the LoRa coding rate (CR), instead using fixed values for computational energy, despite its variability in each setup. Other publications have introduced adaptive techniques for LoRa parameter tuning [7][8], predominantly relying on transmission energy to identify an appropriate set. With this intention, we present crucial elements to estimate the energy consumption of LoRa end nodes.

The chapter comprises two sections: one about the communication system function and one about the energy consumption calculation of LoRa end nodes. The first part delves into the elements of the LoRa transceiver, providing a comprehensive overview of the communication system. This includes operations like channel encoding/decoding, modulation/demodulation, whitening/de-whitening, and interleaving/de-interleaving, considering the impact of each transmission choice on all blocks. The second section details an energy model highlighting the primary factors contributing to energy consumption in the LoRa communication system. This model is centered on chirp spread spectrum (CSS) modulation and spreading factors (SF) to hybrid settings with coded transmissions under the additive white Gaussian noise (AWGN) communication channel. To construct a fully operational LoRa communication system, the model integrates the Hamming channel coding method and the CSS modulation approach, offering intricate insights into the functionality of each system block. Additionally, the provided energy model serves to prevent the energy consumption of a given end node regarding its position in the network.

8.2 LoRa/LoRaWAN components and elements

IoT LPWANs are a type of wireless communication technology that is specially developed for IoT applications that require a long-range, low-power connection. These networks provide broad coverage, allowing devices to send and receive data over vast distances while spending little energy. Additionally, these networks can be identified by their ability to function in low-bandwidth settings, making them ideal for applications involving sporadic data transfer and battery-powered devices. These networks offer a low-cost alternative for connecting many IoT devices across huge geographic areas [1].

There are various LPWA technologies available, each with its own set of benefits and drawbacks. LoRaWAN, narrowband IoT (NB-IoT), and Long-Term Evolution for Machines (LTE-M) are some common LPWA technologies [9]. In this section, we highlight some important LoRa/LoRaWAN elements including how it functions and its architecture elements.

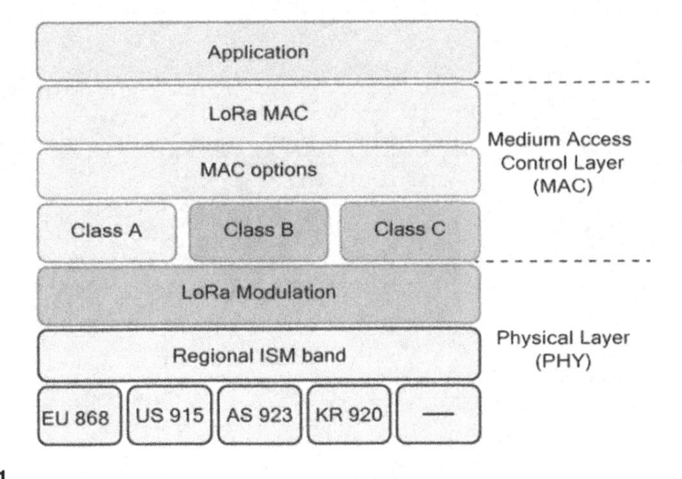

FIGURE 8.1

The LoRaWAN (MAC) protocol stack is implemented on top of LoRa [12].

8.2.1 LoRaWAN

LoRaWAN is an open standard created by Semtech Corporation that acts as a bridge between the proprietary base physical layer and the top levels of communication as shown in Fig. 8.1. LoRaWAN is one of the media access control (MAC) protocols that enable wireless interconnection and time scheduling between end nodes and the network's base station [10]. To connect wirelessly with their gateways (GWs), the nodes use the ALOHA protocol and time division multiple access (TDMA) scheduling access mechanisms. As shown in Fig. 8.2 the nodes are densely dispersed, providing a star-of-star network structure. The received data are sent to backend servers via the GWs. Furthermore, depending on the application's needs, these nodes might be classified as class A, B, or C [11]. Class A is the most frequent because it saves energy by opening two short windows to listen for and prepare for the reception of downlink feedback from the GW following an uplink broadcast. Class B, on the other hand, extends the listening windows to notify the GW of the node's waiting time since it stays open even after transmission occurs. Class C is deemed energy-inefficient since nodes must remain open the majority of the time to provide for continuous and maximum reception windows.

8.2.2 LoRa physical layer

LoRa is the physical layer foundation of Semtech's recently introduced long-range LPWA technology. Fundamentally, it employs a patented CSS-derived approach in which many chirps are distributed throughout the occupied frequency spectrum [13]. The basic chirp changes its frequency values instantly by covering the whole related frequency range. The fundamental chirp is incorporated for LoRa by taking several key characteristics into account including the bandwidth (BW), SFs, CRs, transmis-

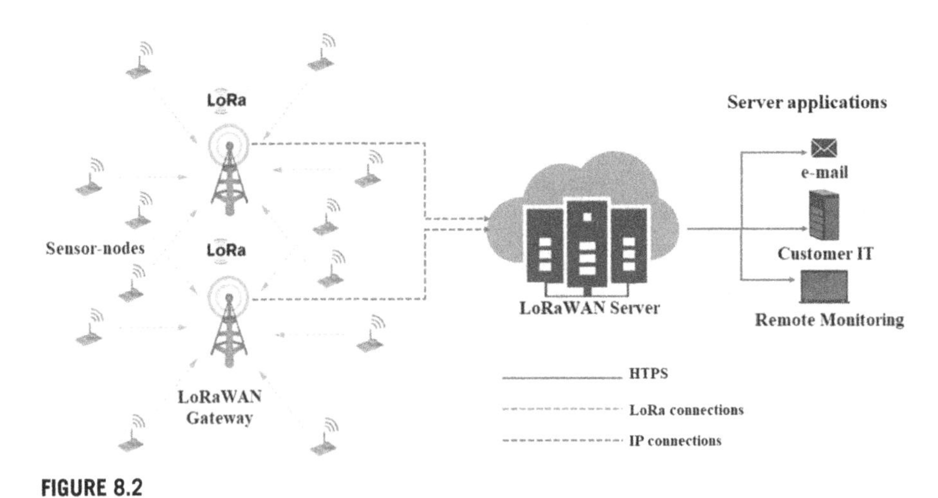

FIGURE 8.2

LoRa network structure.

sion power (TP), and channel signal-to-noise ratio. As they define different features, the SFs $\{7, 8, 9, 10, 11, 12\}$ and various BWs (125, 250, and 500 kHz) are standard.

LoRa employs the CSS modulation technique. Fig. 8.3 illustrates that LoRa symbols are modulated across an up-chirp with a BW of 125 kHz. It is an example of using SF. The figure was generated in MATLAB® using the communication chain described in the below sections.

The modulator generates various chirps (up-chirps, down-chirps) and values. In terms of BW, LoRa may send a sample every $T_c = \left(\frac{1}{BW}\right)$, which may also signify the chirp duration. Each sample contains a subset of the information encoded to the SF bit count. Before being modulated, these data are encoded again into a non-binary symbol with values in $\{0, 1, 2, .., 2^{SF} - 1\}$. As a result, the symbol is superimposed by varying the chirping signal frequencies across the BW by 2^{SF} times T_c. As a result, a symbol is transferred every $T_s = \left(\frac{2^{SF}}{BW}\right)$, and the higher the SF, the longer it takes to send the symbol. LoRa incorporates proportional forward error correction (FEC) codes. These codes can group 4-bit data blocks and encode them by adding adjustable parity bits, varying from one to four bits. This flexibility enables different CRs (CR∈{4/5, 4/6, 4/7, 4/8}) crucial for balancing long-range transmission, low energy usage, and bit rate trade-offs.

Furthermore, the usable bite rate R_b is proportional to the utilized BW, SF, and CR, as described as follows:

$$R_b = SF \times \left(\frac{BW}{2^{SF}}\right) \times CR, \tag{8.1}$$

where $SF \in \{7, 8, 9, 10, 11, 12\}$, $BW \in \{125 \text{ kHz}, 250 \text{ kHz}, 500 \text{ kHz}\}$, and $CR \in \left\{\frac{4}{5}, \frac{4}{6}, \frac{4}{7}, \frac{4}{8}\right\}$.

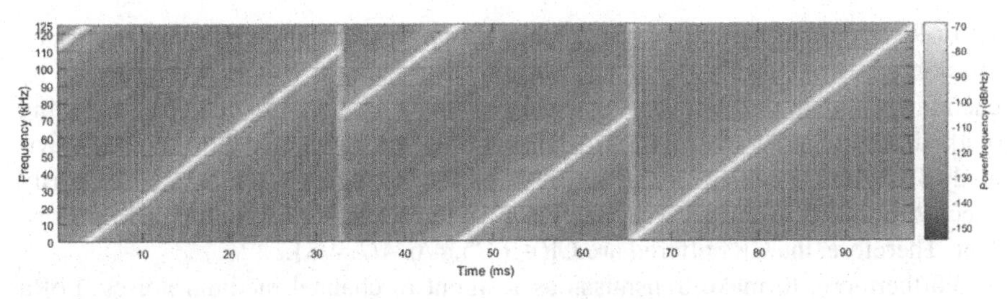

FIGURE 8.3

The description of SF-12 chirp in the 125 kHz frequency band versus time (ms).

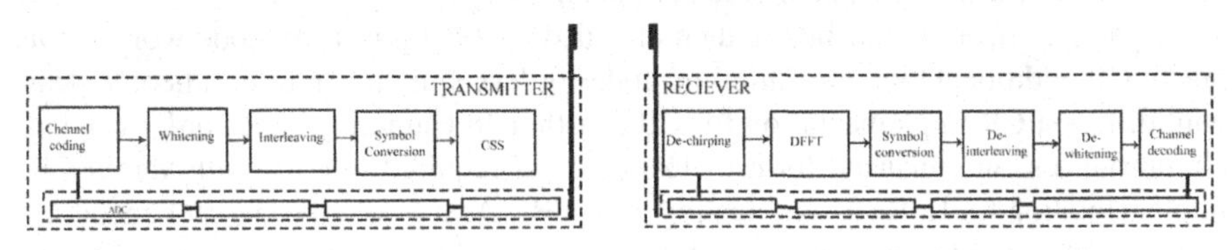

FIGURE 8.4

LoRa transceiver block diagram.

8.2.3 LoRa transceiver functioning principle

In this section, we detail the LoRa transceiver blocks. We give some mathematical equations behind each processing step required by a LoRa device in transmission and reception mode [14]. LoRa relies on physical layer settings and specialized blocks to send data over long distances. The LoRa device transceivers perform a series of concatenating steps prior to every successful LoRa transmission. As shown in Fig. 8.4, the LoRa contains different blocks that are used to encode the transmitted packets. LoRa relies on physical layer settings and then possesses blocks to send small packets over long distances. Before any successful LoRa transmission, a series of concatenating activities is carried out by both the LoRa device transmitter and the receiver. The roles and processes of each unit are described in depth in the following sections. On the transmitter side, the input bits are typically initially encoded using a Hamming code. Then, prior to chirp spreading modulation, whitening, interleaving, and Gray indexing are applied. CSS modulation for the preamble and data is handled by LoRa. Before demodulation, the receiver estimates and compensates for frequency offset. To extract the received information, gray indexing, de-interleaving, de-whitening, and Hamming decoding are used.

As a result, channel decoding approaches improve the likelihood of these mistakes being corrected in the receiver decoding operations. The interleaver typically shuffles the bits in a representative order to prevent errors, but it cannot offer error repair or detection. In the LoRa physical layer, this method is mostly used to rebuild the encoded code words diagonally in the appropriate manner with the chosen SF for CSS modulation.

LoRa integrates adjustable redundancy bits with the sent data to improve transmission robustness against channel medium disturbances. This is accomplished by encoding the data in this model using a Hamming H(k, n) FEC channel encoding method. The code word length is determined by the parameter CRs. It is abbreviated as CR and specified as $CR = (\frac{4}{4+\alpha})$, where α specifies the amount of parity check bits added to the code word, which enables alternative coding speeds to be sent. Therefore, the CRs offered are CR$\in$ {4/5, 4/6, 4/7, 4/8}.

Furthermore, to make transmissions resilient to channel medium noises, LoRa induces controllable redundancy bits into transmitted data. Accordingly, the data in this model are encoded using a Hamming $H(k, n)$ channel encoding scheme with variable code words regulated by the parameter CR.

A piece of data of four bits is then encoded cyclically to form code words of n bits of CR additional bits. As Hamming codes order required conditions, here to deal with different CR implementation for CRs. Although Hamming codes order needed circumstances, an enhanced form of Hamming is required to cope with varied CR implementation for LoRa to suit dealing with CRs. As a result, the channel encoding creates a code word of $n = (4 + \alpha_{cr})$ bits across each block of four bits, where $\alpha_{cr} \in (1, 2, 3, 4)$.

Additionally, the check Hamming matrix for each (n, k) code is a generator matrix of each orthogonal code C denoted by C^T. For instance, if H is the check matrix for C, H is an $(n - k) \times k$ matrix; the rows are orthogonal to C. The resulting block of encoded code words $(4 + \alpha_{cr}) \times q$ is then reordered to prepare symbol generation using the LoRa SFs and add some extra resilience to those code words by interleaving them. The length of the whitened vector and the encoded vector must be equal. In general, this procedure is used to introduce randomness into symbols in order to improve clock recovery at the receiver.

Each queued code word at the output of the FEC coding block is scattered across time during an interleaving process. This technique is used to separate and reorganize the places of potential mistakes. As a result, channel decoding approaches increase the probability of fixing these faults in the receiver decoding operations. Typically, the interleaver only scrambles the bits in a representative sequence to prevent errors from occurring, but it cannot provide error repair or detection; therefore, no benefit is associated with the operations. In the LoRa physical layer, this method is mostly used to rebuild diagonally encoded code words in the appropriate form with the specified SFs for CSS modulation. We have

$$S_m = \sum_{p=0}^{SF-1} V_{\cdot m} \times 2^{SF}. \tag{8.2}$$

Each symbol takes values in $\{0, 1, 2, 3, \ldots, 2^{SF} - 1\}$. Consequently, total generated symbols may be surveyed as a vector of M elements.

The symbol generating procedure is initiated by gray indexing interleaver binary code words of length SF into non-binary symbols. These symbols can have values ranging from 0 to $(2^{SF} - 1)$. Thus, each symbol that may be sent in a given time period is also determined by the SF and T_c. For example, if an SF of length 7 is used, the value of a symbol can range from 0 to 255.

Following that, each created symbol is assigned to a CSS modulator. The modulation technique employs precise criteria to provide each transmitted packet with a distinct shape that is resistant to channel medium interference. As a result, non-binary symbol information is disseminated throughout a time-frequency spectrum utilizing CSS chirps. The CSS modulator normally depends on the on-base chirp waveform to transport each symbol by spreading it throughout the BW within T_s; therefore, the base-band chirp carrying the symbols is represented in time domain as

$$C_0(t) = A(t)\, e^{\frac{\pi j (BW) t^2}{T_s} + \varphi_0}. \tag{8.3}$$

In general, the CSS modulation for each symbol $S_{0<m<(M-1)}$ is indicated as follows:

$$C_{S_m}(t) = A(t)\, e^{2\pi j\left(\frac{(BW)}{2T_s} t^2 + \frac{S_m}{T_s} t + \varphi\right)}. \tag{8.4}$$

The expression of the previous chirping process in discrete domain is obtained respecting the Shannon sampling theorem by taking the frequency sampling equal to BW. Now we have

$$C_{S_m}(kT_c) = A(kT_c)\, e^{2\pi j\left(\frac{(BW)}{2T_s}(kT_c)^2 + \frac{S_m}{T_s} kT_c\right)}, \ \forall k \in \mathbb{Z}. \tag{8.5}$$

Therefore, the discrete time of the spread chirps involving the packet of the encoded symbols is denoted as

$$T_X^{S_m}(kT_c) = \sum_{m=0}^{M-1} C_{S_m}\left(k - m2^{SF}\right). \tag{8.6}$$

Following that, the modulated symbols are successively transmitted over the air through channel medium. The total amount of data acquired at the receiver input is stated as

$$R_X[n] = h \times T_X^{S_m}[n] + Z[n], \tag{8.7}$$

where h, $R_X[.]$, $T_{X_S}^{S_m}[.]$, and $Z[.]$ denote the block-fading channel, received signal, transmitted signal, and zero-mean Gaussian noise with valued variance, respectively.

The demodulation approach is primarily handled in two stages: de-chirping and extraction of the most likely received symbols. The purpose of the de-chirping operation is to extract information from the analog received signal by multiplying it by the base of the broadcast chirp signal. It has little effect on the received signal in most cases; it just takes the section holding the sent symbols, ready to be addressed in subsequent runs. The resulting signal following the de-chirping process is an M-ary FSK modulated signal. We have

$$R_X^D[k] = R_X[k] e^{-\pi j\left(\frac{(BW)}{T_s}(kT_c)^2\right)}, \ \forall k \in \mathbb{Z}. \tag{8.8}$$

LoRa demodulation seeks to locate a collection of symbols on de-chirped signals. The coherent demodulation strategy is an effective method for dealing with M-FSK form signals that have penetrated an AWGN channel. We have

$$\hat{S}_m = Arg_{max}\left\{Re\left\{DFT\left\{R_X^D\left(k + m \times 2^{SF}\right) \times S_m\right\}\right\}\right\}. \tag{8.9}$$

The recovered non-binary symbols $\hat{S}_m$ that are retrieved by the discrete Fourier transform procedure are converted into binary frames using the gray indexing approach. By re-XORing the binary stream comprising binary expressions of the received symbols, the de-whitening operations are used to remove the bit stream form. De-whitening is expressed by XORing with the same applied whitening sequence of bits. The data are then stored in the same rectangular array arrangement as before, but row-wise, one code word at a time, using an inverse of the interleaving performed on the transmission side. More specifically, the vector array of the incoming bit stream is reconstructed to be ready at the input of the decoding block to examine the number of the occurred errors. The Hamming decoding technique is used for FEC decoding of de-interleaved bits.

8.3 LoRa energy modeling

This section focuses on the energy modeling of LoRa devices. The provided parts estimate the overall LoRa network energy drain; nevertheless, a full examination must include all power consumption factors in each node.

8.3.1 Communication energy

LoRa nodes are enabled for a limited duration T_{active} and then deactivated during T_{off}, which is commonly denoted by sleeping mode. In standby mode, the node listens to the radio with clocks that do not consume as much energy as they do in active mode, when it performs many operations that deplete the node's energy supply. To comply with a complete transmission, each node in the network begins by sending acknowledgment messages that are contained in the preamble, and then the payload is transferred following an agreement between the node and the GW. As a result, the total length of the packet transmission may be T_{packet}, represented as

$$T_{packet} = T_{preamble} + T_{payload}, \tag{8.10}$$

where T_{packet}, $T_{preamble}$, and $T_{payload}$ signify the complete packet duration, preamble duration, and physical payload duration, respectively. The preamble time is typically determined by the device type, but the payload duration is determined by the quantity of symbols containing critical data. Their values are articulated as follows:

$$T_{preamble} = (4.25 + N_{pr}).T_s, \tag{8.11}$$

$$T_{payload} = N_{phy}.T_s, \tag{8.12}$$

where N_{pr} is the number of symbols carried on the preamble, T_s is the symbol duration, and N_{phy} represents the length of the transmitted physical payload.

Uplink transmission and downlink reception are both supported by the communication module. The RF oversees end node data transmission and reception. To transmit a packet, the emitter node typically begins by sending two uplink alerts into the GW in a predefined time interval; then the node prepares to receive input from the GW by establishing two listening windows consecutively. In the absence of detection during the initial opening, the second window is normally opened. If no notification is received, the node will wait until the next duty cycle before retransmitting the message until the link exchange is successful. After the ED frees two practical windows to admit likely input from the GW, the transmitter node closes the authorized number of uplink tries inside $T_{preamble}$. If nothing is detected after the initial listening window of T_{W1}, the node tries again by establishing a second window of T_{W2} length. The node is then forced to retransmit the preamble in the following duty cycle if nothing is recognized. We have

$$E_{RF}(t) = E_{tx}(t) + E_{rx,w1}(t) + E_{rx,w2}(t), \tag{8.13}$$

where $E_{rx,w1}$ and $E_{rx,w2}$ are the reception energy during windows 1 and window 2, respectively, and E_{tx} is the transmission energy. We have

$$E_{rx,w1}(t) = P_{rx,w1} \times N_{sym} \times T_s, \tag{8.14}$$

where N_{sym} corresponds to the number of symbols associated with up and down communications, its value depending on the selected SF, namely, 8 symbols for SF-11 and SF-12 and 8 for other SFs, T_s is the symbol duration, and $P_{rx,w1}$ is the required power for window 1 opening. We also have

$$E_{rx,w2}(t) = P_{rx,w1} \times \left(\frac{32 + 2^{SF}}{BW}\right), \tag{8.15}$$

where $P_{rx,w2}$ is the required power for window 2 opening. Moreover, we have

$$E_{tx}(t) = P_{tx} \times T_{packet}, \tag{8.16}$$

where N_{sym} corresponds to the number of symbols associated with up and down communications, its value depending on the selected SF, namely, 8 symbols for SF-11 and SF-12 and 8 for other SFs.

The transmission power is expressed by

$$P_{tx} = \rho \times T \times K \times BW \times 10^{\left(\frac{N_F}{10}\right)} \times \left(\frac{4\pi f_c}{c}\right)^2 \times \left(\frac{d}{d_0}\right)^{\alpha}, \tag{8.17}$$

where ρ, N_F, c, f_c, d, and d_0 signify the spectral efficiency of the system, the noise figure, the celerity, carrier frequency, distance, and the initial distance, which is fixed to 1 m.

8.3.2 Computation energy

In addition to the energy expended by radio components, the node's processing unit adds to energy consumption while processing data through various sequences such as channel coding, whitening, interleaving, and gray indexing, and most notably while modulating the signal. The energy required by the processing unit is primarily ascribed to two components: energy from switching sequences and energy losses due to leakage current. Thus, the entire time required by the MCU physical layer block to perform packet postprocessing is estimated as follows:

$$T_{mcu} = T_{sc} + T_{cc} + T_{whi} + T_{int} + T_{gr} + T_{css}, \tag{8.18}$$

where the time needed by the physical block source coding, channel coding, whitening, interleaving, gray conversion, and CSS modulation block is denoted as T_{sc}, T_{cc}, T_{whi}, T_{int}, T_{gr}, and T_{css}, respectively. The energy expended by a single node during the active stage, i.e., the processing unit's energy consumption E_{mcu}, can be written as

$$E_{mcu}(t) = E_{leakage}(t) + E_{mcu,switch}(t). \tag{8.19}$$

8.3.3 Sensing and circuitry energy

The sensor unit drains the energy resource for data acquisition and digital-to-analog converters responsible for the information collection and digital conversion, which is estimated by

$$E_{sc}(t) = E_c(t) + E_s(t). \tag{8.20}$$

The energy wasted by the node circuitry and sensing unit are indicated respectively by E_c and E_s .

8.3.4 Network total consumed energy

The total energy wasted in the network with a number of N_{node} LoRa EDs after the occurrence of the total number of duty cycles N_{cycle} is the sum of the total energy dissipated by each node identified D_{id} realizing the same number of cycles:

$$E_{total}(t) = \sum_{id=1}^{N_{node}} \sum_{n=1}^{N_{cycle}} E_n(D_{id}). \tag{8.21}$$

Furthermore, we can also use the following formula to predict the residual energy at a specific time point in a given ED equipped with initial power capacity E_{init} as a function of the number of the occurred transmission cycles:

$$E_{total}(t) = \sum_{id=1}^{N_{node}} \sum_{n=1}^{N_{cycle}} E_n(D_{id}). \tag{8.22}$$

8.4 Methodology

Consider a wide area network with hundreds of densely placed wireless end point devices. These devices can gather and share observed data with a remote monitoring base station on their own. Each individual node uses the CSS modulation approach independently, without relying on contact with other nodes in the network. A TDMA slot mechanism is used to arrange exchanges between sensor nodes and the base station to ensure coordinated wireless communication. Considering the nodes rely on limited battery power, it is critical that they optimize their energy consumption. Furthermore, each node is strategically placed within the monitoring region and is responsible for selecting the optimal physical layer parameter to utilize to achieve effective communication while consuming the least amount of energy. LoRa devices are supposed to adapt their transmission to the channel model and distance to succeed in a given transmission. Those parameters include SFs that vary from 7 to 12. In addition, CRs are typically fixed to four different settings {4/5, 4/6, 4/7, or 4/8}.

Next, one must determine the distance at which each specified parameter ensures better performance in terms of energy efficiency and dependability. We present an algorithm that is designed to regulate optimal transmission parameters, ensuring both energy efficiency and network dependability for each LoRa ED on the network. In addition, to determine the amount of energy spent by any LoRa sensor at a certain point per communication, we assume that the SF7 is the highest energy consumption level, which is then used as the baseline reference.

We employ Eq. (8.23) to study the trade-off between energy saved and energy spent by processing activities to estimate the efficiency of such a parameter. Therefore, to give an insightful perspective on the enhanced selectivity of an adaptive parameter while estimating the life of each node on the network. The selection is efficient in either case if the energy saved during transmission between two alternative settings is more than the processing energy required for the equivalent activity. Typically, to determine if the chosen configuration is useful in terms of dependability and energy efficiency a heuristic algorithm is adopted. The algorithm shown in Fig. 8.5 explains the functioning of LoRa adaptive parameters regarding the energy efficiency conditions.

We compare several settings that can perform the same work in the same scenario, as well as distance and packet size. In addition, we calculate the energy efficiency ratio, which considers both necessary transmission and computing energy. However, if the amount of overall transmission energy is strictly larger than the amount of processing energy [15], the difference between two settings $E_{tx}(ED)$ for a specific ED is significant. As a result, the efficiently obtained energy average is written as

$$E_e(ED) = \frac{E_{tx}(ED)}{E_{pros}} \times 100\%. \tag{8.23}$$

The critical distance between two different SF coverages should be zero; therefore, $\Delta E_{tx} = 0$ if SF_j and SF_i are examined at this point. The critical distance d_c for the first SF is the greatest distance that can be achieved with these parameters, whilst

```
1: For a distance d
2: ∀ 1 ≤ i ≤ 6 , SF_i ∈ {7, 8, 9, 10, 11, 12}
3: ∀ 1 ≤ j ≤ 4 , CR_j ∈ {4/5, 4/6, 4/7, 4/8}
4: for i = 1 → 5 do
5:     if d > d_c(SF_i, CR_j) then
6:         SF ← SF_{i+1}
7:         d_c ← d_c(SF_{i+1}, CR_j)
8:     else
9:         SF ← SF_i
10:        d_c ← d_c(SF_i, CR_j)
11:    end if
12:    for j = 1 → 3 do
13:        if Ee(SF, CR_{j+1}) > Ee(SF, CR_j) then
14:            CR ← CR_{j+1}
15:        else
16:            CR ← CR_j
17:        end if
18:    end for
19: end for
```

FIGURE 8.5

LoRa adaptive transmission parameter selection algorithm.

the critical distance d_C for the second is the lowest distance that can be reached. We use the following equation to calculate this distance:

$$d_c(SF_j) = \left(\frac{E_{tx}(SF_j) \times 10^{\left(\frac{-G_{CSS}(SF_j) - N_F}{10}\right)}}{T_{packet}(SF_j) \times \rho \times T \times K \times BW} \left(\frac{1}{4\pi f_C}\right)^2 \right)^{\frac{1}{\alpha}}. \tag{8.24}$$

8.5 Case study and experimental setup

In this section, we assess the energy consumption, energy efficiency, and transmission reliability of LoRa EDs that use CSS modulation in the context of LPWANs. Using MATLAB, we run comprehensive simulations for an ED in a star-of-star topology network. We investigate several propagation scenarios based on physical layer factors. Table 8.1 shows average values of channel model parameters utilized and electrical parameters selected from various genuine LoRa transceiver devices.

To assess the energy calculations, we investigate the described communication chain operations for the chosen coding and modulation gains. The BER is calculated by executing numerous transmissions of various data streams of varying sizes. The

Table 8.1 Simulation parameters and their values.

Parameter	Value
BW	125 kHz
CR	4/5, 4/6, 4/7, 4/8
SF	7, 8, 9, 10, 11, 12
Path loss exponent	$\alpha = 3$, $\alpha = 4$
Voltage	3.3 V
Carrier frequency	858 MHz
Sensor unit voltage	2 V
Processing unit voltage	3.3 V
$I_{leakage}$	10 ηA
I_{rx}	11 mA
I_{sleep}	1.5 μA
N_F	10 dB

simulations are done over 50,000 iterations using an AWGN channel medium with various initial values for the channel medium. Fig. 8.6 shows the SNR thresholds that indicate the performance of the SFs with CR 4/7. As a result, the coding and modulation BER values (10^{-3}, 10^{-4}, and 10^{-5}) are utilized to estimate the SNR gain for any SFs and encoding scheme CRs [16].

We employed a carrier frequency of 868 MHz, multiple potential CRs for the Hamming channel encoding technique, and a unique BW of 125 kHz to explore the behavior of wireless LoRa EDs at long distances. Furthermore, we assumed that all processing unity actions are executed at a processing frequency of 4 MHz. Concerning the route loss condition, we aimed for a high path-loss exponent of $\alpha = 3$ for urban and $\alpha = 4$ for suburban settings to resemble urban and suburban environments. To address the issue of energy efficiency and to validate our assumptions regarding LoRa node energy efficiency, we built a whole LoRa communication chain in the MATLAB simulation framework, as detailed in earlier sections. This method enabled us to select an appropriate BER gain for both CSS and CR to target diverse audiences.

The number of transmissions that every ED may go through is cyclically limited. In most circumstances, the highest duty cycle required is 1%. This means that after a T_{OA} broadcast over the radio, the end node must stay silent for 99% of the time. The node sends a packet for a time length T_{OA} and waits for the time specified by the duty cycle term. One communication delay corresponds to the entire delay between silent mode and the reception time necessary for the node's acknowledgment permission for the first accomplished transmission and subsequent retransmission demands. As a result, greater packets require more T_{OA}; therefore, the associated idle mode is higher, but with less feasible transmission before the whole drain.

Duty cycle compliance must be considered in accordance with any government requirements for public networks. To comprehend the effect of duty cycle on end

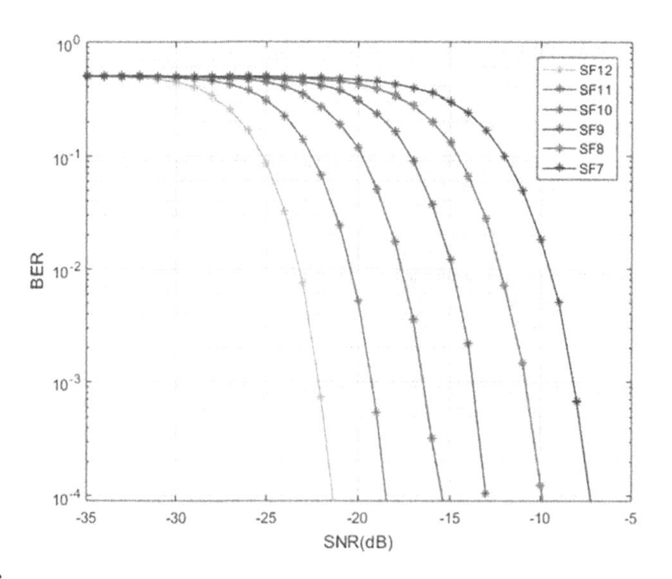

FIGURE 8.6

CSS modulation BER versus SNR performance in an AWGN channel model.

node lifespan length utilizing a chosen configuration based on parameter choices for the researched cases, we calculate the lifespan of an ED based on a duty cycle of 1% and 0.1%. The data size for a typical transmission may vary; hence, we illustrate in Fig. 8.7 the behavior of the node transferring data packets of varying sizes using the configuration (SF7, CR = 4/8). The figure shows that the total number of transmissions that can be realized by the node with a battery resource of 2600 mAh decreases with increasing data size. The node life may be longer for small transmitted packets, which increase the total number of predictable communications.

To predict the life duration the nodes are equipped with different battery source capacities (500 mAh, 2600 mAh, and 3500 mAh) of a node that transmits every duty cycle the same data size equal to 10 bytes. Moreover, we deduce from Fig. 8.8 that the duty cycle is a fundamental parameter that governs the lifespan of a typical node. Therefore, using a reduced duty cycle increases the node's energy drain because it encourages nodes to complete multiple transmissions in a short amount of time; in other words, the quiet time gap between two subsequent transmissions is short. Furthermore, the lifetime of a node is measured not only by the total time a node spends in the network before dying, but also by the total number of realized transmissions, because in some cases the duration of a packet is long, resulting in a longer silent period and a lower number of transmissions. As a result, by lowering the duty cycle ratio, the number of transmissions per day is lowered, and the life of an ED is defined as the set of compromises between the number of happened transmissions by a given setting and the total duration between the active and idle modes of the concerned node.

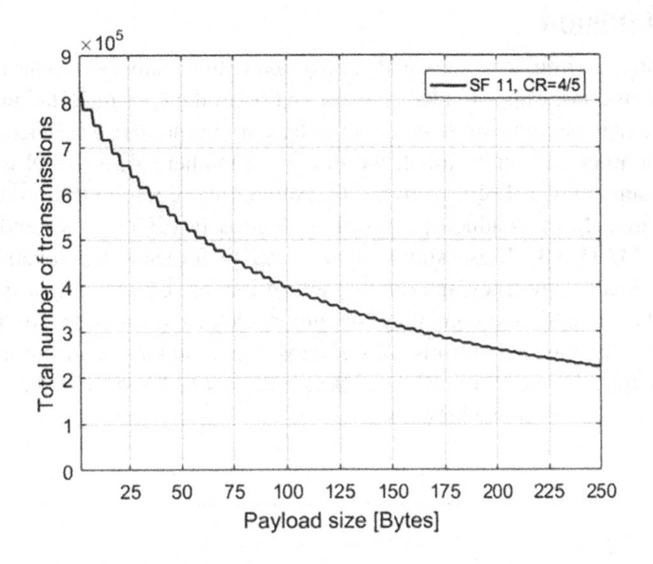

FIGURE 8.7

Total transmissions of LoRa EDs using (SF7, CR = 4/8).

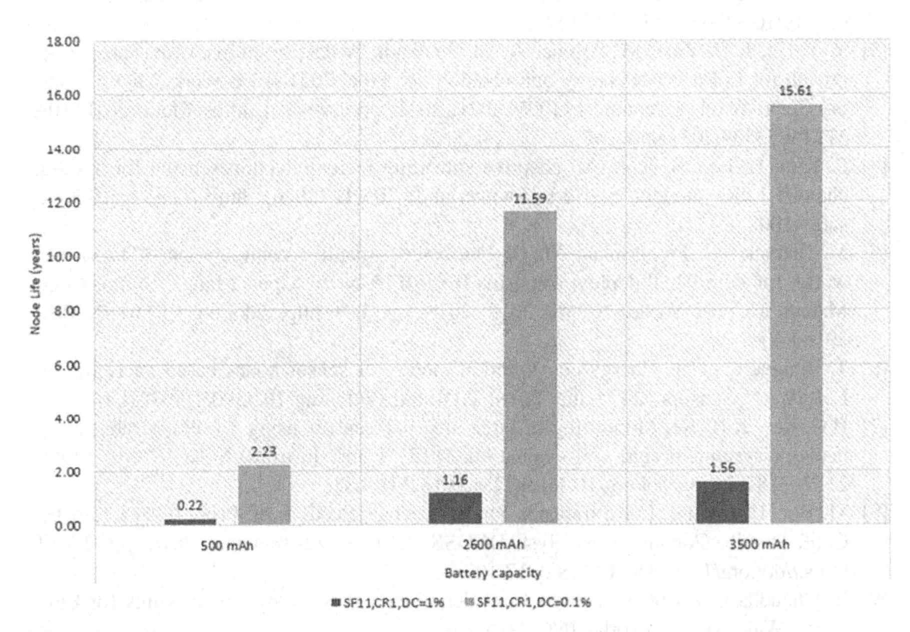

FIGURE 8.8

The lifetime of an ED using different battery capacities.

8.6 Conclusion

In this chapter, we initially discussed the communication chain operations involved in LoRa transmissions. This included an examination of the CSS modulation technique and the channel encoding strategy. Leveraging our comprehensive understanding of the LoRa communication protocol, we devised a mathematical model to calculate energy consumption in LoRa devices. To evaluate the communication and energy estimation model, we conducted multiple scenarios involving LoRa end node operations in MATLAB. These simulations aimed to forecast their overall lifespan concerning energy efficiency and reliable transmissions. By selecting a specific battery capacity for LoRa and regulating the duty cycle of transmission, we determined the estimated number of transmissions feasible using suitable transmission parameters. Consequently, we were able to estimate the entire lifespan of the specialized LoRa sensor node inside the network.

References

[1] B.S. Chaudhari, M. Zennaro, S. Borkar, LPWAN technologies: emerging application characteristics, requirements, and design considerations, Future Internet 12 (3) (2020), https://doi.org/10.3390/fi12030046.

[2] J. Haxhibeqiri, A survey of LoRaWAN for IoT: from technology to application, https://doi.org/10.3390/s18113995, 2018.

[3] Y. Yazid, I. Ez-Zazi, M. Arioua, A. El Oualkadi, A deep reinforcement learning approach for LoRa WAN energy optimization, in: Proc. 2021 IEEE Work. Microw. Theory Tech. Wirel. Commun. MTTW 2021, 2021, pp. 199–204, https://doi.org/10.1109/MTTW53539.2021.9607147.

[4] S. Kim, H. Lee, S. Jeon, An adaptive spreading factor selection scheme for a single channel LoRa modem, Sensors (Switzerland) 20 (4) (2020), https://doi.org/10.3390/s20041008.

[5] M. Slabicki, G. Premsankar, M. Di Francesco, Adaptive configuration of LoRa networks for dense IoT deployments, in: IEEE/IFIP Netw. Oper. Manag. Symp. Cogn. Manag. a Cyber World, NOMS 2018, 2018, pp. 1–9, https://doi.org/10.1109/NOMS.2018.8406255.

[6] T. Bouguera, et al., Energy consumption model for sensor nodes based on LoRa and LoRaWAN, Sensors 18 (7) (Jun. 2018) 2104, https://doi.org/10.3390/s18072104.

[7] H.C. Lee, K.H. Ke, Monitoring of large-area IoT sensors using a LoRa wireless mesh network system: design and evaluation, IEEE Trans. Instrum. Meas. 67 (9) (2018) 2177–2187, https://doi.org/10.1109/TIM.2018.2814082.

[8] M. Bor, U. Roedig, LoRa transmission parameter selection, in: Proc. - 2017 13th Int. Conf. Distrib. Comput. Sens. Syst. DCOSS 2017, vol. 2018-Janua, 2018, pp. 27–34, https://doi.org/10.1109/DCOSS.2017.10.

[9] B. Chaudhari, S. Borkar, Design Considerations and Network Architectures for Low-Power Wide-Area Networks, INC, 2020.

[10] N. Nurelmadina, et al., A systematic review on cognitive radio in low power wide area network for industrial IoT applications, Sustainability 13 (1) (2021) 1–20, https://doi.org/10.3390/su13010338.

[11] A. Augustin, J. Yi, T. Clausen, W.M. Townsley, A study of LoRa: long range & low power networks for the Internet of Things, Sensors (Switzerland) 16 (9) (2016) 1–18, https://doi.org/10.3390/s16091466.

[12] D.H. Kim, E.K. Lee, J. Kim, Experiencing LoRa network establishment on a smart energy campus testbed, Sustainability 11 (7) (2019), https://doi.org/10.3390/su11071917.

[13] C.F. Dias, E.R. De Lima, G. Fraidenraich, Bit error rate closed-form expressions for LoRa systems under Nakagami and rice fading channels, https://doi.org/10.3390/s19204412, 2019, pp. 1–11.

[14] Y. Yazid, A. Guerrero-González, I. Ez-Zazi, A. El Oualkadi, M. Arioua, A reinforcement learning based transmission parameter selection and energy management for long range Internet of Things, Sensors 22 (15) (2022), https://doi.org/10.3390/s22155662.

[15] I. Ez-Zazi, M. Arioua, A. El Oualkadi, Adaptive joint lossy source-channel coding for multihop IoT networks, Wirel. Commun. Mob. Comput. 2020 (2020) 2127467, https://doi.org/10.1155/2020/2127467.

[16] Y. Yazid, I. Ez-Zazi, M. Arioua, A.E.L. Oualkadi, On the LoRa performances under different physical layer parameter selection, in: 2020 Int. Symp. Adv. Electr. Commun. Technol. ISAECT 2020, 2020, pp. 1–6, https://doi.org/10.1109/ISAECT50560.2020.9523690.

CHAPTER

TinyML for 5G networks

Mamoon M. Saeed[a], Rashid A. Saeed[b], and Zeinab E. Ahmed[c,d]

[a]*Department of Communications and Electronics Engineering, Faculty of Engineering, University of Modern Sciences (UMS), Sana'a, Yemen*

[b]*Department of Computer Engineering, College of Computers and Information Technology, Taif University, Taif, Saudi Arabia*

[c]*Department of Electrical and Computer Engineering, International Islamic University Malaysia, Kuala Lumpur, Malaysia*

[d]*Department of Computer Engineering, University of Gezira, Madani, Sudan*

9.1 Introduction

Low-latency, high-speed communication between devices and the network is made possible by the 5G network architecture. Three primary parts make up the 5G architecture: the radio access network (RAN), the core network, and the edge network. The wireless access to the network is provided by the RAN, and data processing and management are handled by the core network. The multi-access edge computing (MEC)-enabled edge network provides computation and data storage at the edge of the network, closer to users and devices [1].

To link a variety of devices, including autonomous vehicles, critical infrastructure, and Internet of Things (IoT) devices, 5G networks must be secure. To guard against dangers including unauthorized access, data breaches, and denial-of-service assaults, 5G networks use a variety of security techniques, including encryption, authentication, and access control. In 5G networks, machine learning can be used to improve resource allocation, network performance, and security. To forecast network traffic patterns, spot anomalies, and spot potential security concerns, machine learning models can be trained using network data. These models can then be applied to enhance security, boost resource allocation, and optimize network performance [2].

The deployment of machine learning models on low-cost, low-power edge devices, like microcontrollers, is the focus of the developing discipline known as tiny machine learning (TinyML). TinyML can be used for edge intelligence in the context of 5G networks, allowing devices to carry out in-the-moment data analysis and processing at the network's edge, nearer the data source [3]. Since TinyML models are made to be small and effective, they can be used on edge devices without consuming a lot of processing power. Applications that need real-time or low-latency processing can benefit from computation being offloaded from the cloud or central server by

TinyML for Edge Intelligence in IoT and LPWAN Networks. https://doi.org/10.1016/B978-0-44-322202-3.00014-2

using TinyML models deployed on edge devices. This reduces network latency and speeds up application response times [4].

Overall, 5G networks, machine learning modeling, and TinyML for edge intelligence have the potential to enable a variety of new applications and enhance the functionality of current apps. It is possible to develop new applications that were not previously possible while also enhancing the performance, efficiency, and security of current applications by utilizing the low-latency communication capabilities of 5G networks and deploying machine learning models and TinyML at the edge of the network [5–7].

9.2 5G network architecture

With the network architecture of the 3rd Generation Partnership Project (3GPP), which is deprived of a structure or method for workers to be able to market the numerous groups of capabilities and services, 5G networks would not be feasible. Several 5G specifications, including Release 15 and Release 16 documents, have been released by the 3GPP. According to Fig. 9.1, showing the 5G network architecture, which was released by the 3GPP in the technical standard paper TS 23.501 [8], the core network's components can be instantiated numerous times to facilitate network slicing and virtualization. The design was developed to get rid of the data overlay that had been employed in earlier mobile network generations.

To support supplying a huge number of devices (massive IoT) and generating sporadic traffic while managing the enormous volume of video traffic via mobile networks predicted for 5G networks, architectural improvements were required. Architectural changes were required to reduce user plane end-to-end (E2E) latency to less than 5 ms and reduce the amount of core network traffic carried over expensive connections by only involving related network functions during individual data sessions. The fact that 5G is supported for specific use cases, such as driverless vehicles, results in lower latency requirements. In comparison to 4G, 5G is anticipated to have 5–10 times reduced latency. The architecture's separation of control and user plane activities is one of its key characteristics [9].

This effectively allows a control plane entity, such as the session management function (SMF), to dynamically create user plane functions (UPFs). This makes it possible for various UPFs to be centrally programmed, distributed in a wide area topology, and instantiated as needed [10]. Through the N9 interface, these UPF components can also be linked together to create a chain of things that can each be a user plane processing entity devoted to carrying out tasks. The access management function (AMF) now handles all other control plane operations, allowing for quicker bearer construction and change as well as simpler context management in the core network. To allow network slicing operations and to distribute network slices to user traffic flows, a specialized component known as the network slice selection function (NSSF) has been developed to work in conjunction with the control plane function AMF [11].

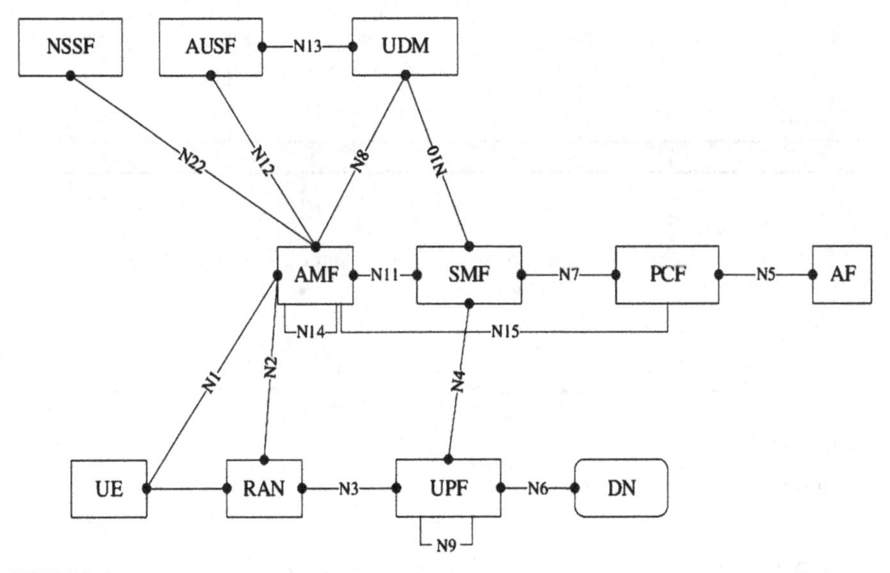

FIGURE 9.1

The non-roaming 3GPP 5G network architecture.

The steps for the 5G system are outlined in a different specification called TS 23.502 [12]. The architecture of the 5G network is service-based. The introduction of function interfaces using the REST API [13] makes it simpler for control plane functions to interact than was possible with earlier generations. On their clearly defined interfaces, control plane functions can communicate with one another utilizing HTTP messages and REST API requests, as illustrated at the top of Fig. 9.2. Compared to previous systems, this design allows the development of a much more modular 5G core and supports various network slicing methods through mobile core network orchestration. Now, freshly instantiated or existing functions can find other functions so that a communication session can be set up among them dynamically. The network exposure function (NEF) and network repository function (NRF) are the names of these new functions [14].

By defining modular components, the architecture based on services successfully gets rid of massive data pipes and lessens system complexity.

9.3 Requirements for 5G use case

Massive machine type communication (mMTC), enhanced mobile broadband (eMBB), and ultrareliable and low-latency communication (uRLLC) are the three categories of 5G use cases established by ITU [15]. 3GPP has also identified these use cases as:

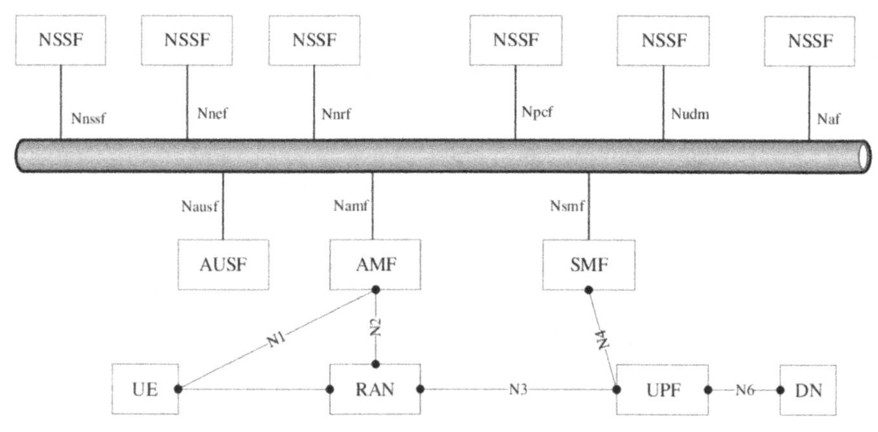

FIGURE 9.2

The service-based interfaces for the 3GPP 5G system architecture.

Table 9.1 The 5G use cases by ITU and 3GPP categories.

Case type to use	Performance objectives
eMBB	Low latency and high data rate
mIoT/mMTC	Ultralow energy usage and ultrahigh device density
Critical communications, extremely uRLLC	High reliability, minimal latency, and robust security

1. enhanced mobile broadband [16],
2. vital communications [17],
3. massive Internet of Things (mIoT) [18].

The performance objectives for each 5G use case are listed in Table 9.1 along with the ITU and 3GPP classifications [16,18].

Table 9.2 provides a summary of the quantitative performance requirements for these application scenarios.

In addition to these performance requirements, 5G systems are anticipated to offer perceptions of increased availability and coverage compared to legacy systems, lower network deployment costs, and support increased mobility profiles, which are divided into four categories: stationary (0 km/h), pedestrian/low mobility (0–10 km/h), vehicular (10–120 km/h), and high speed (120–500 km/h) [19]. Conventional networking techniques cannot realize the 5G ambition; instead, new system innovation that connects numerous systems will be needed. New technologies are therefore required to provide secure interoperable networks or secure 5G services that operate seamlessly across various network borders. This will put pressure on network operators and infrastructure suppliers [20].

Table 9.2 5G use cases' quantitative performance design objectives.

Use case	Category	Requirement
mMTC/mIoT	Linked devices per square kilometer	1 million
URLLC, critical communications	Controlling plane lag	10–20 ms
	Reliability	99.999%
	User plane lag	1 ms
eMBB	User throughput on the uplink	50 Mbps
	Downlink capacity per user	100 Mbps
	Rural mobility	Up to 500 km/h
	(Dense urban) mobility	Up to 30 km/h
	Controlling plane lag	10–20 ms
	User plane lag	4 ms
	Per-m^2 throughput indoors	10 Mbps per m^2
	Throughput per connection	1–10 Gbps
	Aggregate throughput per cell	10 Gbps
	Aggregate throughput for cell downlinks	10 Gbps

9.4 Security in 5G wide scope

For integrated 5G ecosystems made up of numerous sites, such as RANs, mobile core networks, and non-3GPP access networks, with various types of user devices and things connected to the ecosystem, it is necessary to examine the system's multiple security layers from a security perspective. Every 5G network has numerous security layers where different security flaws could potentially appear and need to be managed [21].

The increase in information surface area is a characteristic of all the layers. In comparison to the systems and subsystems used in 4G, each layer provides a larger potential surface area for an attacker. Each one adds new opportunities for the 5G network's purpose and effectiveness to be compromised. However, not all of them are equally important in terms of compromising the network's availability and integrity. Therefore, it is crucial to prioritize security operations to provide 5G networks with adequate security [22].

It is useful to think about a security matrix that spans the transport network, user equipment, IT infrastructure, and applications, as well as the core network and RAN to analyze various places where security vulnerabilities in 5G networks may develop. Table 9.3 summarizes this information by listing the five security layers shown in Fig. 9.3 along with a list of vulnerability topics for each. Then, each subject is flagged for various 5G network difficulty regions. The list is not exhaustive and is likely to be updated when additional possible threats and problems are discovered throughout the DCMS 5G T&T program [23].

In Table 9.3 vulnerabilities are briefly outlined for each of the five security layers in the paragraphs that follow.

Table 9.3 The security matrix for possible flaws in 5G networks. HW, hardware; SW, software; SYS, systems.

5G platform security matrix		Affected areas											
Security layer	Vulnerability topic	Radio network and air interface			Mobile core network			Transport network (backhaul and fronthaul connectivity)			User equipment device		
		HW	SW	SYS	HW	SW	SYS	HW	SW	SYS	HW	SW	SYS
Services, applications, and use case	QoS	✓	✓	✓	✓	✓	✓	✓	✓	✓			
	Access rights to network slices	✓	✓	✓	✓	✓	✓	✓	✓	✓			
	Vertical use cases											✓	✓
	Data confidentiality	✓	✓	✓	✓	✓	✓	✓	✓	✓	✓	✓	✓
	Genuineness, safety, and dependability of services and applications										✓	✓	✓
	Vulnerability of edge computing and services				✓	✓		✓	✓				
Users and things	The authenticity of the connection and device		✓	✓		✓	✓				✓	✓	✓
	Resource restrictions for M2M and IoT devices										✓	✓	✓
	Identification of devices for IoT and M2M					✓	✓						
Internetworking	Model operators			✓			✓			✓			
	Diversified core				✓	✓	✓						
	Using different RAN technologies	✓	✓	✓		✓	✓	✓	✓	✓			
	Separate RAN ownership for enterprise and rural use cases	✓	✓	✓		✓	✓	✓	✓	✓			

continued on next page

Table 9.3 (*continued*)

5G platform security matrix		Affected areas											
Security layer	**Vulnerability topic**	**Radio network and air interface**			**Mobile core network**			**Transport network (backhaul and fronthaul connectivity)**			**User equipment device**		
		HW	**SW**	**SYS**	**HW**	**SW**	**SYS**	**HW**	**SW**	**SYS**	**HW**	**SW**	**SYS**
5G mobile network and virtualization systems	Legacy core network vulnerabilities			✓			✓			✓			
	Functional split in the new RAN	✓	✓	✓									
	Software-based operations						✓						
	Multi-attribute context authentication					✓	✓				✓	✓	✓
	Multi-network latency				✓			✓					
	Network slicing				✓	✓	✓	✓	✓	✓			
	System restoration after failover						✓						
Physical infrastructure	NFV and SDN controllers					✓		✓	✓				
	5G new radio and RAN	✓	✓	✓									
	Active antenna management		✓	✓									
	Commodity hardware vulnerabilities	✓			✓			✓					
	Hardware performance deterioration							✓		✓			
	Physical security of base station, computing systems, and core networks	✓		✓	✓		✓	✓		✓			

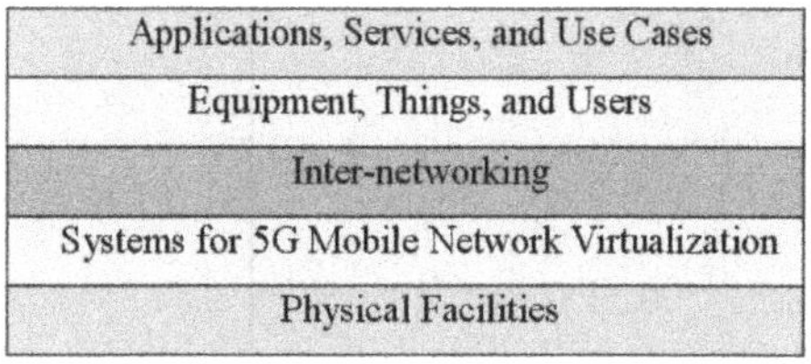

Applications, Services, and Use Cases
Equipment, Things, and Users
Inter-networking
Systems for 5G Mobile Network Virtualization
Physical Facilities

FIGURE 9.3

The layers of security in a 5G system.

9.4.1 Applications and services

Any application that utilizes a 5G network or networks falls under the services and applications layer. User equipment (UE), users, devices, systems, individual customers, business partners, business customers, third-party networks, etc., are some of the names used to describe these applications.

In the OSI networking stack model, services and applications frequently run at the top. From a security standpoint, a service or application's interactions with (i) other services or applications, (ii) the hardware it runs on, (iii) the networks that carry messages generated by or intended for the service or application, (iv) the organizational boundaries that the service or application covers or deals with, and (v) the infrastructure itself are all significant [24].

Common security concerns relate to the use of services, including what they do, who uses them, and how they impact the system [25]. There are additional factors to consider as well, which include the following:

- **Quality of service (QoS)**

In 5G, strict QoS assurances will be necessary for critical communications use cases. For URLLC and mMTC, this necessitates a strong and secure design. The services that will be connected to each network service in 5G will be defined by a unique set of QoS characteristics. User and device context in 5G must have high granularity to efficiently assign network slices to users and control access by users and devices to running slices [26].

- **Control over network slices**

Since many 5G services will operate as network slices, management of access privileges for users and/or devices to these services, and hence the slices, is necessary to prevent unauthorized access to the slices. This is especially important for network slices that are running important applications. Since different flows of multiple devices are handled by a single logical instance of a protocol stack layer, access permissions must be properly maintained [27].

- **Aspects of vertical use**

To what extent integration and interoperability applications and services must adhere to meet the requirements of vertical use cases is a question that must be answered. The numerous configuration possibilities for applications for users and devices, such as IoT devices, autonomous vehicles, smart homes, industrial control systems, or smart cities, may lead to security vulnerabilities. Financial services and information services related to medicine and healthcare are among the more difficult industries. Data security, privacy, integrity, and regulatory compliance are challenging but essential facets of these verticals [28].

- **Confidentiality of data**

One of the most important needs from the user's perspective is to make sure that data confidentiality can be guaranteed. To put it another way, solid procedures must be in place to stop breaches that could compromise the security and integrity of data, as well as the E2E risk appetite of users. This entails protecting not only the actual data content but also any system guarantees provided to businesses, such as business key performance indicators (KPIs) and service stability and dependability [29].

Early in the design phase, if these needs are assessed, there is the greatest chance of simultaneously meeting both law enforcement requirements and data confidentiality principles. Along with the legal and regulatory requirements (such as lawful interception [LI] and data retention) for law enforcement access to data, the demand for data confidentiality must be met. Please see NFV-SEC-011 Clause 4, NFV-SEC-009 Clause 4.5, and NFV-SEC-010 [30], [31], and [32] for more information.

- **The sincerity of the application and service**

Systems must make sure that services and applications are legitimate and only interact with the intended end points. This mostly concerns Industry 4.0 applications and has to do with testing, ensuring, and maintaining these applications. These programs need to be dependable and safe to use [33].

- **Service and edge computing vulnerabilities**

By making applications and services more easily accessible to users and devices at the network's edge, where user devices connect to networks, or at adjacent local data centers, edge computing aims to reduce the latency of applications and services. The updated 5G design now includes dedicated UPF nodes near the network edge, as well as distributed UPF nodes, spread out across the deployment area. As a result, MEC servers running specialized MEC services are closer to consumer devices. The communication between edge UPF machines and MEC servers needs to be safeguarded to prevent traffic sniffing attacks [34].

The UPF devices must also be physically protected to prevent tampering, which could result from a mobile network's control, and user planes to/from such UPFs being vulnerable. Application security may be compromised as a result of edge computing. To ensure the security of such data centers or box units, low-touch environments and less secure locations of edge computing centers are possible [35].

A further difficulty that adds to this is MEC application administration and provisioning, which necessitates that management and monitoring sessions end at the MEC application services provided by application providers. Edge computing services need to use API security to guarantee that MEC apps can only access data that have been authorized for their use. Applications and security protocols must also consider the fact that, in some cases, the network edge may only have a limited amount of edge resources (computing, networking, and storage) [14]. This influences the security mechanisms' policy decisions and might impede some complex procedures.

9.4.2 Internet of Things users

To enable several new application scenarios and use cases, 5G networks will need to connect a variety of devices to the mobile infrastructure via various access technologies. This is a problem because every device has inherent security features built into its systems, hardware, and software. Additionally, the traffic produced by 5G-connected devices may differ greatly from that produced by legacy networks' traditional user equipment. It may not be possible to provide common security methods for all devices, necessitating a case-by-case review of each device type [36].

- **Device authenticity**

Device dependability and trustworthiness must be ensured, much like with services and applications. A fully secure automation system requires the safety, security, and reliability of devices, especially when it is connected to networks. Important tasks include testing and ensuring the sensors used in automated production equipment. Malware on various kinds of devices creates a new security threat surface.

Equally dangerous are firmware breaches that compromise devices. Certain sensors could be more prone to changes and accessible to adaptations. A distributed denial-of-service (DDoS) assault is one clear example of an attack that degrades business KPIs for many, if not all, users and has an overall negative impact on the system [37].

DDoS defense measures are necessary for networks to thwart external attacks, which frequently come from Internet servers. To detect and counteract coordinated DDoS attacks that might be conducted from rogue or even legitimate devices, safeguards must be put in place at various points across the system. Man in the Middle (MitM) Trivial File Transfer Protocol (TFTP) attacks may result in third devices listening in on ongoing connections. Authentic, tamper-proof UE hardware is required for the secure processing and storing of user credentials [38].

- **User approval and data processing procedures**

There is a need for additional system integration including new building blocks that offer user privacy and data security techniques, as well as data handling and permission procedures that are integrated with mobile communication protocols. New billing and value chain distribution management systems must be connected to these blocks.

Building a new, safe environment that supports consumer and corporate trust in 5G networks and services requires such integration. Building a knowledge of the danger and responsibility of stakeholders and risk owners in relationships will also be facilitated by the integration of security systems with data, permission, and billing processes [38].

- **Devices that connect machines must use limited resources**

Compared to typical data centers and communications equipment, machine-to-machine (M2M) devices use substantially less electricity and transmit data in a variety of ways. Common UEs, such as smartphones, laptops, personal digital assistants (PDAs), and other mobile devices, are supplied for security architectures for ordinary mobile networks. To avoid taxing low-power devices, protocols must be designed specifically for them [39].

Because of this, security designs must address M2M use cases, or mIoT in 3GPP, which allows for lightweight solutions to traditional security issues without jeopardizing security operations or leaving backdoors that can jeopardize the security of the system. Security measures need to include several crucial M2M communication applications that demand high levels of reliability and latency, especially on the user plane (4 ms RTT). To balance energy efficiency, battery life, and high-security assurance, security methods and practices are required for M2M/IoT devices (or gateways for these devices) [34].

- **Identifying M2M/IoT gadgets**

Since non-access stratum (NAS) is defined abstractly in 5G networks, SIM-based device identification and authentication will no longer be the only method used. Especially for IoT devices that are anticipated to use flexible identifying mechanisms, such as soft-SIM, this could lead to new sorts of security concerns in the operation and maintenance of networks. Therefore, identity management processes should be in place for IoT/M2M devices as well as traditional user equipment [40].

9.4.3 Crossing organizational boundaries in networking

Distinct network access technologies must be connected to 5G networks to accommodate new 5G application types, each of which may be handled by a distinct operator domain. The 5G ecosystem is currently more complex than legacy systems due to the entry of new participants; vendors of virtual applications, virtual network service providers, and virtual infrastructure providers are all included in E2E services [41].

- **Models for operator**

In a neutral host paradigm, it is necessary to identify who is in charge of maintaining security. From radio access to the virtual network core, this crosses several network domains. New potential security attack surfaces are introduced by network-as-a-service (NaaS) and public cloud network hosting [42].

- **Dispersed core**

To decrease user plane latency to devices, the core network is now dispersed across the network topology. This is a highly desirable quality from a performance standpoint, but it also implies that the core must maintain security controls at the networking components that make up its edge, which will be situated apart from the data center for the main core network. Therefore, as a component of a distributed core management system, procedures and features for device authentication must be in place [43].

- **Using different RAN technologies**

Different RAN technologies, such as mmWave, WiFi, LiFi, and NB-IoT, are being merged with 5G as it envisions a wide range of application scenarios. Networks run the risk of such integration when systems are not tested or tried to identify their known or unknown vulnerabilities. MitM attacks, which include listening in on ongoing communication sessions, are another danger. These assaults may become more common as 5G networks integrate various access technologies. Various access networks operating on various frequency bands require systems to identify and mitigate such assaults. Access networks are also susceptible to jamming attacks, which can interfere with air interface interactions [44].

- **Owning RAN separately for rural and business use situations**

Many scenarios where there is a division of ownership or interests between the RAN and the core network are made possible by 5G. As infrastructure is added or removed, the core must be able to handle large volumes of transfers and changes while upholding strict security, QoS, and control protocols. A large cell's and a micro cell's environment may vary by thousands of times per day. Furthermore, the networks must be intelligent to detect any rogue nodes that might be RAN equipment being used maliciously. This is particularly frequent in 5G networks [45] due to the wide range of access technologies used, the extremely dense deployment, and the multiple parties that may possess distinct radios.

- **Backhaul network weaknesses**

To safeguard the backhaul networks that link the sites, standard IT procedures and security measures like DDoS prevention systems and security firewalls toward the public Internet must be utilized. The 5G ecosystem must include capabilities to identify DDoS assaults so that devices or rogue RAN equipment does not produce a lot of malicious traffic and takes down servers for services. Additionally, IMSI catchers that steal user device identities, impersonate network connections, and flood communication interfaces must be protected by mechanisms [46,47].

9.5 5G mobile network and virtualization systems

To achieve more flexibility and scalability in operating network systems and software as well as to support the dynamic and rapid deployment of network services, 5G networks will have integrated NFV solutions or a virtualized E2E mobile system that is run on virtual infrastructure systems. The advantages of operating virtualization services in 4G systems are already being seen by operators, and the roadmap for 5G has placed a strong emphasis on the supporting technologies of NFV and SDN. Virtualization has several benefits over traditional mobile systems that utilize fixed hardware and are fundamentally different, but it also creates several open issues that must be resolved before it can be effectively deployed in live systems. The majority of these issues are connected to performance guarantees and security assurance.

A logical security architecture would allow for greater flexibility and lessen the importance of physically allocating security functions, as was the case in older systems, according to the 5GPPP Phase 1 Security Landscape White Paper [48]. The virtualization of services and the existence of numerous operational domains involving various actors are the driving forces behind this. Flexible trust models, security administration, and dynamic slicing operations would be possible with a logical design.

9.5.1 Computerized operations

The increased software dependence of 5G networks over preceding generations creates significant operational hazards. Networks, for instance, must make sure that such software is not compromised or exposed, so they must have powerful virus prevention systems that are regularly updated and maintained. Normative standards for NFV security are being developed by the European Telecommunications Standards Institute (ETSI), drawing on the report NFV-SEC-12 [49].

One such draft standard is GS NFV-SEC-019. If hypervisors are hostile or compromised, there may still be some significant and structural security vulnerabilities regarding the isolation of critical functions. ETSI also has draft standards (NFV-IFA-026 [50] and NFV-IFA-033 [51]) for developing and testing security monitoring and management for NFV. These need to be transformed into tried-and-true solutions (NFV-SOL and NFV-TST groups) and put into use as open-source solutions.

9.5.2 Slicing the network

As mentioned in Section 9.1, 5G networks are expected to leverage new technologies including NFV, SDN, MEC, network slicing, and distributed core to achieve the aims of 5G. This implies that from a security standpoint, any element that lowers the performance of any of these systems is of interest. The mobile core of 5G networks will be fully virtualized, with many network slices operating concurrently and dedicated to distinct consumer domains or vertical markets.

The use of NFV technology, which provides efficient and scalable support for running virtual machines in commercially available servers, is crucial to the development of 5G network slices. Instead of taking weeks or months, operators will benefit from the flexibility and efficiency of swiftly building virtual mobile core networks. The matching of core network slices with corresponding transport slices will be made possible by the programmability support provided by the use of SDN in the transport network, i.e., control of SDN switches via protocols like OpenFlow. Additionally, network slices will support cloud-RAN principles if running slices and RAN sections are dynamically linked following demand and load conditions [52]. However, complete security solutions that would guarantee cross-slice security must be equally backed by the flexibility and effectiveness of network slicing. What is needed for this is:

- ✓ Complete multi-tenancy support in an E2E network to enable customized domains of operations for security processes and complete isolation of slices in virtualization systems. The themes from ETSI NFV-SEC-009 [30] and NTV-SEC-012 [31] should be considered.
- ✓ Preventing unauthorized access between these domains using authentication techniques that would only allow slice access to authorized users.
- ✓ Effective security management and monitoring, such as that provided by the NFV-SEC-13 [53], NFV-IFA-026 [50], and NFV-IFA-033 [51] standards, will reduce the risk and impact of attacks.
- ✓ A slice's allotted resources must be inaccessible to others.
- ✓ Network slice management and monitoring must be segregated.
- ✓ Any potential side channels resulting from shared virtualization infrastructure, such as sequential memory location usage by several virtual machines belonging to distinct network slices, must be avoided. gNB functions may be exposed to attacks that are normally anticipated on virtualization infrastructure when they are implemented as virtualized network functions (VNFs) on shared virtualization infrastructure.
- ✓ Such infrastructure needs to be shielded from both local and remote (software-based) hacking attempts. On network slices and the virtualization infrastructure, there are additional potential security vulnerabilities [54,55]. Such flaws result from network slicing's insufficiently widespread implementation in commercial systems and incomplete assessment of its security flaws. Attacks may target the following:
  - ✓ shared network interfaces between network slices,
  - ✓ slice selection and management,
  - ✓ interslice interfaces,
  - ✓ management interfaces for network slices.

9.5.3 Restoring the system following hardware failure

Unexpected and unplanned power-down occurrences, a common problem in today's data centers, pose a risk to the ongoing and dependable operation of services. In data centers, there are standard procedures to avoid such circumstances, restart the system, and then restore the service context and system context.

Due to the addition of NFV-based infrastructures, the orchestration capabilities provided by Management and Orchestration (MANO) systems, and dynamic programmability with SDN, it is now also important to have NFV, SDN, and MANO in place to have effective recovery mechanisms [56]. The following must be guaranteed by the recovery methods in the virtualization systems:

✓ It is necessary to fully restore the virtualization infrastructure and make the appropriate adjustments to all configurations and settings. This entails reloading settings with the appropriate parameters, pointing controller nodes to the appropriate group of components, and restoring complete interoperability. The automated restoration of NFV, SDN, and MANO system interoperability without the need for manual reconfiguration to make these systems function once more is crucial [57].

✓ The entire service context must be recovered. This entails appropriately establishing the final state and reloading virtual machines and network services. The many designs of 5G networks, which also include virtualization systems and technologies, must be verified for both performance and any potential attack surfaces they may expose [58].

The new RAN's functional division

Understanding the new RAN functional divides [59], including potential attack pathways to restrict or deny service layer functions, is important. The division and distribution of the RAN protocol stack to various components may have an impact on security. Open disaggregated V-RAN should be addressed by functional splits in the RAN. It may be difficult to ensure the security and integrity of the RAN system given how RAN systems are operated and managed because the radio stack has been divided into three distinct components: the radio unit (RU), distributed unit (DU), and central unit (CU) [58].

9.5.4 Context authentication on multiple levels

For context-aware operations in intelligent networks, numerous user context components and components of the device context, such as apps and usage habits, as well as position and velocity, are currently being specified and enabled [60]. User and device context will be very helpful for 5G networks so that network slicing activities can be automatically customized according to patterns and demands for usage. The capacity to effectively handle and process various context-related pieces of information and correlate them with threats to networks and systems in general and potential security breaches, in particular, is a requirement for increased authentication, ongoing security checks, and the actual delivery of security assurance to users and businesses [61].

9.5.5 The difficulty is in minimizing false positives while maintaining uninterrupted and secure operations

- **Multiple-network lag**

The mobile core network, RAN, transport network, and other access technologies are all part of the 5G networks, which are more complicated than earlier generations. Deterministic latency levels must therefore be determined, measured, and set up using efficient procedures. While achieving latency goals should not compromise security assurance levels, it is important to remember that not all application circumstances call for the same level of security; some methods may be overly complex while others may be insufficient. As with any other communication system, there is a trade-off between performance and security assurance.

- **Flaws in earlier iterations of the 3GPP core network systems**

Backward compatibility is supported by the new 3GPP architecture for 5G networks [62]. Additionally, this means that the system will come with built-in 4G vulnerabilities that need to be fixed, especially if a non-standalone system is used.

9.5.6 Physical infrastructure

It is crucial to make sure that parts of the system and physical equipment are not altered, are securely installed, and can only be accessed by authorized individuals. This is in addition to the software, systems, users, items, and services. This covers both access to the systems managing the infrastructure and physical access to the infrastructure by human operators.

- **Controllers for NFV and SDN**

NFV functions handle the control plane operations about network slices, while SDN-based transport layer support now separates the control of switching and routing from the data plane. Network operations are made flexible, elastic, and reconfigurable via NFV and SDN. The controllers of NFV and SDN systems are weak areas in network operations; therefore, using these technologies also requires using them.

It is crucial to make sure that virtualization controllers are safe, secure, and guarded against unauthorized access. This refers to access through the controller API to services like virtual infrastructure controller services. Testing new radio technologies and systems is essential to ensure that they are secure and that they do not have any vulnerabilities [63]. This includes testing for security leaks as well as for system, software, and hardware flaws.

- **New radio 5G**

The new radio in 5G networks is designed to deliver faster data rates at the air interface between UEs and the RRH than previous generations can. mmWave technology, which is presently employed in some RRHs, is one example of a novel air interface technology that may be more brittle. This raises reliability concerns, which could impact security. Pico-cell deployments, which are required for ultradense 5G RAN deployments, present their own set of difficulties [64]. The most notable change is that air interface attacks may now affect a broader area and have access to more cells. This can involve physically impeding a cell's operation or interfering with it to disrupt or listen in on ongoing communication sessions.

Typical RAN attacks include traffic interception, mobile device impersonation, location monitoring, network spoofing, false base stations (primarily IMSI catchers), and denial-of-service by jamming, especially on control channels [65]. An educational annex (E.1) to 3GPP TS 33.501 [66] covers network-based bogus base station detection with UE assistance. This claims that SUPI/5G-GUTI catchers can be identified using security factors in UE measurement data.

- **Control of active antenna**

New active antenna management techniques or comparable external activities that manage or configure antenna functioning in real time should not harm or interfere with antenna operations. These management procedures need to be carried out by authorized users or automation engines. In addition, control plane information-based transmission beam control towards devices may be open to interference or breaches [67].

In addition to the hazards associated with the new radio and the virtualization controllers, data centers and equipment locations must constantly guard against threats to common hardware equipment and systems.

- **Commodity hardware weaknesses**

Physical hardware equipment contains weaknesses that must be considered, just like any system. It is crucial to have the system restored in cases of system failure or meltdown and, while doing so, to reload any security systems so that they are once again fully functional [18].

- **Decreasing hardware performance**

Traditional IT network hardware, such as sluggish firewalls, may cause latency problems. Defending the network, users, and services from external threats, this puts a risk on maintaining customer SLAs [19].

- **The base station, computer, and core network physical security**

Due to the proximity of physical components, systems, and subsystems, this element, like any system, is connected to the physical security of these locations. This includes the accessibility of any features. The difficulty with 5G networks is their distributed architecture, which may include many data centers hosting the essential network elements as well as systems to enable MEC operations. Additionally, 5G networks deploy RAN equipment densely, resulting in more base stations per square km [20].

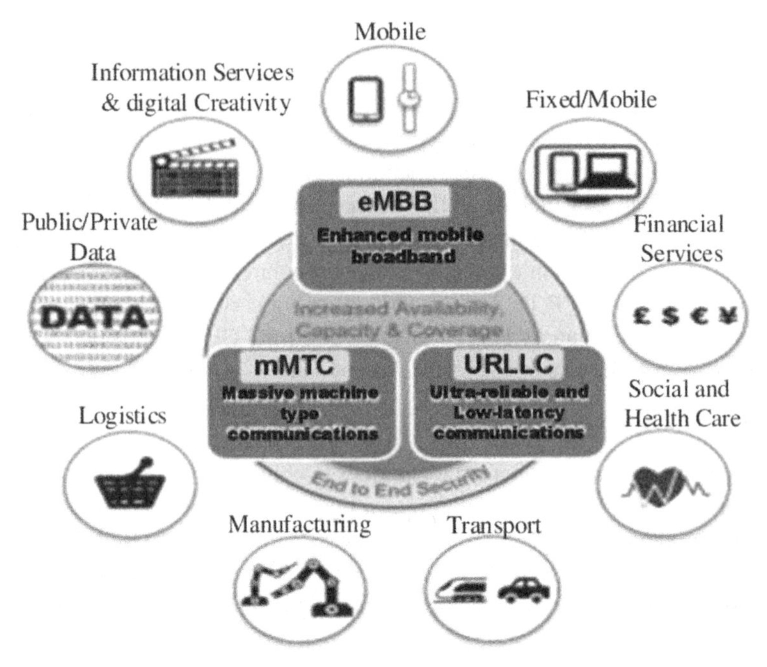

FIGURE 9.4

Vertical industries in 5G application scenarios.

9.5.7 Security layers summary

Different vulnerabilities exist in each of the five security layers that make up the enormous attack surface of 5G networks and systems, as described in this section. To build a 5G network that not only meets the ITU performance standards but also allays the security worries associated with these vulnerability areas, businesses and organizations from numerous industry verticals must cooperate through standardization bodies and industry forums. The support of multiple standard developing organizations (SDOs) is necessary for the various 5G verticals, as shown in Fig. 9.4, to produce security requirements [21].

As well as having specific and challenging security requirements, industry-focused fields like robotics and automation, emergency and critical communications applications and systems on autonomous vehicles, remote healthcare, and user- and community-focused new technologies like augmented reality (AR), virtual reality (VR), and mixed reality (MR) all have their particular challenges. A more complicated network will be needed for these applications than for legacy systems [22].

It is crucial to bring all SDOs together and align specifications to ensure interoperability of security across all components of the mobile core, transportation, access technologies, devices, services, and applications of the 5G system. This is necessary to ensure that a single SDO's security solutions do not have an excessively narrow scope and leave security gaps when all components need to join [23].

9.6 Multi-access edge computing for 5G

MEC is a technology that allows data processing and storage closer to users and devices at the edge of the network. Because it provides low-latency communication and enhances the performance of applications that need real-time or low-latency processing, MEC is a crucial part of 5G networks [24].

MEC operates by placing servers, storage, and networking hardware on the network's edge, where they can be situated nearer to users and gadgets. Applications that need real-time or low-latency processing may benefit from this by having their latency reduced and their performance increased. By keeping data near the source and lowering the likelihood of data breaches, MEC can also increase security and privacy [25].

The versatility of MEC's applications and use scenarios is one of its main advantages. Applications for MEC include:

- Applications for AR and VR that need real-time processing and low-latency connectivity can be made possible using MEC. It is feasible to decrease network latency and enhance the performance of these applications by installing computation and storage resources at the network's edge [26].
- In autonomous automobiles, real-time sensor data processing is necessary to make steering, braking, and acceleration decisions. It is possible to decrease network latency and raise the safety and effectiveness of autonomous vehicles by placing compute and storage resources at the network's edge [27].

Smart cities: In a smart city, there may be thousands of sensors deployed throughout the city to collect data on traffic flow, air quality, and other variables. By deploying compute and storage resources at the edge of the network, it is possible to analyze the data locally and identify patterns or anomalies in real time. This can help city planners optimize traffic flow, reduce air pollution, and improve the overall quality of life in the city [28].

Industrial automation: In a factory or industrial setting, there may be hundreds or thousands of sensors monitoring equipment and processes. By deploying compute and storage resources at the edge of the network, it is possible to detect anomalies or predict failures before they occur. This can increase the general effectiveness of industrial processes while lowering maintenance costs and downtime [29].

Healthcare: Sensors may be used in a healthcare environment to track patient behavior or to monitor vital signs. Local data analysis and real-time pattern recognition are made possible by placing compute and storage resources at the network's edge [30]. This can enhance patient outcomes by enabling healthcare professionals to make better decisions about patient care.

MEC is a crucial part of 5G networks overall because it makes low-latency communication possible and enhances the performance of applications that need real-time or low-latency processing. A wide variety of applications and use cases can be supported, and the effectiveness, security, and privacy of these applications can all be enhanced by placing computing and storage resources at the network's edge [31].

9.6.1 Connected multi-access edge computing

The ETSI MEC specifications [32] and a new draft merging NFV with MEC [33] are the foundations for the key ideas surrounding MEC. The definitions from [34] have been updated in line with the shift in terminology from mobile edge computing to MEC. Multi-access edge applications (MEAs) are programs that can be launched on multi-access edge hosts in a multi-access edge system and that can either produce or consume multi-access edge services [35].

The Multiple-access Edge Application Orchestrator (MEAO) performs similar tasks to the multiple-access edge orchestrator (MEO), with the exception that it should employ the NFVO to create the virtual resources for both the MEA and the multi-access edge platform (MEP) [36]. The multiple-access edge host (MEC host), which interacts with mobile network entities via the MEP platform to supply MES and offload data to MEA, provides the virtualized environment required to run MEC applications. MEA orchestration and instantiation are handled by an MEO [37].

It oversees managing the deployed MEA's lifecycle. This platform is called the Multiple-access Edge Platform Manager (MEPM). The MEP setup is the responsibility of the MEPM, which also handles DNS redirection, traffic types that must be offloaded to the MEC application, and MEC application authorization. A virtualized version of the MEPM keeps the MEP configuration while assigning the LCM of MEA to one or more VNFMs [38].

A collection of functionalities known as a MEP is needed, which may both deliver and receive multi-access edge services by running MEAs on a specific multi-access edge host virtualization infrastructure. A range of multi-access edge services is also available through the MEP [39]. An MEC application or the MEC platform itself may provide an MES via the multi-access edge platform. The location service and radio network information service are two examples of MES offered by the MEP [40].

Some of the MEC ideas have NFV analogs in terms of orchestration; for example, similar to the NFVO, the MEAO and MEO also coordinate virtual functions. These tasks are performed by MEAs in MEC [41], whereas VNFs are used in NFV. MEC offers preconfigured services for applications connected to mobile devices in addition to these associated ideas.

9.6.2 Energy-efficient radio access networks and multi-access edge computing

One of the main responsibilities of the SD-RAN controller is radio resource management control, which a collection of RRM algorithms performs. These algorithms frequently involve a scheduler, admission/congestion control, and other MAC layer-related operations [42]. The scheduler assigns physical resource blocks in LTE-like systems dependent on the underlying channel quality and the type of traffic (or QoS criterion). Many special design issues must be overcome to effectively manage radio resources for the virtualized RANs, also referred to as the edge cloud, which can virtualize several different parts of the conventional wireless stack [43]. These challenges are listed below:

- ✓ The main goals of such centralized resource control are to ensure the target 5G KPIs, take into account the mobile network operator's (MNO) need to reduce costs (OPerational EXpenditure [OPEX] and Capital EXpenditure [CAPEX]) [44], and attend to the needs of energy optimization.
- ✓ Centralized virtualization of all or parts of the wireless stack's baseband processing allows for resource pooling and right-sizing to reduce overprovisioning.
- ✓ The various solutions use SDN technology (two-tier virtualization) to overcome regional restrictions and technological gaps in both vertically tier-based and horizontally celled heterogeneous networks to deliver ubiquitous and universal network services in a variety of use cases [45]. Additional (in-depth) specifications for the construction of a centralized scheduler are offered in this part, primarily based on [46]:

    1. using general-purpose platforms (GPPs) with improvements for real-time virtualization;
    2. cloud-RANs' virtualization and resource sharing;
    3. determination of the 3GPP radio stack's effective function division across network functions (virtual or physical);
    4. the need for efficient interference mitigation strategies, particularly in the transition to "cell-less" architecture;
    5. considering the necessity to mitigate problems connected to handover, load balancing in cellular networks, and network convergence to handle the needs of 5G networks [47].

The next sections go over each of the design issues for the centralized scheduler one by one.

Application of virtualization using general-purpose platforms: It is challenging to provide real-time operation so that millisecond response times may be achieved when using GPPs for RAN task requirements for the Layer-1 function split may be met. However, given extremely tight time boundaries (lower than 100 μs), GPP can be used for other compute-intensive baseband components, such as the MAC scheduler. The scheduler and MAC must carry out a specific set of operations during each transmission time interval (TTI) [48]; however, specialized patches, such as "run to completion," which get rid of the unpredictable nature of kernel interruptions to wake up jobs during each TTI, may be required. Additionally, the DPDK environment [49] can be utilized to accelerate packet processing, by using either user-space run-to-completion or pipeline models, which is an open-source framework on Intel x86 CPUs [73]. A virtual machine's virtual CPUs can be tied to physical CPUs. Such adjustments have the effect of establishing an effectively exclusive environment for RAN operations [50].

Since cloud-RAN enables dynamic right-sizing of the processing resources based on workload [51], a mapping between virtualized entities (virtual machines, containers), cells, and CPU cores must be constructed. The possibility of dividing RAN modules into "per-user" and "per-cell" activities is particularly important to consider. Since the scheduler for a cell must consider the channel status, scheduling metrics, and available resources to make the best resource allocations, scheduling is a "per-cell" process [52]. When choosing whether to allocate resources on a per-user or per-cell basis, for example, the trade-off between the benefits of virtualization and implementation complexity should be taken into account.

A virtual passive optical network (VPON) of cloud/centralized-RAN can be reconfigured to accommodate changes in traffic volume. Resource sharing and base station (BS) cooperation are both made possible by VPON creation. By forming VPONs, the entire radio access region can be divided into several service areas. Resource sharing and BS coordination are both made possible by VPON formation. A VPON can be formed of nearby RUs and managed by a single DU [53].

Finding the effective split at the 3GPP radio stack level: The fronthaul link's latency and bandwidth requirements are largely determined by the functional splits of the radio stack, which have all been examined by NGMN [54] and in the 5G New Radio (NR) specification [55]. For example, to ensure peak throughput and enhance the effects of centralization, the split low at the physical layer (PHY), which breaks the hybrid automatic repeat request (HARQ) loop, a delay of 0.5 ms on the fronthaul link roundtrip time is required. As baseband processing (PHY) must be installed close to the remote radio head (RRH), splitting at the non-real-time Layer 2/3 (like MAC-MAC) [55] provides more scale centralization and lower latency/bandwidth requirements on the fronthaul while pooling gains are decreased. According to the ETSI MANO framework, the VNF Manager or Orchestrator should collaboratively optimize the placement of certain PHY and MAC functions. Currently, an agreement has not been reached on how the centralized baseband processing units and network resources from a radio network segment will be virtualized and how fronthaul traffic will be carried between RUs and DUs [56].

Techniques for mitigating interference: With more cell sites, there is a greater likelihood of interference as a result of poor design (more sites require more work to optimize site parameters). The Coordinated Scheduling and Dynamic Point Selection (CSDPS), the Joint Reception (JR), the enhanced ICIC (eICIC), the uplink's low-latency Joint Reception (JR), and the uplink's fast-latency Coordinated Multi-Point (CoMP) are a few techniques that need to be taken into consideration to combat interference. The performance of multi-cell coordination is significantly influenced by the bandwidth available for coordination and the latency of information flow between cells [57,58].

Distributed BSs in RANs (DRANs) lack the processing power necessary for CoMP, and backhaul lines connecting BSs to the main network experience significant signaling delays (4–15 ms); therefore, with their virtualization mechanism in place, centralized RAN (CRAN) solutions are required. CoMP offers fluid communication when a user is moving about by reorganizing dynamic clusters of RUs that can collectively communicate signals to the user. A cross-layer optimization methodology for resource allocation in SD-RAN is required by edge cell design concepts. E2E resource allocation is accomplished by this system, which allows (baseband unit) processing, transmission, and radio resources to each user [60].

Addressing issues with mobility and 5G traffic requirements: The random fluctuation in the space domain can be smoothed out by macro cells, but when cells get smaller in 5G networks, the issue of traffic load balance arises. Rapidly moving terminals cause frequent handovers and more latency is unavoidably introduced because 5G cellular networks have cell sizes reduced to tens of meters [61].

The high overhead will reduce the effectiveness of data exchange when handovers take place between various types of heterogeneous wireless networks. Enabling "zero" interference is the ultimate goal of constructing an efficient scheduler. The term "cell-less architecture" is used in the literature to describe this tendency [62]. The suggested plan permits adaptively adjusting the number of baseline schedulers for access points (BSs/APs), which is determined by the requirements of mobile terminals and the state of wireless channels in diverse environments.

Standard scheduler: Simple round-robin (designed for benchmarking) and complicated channel-aware proprietary scheduling are the two configurable scheduling algorithms that will be used to develop the baseline scheduler for an LTE eNB network. It is advised to communicate with the LTE eNB protocol stack using the Functional Application Platform Interface (FAPI), which has been modified to accommodate carrier aggregation and is compliant with the Small Cell Forum. Standardized FAPI interface usage makes it possible to collaborate with well-liked open-source protocol stacks (like OpenAirInterface) right out of the box [63].

The criteria considered by LTE eNB Scheduler when generating scheduling decisions will serve as a solid foundation for establishing a central scheduler that takes the abovementioned requirements and capabilities into account. Functionalities of LTE PHY Lab expand the 3GPP specification by including prospective 5G technologies like the Universal Filtered Multicarrier Modulation (UFMC) [64] modulator and demodulator because it is focused on 5G experiments. LTE PHY Lab [65] can be used as a foundation for in-depth 5G study in our 5G projects because of its modular software architecture, allowing for the validation and verification of any novel waveforms or algorithms that might be included in the next 3GPP releases. Many projects and tests successfully used LTE PHY Lab after successfully verifying it.

In phase 3 of the eWINE project [66], where orthogonal frequency division multiplexing (OFDM) and generalized frequency division multiplexing (GFDM) are two methods being investigated, LTE PHY Lab is one of the essential components.

9.7 Open5GS, srsRAN/srsLTE, and OpenAirInterface

srsRAN/srsLTE, OpenAirInterface, and Open5GS are all open-source software initiatives for 5G networks. Open5GS is a 3GPP Release 15-compatible open-source implementation of the 5G core network and 5G base station (gNodeB). It offers a feature-rich and adaptable platform with support for a variety of use cases and deployment scenarios for developing 5G networks. For network management and monitoring, Open5GS comes with a web-based user interface, as well as APIs for system integration [67].

An open-source software package known as srsRAN/srsLTE implements a 5G RAN and an LTE base station (eNodeB). It may be used on a variety of hardware platforms and is made to be adaptable, effective, and simple to use. Support for several features and protocols, such as carrier aggregation, beamforming, and MIMO, is provided by srsRAN/srsLTE [68].

The physical layer and protocol stack of 4G and 5G RANs are both implemented in OpenAirInterface, an open-source solution. It supports a variety of platforms and deployment scenarios and is made to be versatile and adaptable. Carrier aggregation, beamforming, and MIMO are just a few of the capabilities and technologies that are supported by OpenAirInterface [69].

Overall, these open-source software initiatives offer adaptable and strong platforms for developing 5G networks, allowing developers and operators to test new ideas and personalize their 5G deployments. Building 5G networks that are customized to certain use cases and deployment situations is achievable by utilizing these open-source technologies, and you can also gain access to the community-driven development and support that come with open-source software [70]. Of course, the following information on Open5GS, srsRAN/srsLTE, and OpenAirInterface is also included:

Open5GS: The 5G core network and 5G base station (gNodeB) are both implemented using Open5GS, an adaptable and feature-rich open-source solution. It contains support for a variety of use cases and deployment situations and is built to comply with the 3GPP Release 15 guidelines [71]. For network management and monitoring, Open5GS comes with a web-based user interface, as well as APIs for system integration. It is written in C and is compatible with Linux and FreeBSD, among other operating systems [72].

A 5G RAN and an LTE base station (eNodeB) are implemented using the open-source software package srsRAN/srsLTE. It may be used on a variety of hardware platforms and is made to be adaptable, effective, and simple to use. Support for several features and protocols, such as carrier aggregation, beamforming, and MIMO, is provided by srsRAN/srsLTE. It is developed in C++ and may be built on a variety of operating systems, including Windows, Linux, and macOS [73].

OpenAirInterface: An open-source implementation of an RAN for 4G and 5G, OpenAirInterface includes both the physical layer and the protocol stack. It supports a variety of platforms and deployment scenarios and is made to be versatile and adaptable. Carrier aggregation, beamforming, and MIMO are just a few of the capabilities and technologies that are supported by OpenAirInterface [74]. It is written in C and is compatible with Linux and macOS, among other operating systems.

Open-source 5G software advantages: The flexibility to modify and adapt the software to particular use cases and deployment circumstances is one of the key advantages of adopting open-source 5G software [75]. Because it is frequently offered for free or at a reduced price, open-source software can also be more affordable than proprietary software. Additionally, community-driven development and support for open-source software are advantageous since they can result in quicker development, more frequent updates, and greater quality control [76].

Using open-source 5G software has some drawbacks. The need for specialized knowledge and resources to build, implement, and maintain the software is one of the key obstacles to using open-source 5G software. Furthermore, open-source software could not be as well supported or popular as proprietary software, which might limit its use in some situations [77].

srsRAN/srsLTE, OpenAirInterface, and Open5GS are all effective open-source software tools for developing 5G networks. These technologies allow for the customization and tailoring of 5G networks to certain use cases and deployment scenarios while also gaining access to the open-source software's community-driven development and support.

9.8 Low-power communication, massive machine type communication (mMTC), and IoT

The development of extensive, low-power, and economically viable IoT networks depends on the interconnected ideas of low-power communication and mMTC, along with IoT [78]. The usage of wireless communication methods is known as low-power communication, which utilizes a very small amount of power, allowing devices to run for extended periods on a single battery or other power source. Technologies for low-power communication include Zigbee, LoRaWAN, and Bluetooth Low Energy (BLE) [79].

mMTC describes an IoT network's capacity to accommodate a high number of devices that communicate sparse data at low data rates. The development of large-scale IoT networks that can support a variety of applications, from smart homes to industrial automation, depends on mMTC [80].

The network of networked objects known as IoT is equipped with sensors, electronics, and software, allowing it to collect and exchange data. To construct large-scale, low-power, and affordable IoT networks, low-power communication technologies and support for mMTC are needed [81].

Low-power communication, mMTC, and IoT work together to make it possible to construct large-scale, low-power, and affordable IoT networks that can support a variety of applications. It is feasible to lower the power requirements of IoT devices, increase their battery life, and enable the deployment of several devices in a single network by utilizing low-power communication methods and enabling mMTC. This can make data collection and exchange easier for a variety of applications, including smart cities, agribusiness, industrial automation, and smart homes [8][82].

For instance, in a smart home, smart devices like lights, thermostats, and security systems can be connected and controlled using low-power communication protocols like Zigbee or BLE. These devices can operate for extended periods on a single battery because they can transfer small amounts of data at low data rates [83].

To connect industrial sensors and transfer data over long distances, low-power communication technologies like LoRaWAN can be utilized. This allows the implementation of IoT networks in huge factories or outdoor settings [84].

The deployment of expansive IoT networks for industrial automation is made possible using mMTC, which can support a high number of sensors and devices in a single network. In general, mMTC, IoT, and low-power communication are crucial elements of the contemporary linked world, enabling the construction of extensive, low-power, and affordable IoT networks that may serve a variety of applications [85].

Technologies for low-power communication: Low-power communication technologies are created so that devices can run for extended periods using just one battery or other power source. These technologies include Narrowband IoT (NB-IoT), Zigbee, LoRaWAN, and BLE. Smart homes, industrial automation, and agriculture are just a few of the applications that can make use of low-power communication technology [86].

mMTC: The ability of IoT networks to support numerous devices that send sparse amounts of data at low data rates is referred to as mMTC. A large-scale IoT network's ability to install and support a variety of applications depends on mMTC. Long-range (LoRa) communication and NB-IoT technologies support mMTC [87].

The IoT is a network of networked objects that have sensors, electronics, and software built into them so they can gather and exchange data. IoT gadgets can range from wearables and smart thermostats to industrial sensors and self-driving cars. To construct large-scale, low-power [88], and affordable IoT networks, low-power communication technologies and support for mMTC are needed.

Low-power communication, mMTC, and IoT applications include smart homes, cities, agriculture, logistics, and healthcare and industrial automation. These applications all depend on these technologies [89]. Smart gadgets like lights, thermostats, and security systems can be connected to and controlled by smart houses using low-power communication technology. A significant number of smart devices can be installed in a smart home thanks to the usage of mMTC, which supports numerous devices on a single network.

Precision irrigation and fertilization in agriculture are made possible using IoT sensors to monitor soil moisture, temperature, and other environmental conditions. IoT networks can be deployed in large-scale agriculture thanks to low-power communication technologies like LoRaWAN that can carry data over vast distances [90].

IoT devices can be utilized in industrial automation to monitor and manage production processes, enabling in-the-moment data collection and analysis. Industrial sensor connections and data transmission via cellular networks are made possible by low-power communication technologies like NB-IoT, allowing the implementation of IoT networks in huge factories or open spaces [91].

IoT devices can be utilized in the healthcare industry to manage medication adherence and monitor patient vital signs. Wearable devices can be linked to smartphones or other devices using low-power communication technologies like BLE, enabling remote monitoring and data collection [92].

In general, mMTC, IoT, and low-power communication technologies work together to enable the implementation of large-scale, low-power, and affordable IoT networks that can support a variety of applications in many industries. These technologies make it possible to increase the battery life of IoT devices, decrease their power consumption, and allow for the deployment of many devices in a single network [93].

This can make it easier to gather and share data for use in a variety of applications, resulting in increased effectiveness, cost savings, and better decision-making. Additionally, improvements in machine learning and AI can aid in the analysis and interpretation of the enormous amounts of data produced by IoT networks, providing fresh perspectives and opening new avenues for innovation [94].

MTC is a third term to consider since M2M and IoT have already been defined. You may think of this as the all-encompassing phrase that includes the other two phrases. It defines, generally, how two or more robots can speak to one another without coming into direct contact with a person [95]. No paradigm relating to the devices' media or communication mechanisms is mentioned in this term. The following will concentrate on M2M communication through the Internet, specifically communication with the cloud. IoT is the term used when such devices are connected in huge numbers to the cloud.

There are numerous potential applications for IoT hardware and software. These domains are the most obvious to most people since they start in the region of every person's house, where technologies like home automation or smart buildings are used. Applications could include using a smart doorbell or fridge as well as highly automated home installations and building management systems for larger structures. Other possible IoT application scenarios include the management of automated manufacturing processes that enable speedier and more efficient operations by tracking all parts and process steps to enhance workflow and resource allocation.

Countless further applications for IoT might be thought of. Since the requirements for these various scenarios vary, various groupings of such settings are outlined in the following section.

9.8.1 Various IoT application types

The first category is the collection of mobile broadband applications for IoT that are utilized by everyone. Then there is IoT for industrial automation, which applies the available services via wireless communication by integrating with the already existing Ethernet infrastructure of the automation. Additionally, IoT for industrial automation needs to have the same characteristics as critical IoT, which are outlined in Section 9.8.2.

The first two categories are the least common; in contrast, big IoT and critical IoT are more common and well known. The emphasis will be on critical and enormous IoT in the next chapters because industrial automation IoT is fundamentally a subclass of crucial IoT, and broadband IoT does not have needs that are noticeably different from mobile broadband applications. Applications in the massive IoT category make use of millions or even billions of connected devices.

These devices must be affordable to enable these large constellations. Concerning connectivity, they update their data sporadically, with little significance and low priority. These highly scalable systems aim to provide low-cost, energy-efficient devices that can be disseminated over a wide area without the need for a power source, which may be a town or a whole country. Such systems include, for instance, those that monitor the land or the climate for research or agricultural purposes. To get a better overview of the process and identify bottlenecks, a shipping environment for parcels is another conceivable scenario [96].

In this environment, all components and processes are followed and tracked. However, in the second major group of critical IoT, there are applications in traffic control and healthcare that have different requirements, such as a need for much higher availability and reliability as well as very safe and secure communication with guaranteed low and bounded latency.

These requirements are in place in this industry to prevent failures, especially those that could result in a dangerous situation or potentially damage people.

9.8.2 Requirements

The above discussion covered the special criteria for huge and crucial IoT. Additionally, there are six universal requirements for IoT applications that will be described below.

Low-cost devices: The operational margin is low owing to a single device's minimal complexity and cost, but the cost of a single device must be low to deploy the devices in a large number.

Low network cost: Low network and communication costs are necessary to enable a large-scale system [97].

Long battery life: This is necessary due to the substantial number of devices and the dispersal-based deployment strategy, which involves scattering the devices.

Extended coverage: Because many scenarios take place indoors, the network's coverage must also account for indoor use.

Scalability: The new network must be able to support all connected devices once to handle a large installation of devices.

Security and privacy: Protecting a single IoT user is just one illustration of how security and privacy should be a priority for IoT systems [98].

9.8.3 Current IoT situations

The decision made about the communication technology and network to be employed will affect all of the aforementioned needs. Many protocols built on the various technologies already in use have been established in recent years. Fig. 9.5 contrasts the various methods based on the features of the putative peak rate (vertical axis) and the maximum range (horizontal axis). In the diagram, a logarithmic scale is employed. Five primary sections of the graph can be distinguished, primarily by their maximum range.

These so-called short-range and medium-range wireless networks are used mostly in local installations like single-family homes or factories and are limited to an area of roughly 5 km^2. Multiple technologies in this field call for the consideration of var-

Massive IoT	Broadband IoT	Industrial Automation IoT	Critical IoT
Low-cost gadgets Miniature data volumes Broad coverage	Greater data rates Large quantities of data Little latency	Integration of the Ethernet protocol Service for synchronizing clocks The network that reacts quickly	Extremely dependable data delivery Extremely low latency Constraint latencies
Device battery life, network exposure, network data analysis, network slicing, and device positioning			

FIGURE 9.5

Categorization of all potential IoT usage situations into four key usage groups.

ious factors. First, there are WiFi networks, which can achieve relatively high peak data speeds, but they also use a great deal of energy when operating and communicating. Other alternative technologies, such as ZigBee or Z-Wave, that achieve this utilizing other ideas and methodologies exist at much lower data rates in this industry [99].

They are currently largely utilized in smart home systems, as their primary value is that they communicate while consuming a negligible amount of energy. All of these technologies' shortcomings in this initial sector include their inability to, without a lot of work and expense, cover anything greater than a single residence. There are satellite networks on the other side of the range with a range of more than 100 km. The key issue is the high expense of setting up such a network because it takes a lot of money to conceive, build, and launch a satellite into orbit. The distance presents yet another issue but also the main benefit [100].

The network can be accessed from a large area, but on the other side, the signal takes a very long time to reach the satellite from the device. There are still two other alternatives in addition to the local network, which may include a mesh network and the worldwide satellite network. The first is the cellular network, which contains the existing 4G, 3G, and 2G standards as well as the currently underdeveloped 5G standard, which allows for the use of high bandwidth across a wide region [101]. Low-power wide area networks (LPWANs) are used in wide area networks with lesser bandwidth as an alternative to high peak data rates. Examples of this are LoRaWAN or SigFox to consume less power, which combines a wide maximum range with a narrow bandwidth technique. Each of these technologies has advantages and disadvantages. All of these approaches have the flaw that they are incompatible with one another and frequently rely on separate protocols and technology [102]. A specialized middleware architecture for interoperability that facilitates switching between them or even permits the use of multiple technologies at once to improve the availability and utilization of these networks is also lacking at the software layer.

9.8.4 5G structure

Varied use cases have varied requirements; therefore, not every scenario will require all the ones previously illustrated. There is a wide range of potential applications for this, including mining, factory automation, and public safety, so countless distinct configurations are conceivable [103]. These various settings must deal with various parameters. For instance, application characteristics in the area of farming and agriculture are very different from those found in the area of healthcare and well-being. For instance, in a healthcare setting, equipment must be focused on safety and security, yet in an agricultural setting, communication must be feasible in extremely remote locations.

The 5G standard has been split into three primary capability areas to better accommodate the specifics of each segment while focusing on specific usage situations [104]. The first is enhanced mobile broadband (eMBB), which enables applications that need a lot of bandwidth, a lot of users, and a small amount of space – like they would in a densely populated city center. It also allows for applications that demand a lot of mobility. mMTC, which enables M2M communication, for example, by enabling a sizable number of linked devices to communicate at low data rates and with little energy consumption, is the second method. The third feature, ultrareliable and low-latency communication (URLLC), is made "for safety-critical and mission-critical applications" [105]. The last two are mostly used for M2M communication and do not include human contact like IoT. The eight crucial factors listed below apply to mobile 5G telecommunication [106]:

- ✓ The highest data rate that is feasible is called the peak data rate.
- ✓ The "user-experienced data rate" is the maximum data rate a user can attain with their devices in a practical setting.
- ✓ The data rate with the bandwidth is known as spectrum efficiency.
- ✓ Mobility refers to the pace at which mobile devices may still switch between cells.
- ✓ The duration of data transmission between two end points is known as latency.
- ✓ The number of linked devices per area is known as the connection density.
- ✓ The effectiveness of a network's energy use in comparison to that of other networks is known as network energy efficiency.
- ✓ Area traffic capacity determines the pace of traffic in each area.

Fig. 9.6 illustrates how these eight characteristics should be prioritized among the three previously indicated competencies. The data rate, mobility, and area traffic capacity are the three areas where eMBB places the highest priority, although latency is less important because most applications involve human interaction, and connection density is less important because few devices demand high bandwidth [107]. However, all other issues are very small in comparison to latency and mobility within URLLC due to its crucial use. Due to the requirements of large IoT setups, mMTC also emphasizes the necessity for high connection density and medium energy efficiency [108].

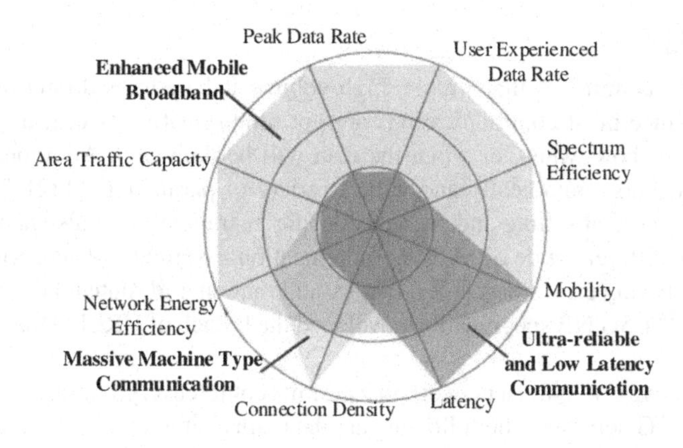

FIGURE 9.6

Characterization of the three components of the new 5G standard based on key mobile telecommunication characteristics.

9.8.5 mMTC's specialties

The following will put less emphasis on eMBB and more on mMTC and URLLC because the theme is M2M communication. IoT has become a hot topic in all communication disciplines due to its possible application scenarios, so the new cellular network standard of 5G was required to combine communication techniques to enable IoT communication. The main use cases for mMTC are "where a large number of machine-type devices deployed in a wide area communicate sporadically without specific latency requirements, ignoring reliability guarantees" [109]. This is the 5G new core network setup's first focus on machine communication.

Nevertheless, it allows for cheap production and operating costs over its 10-year operational lifespan. LPWANs were already described in Section 9.13 on an equal footing with the connectivity of IoT devices using cellular networks. This technique, which mMTC is now introducing as a part of 5G, is specifically made to deal with the needs of these IoT applications while allowing for continued use of the 5G network. The following are the main points the mMTC does offer [110]:

- ✓ long battery life,
- ✓ 10^6 devices per square mile, and
- ✓ widespread coverage.

These specifications were developed to support IoT setups that have many devices dispersed across a wide region. Additionally, sockets cannot be used to connect devices of this size and area. As a result, these battery-operated gadgets must be durable to considerably lower maintenance expenses [111].

9.8.6 URLLC

The mMTC recommends that simple, high-volume IoT systems do not require any warranties, since most communications are not time-sensitive or crucial, making it irrelevant as to how, when, or when the data will be delivered. Additionally, communication occurs sporadically and with a variety of parameters [112]. To address problems in a crucial setting and provide specific guarantees, it is also necessary for a completely different style of M2M communication. Examples of scenarios for this include self-driving cars, smart cities, factory automation, rail route management, and many more. The 5G NR specification developed the branch of URLLC for diverse use cases.

The objective is to build a network that permits complete cryptographic communication on the 5G network. The following are the main characteristics for critical communication in 5G NR using URLLC: low latency (guaranteed 1 ms) with >99.999% availability [113]. The utilization situations called for these numbers. For instance, self-driving cars need to be able to communicate with one another quickly and have a highly available network so that they may, for example, avert an accident or contact a remote driver in the event of a breakdown.

URLLC also permits the creation of non-public networks (NPNs) that are designed for a single task, such as the creation of an entire factory in the field of industrial automation IoT that automates the entire production and requires several conditions to ensure time-sensitive communication as well as safety and security for all participants [114].

9.9 5G edge intelligence features

The ability of 5G networks to provide edge computing and AI capabilities at the network edge is referred to as 5G edge intelligence features. Examples of 5G edge intelligence features include the following:

Edge processing: Edge computing, which entails processing and analyzing data at the network edge, is a capability of 5G networks. Numerous applications, such as automated manufacturing, smart cities, and driverless vehicles, can benefit from edge computing [115].

Edge devices with AI capabilities: Edge devices having AI capabilities, such as machine learning or computer vision, can be supported by 5G networks. At the network edge, these devices can collect and analyze data, enabling automation and decision-making in real time [116]. Wearables, driverless vehicles, and smart homes are just a few examples of the many applications that AI-enabled edge devices can be employed in.

Network slicing: The ability to create virtual networks tailored to particular use cases or applications is supported by 5G networks. As a result, it may be possible to deploy customized edge computing and AI capabilities for certain applications or industries, such as industrial automation or healthcare [117].

Multi-access edge computing: MEC involves placing computing assets and services closer to end users and devices at the network edge, which is supported by 5G networks. Applications like VR and AR, gaming, and video streaming can all benefit from MEC's low latency and high bandwidth capabilities [118].

5G networks can lower latency and boost network performance for these applications by introducing MEC capabilities at the network's edge.

Network automation: The use of AI and machine learning techniques to automate network administration and operations is supported by 5G networks. This can make network configuration, upkeep, and troubleshooting quicker and more effective, lowering the need for manual intervention and enhancing service quality.

Overall, 5G edge intelligence features enable the deployment of fresh and cutting-edge applications with quick response times, high bandwidth, and low latency [119].

5G networks can offer a platform for innovation and development in a variety of industries, including healthcare, transportation, and manufacturing, by supporting edge computing, AI-enabled edge devices, network slicing, MEC, and network automation. It is crucial to keep in mind that putting these features into practice might necessitate specialized knowledge and resources, such as edge computing infrastructure and AI, as well as careful consideration of data privacy and security issues [120].

A distributed computing paradigm called "edge computing" puts computation, storage, and application services closer to the device or end user. Edge computing enables new applications and services while reducing latency and enhancing network performance by allowing data to be processed and analyzed in real time at the network edge. For instance, edge computing can be utilized in a smart city to interpret data from multiple IoT sensors and devices to manage energy usage or improve traffic flow. Edge devices having AI capabilities, such as computer vision or machine learning algorithms, are referred to as AI-enabled edge devices [121].

At the network edge, these devices can collect and analyze data, enabling automation and decision-making in real time. Wearables, driverless vehicles, and smart homes are just a few examples of the many applications that AI-enabled edge devices can be employed in. An AI-enabled thermostat in a smart home, for instance, can learn user preferences and modify temperature settings accordingly [122].

Network slicing: This 5G technology enables the construction of virtual networks that are specialized for particular use cases or applications. Every network slice can be tailored to focus on traits like latency, bandwidth, security, and dependability. As a result, it is possible to deploy specialized edge computing and AI capabilities for certain applications or sectors [123]. For instance, a network slice can be developed in the healthcare industry to facilitate remote monitoring and diagnostics, together with specific edge devices and AI algorithms for examining patient data. MEC is a 5G feature that entails placing computing tools and services closer to end users and devices, at the network edge.

Applications like VR and AR, gaming, and video streaming can all be made possible with MEC. 5G networks can lower latency and boost network performance for these applications by introducing MEC capabilities at the network's edge. MEC can be used, for instance, in gaming applications to offload computationally heavy operations to the network edge, such as rendering and physics calculations [124]. Network automation: Automating network administration and operations entails the use of AI and machine learning algorithms. This can make network configuration, upkeep, and troubleshooting quicker and more effective, lowering the need for manual intervention and enhancing service quality. For instance, real-time network fault detection and diagnosis, as well as automated network reconfiguration for service continuity, are all possible using AI algorithms [125].

Overall, 5G edge intelligence skills are crucial for enabling the deployment of novel and cutting-edge applications that demand high bandwidth, low latency, and real-time decision-making abilities. 5G networks can offer a platform for innovation and development in a variety of industries, including healthcare, transportation, and manufacturing, by supporting edge computing, AI-enabled edge devices, network slicing, MEC, and network automation. It is crucial to remember that putting these features into practice might call for specific knowledge, equipment, and infrastructure, such as edge computing infrastructure and AI, as well as careful consideration of data privacy and security issues [126]. To guarantee interoperability and compatibility amongst various 5G ecosystem components, network operators, device manufacturers, and application developers must cooperate closely together to deploy 5G edge intelligence features.

9.10 Machine learning modeling for 5G

Service management using zero-touch (ZSM): ZSM is a new horizontal and vertical E2E architecture framework designed for closed-loop automation and improved for machine learning and AI algorithms that are data-driven, and it is referred to by the ETSI [127] specification. The goal is to create largely autonomous networks that are guided by high-level policies and norms while also being able to self-configure, self-monitor, self-heal, and optimize without further human intervention [128].

The 5G Core Network function known as network data analytics services is provided by the Network Data Analytics Function (NWDAF) [129]. Data collection from Network Functions, Application Functions, or OAM, service registration and metadata exposure to Network Functions and Application Functions, the provision of analytics information, and support for machine learning model training are just a few of the functionalities that the NWDAF includes. The NWDAF functions are described in [130] in detail.

Management data analytics (MDA): MDA provides the capacity to process and analyze the raw data about network and service events and status (e.g., performance measurements, trace reports, quality of experience reports, alarms, configuration data, network analytics data, service experience data from Application Functions, etc.) to provide analytics reports, including recommended actions, to enable the necessary actions for network and service operations [131].

The AI/Machine Learning Platform (AIMLP) is an architectural component that manages the lifespan of AI/machine learning (AIML) models and makes them available to other elements of the network architecture via clearly defined interfaces [132].

AIML-as-a-service (AIMLaaS): AIMLaaS provides AIML capabilities through clearly defined APIs to any network object or service that might require them, such as to efficiently utilize resources. Offline learning with batch processing of a predetermined dataset is the basis for the training and evaluation of machine learning models. Inference (predictions) can then be made on the spot using the pretrained machine learning model [133].

Real-time/online training: An online learning-based approach to training and evaluating a machine learning model (also known as incremental learning). Machine learning models are tested (inference) on-the-fly with fresh data while being trained in real time. In self-managing systems (nodes, networks, and/or systems), the process of developing (acquiring and sustaining) knowledge of the operational environment is known as self-awareness. Automated configuration of systems (nodes, networks, and/or services), as well as resource adaptation to constantly changing environmental conditions, is known as self-configuration [134].

Processes in nodes, networks, and/or services that enable problem identification through fault detection, diagnosis, and launching necessary actions to prevent interruptions are collectively referred to as self-healing. Self-management is a process by which systems (nodes, networks, and/or services) continuously attempt to optimize their load and resource utilization to boost their effectiveness and performance [135].

Machine learning can be applied in a variety of ways to model and improve 5G networks, for example by optimizing a network by enhancing network performance and lowering latency or by enhancing the architectures and configurations of 5G networks. Based on real-time network data, machine learning algorithms can be used to optimize network parameters like handover thresholds, packet size, and resource allocation.

Predictive maintenance: Based on information from sensors and other devices, machine learning algorithms can be used to forecast equipment failures and maintenance requirements in 5G networks. As a result, preventive maintenance may be possible, lowering downtime and enhancing network dependability [136]. Machine learning algorithms can be used to forecast traffic patterns and manage network resources accordingly.

Traffic prediction and management: Algorithms based on machine learning, for instance, can be used to forecast network congestion and dynamically distribute network resources to improve network performance [137].

Security: In 5G networks, machine learning algorithms can be used to identify and counteract security threats including malware identification and denial-of-service assaults. Additionally, machine learning algorithms can be used to spot unusual network traffic patterns that might be signs of security flaws or other anomalies [138].

Network slicing: Machine learning algorithms can be applied to network slicing to improve it and make it possible to build virtual networks that are suited to particular use cases or applications. Based on elements including network traffic, user behavior, and application needs, machine learning algorithms can be used to optimize network slicing [139].

Overall, network performance, maintenance, traffic management, security, and network slicing are just a few of the different facets of 5G networks that may be modeled and optimized using machine learning. The massive volumes of data produced by 5G networks may be utilized by machine learning algorithms to enable new levels of automation, efficiency, and optimization, improving network performance, reducing costs, and improving user experience. It is crucial to remember that machine learning algorithms may need specific knowledge and resources to construct and maintain, in addition to a sizeable amount of data to train them properly [140]. As a result, individual use cases and applications for which machine learning is being employed in 5G networks should be carefully considered.

IoT network data can be analyzed in a variety of ways using AI and machine learning. Here are a few instances:

Prevention-based maintenance IoT devices: Prevention-based maintenance IoT devices can be used to track and gather information about the condition and operation of equipment, such as commercial HVAC systems or industrial machines. Based on these data, machine learning algorithms can be trained to find patterns and anomalies that point to potential failures or maintenance requirements. By servicing or repairing equipment before it breaks down, predictive maintenance can be made possible, cutting down on both maintenance costs and downtime [141].

Finding anomalies: IoT networks produce a ton of data, including logs, sensor data, and other operational data. To find abnormalities or trends in these data that can point to security breaches, equipment problems, or other problems, machine learning methods can be applied. This can help in the early identification and correction of any issues before they worsen.

Optimization: Data from IoT networks can be analyzed using machine learning algorithms to spot chances for optimization and productivity gains. For instance, in a smart city, traffic flow data can be analyzed using machine learning algorithms to improve the timing of traffic signals, lowering congestion, and enhancing traffic flow.

Personalization: IoT devices, such as wearables or smart homes, can gather information on user behavior and preferences. These data can be analyzed by machine learning algorithms to customize the user experience, such as modifying temperature settings or recommending unique fitness regimens [142].

Analysis of images and videos: IoT devices like security cameras and drones can produce a lot of image and video data. To identify intrusive individuals or recognize facial expressions, machine learning algorithms can be employed to examine these data and detect things, persons, or events of interest. This can enable automatic responses to potential security risks or other events as well as real-time decision-making.

In general, IoT network data may be analyzed using AI and machine learning to produce insights, spot trends, and anomalies, streamline operations, and enhance decision-making. Machine learning algorithms can enable new levels of automation, personalization, and efficiency by utilizing the massive volumes of data produced by IoT networks [143]. This will increase performance, result in cost savings, and improve user experiences.

9.11 TinyML for edge intelligence in 5G

TinyML is a machine learning tool or technique that can conduct on-device analyses for several sensory modalities, including vision, audio, and voice. TinyML has a very low power/energy consumption, making it appropriate for embedded, battery-powered systems. In the context of the IoT network framework, TinyML can also be applied for large-scale applications [144]. Cloud-enabled machine learning systems now face a variety of challenges, such as excessive power consumption and problems with security, privacy, dependability, and latency. Preinstalled models are therefore now implemented on hardware–software systems (such as edge impulse) [145].

Sensors collect raw data, simulating the physical world, which are then processed by a CPU or microprocessor unit (MPU). The MPU supports the machine learning-aware analytic support made possible by edge-aware machine learning networks. Keep in mind that edge machine learning exchanges knowledge with any distant cloud machine learning through communication. The physical world will become substantially more intelligent than it is now because of the integration of TinyML into systems [146].

A system like this can aid edge devices in making important decisions even without help from edge AI or cloud AI. Notably, the system's performance may increase on many fronts, including delay, effective data privacy, and energy economy. Overall, TinyML is envisioned as a combination of algorithms, hardware, and software. Hardware-wise, IoT devices, which might or might not include hardware accelerators, might need analog and memory computing to offer a useful educational experience. Software-wise, TinyML applications can be run on a variety of platforms, including Linux/embedded Linux, or over cloud-enabled software [147].

Finally, to prevent excessive memory usage, new methods for tiny machine learning systems must be developed. Overall, TinyML systems need to optimize machine learning with a small software design when working with high-quality data. Then, using binary files produced by the models that were learned on a larger machine, these data must be flashed [148]. Additionally, for TinyML implementation to provide low power consumption, compact software is needed. As a result, systems made possible by TinyML must work within strict limitations while still offering great accuracy. To support its functioning and/or enable battery-operated embedded edge devices, TinyML may frequently rely on energy harvesting at the edge devices [149].

Fundamental criteria for TinyML include (a) scaling to billions of inexpensive embedded devices and (b) storing codes in a finite number of kBs on device RAM. Finally, it has been shown that TinyML 4 may also be utilized within a conventional pipeline over the edge, which can be modified as needed by cross-section data [150]. Numerous sectors of the economy, specialized developers, and research organizations are working to improve TinyML frameworks. Except for the AlfES and NanoEdge

AI Studio architectures created by Fraunhofer IMS and Cartesiam, many of the architectures are publicly accessible [151].

A few of these frameworks also support Arduino and Raspberry Pi; however, most of them support the ARM Cortex group. Other hardware groups supported by these frameworks include the ESP8266 and ESP32 groups. It is clear from these architectures that C and C++ are the most often used languages, but the frameworks also support a variety of additional libraries, like TensorFlow Lite and TensorFlow.

Several frameworks are now being launched by numerous research organizations across the globe in preparation for the implementation of TinyML models on devices with limited resources. To decrease delay and enhance privacy, the work in [152] focuses on the deployment of TinyML models over edge devices. For use in IoT-enabled CPUs, the authors have presented a parallel ultralow-power (PULP) architecture. Non-neural machine learning kernels, which outperform neural networks in terms of accuracy, can be used with PULP. It has been shown that PULP-OPEN hardware is 12.87 times faster than an ARM Cortex-M4 MCU. Fast Artificial Neural Network (FANN)-on-MCU, a PULP-based open-source toolkit, was created to reduce energy usage [153].

The suggested toolkit can operate the architecture, as implemented in the InfiniWolf prototype, over light IoT-aware devices thanks to the FANN library. The authors show that when the RISCV octacore processor is compared to FANN-on-MCU, FANN-on-MCU is 22 times faster and uses 69% less energy. The creators of [154] have introduced the hls4ml TinyML architecture to cut down on energy use. The suggested framework makes it possible and simple to accelerate the development of machine learning-aware FPGA and ASIC implementation, and it offers Python-based APIs to take advantage of the framework's scientific advantages. For low-power embedded systems, it additionally offers quantization and pruning-aware training.

In [155], PhiNets is described, a scalable DNN backbone architecture intended to enable image processing applications for resource-constrained edge IoT devices. To decouple cost, memory, and overprocessing, the suggested structure is built on inverted residual blocks. The findings showed that, when compared to conventional architectures, PhiNets reduces the number of parameters by 85% to 90%. The HANNAH framework, which attempts to automate co-optimization processes of a neural network framework for effective E2E DNN training and application over edge devices, is described in [156] to address the hardware/software co-design.

Three steps are taken to implement HANNAH using the Ultra-Trail NN unit. According to [157], each sector has its own software and machine learning model to enable TinyML apps to run on embedded system boards. With TinyML, though, there might be serious problems. TinyML is a method for using incredibly few computer programs for jobs like motion and voice recognition. TinyML will be difficult to use across several devices without sacrificing accuracy, as each sector has its own software. Therefore, offering a standardized method for adopting TinyML will be essential. This calls for the creation of a universal framework that can function on a variety of hardware produced by various companies [158].

Once successful, this will guarantee that TinyML can offer a variety of everyday applications. From the foregoing, it can be concluded that hardware platforms represent the main restrictions for TinyML's high performance among the many resource constraints, including dataset creation, execution time, etc. Therefore, it is necessary to promote the creation and training of any TinyML model using software, libraries, or frameworks and then deploy the taught model on the enabling hardware platform. We also include a list of the software and libraries that can be integrated with the relevant hardware platforms to enable a variety of additional applications and use cases [159].

According to the results of our poll, roughly (i) 59% of participants use TensorFlow, (ii) 17% use PyTorch, and (iii) 24% utilize other frameworks. Furthermore, vision, motion, and sound data are the top three data types used by machine learning users. In terms of hardware boards, (i) Raspberry Pi, (ii) Arduino Nano 33 BLE Sense, (iii) ESP32, and (iv) Raspberry Pi Pico and NVIDIA Jetson Nano are the most frequently utilized for creating TinyML projects [160]. Applications requiring low-latency processing and effective use of computational resources could be made possible by the combination of TinyML and 5G networks. Here are a few instances of how TinyML can be applied to 5G networks [161]:

Modern cities: Thousands of sensors may be placed throughout a smart city to gather information on air quality, traffic movement, and other factors. It is feasible to examine the data locally and spot trends or anomalies in real time by deploying TinyML models on these sensors. This can assist city planners in streamlining traffic, lowering air pollution, and enhancing city life in general [162].

Autonomous vehicles: To make decisions regarding steering, braking, and accelerating, autonomous vehicles need to process sensor data in real time. TinyML models can be installed on the vehicles themselves, enabling local data analysis and real-time decision-making. This can increase the effectiveness and safety of autonomous vehicles while lowering the volume of data that must be sent to the cloud for processing [163].

Industrial automation: Hundreds or thousands of sensors may be watching equipment and procedures in a factory or other industrial setting. It is feasible to spot anomalies or foresee problems by deploying TinyML models on these sensors. This can increase the general effectiveness of industrial processes while lowering maintenance costs and downtime.

Healthcare: In a healthcare setting, sensors may be used to collect information on patient behavior or to monitor patients' vital signs. It is feasible to examine the data locally and spot trends or anomalies in real time by deploying TinyML models on these sensors. This can enhance patient outcomes by enabling healthcare professionals to make better decisions about patient care [164].

TinyML and 5G networks collectively have the potential to enable a variety of applications that call for low-latency processing and effective use of computational resources. It is possible to decrease network latency, enhance reaction times, and minimize the volume of data that must be sent to the cloud for processing by implementing machine learning models on edge devices. As a result, resources may be

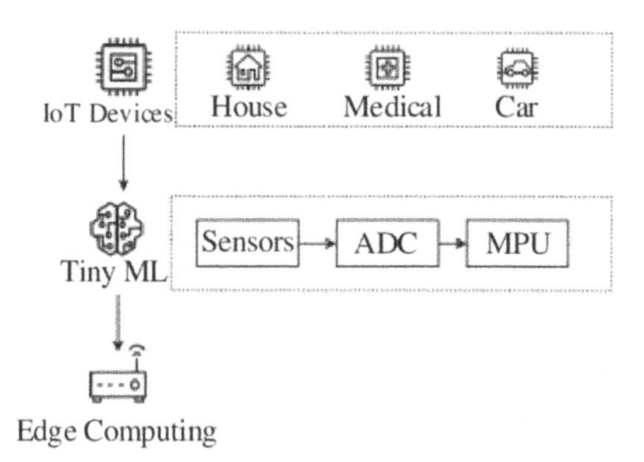

FIGURE 9.7

Edge computing with TinyML and edge devices.

used more effectively and efficiently, and previously impractical new and creative uses may now be made available [165].

TinyML has a variety of uses, such as applications for voice and vision, autonomous vehicles, phenomics, ecosystem monitoring, edge computing, brain–computer interface (BCI), data pattern categorization and compression, and health diagnosis. The innovative applications of TinyML that utilize various innovative technologies are discussed in this section [166].

A. Speech-based applications
- Speech communications
- A hearing aid (HA)

B. Applications based on vision
C. Data pattern classification and compression
D. Medical diagnosis
E. Edge computing

Setting up edge devices is urgently needed to lessen the burden on the cloud given the enormous growth in IoT devices connecting to the global network. These edge devices contain independent, high-performance data centers, resulting in increased security and decreased reliance on the cloud, latency, and bandwidth. The power, memory, and processing time limits as depicted in Fig. 9.7 will be met by the edge devices enhanced with TinyML algorithms. The authors of [167] described the energy efficiency issues that arose during the actual use of edge unmanned aerial vehicle (UAV) equipment. The MCU, which serves as the host controller for the UAV, is interfaced with TinyML to construct an energy-efficient device with minimal latency.

Sensors for data collection are required for a variety of edge computing tasks. For the data collection process during edge computing, it is common practice to use edge sensors including blood pressure sensors, accelerometers, glucose sensors, electrocardiogram (ECG) sensors, motion sensors, and electroencephalogram (EEG) sensors [168].

F. BCI
G. Autonomous vehicles
H. Phenomics and ecological preservation
I. Anomaly detection
J. Predictive maintenance

9.11.1 Challenges and prospects for research

We outline the many difficulties and problems facing research on the numerous TinyML applications in this section. We specifically summarize the main ideas from the in-depth discussions throughout the chapter that relate to TinyML. To give scholars future opportunities to contribute to finding solutions to the many problems and obstacles, we also outline potential solutions. Difficulties major obstacles that TinyML faces prevent growth. The following are the main obstacles:

- As of right now, battery power consumption for embedded edge IoT devices is anticipated to be over 10 years [169]. For instance, given that the power consumption is less than 12 Watt, a battery with a capacity of 2 Ah should have a lifespan of more than 10 years under ideal circumstances. However, the total current consumption of a straightforward edge IoT device's circuit, which combines an MCU, a temperature sensor, and a WiFi module, is about 176.4 mA [170]. The 2 Ah battery's life cycle will inevitably be shortened to about 11 h and 20 min. The TinyML ecosystem faces significant difficulties in this area. Most edge devices run at clock speeds between 10 and 1000 MHz, which limits their ability to execute complicated learning models efficiently at the edge [171].

There is also a memory restriction. Current TinyML edge devices function with onboard flash memory that is less than 1 MB [172]. This limits the performance of the models and poses a serious problem considering the MCU's accommodation.

- Even when the cost per device is minimal, the cost of large-scale deployments can be substantial. Therefore, financial concerns must be resolved for low-cost edge platforms to succeed [173].

A. Lessons discovered

Through our thorough survey, we were able to pinpoint the main problem with TinyML research – low power availability in edge devices and the need for TinyML system designs that are energy-efficient. To properly allocate energy to machine learning-related tasks, effective energy harvesting strategies must be created for powering smart devices [174]. Another issue impeding TinyML's expansion is memory constraints. Therefore, research for TinyML systems must concentrate on edge hardware with a small memory footprint. Much attention must be given to finding the best answer to difficulties relating to clock speeds to resolve processor capacity problems.

Additionally, the limited CPU capability of MCUs makes it challenging to execute complex machine learning algorithms on them. When considering the adoption of a particular learning mechanism and deployment approach, the variability in the hardware/software infrastructures is a significant challenge for TinyML systems. Additionally, the edge computing industry is still in its infancy, making it impossible to adapt resources that fluctuate dynamically within edge devices [175]. So, when deploying machine learning models, it will be necessary to take device mobility and reliability into account [176].

The main reliability problems will be variances in the process, hard and soft faults, and aging. Therefore, it will be crucial to make sure that each edge device is subjected to a reliability assessment before being used in any application [177]. New machine learning models will need to be developed to be introduced to the TinyML ecosystem with the implementation of TinyML [178].

For the model design, which must deliver real-time solutions, strategies like federated learning, transfer learning, and reinforcement learning might be applied. For edge-based solutions to handle multiple-level dynamics, the edge infrastructure must be built using virtual optimization approaches. Such a design will need specialist skillsets when it comes to edge software. The blending of edge hardware and software presents another difficulty. Overall, the edge intelligence framework must offer cutting-edge services like 5G/6G wireless networking, cooperative intelligence, energy management, machine learning-as-a-service, etc. [179].

Finally, a major issue that must be rectified for TinyML research to proceed is the dearth of benchmarking tools, datasets, and acknowledged models. One of the key issues making it difficult for developers to produce cross-platform compatible solutions is the lack of standardization.

9.12 TinyML for 5G security and data anomaly

The term "TinyML" refers to a young field where machine learning models are applied to inexpensive, low-power edge devices like microcontrollers. Numerous applications, particularly those relating to security and data anomaly detection, may be made possible by the union of TinyML and 5G networks.

TinyML can be used to identify and counteract numerous security risks, including malware and intrusion attacks, in the context of 5G security. On edge devices, for instance, TinyML models can be used to monitor network traffic and spot unusual behavior that can point to a security breach [180]. By examining network traffic patterns and locating the sources of unexpected traffic, these models can also be used to recognize and stop DDoS assaults.

TinyML can be used in 5G networks to find data abnormalities in addition to security. Finding trends and anomalies that could point to problems can be difficult given the enormous amounts of data that 5G networks generate. TinyML models can be trained to examine network data in real time and spot trends or abnormalities that might point to problems like hardware failures, network congestion, or other concerns [181].

Overall, a variety of security and data anomaly detection applications may be made possible by the union of TinyML and 5G networks. We may anticipate more use cases for TinyML in this space as the rollout of 5G networks increases [182]. TinyML is a technology that makes it possible to use machine learning on small, low-power gadgets like microcontrollers. In 5G networks, TinyML can be used to enhance security and find data anomalies. Here are a few instances:

Finding anomalies: TinyML can be used to identify irregularities in 5G network traffic, such as strange data transfer patterns that could point to security breaches or other problems. TinyML algorithms can be installed on edge devices like routers or gateways to enable real-time anomaly detection and response, enhancing network security and lowering the possibility of data breaches.

Intrusion detection: TinyML can be used to identify intrusions or attacks on 5G networks, such as malware infections or denial-of-service assaults. TinyML algorithms are used to analyze network traffic in real time so that suspicious activity can be immediately identified and dealt with, minimizing the chance of network outages or data loss.

Network optimization: By examining data traffic patterns and modifying network resources as necessary, TinyML can be used to improve the performance of 5G networks. TinyML algorithms can be installed on edge devices like base stations and routers to optimize network traffic in real time while lowering latency and enhancing network performance [183].

Privacy protection: TinyML can be used to analyze user behavior and preferences in wearables or smart homes to safeguard user privacy in 5G networks. Data transmission to centralized servers can be minimized by implementing TinyML algorithms on edge devices, such as smart speakers and smartwatches, to evaluate and process user data locally. Lowering the possibility of data breaches or unwanted access to personal information can enhance data privacy.

Threat intelligence: TinyML can be used to evaluate threat intelligence data to find potential security threats to 5G networks, such as security warnings or threat feeds. Threat intelligence data can be processed in real-time, enabling real-time threat detection and response, by implementing TinyML algorithms on edge devices, including security gateways or firewalls [184].

Overall, by evaluating data traffic patterns, spotting intrusions or assaults, enhancing network performance, safeguarding user privacy, and assessing threat intelligence data, TinyML can be utilized to enhance security and detect data abnormalities in 5G networks. These features can be implemented at the network edge by installing TinyML algorithms on edge devices like routers or gateways, allowing for real-time detection and reaction to potential security threats or network anomalies [185]. The implementation of TinyML on edge devices may, however, necessitate specialized knowledge and resources, such as infrastructure for AI and machine learning, as well as careful consideration of data privacy and security issues. To ensure interoperability and compatibility amongst various 5G ecosystem components, network operators, device manufacturers, and application developers must cooperate closely together to deploy TinyML for 5G security and data anomaly detection.

9.13 Design and modeling for LPWAN-based 5G

Low-power devices with low data rates can connect over long distances using LP-WANs, a form of wireless network. LPWANs can offer improved capabilities for IoT applications when paired with 5G networks, allowing devices to interact over greater distances and use less power [186].

The steps listed below can be used to develop and model an LPWAN-based 5G network:

- **Define the IoT use case:** Determining the use cases that the network will support is the first stage in creating a 5G LPWAN-based network. This can be used to determine the network's requirements, including its range, data throughput, and power consumption [187].
- **Pick an LPWAN technology:** There are several LPWAN options, including LoRaWAN, Sigfox, and NB-IoT. Depending on the needs of the use case, such as range, data rate, and power consumption, a particular technology will be chosen.
- **Establish the network architecture:** The LPWAN-based 5G network architecture will be determined by the number of devices that need to be linked and the network's coverage area [188]. The network can be built using a mesh topology, a star topology, or a combination of the two.
- **Model the network:** After deciding on the network architecture, the following step is to use a network simulation tool to model the network. This can assist in identifying the ideal network characteristics, such as the required number of gateways, their location, and the transmission power settings.
- **Optimize the network:** Following modeling, the network can be improved for maximum effectiveness and performance. This can involve choosing the best frequency band, placing the gateway optimally, and modifying the transmission power parameters.
- **Test and validate the network:** The network should be tested and validated once it has been created and optimized to make sure it satisfies the use case's criteria. This may entail testing the network in various scenarios, such as at various data rates and ranges, and assessing the effectiveness and performance of the network.

In general, careful consideration of the needs of the use case, the selection of LPWAN technology, the network architecture, and the optimization of the network parameters are necessary when developing and modeling an LPWAN-based 5G network. These techniques can be used to build a powerful LPWAN-based 5G network for IoT applications.

Low-power, low-bandwidth devices can be connected across long distances with the help of LPWANs, a form of wireless network. LPWANs are highly suited for use cases where devices need to be connected over vast geographic areas with little power consumption, such as smart cities, agriculture, and industrial IoT [189]. Considerations for the design and modeling of LPWAN-based 5G networks are listed below:

Network architecture: Low-power, low-bandwidth devices, such as sensors and actuators, should be supported by the network architecture for LPWAN-based 5G networks. The network should be designed with low power consumption and long-range communication in mind. Long-range, low-power communication can be made possible with the adoption of LPWAN technologies like LoRaWAN or Sigfox, which have minimal infrastructure requirements.

Resource allocation: For LPWAN-based 5G networks, resource allocation is a crucial factor. The network should be designed to optimally distribute resources depending on the unique requirements of various applications and devices. Network slicing, which enables the development of virtual networks tailored for certain use cases or applications, can be used to do this.

Security: Given the numerous devices that could be linked to the network, security is a crucial factor for LPWAN-based 5G networks. Strong security elements for the network should be included in its design, such as access control, authentication, and encryption. The usage of blockchain technology, which can offer a secure and decentralized framework for managing device IDs, access control, and data privacy, may also be advantageous for LPWAN-based 5G networks [190].

Edge computing: Data transmission over vast distances can be minimized by using edge computing to process and analyze data locally at the network edge. This can enable real-time data processing and analysis while lowering latency and enhancing network performance. Predictive maintenance, anomaly detection, and optimization are just a few of the uses for edge computing.

Machine learning: Machine learning techniques can be used to evaluate data from 5G networks built on LPWAN, looking for trends and anomalies that can point to impending breakdowns or repairs that are required. By examining data traffic patterns and modifying network resources as necessary, machine learning can also be used to improve network performance. Real-time data processing and analysis can be facilitated, enhancing network performance and lowering latency, by putting machine learning algorithms on edge devices, such as gateways or routers.

Overall, LPWAN-based 5G networks should focus on resource allocation, security, edge computing, and machine learning capabilities to serve low-power, low-bandwidth devices. The deployment of network slicing, blockchain, edge computing, and machine learning algorithms along with the use of LPWAN technologies like LoRaWAN or Sigfox allows LPWAN-based 5G networks to enable a variety of innovative applications and services while reducing power consumption and infrastructure needs. LPWAN end points, LPWAN network servers, and LPWAN gateways may all be necessary for LPWAN-based 5G networks, which is another crucial point to remember.

To guarantee interoperability and compatibility amongst various elements of the LPWAN-based 5G ecosystem, network operators and device manufacturers should collaborate closely. Additionally, strong encryption, authentication, and access control techniques should be implemented in LPWAN-based 5G networks to ensure data security and privacy. Finally, LPWAN-based 5G networks should be built to be flexible and scalable, with the capacity to adjust to evolving device and application requirements [191].

9.14 5G MEC latency standards

MEC is a technology that allows processing and data storage closer to users and devices at the edge of the network. By doing this, latency can be decreased and applications that need real-time or low-latency processing can function better. MEC is a crucial element in the context of 5G networks that can facilitate a variety of cutting-edge applications [192].

Several standards bodies have established latency criteria for MEC to guarantee that it complies with the 5G networks' latency requirements. The following are some of the major 5G MEC latency standards:

ETSI MEC: In its MEC standards, the ETSI outlined some latency requirements for MEC. User plane latency for MEC services must not exceed 20 ms and control plane latency must not exceed 10 ms, according to ETSI MEC [193].

3GPP MEC: In its MEC standards, the 3GPP has outlined several latency criteria for MEC. User plane latency for MEC services must not exceed 50 ms, and control plane latency must not exceed 20 ms, according to 3GPP MEC [194].

OpenFog: The OpenFog Consortium has established a set of MEC-like latency specifications for fog computing. Fog services must have a maximum user plane latency of 10 ms and a maximum control plane latency of 1 ms to be compliant with OpenFog [195].

IEEE: In its IEEE 1934 standard, the Institute of Electrical and Electronics Engineers (IEEE) also outlined a set of latency requirements for edge computing. User plane latency for edge services must not exceed 5 ms and control plane latency must not exceed 1 ms, according to IEEE 1934 [196].

These standards are crucial for ensuring that MEC services deliver the performance needed for applications that need real-time or low-latency processing and adhere to the latency specifications of 5G networks. Network operators and service providers can make sure that their MEC implementations are performant and satisfy the expectations of their clients by adhering to these guidelines. A crucial component of 5G networks is MEC, which enables the deployment of computing resources and services closer to end users and devices at the network edge.

For low-latency applications like gaming, streaming, and VR and AR, MEC can lower latency and boost network performance. The following 5G MEC latency norms should be considered:

E2E latency: The amount of time it takes for a data packet to go from a device to a server and back is known as E2E latency. When using MEC, E2E latency should be reduced by putting MEC capabilities closer to end users and devices at the network's edge. For E2E connectivity, the 5G standard sets a target latency of 1 ms, which is essential for real-time applications like industrial automation and driverless vehicles.

Round-trip latency: Round-trip latency, which includes processing time at the server, is the length of time it takes for a data packet to travel from a device to a server and back. When using MEC, round-trip latency should be kept to a minimum by positioning computing tools and services closer to end users and devices at the network edge. To support applications like gaming and VR, the 5G standard sets a goal round-trip latency of 10 ms.

Network latency: Network latency refers to the amount of time it takes a data packet to go from one point in the network to another, and it can be impacted by things like routing and network congestion. By putting MEC capabilities at the network's edge, where they are closer to end users and devices, network latency can be lowered in the case of MEC. For real-time applications like autonomous vehicles and industrial automation, the 5G standard sets a target latency of less than 5 ms for networks.

Processing latency: The length of time it takes for a computing resource or service to process a data packet is referred to as processing latency. By positioning computing facilities and services nearer to end users and other devices at the network edge in the case of MEC, processing latency can be reduced. For real-time applications like autonomous vehicles and industrial automation, the 5G standard sets a target latency of less than 1 ms for processing delay.

All things considered, 5G MEC latency standards are essential for allowing low-latency, high-bandwidth applications like gaming, VR and AR, and streaming. Latencies can be reduced, network performance can be enhanced, and real-time decision-making is made possible by installing MEC capabilities at the network edge, nearer to end users and devices. The development of a wide range of cutting-edge applications and services is made possible by the adoption of the 5G MEC latency standards, which are crucial for ensuring interoperability and compatibility between various elements of the 5G ecosystem.

There are many reasons why it may be difficult to meet the 5G MEC latency standards. Some of the major difficulties are listed below:

Infrastructure on the network: MEC calls for the deployment of computing tools and services at the network edge, which can be difficult given the sporadic infrastructure in some places. Network operators may have to make extra investments in infrastructure, including tiny cells and edge data centers, to meet the 5G MEC latency standards.

Network design: MEC demands a dispersed network design with computer resources and services provided at the network edge. The implementation of this can be difficult, particularly in older networks that were built for centralized computing. It may be necessary for network operators to restructure their networks to enable MEC, which can be time-consuming and costly.

Interoperability: To meet the 5G MEC latency criteria, various 5G ecosystem elements, such as devices, networks, and apps, must work together seamlessly. Given the wide variety of devices and applications that might be connected to the network, ensuring compatibility can be difficult.

Standardization: Standardization is essential for meeting the 5G MEC latency standards since it makes sure that all the ecosystem's many parts can function together in harmony. Despite this, standardization can be difficult to establish, especially considering the numerous stakeholders in the 5G ecosystem.

Security: As computing resources and services are delivered at the network edge, closer to end users and devices, MEC brings new security threats. Given the numerous devices that might be connected to the network, it can be difficult to ensure MEC's security. To ensure the security of MEC, network operators may need to adopt extra security mechanisms including encryption, access control, and threat detection.

Power usage: MEC calls for the deployment of computing tools and services at the network edge, which may result in higher power usage. However, power consumption must be kept to a minimum to meet 5G MEC latency standards. To achieve the necessary latency while reducing power consumption, network operators may need to invest in low-power computing technologies like edge AI and machine learning.

Scalability: To handle the numerous devices and applications that might be linked to the network, MEC must be scalable. However, establishing scalability can be difficult, especially given some regions' poor infrastructure. To achieve scalability, network operators might need to make investments in extra infrastructure, such as edge data centers and tiny cells.

Overall, several variables, including network infrastructure, network architecture, interoperability, standardization, security, power consumption, and scalability, can make it difficult to meet the 5G MEC latency standards. To successfully deploy MEC and realize the advantages of 5G networks, such as improved network performance, reduced latency, and the ability to support a wide range of cutting-edge applications and services, network operators and device manufacturers must collaborate closely to address these challenges.

A concerted effort by network operators, device manufacturers, application developers, and other stakeholders in the 5G ecosystem is also necessary to meet the 5G MEC latency standards. The 5G MEC latency standards can be met by collaborating to address these issues, enabling the deployment of a variety of cutting-edge applications and services that can revolutionize industries and enhance our daily lives.

9.15 TinyML for 5G computation complexity

The term "TinyML" refers to a young field where machine learning models are applied to inexpensive, low-power edge devices like microcontrollers. Numerous applications, particularly those involving processing difficulty, may be made possible by the union of TinyML and 5G networks. TinyML can be utilized in the context of 5G networks to lessen the computational complexity of machine learning models that are implemented on edge devices. This is significant because running complicated machine learning models on edge devices may be difficult due to their often low processing and memory capacities.

Due to the small weight and efficiency of TinyML models, they can be used on edge devices without consuming a lot of processing power. For instance, a TinyML model for classifying images might only need a few kilobytes of memory and be able to function on a microcontroller with a modest amount of computing power.

TinyML models can be installed on edge devices to offload computation from a cloud or central server, lowering network latency and speeding up response times for applications that need real-time or low-latency processing. As there is no need to transport data to the cloud for processing, this can help lower the network's bandwidth needs. TinyML and 5G networks collectively have the potential to enable a variety of applications that call for low-latency processing and effective use of computational resources. We may anticipate new use cases for TinyML in this space, including in smart cities, driverless vehicles, and industrial automation, as the implementation of 5G networks spreads.

Here are some more specifics on TinyML and 5G networks:

TinyML models are made to be compact and effective: TinyML models for deployment on low-cost, low-power edge devices like microcontrollers. These models can be used on edge devices because they do not consume a lot of processing power and are lightweight and effective. By putting TinyML models on edge devices, the computation can be offloaded from the cloud or a central server, lowering network latency and reducing the response time for applications that need real-time or low-latency processing. This can be crucial in situations where decisions need to be made fast, such as industrial automation or autonomous vehicles.

Low-latency communication is made possible by 5G networks: Low-latency communication, with latency as low as 1 ms, is made possible by 5G networks. Emerging applications such as AR or remote surgery can benefit from real-time or low-latency processing.

5G networks and TinyML open new applications: Smart cities, autonomous vehicles, and industrial automation are just a few of the new applications that could be made possible by the union of TinyML with 5G networks. It is conceivable to develop new applications that were not previously feasible by installing machine learning models on edge devices and utilizing the low-latency communication capabilities of 5G networks.

It is safer to use machine learning at the edge: Data can be preserved more securely by executing computing at the edge rather than in the cloud. This is because processing data locally instead of in the cloud lowers the risk of data breaches.

Additionally, TinyML models are made to be compact and effective, enabling their deployment on hardware with processing and memory constraints. By lowering the device's attack surface, attackers may find it more challenging to exploit flaws as a result.

Overall, TinyML and 5G networks have the potential to enable a variety of new applications while enhancing the performance of already existing ones. It is conceivable to develop new apps that were not previously possible while simultaneously enhancing the performance and security of current applications by putting machine learning models on edge devices and utilizing the low-latency communication capabilities of 5G networks. TinyML is a technology that makes it possible to use machine learning on small, low-power gadgets like microcontrollers. The issues of 5G networks' computational complexity can be solved with TinyML. Here are a few instances:

Edge processing: Data transmission across long distances can be avoided by using edge computing to process and analyze data locally at the network edge. Enabling real-time data processing and analysis on edge devices, such as gateways or routers, can lower the computing complexity of 5G networks. Real-time data processing and analysis can be made possible by installing TinyML algorithms on edge devices like microcontrollers or low-power processors, which will enhance network performance and lower latency.

Resource allocation: Resource allocation is a crucial factor to consider for 5G networks, especially considering the numerous devices that could be linked to the network. By examining the patterns of data traffic and modifying network resources accordingly, TinyML can be used to optimize resource allocation. Network traffic can be optimized in real time, enhancing network speed and lowering latency, by putting TinyML algorithms on edge devices like base stations or routers.

Anomaly detection: TinyML can be used to find irregularities in 5G network traffic, such as strange data transmission patterns that might point to security breaches or other problems.

TinyML algorithms: TinyML algorithms can be installed on edge devices like routers and gateways to enable real-time anomaly detection and response, enhancing network security and reducing the computational strain on the network core. TinyML can be used to forecast maintenance requirements for 5G network equipment, including base stations and antennas. TinyML algorithms can analyze data from sensors and other sources to find patterns that could point to upcoming failures or maintenance requirements. This can enable preventative upkeep and repair, minimizing downtime and enhancing network efficiency.

Network optimization: By examining data traffic patterns and modifying network resources as necessary, TinyML can be used to improve the performance of 5G networks. TinyML algorithms can be installed on edge devices like base stations and routers to optimize network traffic in real time while lowering latency and enhancing network performance.

Overall, by enabling real-time data processing and analysis on edge devices, optimizing resource allocation, spotting abnormalities, anticipating maintenance requirements, and enhancing network performance, TinyML can be utilized to address the computational complexity concerns of 5G networks. Network administrators can reduce the computational strain on the network core, enhancing network speed and lowering latency, by putting TinyML algorithms on edge devices. TinyML implementation on edge devices, it should be noted, may necessitate specific knowledge and resources, including infrastructure for AI and machine learning as well as careful consideration of data privacy and security issues.

To guarantee interoperability and compatibility amongst various 5G ecosystem components, network operators, device manufacturers, and application developers must cooperate closely to deploy TinyML for 5G networks.

References

[1] I. Parvez, A. Rahmati, I. Guvenc, A.I. Sarwat, H. Dai, A survey on low latency towards 5G: RAN, core network and caching solutions, IEEE Communications Surveys and Tutorials 20 (4) (2018) 3098–3130.

[2] R. Barona, E.M. Anita, A survey on data breach challenges in cloud computing security: issues and threats, in: 2017 International Conference on Circuit, Power and Computing Technologies (ICCPCT), IEEE, 2017, pp. 1–8.

[3] N. Schizas, A. Karras, C. Karras, S. Sioutas, TinyML for ultra-low power AI and large scale IoT deployments: a systematic review, Future Internet 14 (12) (2022) 363.

[4] K. Zhang, S. Leng, Y. He, S. Maharjan, Y. Zhang, Mobile edge computing and networking for green and low-latency Internet of Things, IEEE Communications Magazine 56 (5) (2018) 39–45.

[5] S. Zaidi, A.M. Hayajneh, M. Hafeez, Q. Ahmed, Unlocking edge intelligence through tiny machine learning (TinyML), IEEE Access 10 (2022) 100867–100877.

[6] L. Dutta, T. Bharali, TinyML meets IoT: a comprehensive survey, Internet of Things 16 (2021) 100461.

[7] H. Liu, Z. Wei, H. Zhang, B. Li, C. Zhao, Tiny machine learning (Tiny-ML) for efficient channel estimation and signal detection, IEEE Transactions on Vehicular Technology 71 (6) (2022) 6795–6800.

[8] 3GPP, Technical Specification 23.501, System architecture for the 5G system (5GS), 2020.

[9] A. Muthana, M. Saeed, A. Ghani, R. Mahmod, Enhancing privacy of paging procedure in LTE, International Journal of Engineering Science Invention 7 (2) (2018).

[10] K. Samdanis, T. Taleb, The road beyond 5G: a vision and insight of the key technologies, IEEE Network 34 (2) (2020) 135–141.

[11] J. Ni, X. Lin, X. Shen, Efficient and secure service-oriented authentication supporting network slicing for 5G-enabled IoT, IEEE Journal on Selected Areas in Communications 36 (3) (2018) 644–657.

[12] F. Mademann, The 5G system architecture, Journal of ICT Standardization 6 (1–2) (2021) 77–86, https://doi.org/10.13052/jicts2245-800X.615.

[13] U. Riaz, S. Hussain, H. Patel, A comparative study of rest with soap, in: Multimedia Technology and Enhanced Learning: Third EAI International Conference, ICMTEL 2021, Virtual Event, Proceedings, Part I 3, April 8–9, 2021, Springer, 2021, pp. 485–491.

[14] S. Shah, M.A. Gregory, S. Li, Cloud-native network slicing using software defined networking based multi-access edge computing: a survey, IEEE Access 9 (2021) 10903–10924.

[15] K. Husenovic, et al., Setting the scene for 5G: opportunities & challenges, International Telecommunication Union (ITU) Report, vol. 56, https://www.itu.int/pub/D-PREF-BB.5G_01, 2018.

[16] T. Norp, 5G requirements and key performance indicators, Journal of ICT Standardization 6 (1–2) (2018) 15–30.

[17] G. Akpakwu, B. Silva, G. Hancke, A. Abu-Mahfouz, A survey on 5G networks for the Internet of Things: communication technologies and challenges, IEEE Access 6 (2017) 3619–3647.
[18] F. Liu, J. Peng, M. Zuo, Toward a secure access to 5G network, in: 2018 17th IEEE International Conference on Trust, Security and Privacy in Computing and Communications/12th IEEE International Conference on Big Data Science and Engineering (TrustCom/BigDataSE), IEEE, 2018, pp. 1121–1128.
[19] NGMN Alliance Report, white paper, "5G white paper," vol. 1, no. 2015, https://www.ngmn.org/wp-content/uploads/NGMN_5G_White_Paper_V1_0.pdf, 2015.
[20] P. Sharma, S. Jain, S. Gupta, N. Chamola, Role of machine learning and deep learning in securing 5G-driven industrial IoT applications, Ad Hoc Networks 123 (2021) 102685.
[21] L. Tawalbeh, F. Muheidat, M. Tawalbeh, M. Quwaider, IoT privacy and security: challenges and solutions, Applied Sciences 10 (12) (2020) 4102.
[22] B. Silva, M. Khan, K. Han, Towards sustainable smart cities: a review of trends, architectures, components, and open challenges in smart cities, Sustainable Cities and Society 38 (2018) 697–713.
[23] J. del Peral-Rosado, R. Raulefs, J. López-Salcedo, G. Seco-Granados, Survey of cellular mobile radio localization methods: from 1G to 5G, IEEE Communications Surveys and Tutorials 20 (2) (2017) 1124–1148.
[24] R. Rojas, M. Ruiz Garcia, Implementation of industrial internet of things and cyber-physical systems in SMEs for distributed and service-oriented control, in: Industry 4.0 for SMEs: Challenges, Opportunities and Requirements, 2020, pp. 73–103.
[25] M. Khan, K. Salah, IoT security: review, blockchain solutions, and open challenges, Future Generation Computer Systems 82 (2018) 395–411.
[26] S. Guo, B. Lu, M. Wen, S. Dang, M. Saeed, Customized 5G and beyond private networks with integrated URLLC, eMBB, mMTC, and positioning for industrial verticals, IEEE Communications Standards Magazine 6 (1) (2022) 52–57.
[27] Y. Wu, Y. Ma, H.-N. Dai, H. Wang, Deep learning for privacy preservation in autonomous moving platforms enhanced 5G heterogeneous networks, Computer Networks 185 (2021) 107743.
[28] L. Babun, K. Denney, Z. Celik, P. McDaniel, A. Uluagac, A survey on IoT platforms: communication, security, and privacy perspectives, Computer Networks 192 (2021) 108040.
[29] S. Sicari, A. Rizzardi, A. Coen-Porisini, 5G in the internet of things era: an overview on security and privacy challenges, Computer Networks 179 (2020) 107345.
[30] ETSI report, Network Functions Virtualisation (NFV); Management and Orchestration, vol. ETSI GS NFV-MAN 001 V1.1.1 (2014), https://www.etsi.org/deliver/etsi_gs/NFV-MAN/001_099/001/01.01.01_60/gs_NFV-MAN001v010101p.pdf.
[31] ETSI report, Network Functions Virtualisation (NFV); NFV Security; Problem Statement, ETSI GS NFV-SEC 001 V1.1.1 (2014-10), https://www.etsi.org/deliver/etsi_gs/nfv-sec/001_099/001/01.01.01_60/gs_nfv-sec001v010101p.pdf.
[32] ETSI report, Network Functions Virtualisation (NFV); NFV Security; Security and Trust Guidance, ETSI GR NFV-SEC 003 V1.2.1 (2016-08), https://www.etsi.org/deliver/etsi_gr/NFV-SEC/001_099/003/01.02.01_60/gr_nfv-sec003v010201p.pdf.
[33] A. Theorin, et al., An event-driven manufacturing information system architecture for Industry 4.0, International Journal of Production Research 55 (5) (2017) 1297–1311.
[34] F. Spinelli, V. Mancuso, Toward enabled industrial verticals in 5G: a survey on MEC-based approaches to provisioning and flexibility, IEEE Communications Surveys and Tutorials 23 (1) (2020) 596–630.

[35] M.M. Saeed, et al., A novel variable pseudonym scheme for preserving privacy user location in 5G networks, Security and Communication Networks 2022 (2022), https://doi.org/10.1155/2022/7487600.
[36] I. Ali, S. Sabir, Z. Ullah, Internet of things security, device authentication and access control: a review, arXiv preprint, arXiv:1901.07309, 2019, https://doi.org/10.48550/arXiv.1901.07309.
[37] B. Sinha, R. Dhanalakshmi, Recent advancements and challenges of Internet of Things in smart agriculture: a survey, Future Generation Computer Systems 126 (2022) 169–184.
[38] S. Wani, M. Imthiyas, H. Almohamedh, K. Alhamed, S. Almotairi, Y. Gulzar, Distributed denial of service (DDoS) mitigation using blockchain—a comprehensive insight, Symmetry 13 (2) (2021) 227.
[39] J. Téllez, S. Zeadally, Mobile Payment Systems, Springer, 2017.
[40] M. Lanoue, C. Bollmann, J. Michael, J. Roth, D. Wijesekera, An attack vector taxonomy for mobile telephony security vulnerabilities, Computer 54 (4) (2021) 76–84.
[41] F. Yousaf, M. Bredel, S. Schaller, C. Schneider, NFV and SDN—key technology enablers for 5G networks, IEEE Journal on Selected Areas in Communications 35 (11) (2017) 2468–2478.
[42] M. Geller, P. Nair, 5G security innovation with Cisco, Whitepaper, Cisco Public2018, pp. 1–29, https://www.cisco.com/c/dam/en/us/solutions/collateral/service-provider/service-provider-security-solutions/5g-security-innovation-with-cisco-wp.pdf.
[43] K. Bilal, O. Khalid, A. Erbad, S. Khan, Potentials, trends, and prospects in edge technologies: fog, cloudlet, mobile edge, and micro data centers, Computer Networks 130 (2018) 94–120.
[44] S. Abdel Hakeem, H. Hussein, H. Kim, Security requirements and challenges of 6G technologies and applications, Sensors (Basel) 22 (5) (2022) 1969.
[45] B. Farroha, D. Farroha, J. Cook, A. Dutta, Exploring the security and operational aspects of the 5th generation wireless communication system, in: Open Architecture/Open Business Model Net-Centric Systems and Defense Transformation 2019, vol. 11015, SPIE, 2019, pp. 60–79.
[46] P. Krishnan, K. Jain, A. Aldweesh, P. Prabu, R. Buyya, OpenStackDP: a scalable network security framework for SDN-based OpenStack cloud infrastructure, Journal of Cloud Computing 12 (1) (2023) 26.
[47] M. Saeed, et al., Preserving privacy of user identity based on pseudonym variable in 5G, Computers, Materials & Continua 70 (3) (2022) 5551–5568.
[48] 5G PPP Security WG, 5G PPP phase1 security landscape, 5G-ENSURE project, 2017.
[49] M. De Benedictis, A. Lioy, On the establishment of trust in the cloud-based ETSI NFV framework, in: 2017 IEEE Conference on Network Function Virtualization and Software Defined Networks (NFV-SDN), IEEE, 2017, pp. 280–285.
[50] B. Martini, et al., Pushing forward security in network slicing by leveraging continuous usage control, IEEE Communications Magazine 58 (7) (2020) 65–71.
[51] N. Rajatheva, et al., White paper on broadband connectivity in 6G, arXiv preprint, arXiv:2004.14247, 2020, https://doi.org/10.48550/arXiv.2004.14247.
[52] N. Cardona, E. Coronado, S. Latré, R. Riggio, A. Marquez-Barja, Software-defined vehicular networking: opportunities and challenges, IEEE Access 8 (2020) 219971–219995.
[53] A. Alwakeel, A. Alnaim, E. Fernandez, A survey of network function virtualization security, in: SoutheastCon 2018, IEEE, 2018, pp. 1–8.

[54] T. Wichary, J. Mongay Batalla, C. Mavromoustakis, J. Żurek, G. Mastorakis, Network slicing security controls and assurance for verticals, Electronics 11 (2) (2022) 222.
[55] M.M. Saeed, R.A. Saeed, E. Saeid, Identity division multiplexing based location preserve in 5G, in: 2021 International Conference of Technology, Science and Administration (ICTSA), IEEE, 2021, pp. 1–6.
[56] A. Gonzalez, G. Nencioni, A. Kamisiński, B. Helvik, S. Heegaard, Dependability of the NFV orchestrator: state of the art and research challenges, IEEE Communications Surveys and Tutorials 20 (4) (2018) 3307–3329.
[57] J. Baranda, et al., Orchestration of end-to-end network services in the 5G-crosshaul multi-domain multi-technology transport network, IEEE Communications Magazine 56 (7) (2018) 184–191.
[58] D. Je, J. Jung, M. Choi, Toward 6G security: technology trends, threats, and solutions, IEEE Communications Standards Magazine 5 (3) (2021) 64–71.
[59] I. GSTR-TN5G, Geneva, Switzerland, Oct, Transport network support of IMT-2020/5G, http://handle.itu.int/11.1002/pub/810db8e7-en, 2018.
[60] O.S. Peñaherrera-Pulla, C. Baena, S. Fortes, E. Baena, R. Barco, KQI assessment of VR services: a case study on 360-video over 4G and 5G, IEEE Transactions on Network and Service Management 19 (4) (Dec. 2022) 5366–5382, https://doi.org/10.1109/TNSM.2022.3192762.
[61] M.M. Saeed, R.A. Saeed, E. Saeid, Preserving privacy of paging procedure in 5^{th}G using identity-division multiplexing, in: 2019 First International Conference of Intelligent Computing and Engineering (ICOICE), IEEE, 2019, pp. 1–6.
[62] ETSI Report, 5G; System Architecture for the 5G System (3GPP TS 23.501 version 15.2.0 Release 15), ETSI TS 123 501 V15.2.0 (2018-06).
[63] H. Kim, 5G core network security issues and attack classification from network protocol perspective, Journal of Internet Services and Information Security 10 (2) (2020) 1–15.
[64] W. Sun, Q. Wang, N. Zhao, H. Zhang, C. Shen, L. Wong, Ultra-Dense Heterogeneous Networks, CRC Press, 2022.
[65] R.A. Saeed, M.M. Saeed, R.A. Mokhtar, H. Alhumyani, E. Abdel-Khalek, Pseudonym mutable based privacy for 5G user identity, Computer Systems Science and Engineering 39 (1) (2021) 1–14.
[66] J. Konečný, et al., Federated optimization: distributed machine learning for on-device intelligence, arXiv:1610.02527, 2016, https://doi.org/10.48550/arXiv.1610.02527.
[67] V. Chamola, P. Kotesh, A. Agarwal, N. Gupta, M. Guizani, A comprehensive review of unmanned aerial vehicle attacks and neutralization techniques, Ad Hoc Networks 111 (2021) 102324.
[68] M.A.M. Ali, A.S.A. Gaid, M.M. Saeed, R.A. Saeed, Design and performance analysis of a 38 GHz microstrip patch antenna with slits loading for 5G millimeter-wave communications, in: 2023 3rd International Conference on Emerging Smart Technologies and Applications (eSmarTA), Taiz, Yemen, 2023, pp. 1–6, https://doi.org/10.1109/eSmarTA59349.2023.10293289.
[69] A. Aasheed, et al., An overview of mobile edge computing: Architecture, technology and direction, KSII Transactions on Internet and Information Systems (TIIS) 13 (10) (2019) 4849–4864, https://doi.org/10.3837/tiis.2019.10.002.
[70] G. Baldoni, et al., Edge computing enhancements in an NFV-based ecosystem for 5G neutral hosts, in: 2018 IEEE Conference on Network Function Virtualization and Software Defined Networks (NFV-SDN), IEEE, 2018, pp. 1–5.
[71] L. Gavrilovska, V. Rakovic, D. Denkovski, From cloud RAN to open RAN, Wireless Personal Communications 113 (2020) 1523–1539.

[72] W. Xiang, K. Zheng, X. Shen, 5G Mobile Communications, Springer, 2016.

[73] O. Newton, S. Saadat, J. Song, S. Fiore, G. Sukthankar, EveryBOTy counts: examining human–machine teams in open source software development, Topics in Cognitive Science (2022), https://doi.org/10.1111/tops.12613.

[74] N. Zhan, C. Gan, J. Hui, Y. Guo, Fair resource allocation based on user satisfaction in multi-OLT virtual passive optical network, IEEE Access 8 (2020) 134707–134715.

[75] M.E. Diago-Mosquera, A. Aragón-Zavala, G. Castañón, Bringing it indoors: A review of narrowband radio propagation modeling for enclosed spaces, IEEE Access 8 (2020) 103875–103899, https://doi.org/10.1109/ACCESS.2020.2999848.

[76] R. Solozabal, et al., Design of virtual infrastructure manager with novel VNF placement features for edge clouds in 5G, in: Engineering Applications of Neural Networks: 18th International Conference, EANN 2017, Proceedings, Athens, Greece, August 25–27, 2017, Springer, 2017, pp. 669–679.

[77] S. Gulati, S. Kalyanasundaram, P. Nashine, B. Natarajan, R. Agrawal, A. Bedekar, Performance analysis of distributed multi-cell coordinated scheduler, in: 2015 IEEE 82nd Vehicular Technology Conference (VTC2015-Fall), IEEE, 2015, pp. 1–5.

[78] D. Pengoria, S. Nagaraj, R. Agrawal, Performance of co-operative uplink reception with non-ideal backhaul, in: 2015 IEEE 81st Vehicular Technology Conference (VTC Spring), IEEE, 2015, pp. 1–5.

[79] T. Han, X. Ge, L. Wang, K. Kwak, Y. Han, M. Liu, 5G converged cell-less communications in smart cities, IEEE Communications Magazine 55 (3) (2017) 44–50.

[80] T. Šolc, M. Mohorčič, C. Fortuna, A methodology for experimental evaluation of signal detection methods in spectrum sensing, PLoS ONE 13 (6) (2018) e0199550.

[81] Y. Chun, M. Mokhtar, A. Rahman, A. Samingan, Performance study of LTE experimental testbed using OpenAirInterface, in: 2016 18th International Conference on Advanced Communication Technology (ICACT), IEEE, 2016, pp. 617–622.

[82] F. Kaltenberger, A. Silva, A. Gosain, L. Wang, N. Nguyen, OpenAirInterface: democratizing innovation in the 5G era, Computer Networks 176 (2020) 107284.

[83] H. Khalili, et al., Implementation of 5G experimentation environment for accelerated development of mobile media services and network applications, in: 2023 26th Conference on Innovation in Clouds, Internet and Networks and Workshops (ICIN), IEEE, 2023, pp. 153–160.

[84] L. Bolivar, C. Tselios, D. Area, G. Tsolis, On the deployment of an open-source, 5G-aware evaluation testbed, in: 2018 6th IEEE International Conference on Mobile Cloud Computing, Services, and Engineering (MobileCloud), IEEE, 2018, pp. 51–58.

[85] H. Wang, A. Fapojuwo, A survey of enabling technologies of low power and long range machine-to-machine communications, IEEE Communications Surveys and Tutorials 19 (4) (2017) 2621–2639.

[86] M.M. Saeed, R.S. Saeed, A.S. Gaid, R.A. Mokhtar, O.O. Khalifa, Z.E. Ahmed, Attacks detection in 6G wireless networks using machine learning, in: 2023 9th International Conference on Computer and Communication Engineering (ICCCE), IEEE, 2023, pp. 6–11, https://doi.org/10.1109/ICCCE58854.2023.10246078.

[87] G. Callebaut, G. Ottoy, L. Van der Perre, Cross-layer framework and optimization for efficient use of the energy budget of IoT nodes, in: 2019 IEEE Wireless Communications and Networking Conference (WCNC), IEEE, 2019, pp. 1–6.

[88] S. Li, M. Li, Y. Xu, Z. Bao, L. Fu, Y. Zhu, Capsules based Chinese word segmentation for ancient Chinese medical books, IEEE Access 6 (2018) 70874–70883.

[89] J. Ren, Y. Zhang, K. Zhang, M. Shen, Exploiting mobile crowdsourcing for pervasive cloud services: challenges and solutions, IEEE Communications Magazine 53 (3) (2015) 98–105.
[90] M. Nakip, A. Helva, C. Güzeliş, V. Rodoplu, Subspace-based emulation of the relationship between forecasting error and network performance in joint forecasting-scheduling for the Internet of Things, in: 2021 IEEE 7th World Forum on Internet of Things (WF-IoT), IEEE, 2021, pp. 247–252.
[91] R. Rajasekar, C. Moganapriya, M.H. Kumar, P.S. Kumar, Integration of Mechanical and Manufacturing Engineering with IoT: A Digital Transformation, John Wiley & Sons, 2023.
[92] A. Muthana, M.M. Saeed, Analysis of user identity privacy in LTE and proposed solution, International Journal of Computer Network and Information Security 9 (1) (2017) 54–63.
[93] M. Ballerini, T. Polonelli, D. Brunelli, M. Magno, L. Benini, NB-IoT versus LoRaWAN: an experimental evaluation for industrial applications, IEEE Transactions on Industrial Informatics 16 (12) (2020) 7802–7811.
[94] H.-T. Chien, Y.-D. Lin, C.-L. Lai, C.-T. Wang, End-to-end slicing as a service with computing and communication resource allocation for multi-tenant 5G systems, IEEE Wireless Communications 26 (5) (2019) 104–112.
[95] D.N. Molokomme, A.J. Onumanyi, A.M. Abu-Mahfouz, Edge intelligence in smart grids: a survey on architectures, offloading models, cyber security measures, and challenges, Journal of Sensor and Actuator Networks 11 (3) (2022) 47.
[96] D. Xu, et al., Edge intelligence: architectures, challenges, and applications, arXiv preprint, arXiv:2003.12172, 2020, https://doi.org/10.48550/arXiv.2003.12172.
[97] V.K. Prasad, S. Tanwar, M.D. Bhavsar, Advance cloud data analytics for 5G enabled IoT, in: Blockchain for 5G-Enabled IoT, Springer, 2021, pp. 159–180.
[98] O. Vermesan, J. Bacquet, Next Generation Internet of Things: Distributed Intelligence at the Edge and Human Machine-to-Machine Cooperation, River Publishers, 2019.
[99] L. Zhang, W. Yang, B. Hao, Z. Yang, A. Zhao, Edge computing resource allocation method for mining 5G communication system, IEEE Access 11 (2023) 49730–49737, https://doi.org/10.1109/ACCESS.2023.3244242.
[100] M.Z. Asghar, S.A. Memon, J. Hämäläinen, Evolution of wireless communication to 6G: potential applications and research directions, Sustainability 14 (10) (2022) 6356.
[101] M.M. Saeed, et al., A comprehensive review on the users' identity privacy for 5G networks, IET Communications 16 (5) (2022) 384–399.
[102] R. Olimid, G. Nencioni, 5G network slicing: a security overview, IEEE Access 8 (2020) 99999–100009.
[103] M.M. Saeed, E.S. Ali, R.A. Saeed, Data-driven techniques and security issues in wireless networks, in: Data-Driven Intelligence in Wireless Networks, CRC Press, 2023, pp. 107–154.
[104] E. Pateromichelakis, D. Dimopoulos, A. Salkintzis, NetApps enabling application-layer analytics for vertical IoT industry, IEEE Internet of Things Magazine 5 (4) (2022) 130–135.
[105] Y. Ouyang, et al., The next decade of telecommunications artificial intelligence, arXiv preprint, arXiv:2101.09163, 2021, https://doi.org/10.48550/arXiv.2101.09163.
[106] J. Baranda, et al., On the integration of AI/ML-based scaling operations in the 5Growth platform, in: 2020 IEEE Conference on Network Function Virtualization and Software Defined Networks (NFV-SDN), IEEE, 2020, pp. 105–109.

[107] M.M. Saeed, E.S.A. Saeed, R. Ahmed, M.A. Azim, Green machine learning protocols for cellular communication, in: Green Machine Learning Protocols for Future Communication Networks, CRC Press, 2023, pp. 15–62, https://doi.org/10.1201/9781003230427.

[108] S. Kaparthi, D. Bumblauskas, Designing predictive maintenance systems using decision tree-based machine learning techniques, https://doi.org/10.1108/IJQRM-04-2019-0131, 2020.

[109] A. Thantharate, R. Paropkari, V. Walunj, C. Beard, DeepSlice: a deep learning approach towards an efficient and reliable network slicing in 5G networks, in: 2019 IEEE 10th Annual Ubiquitous Computing, Electronics & Mobile Communication Conference (UEMCON), IEEE, 2019, pp. 0762–0767.

[110] A. Celesti, A. Galletta, L. Carnevale, M. Fazio, A. Ĺay-Ekuakille, M. Villari, An IoT cloud system for traffic monitoring and vehicular accidents prevention based on mobile sensor data processing, IEEE Sensors Journal 18 (12) (2017) 4795–4802.

[111] R. Asaithambi, S. Venkatraman, D. Venkatraman, C. Computing, Big data and personalisation for non-intrusive smart home automation, Big Data and Cognitive Computing 5 (1) (2021) 6.

[112] R. Kallimani, K. Pai, P. Raghuwanshi, S. Iyer, O. López, TinyML: tools, applications, challenges, and future research directions, Multimedia Tools and Applications (2023) 1–31, https://doi.org/10.1007/s11042-023-16740-9.

[113] S. Gupta, S. Jain, B. Roy, A. Deb, A TinyML approach to human activity recognition, Journal of Physics. Conference Series 2273 (1) (2022) 012025.

[114] P.P. Ray, A review on TinyML: state-of-the-art and prospects, Journal of King Saud University: Computer and Information Sciences 34 (4) (2022) 1595–1623.

[115] H. Tataria, M. Shafi, A.F. Molisch, M. Dohler, H. Sjöland, F. Tufvesson, 6G wireless systems: vision, requirements, challenges, insights, and opportunities, Proceedings of the IEEE 109 (7) (2021) 1166–1199.

[116] R. Sanchez-Iborra, C. Skarmeta, S. Magazine, TinyML-enabled frugal smart objects: challenges and opportunities, IEEE Circuits and Systems Magazine 20 (3) (2020) 4–18.

[117] E. Tabanelli, G. Tagliavini, S. Benini, DNN is not all you need: parallelizing non-neural ML algorithms on ultra-low-power IoT processors, ACM Transactions on Embedded Computing Systems 22 (3) (2023) 1–33.

[118] F. Fahim, et al., hls4ml: an open-source codesign workflow to empower scientific low-power machine learning devices, arXiv preprint, arXiv:2103.05579, 2021, https://doi.org/10.48550/arXiv.2103.05579.

[119] F. Paissan, A. Ancilotto, S. Farella, PhiNets: a scalable backbone for low-power AI at the edge, ACM Transactions on Embedded Computing Systems 21 (5) (2022) 1–18.

[120] B. Ozpoyraz, A.T. Dogukan, Y. Gevez, U. Altun, E. Basar, Deep learning-aided 6G wireless networks: a comprehensive survey of revolutionary PHY architectures, IEEE Open Journal of the Communications Society 3 (2022) 1749–1809, https://doi.org/10.1109/OJCOMS.2022.3210648.

[121] N.N. Alajlan, M. Ibrahim, TinyML: enabling of inference deep learning models on ultra-low-power IoT edge devices for AI applications, Micromachines 13 (6) (2022) 851.

[122] M. Antonini, M. Pincheira, M. Vecchio, F. Antonelli, An adaptable and unsupervised TinyML anomaly detection system for extreme industrial environments, Sensors 23 (4) (2023) 2344.

[123] W. Raza, A. Osman, F. Ferrini, F. De Natale, Energy-efficient inference on the edge exploiting TinyML capabilities for UAVs, Drones 5 (4) (2021) 127.

[124] X. Chen, W. Feng, N. Ge, Y. Zhang, Zero trust architecture for 6G security, IEEE Network (2023), https://doi.org/10.1109/MNET.2023.3326356.

[125] Q.-u. Ain, S. Iqbal, S.A. Khan, A.W. Malik, I. Ahmad, N. Javaid, IoT operating system based fuzzy inference system for home energy management system in smart buildings, Sensors 18 (9) (2018) 2802.

[126] R. David, et al., TensorFlow Lite Micro: embedded machine learning for TinyML systems, in: Proceedings of Machine Learning and Systems, vol. 3, 2021, pp. 800–811.

[127] H. Doyu, R. Morabito, M. Brachmann, A TinyMLaaS ecosystem for machine learning in IoT: overview and research challenges, in: 2021 International Symposium on VLSI Design, Automation and Test (VLSI-DAT), IEEE, 2021, pp. 1–5.

[128] I. Zhou, et al., Internet of Things 2.0: concepts, applications, and future directions, IEEE Access 9 (2021) 70961–71012.

[129] U. Khalil, O.A. Malik, M. Uddin, C. Chen, A comparative analysis on blockchain versus centralized authentication architectures for IoT-enabled smart devices in smart cities: a comprehensive review, recent advances, and future research directions, Sensors 22 (14) (2022) 5168.

[130] A.I. Awad, M.M. Fouda, M.M. Khashaba, E.R. Mohamed, E. Hosny, Utilization of mobile edge computing on the Internet of Medical Things: a survey, ICT Express 9 (3) (2023) 473–485, https://doi.org/10.1016/j.icte.2022.05.006.

[131] A.A. Elnaim, et al., Energy consumption for cognitive radio network enabled multi-access edge computing, in: 2023 3rd International Conference on Emerging Smart Technologies and Applications (eSmarTA), IEEE, 2023, https://doi.org/10.1109/eSmarTA59349.2023.10293270.

[132] E.U. Ogbodo, A.M. Abu-Mahfouz, A.M. Kurien, A survey on 5G and LPWAN-IoT for improved smart cities and remote area applications: from the aspect of architecture and security, Sensors 22 (16) (2022) 6313.

[133] J. Sanchez-Gomez, et al., Integrating LPWAN technologies in the 5G ecosystem: a survey on security challenges and solutions, IEEE Access 8 (2020) 216437–216460.

[134] S. Ugwuanyi, G. Paul, J. Irvine, Survey of IoT for developing countries: performance analysis of LoRaWAN and cellular electronics networks, Electronics 10 (18) (2021) 2224.

[135] B.S. Chaudhari, M. Zennaro, S. Borkar, LPWAN technologies: emerging application characteristics, requirements, and design considerations, Future Internet 12 (3) (2020) 46.

[136] N. Tadrist, O. Debauche, S. Mahmoudi, A. Guttadauria, Towards low-cost IoT and LPWAN-based flood forecast and monitoring system, Journal of Ubiquitous Systems & Pervasive Networks 17 (1) (2022).

[137] C. Colman-Meixner, et al., Deploying a novel 5G-enabled architecture on city infrastructure for ultra-high definition and immersive media production and broadcasting, IEEE Transactions on Broadcasting 65 (2) (2019) 392–403.

[138] D. Sabella, A. Vaillant, P. Kuure, U. Rauschenbach, F. Giust, Mobile-edge computing architecture: the role of MEC in the Internet of Things, IEEE Consumer Electronics Magazine 5 (4) (2016) 84–91.

[139] M. García-Valls, A. Dubey, V. Botti, Introducing the new paradigm of social dispersed computing: applications, technologies and challenges, Journal of Systems Architecture 91 (2018) 83–102.

[140] G. Caiza, M. Saeteros, W. Oñate, M.V. Garcia, Fog computing at industrial level, architecture, latency, energy, and security: a review, Heliyon 6 (4) (2020) e03706.

[141] A.A. Barakabitze, A. Ahmad, R. Mijumbi, A. Hines, 5G network slicing using SDN and NFV: a survey of taxonomy, architectures and future challenges, Computer Networks 167 (2020) 106984.
[142] T. Taleb, et al., On multi-access edge computing: a survey of the emerging 5G network edge cloud architecture and orchestration, IEEE Communications Surveys and Tutorials 19 (3) (2017) 1657–1681.
[143] N.S. Ali, A.S.A. Gaid, M.M. Saeed, R.A. Saeed, B. Hawash, High gain, E-shaped microstrip antenna with two identical slits for 5G applications in the 60 GHz band, in: 2023 3rd International Conference on Emerging Smart Technologies and Applications (eSmarTA), Taiz, Yemen, Springer, 2023, pp. 1–7, https://doi.org/10.1109/eSmarTA59349.2023.10293412.
[144] F. Giust, et al., Multi-access edge computing: the driver behind the wheel of 5G-connected cars, IEEE Communications Standards Magazine 2 (3) (2018) 66–73.
[145] Mona Bakri Hassan Dahab, et al., Artificial intelligence and machine learning approaches in smart city services, in: K. Hemant Kumar Reddy, et al. (Eds.), Handbook of Research on Network-Enabled IoT Applications for Smart City Services, IGI Global, 2023, pp. 339–352, https://doi.org/10.4018/979-8-3693-0744-1.ch019.
[146] Mona Bakri Hassan, Elmustafa Sayed Ali Ahmed, Rashid A. Saeed, Machine learning for industrial IoT systems, in: Jingyuan Zhao, V. Vinoth Kumar (Eds.), Handbook of Research on Innovations and Applications of AI, IoT, and Cognitive Technologies, IGI Global, Hershey, PA, 2021, pp. 336–358, https://doi.org/10.4018/978-1-7998-6870-5.ch023.
[147] Sara A. Mahboub, Elmustafa Sayed Ali Ahmed, Rashid A. Saeed, Smart IDS and IPS for cyber-physical systems, in: Ashish Kumar Luhach, Atilla Elçi (Eds.), Artificial Intelligence Paradigms for Smart Cyber-Physical Systems, IGI Global, Hershey, PA, 2021, pp. 109–136, https://doi.org/10.4018/978-1-7998-5101-1.ch006.
[148] Nagi Faroug M. Osman, Ali Ahmed A. Elamin, Elmustafa Sayed Ali Ahmed, Rashid A. Saeed, Cyber-physical system for smart grid, in: Ashish Kumar Luhach, Atilla Elçi (Eds.), Artificial Intelligence Paradigms for Smart Cyber-Physical Systems, IGI Global, Hershey, PA, 2021, pp. 301–323, https://doi.org/10.4018/978-1-7998-5101-1.ch014.
[149] M.M. Saeed, et al., Task reverse offloading with deep reinforcement learning in multi-access edge computing, in: 2023 9th International Conference on Computer and Communication Engineering (ICCCE), IEEE, 2023, https://doi.org/10.1109/ICCCE58854.2023.10246081.
[150] Rania Salih Abdalla, Sara A. Mahbub, Rania A. Mokhtar, Elmustafa Sayed Ali, Rashid A. Saeed, IoE design principles and architecture, in: Internet of Energy for Smart Cities: Machine Learning Models and Techniques, CRC Press Publisher, 2020, https://doi.org/10.4018/978-1-7998-5101-1.
[151] Zeinab E. Ahmed, et al., Monitoring of wildlife using unmanned aerial vehicle (UAV) with machine learning, in: Applications of Machine Learning in UAV Networks, IGI Global, 2024, pp. 97–120, https://doi.org/10.4018/979-8-3693-0578-2.ch005.
[152] E.S. Ali, M.B. Hassan, R.A. Saeed, Machine learning technologies on Internet of Vehicles, in: N. Magaia, G. Mastorakis, C. Mavromoustakis, E. Pallis, E.K. Markakis (Eds.), Intelligent Technologies for Internet of Vehicles. Internet of Things (Technology, Communications, and Computing), Springer, Cham, 2021, https://doi.org/10.1007/978-3-030-76493-7_7.
[153] Rofida O. Dirar, Rashid A. Saeed, Mohammad Kamrul Hasan, Musse Mahmud, Persistent overload control for backlogged machine to machine communications in long term

evolution advanced networks, Journal of Telecommunication, Electronic and Computer Engineering (JTEC) 9 (3) (Dec. 2017).
[154] Rashid A. Saeed, Mohammed Al-Magboul, Rania A. Mokhtar, Machine-to-machine communication, in: Encyclopedia of Information Science and Technology, third edition, IGI Global, July 2014, pp. 6195–6206, https://doi.org/10.4018/978-1-4666-5888-2.
[155] Fahad A. Alqurashi, F. Alsolami, S. Abdel-Khalek, Elmustafa Sayed Ali, Rashid A. Saeed, Machine learning techniques in the Internet of UAVs for smart cities applications, Journal of Intelligent & Fuzzy Systems 24 (4) (2021) 1–24, https://doi.org/10.3233/JIFS-211009.
[156] Elmustafa Sayed Ali, Mohammad Kamrul Hasan, Rosilah Hassan, Rashid A. Saeed, Mona Bakri Hassan, Shayla Islam, Nazmus Shaker Nafi, Savitri Bevinakoppa, Machine learning technologies for secure vehicular communication in Internet of vehicles: recent advances and applications, Journal of Security and Communication Networks (SCN) 2021 (2021) 8868355, https://doi.org/10.1155/2021/8868355, Wiley-Hindawi.
[157] R.H. Aswathy, P. Suresh, M.Y. Sikkandar, S. Abdel-Khalek, H. Alhumyani, et al., Optimized tuned deep learning model for chronic kidney disease classification, Computers, Materials & Continua 70 (2) (2022) 2097–2111.
[158] R.F. Mansour, N.M. Alfar, S. Abdel-Khalek, M. Abdelhaq, R.A. Saeed, R. Alsaqour, Optimal deep learning based fusion model for biomedical image classification, Expert Systems 39 (3) (2022) e12764.
[159] Elmustafa Sayed Ali Ahmed, Zahraa Tagelsir Mohammed, Mona Bakri Hassan, Rashid A. Saeed, Algorithms optimization for intelligent IoV applications, in: Jingyuan Zhao, V. Vinoth Kumar (Eds.), Handbook of Research on Innovations and Applications of AI, IoT, and Cognitive Technologies, IGI Global, Hershey, PA, 2021, pp. 1–25, https://doi.org/10.4018/978-1-7998-6870-5.ch001.
[160] Z.E. Ahmed, A.A. Hashim, R.A. Saeed, M.M. Saeed, Mobility management enhancement in smart cities using software defined networks, Scientific African 22 (2023) e01932, https://doi.org/10.1016/j.sciaf.2023.e01932.
[161] Mamoon M. Saeed, et al., Anomaly detection in 6G networks using machine learning methods, Electronics 12 (15) (2023) 3300, https://doi.org/10.3390/electronics12153300.
[162] Asif Khan, Jian Ping Li, Mohammad Kamrul Hasan, Naushad Varish, Zulkefli Mansor, Shayla Islam, Rashid A. Saeed, Majid Alshammari, Hesham Alhumyani, PackerRobo: model-based robot vision self-supervised learning in CART, Alexandria Engineering Journal 61 (12) (2022) 12549–12566, https://doi.org/10.1016/j.aej.2022.05.043.
[163] R.A. Saeed, M. Omri, S. Abdel-Khalek, et al., Optimal path planning for drones based on swarm intelligence algorithm, Neural Computing & Applications 34 (2022) 10133–10155, https://doi.org/10.1007/s00521-022-06998-9.
[164] Lina Elmoiz Anatabine, Elmustafa Sayed Ali, Rania A. Mokhtar, Rashid A. Saeed, Hesham Alhumyani, Mohammad Kamrul Hasan, Deep and reinforcement learning technologies on Internet of Vehicle (IoV) applications: current issues and future trends, Journal of Advanced Transportation 2022 (2022) 1947886, https://doi.org/10.1155/2022/1947886.
[165] Othman O. Khalifa, Muhammad H. Wajdi, Rashid A. Saeed, Aisha H.A. Hashim, Muhammed Z. Ahmed, Elmustafa Sayed Ali, Vehicle detection for vision-based intelligent transportation systems using convolutional neural network algorithm, Journal of Advanced Transportation 2022 (2022) 9189600, https://doi.org/10.1155/2022/9189600.
[166] Mamoon M. Saeed, et al., Green machine learning approach for QoS improvement in cellular communications, in: 2022 IEEE 2nd International Maghreb Meeting of the

Conference on Sciences and Techniques of Automatic Control and Computer Engineering (MI-STA), IEEE, 2022, pp. 523–528, https://doi.org/10.1109/MI-STA54861.2022.9837585.

[167] Alaa M. Mukhtar, Rashid A. Saeed, Rania A. Mokhtar, Elmustafa Sayed Ali, Hesham Alhumyani, Performance evaluation of downlink coordinated multipoint joint transmission under heavy IoT traffic load, Wireless Communications and Mobile Computing 2022 (2022) 6837780, https://doi.org/10.1155/2022/6837780.

[168] L.E. Alatabani, E.S. Ali, R.A. Saeed, Deep learning approaches for IoV applications and services, in: N. Magaia, G. Mastorakis, C. Mavromoustakis, E. Pallis, E.K. Markakis (Eds.), Intelligent Technologies for Internet of Vehicles. Internet of Things (Technology, Communications, and Computing), Springer, Cham, 2021, https://doi.org/10.1007/978-3-030-76493-7_8.

[169] M.B. Hassan, S. Alsharif, H. Alhumyani, et al., An enhanced cooperative communication scheme for physical uplink shared channel in NB-IoT, Wireless Personal Communications 120 (2021) 2367–2386, https://doi.org/10.1007/s11277-021-08067-1.

[170] Mona Bakri Hassan, Elmustafa Sayed Ali, Nahla Nurelmadina, Rashid A. Saeed, Artificial intelligence in IoT and its applications, in: Intelligent Wireless Communications (Telecommunications), IET Digital Library, 2021, pp. 33–58, https://doi.org/10.1049/PBTE094E_ch2, Chap. 2.

[171] Rashid A. Saeed, et al., Enhancing medical services through machine learning and UAV technology: applications and benefits, in: Applications of Machine Learning in UAV Networks, IGI Global, 2024, pp. 307–343, https://doi.org/10.4018/979-8-3693-0578-2.ch012.

[172] Mona Bakri Hassan, Elmustafa Sayed Ali, Rania A. Mokhtar, Rashid A. Saeed, Bharat S. Chaudhari, NB-IoT: concepts, applications, and deployment challenges, in: Bharat S. Chaudhari, Marco Zennaro (Eds.), LPWAN Technologies for IoT and M2M Applications, Elsevier, ISBN 9780128188804, March 2020, Book Chapter (Ch 6).

[173] Nada M. Elfatih, Mohammad Kamrul Hasan, Zeinab Kamal, Deepa Gupta, Rashid A. Saeed, Elmustafa Sayed Ali, Md. Sarwar Hosain, Internet of vehicle's resource management in 5G networks using AI technologies: current status and trends, IET Communications (2021) 1–21, https://doi.org/10.1049/cmu2.12315.

[174] Nahla Nurelmadina, Mohammad Kamrul Hasan, Imran Mamon, Rashid A. Saeed, Khairul Akram, Zainol Ariffin, Elmustafa Sayed Ali, Rania A. Mokhtar, Shayla Islam, Eklas Hossain, Md. Arif Hassan, A systematic review on cognitive radio in low power wide area network for industrial IoT applications, Sustainability 13 (1) (2021) 338, https://doi.org/10.3390/su13010338, MDPI.

[175] Rashid A. Saeed, Ahmed A.M. Hassan Mabrouk, Amitava Mukherjee, Francisco Falcone, K. Daniel Wong, WiMAX, LTE and WiFi interworking, Journal of Computer Systems, Networks, and Communications 2010 (2010) 754187, Hindawi Publishing Corporation.

[176] S.N. Ghorpade, M. Zennaro, B.S. Chaudhari, R.A. Saeed, H. Alhumyani, S. Abdel-Khaled, A novel enhanced quantum PSO for optimal network configuration in heterogeneous industrial IoT, IEEE Access 9 (2021) 134022–134036, https://doi.org/10.1109/ACCESS.2021.3115026.

[177] S.N. Ghorpade, M. Zennaro, B.S. Chaudhari, R.A. Saeed, H. Alhumyani, S. Abdel-Khaled, Enhanced differential crossover and quantum particle swarm optimization for IoT applications, IEEE Access 9 (2021) 93831–93846, https://doi.org/10.1109/ACCESS.2021.3093113.

[178] Mamoon Mohammed Ali Saeed, Rashid A. Saeed, Zeinab E. Ahmed, Data security and privacy in the age of AI and digital twins, in: Digital Twin Technology and AI Implementations in Future-Focused Businesses, IGI Global, 2024, pp. 99–124, https://doi.org/10.4018/979-8-3693-1818-8.ch008.
[179] Abbas Alnazir, Rania A. Mokhtar, Hesham Alhumyani, Elmustafa Sayed Ali, Rashid A. Saeed, S. Abdel-khalek, Quality of services based on intelligent IoT WLAN MAC protocol dynamic real-time applications in smart cities, Computational Intelligence and Neuroscience 2021 (2021) 2287531, https://doi.org/10.1155/2021/2287531.
[180] Zeinab E. Ahmed, Hasan Kamrul, Rashid A. Saeed, Sheroz Khan, Shayla Islam, Mohammad Akharuzzaman, Rania A. Mokhtar, Optimizing energy consumption for cloud Internet of things, Frontiers in Physics 8 (2020), https://doi.org/10.3389/fphy.2020.00358.
[181] Zeinab E. Ahmed, Rashid A. Saeed, Sheetal N. Ghopade, Amitava Mukherjee, Energy optimization in LPWANs by using heuristic techniques, in: Bharat S. Chaudhari, Marco Zennaro (Eds.), LPWAN Technologies for IoT and M2M Applications, Elsevier, ISBN 9780128188804, March 2020, Book Chapter (Ch 11).
[182] S.E. Abdelsamad, M.A. Abdelteef, O.Y. Elsheikh, Y.A. Ali, T. Elsonni, M. Abdelhaq, R. Alsaqour, R.A. Saeed, Vision-based support for the detection and recognition of drones with small radar cross sections, Electronics 12 (2023) 2235, https://doi.org/10.3390/electronics12102235.
[183] Mohamed M. Siddik, Thowiba E. Ahmed, Fatima R. Ahmed, Rania A. Mokhtar, Elmustafa S. Ali, Rashid A. Saeed, Development of health digital GIS map for tuberculosis disease distribution analysis in Sudan, Journal of Healthcare Engineering 2023 (2023) 6479187, https://doi.org/10.1155/2023/6479187.
[184] M. Hassan, M. Singh, K. Hamid, R. Saeed, M. Abdelhaq, R. Alsaqour, N. Odeh, Enhancing NOMA's spectrum efficiency in a 5G network through cooperative spectrum sharing, Electronics 12 (2023) 815, https://doi.org/10.3390/electronics12040815.
[185] Elmustafa Sayed Ali, Rashid A. Saeed, Ibrahim Khider Eltahir, Othman O. Khalifa, A systematic review on energy efficiency in the internet of underwater things (IoUT): recent approaches and research gaps, Journal of Network and Computer Applications 213 (2023) 103594, https://doi.org/10.1016/j.jnca.2023.103594.
[186] M.S. Elbasheir, R.A. Saeed, S. Edam, Multi-technology multi-operator site sharing: compliance distance analysis for EMF exposure, Sensors 23 (2023) 1588, https://doi.org/10.3390/s23031588.
[187] Bilal Ur Rehman, Mohammad I. Babar, Gamil Abdel Azim, Muhammad Amir, Hesham Alhumyani, Mohammed S. Alzaidi, Majid Alshammari, Rashid Saeed, Uplink power control scheme for spectral efficiency maximization in NOMA systems, Alexandria Engineering Journal 64 (2023) 667–677, https://doi.org/10.1016/j.aej.2022.11.030.
[188] O.O. Khalifa, A. Roubleh, A. Esgiar, M. Abdelhaq, R. Alsaqour, A. Abdalla, E.S. Ali, R. Saeed, An IoT-platform-based deep learning system for human behavior recognition in smart city monitoring using the Berkeley MHAD datasets, Systems 10 (2022) 177, https://doi.org/10.3390/systems10050177.
[189] L.E. Alatabani, R.A. Saeed, E.S. Ali, R.A. Mokhtar, O.O. Khalifa, G. Hayder, Vehicular network spectrum allocation using hybrid NOMA and multi-agent reinforcement learning, in: G.H.A. Salih, R.A. Saeed (Eds.), Sustainability Challenges and Delivering Practical Engineering Solutions, in: Advances in Science, Technology & Innovation, Springer, Cham, 2023, https://doi.org/10.1007/978-3-031-26580-8_23.
[190] R.A.M. Elnour, et al., Social Internet of Things (SIoT) localization for smart cities traffic applications, in: G.H.A. Salih, R.A. Saeed (Eds.), Sustainability Challenges and

Delivering Practical Engineering Solutions, in: Advances in Science, Technology & Innovation, Springer, Cham, 2023, https://doi.org/10.1007/978-3-031-26580-8_24.
[191] K. Kuna, R.A. Saeed, E.S. Ali, A. Babiker, Self-organizing algorithm for fairness in joint admission and power control for cognitive radio cellular network, in: G.H.A. Salih, R.A. Saeed (Eds.), Sustainability Challenges and Delivering Practical Engineering Solutions, in: Advances in Science, Technology & Innovation, Springer, Cham, 2023, https://doi.org/10.1007/978-3-031-26580-8_11.
[192] M. Barakat, R.A. Saeed, S. Edam, A.A. Elnaim, I. Nasar, Performance evaluation of multi-access edge computing for blended learning services, in: 2024 21st Learning and Technology Conference (L&T), Jeddah, Saudi Arabia, 2024, pp. 197–202, https://doi.org/10.1109/LT60077.2024.10469103.
[193] L.E. Alatabani, E.S. Ali, R.A. Mokhtar, O.O. Khalifa, R.A. Saeed, Robotics architectures-based machine learning and deep learning approaches, in: 8th International Conference on Mechatronics Engineering (ICOM 2022), Kuala Lumpur, Malaysia, 2022, pp. 107–113, https://doi.org/10.1049/icp.2022.2274.
[194] M. Hoque, S.S.B. Farhad, S. Dewanjee, Z. Alom, R.A. Mokhtar, R.A. Saeed, O.O. Khalifa, E.S. Ali, M. Abdul Azim, Green communication in 6G, in: 8th International Conference on Mechatronics Engineering (ICOM 2022), Kuala Lumpur, Malaysia, 2022, pp. 101–106, https://doi.org/10.1049/icp.2022.2273.
[195] T. Ali, R.A. Saeed, O.O. Khalifa, E.S. Ali, N. Odeh, G. Hayder, A.A. Hashim, Agile Enterprise Geographic Information System (AEGIS) from design and development perspective, in: 8th International Conference on Mechatronics Engineering (ICOM 2022), Kuala Lumpur, Malaysia, 2022, pp. 26–31, https://doi.org/10.1049/icp.2022.2260.
[196] M.K. Hasan, M. Junjie, A.K.M. Ahasan Habib, A. Al Mamun, T.M. Ghazal, R.A. Saeed, IoT-based warehouse management system, in: 2022 International Conference on Cyber Resilience (ICCR), Dubai, United Arab Emirates, 2022, pp. 1–6, https://doi.org/10.1109/ICCR56254.2022.9995768.

CHAPTER 10

Non-static TinyML for ad hoc networked devices

Evangelia Fragkou[b] and Dimitrios Katsaros
Department of Electrical and Computer Engineering, Volos, Greece

10.1 Introduction

Nowadays, the availability of data produced by edge devices is high, leading to an increase in techniques able to process these data, so new methods of machine learning/deep learning (ML/DL) have emerged. In the existing literature, ML/DL models are trained in devices without computational limitations like servers or the cloud and then the model can be deployed in any other device, running the known inference task. However, the tremendous success of the Internet of Things (IoT) networks [1] has solemnified the demands of real-time processing of data at the edge. Processing data at the core from which they originate provides data safety, since information remains at the edge device, and low latency, as data do not have to be transmitted to a server for processing, resulting in less total energy consumption.

However, taking into account the lack of resources, even inference tasks are not capable of running on devices with extremely restricted computational resources, like storage or energy consumption limitations, for instance microcontroller units (MCU) allotted with some megabytes of flash memory. This bottleneck in ML/DL deployment led to the development of *tiny machine learning* (*TinyML*) [1–3]. TinyML aims at producing compact models that can satisfactorily run inference tasks in the aforementioned devices. The proposed method includes off-line training of a model, and then, using a suitable tool, e.g., TensorFlow Lite [4], an interpreter is used to convert the pretrained model into a small memory-efficient one, applicable to an MCU, in which it is further applied.

Although progress has been made, taking into account that inference tasks can now be deployed on extremely resource-limited devices in some cases, inference-based methodologies place constraints on the efforts, so the really successful models can surface. This occurs since these methods are not able to adjust to new incoming data. The reason is that in inference tasks we can only modify the classifier of the network, not the core of the model responsible for the final classification (*static TinyML*). So, as mentioned in [5], the processing of data at the edge necessitates

[b] The research work is supported by the Hellenic Foundation for Research and Innovation (HFRI) under the 3rd Call for HFRI PhD Fellowships (Fellowship Number: 5631).

TinyML for Edge Intelligence in IoT and LPWAN Networks. https://doi.org/10.1016/B978-0-44-322202-3.00015-4

the introduction of *reformable TinyML* (non-static TinyML) techniques, or in other words, the development of techniques which reduce the size of neural networks and thus the number of parameters during training, leading to better generalization of the processed data and further performance enhancement. In [6], the techniques called *pruning* and *sparsification* are extensively demonstrated, which are among the most widely known methods of accelerating the learning process of the model by eliminating unnecessary weights/connections/filters.

Despite the fact that non-static TinyML seems to be a productive means of successfully addressing the problem of proper raw data integration into the model, it demands both abundance of data and enormous execution time in order for the model to converge, conditions that are not in line with the TinyML device specifications. This problem is tackled to a great extent by *transfer learning* (TL) [7]. TL is a promising technique in which knowledge of a trained model on a task A (usually on a computationally efficient machine with data availability, like servers) can be used as preknowledge and further applied to an aforementioned device, e.g., an MCU, to learn its own task B. It is deduced in [8] that pretrained weights used for neural network initialization can lead to faster convergence of the model, irrespective of the task they were produced by.

However, spurring the deployment of training procedures at the edge and making endeavors to perform effectually learning "on the fly," it is significant to design and thus construct suitable decentralized neural network topologies (*backbones*) that can support cooperative training (*federated learning* [FL]). Edge devices lack not only computational resources, but also availability of data, making FL, in which a global model is trained by many devices, sending each other, e.g., gradients, instead of the data themselves, a promising method for successful cooperative training. There is a lot of work regarding centralized FL schemes [9], but semi/hybrid-decentralized ones [10–12] or stipulated purely decentralized schemes, like the ones edge learning concepts demand, are still in their infancy [13–17]. The cost overhead and reliability issues incurred by all-to-all nodes communication in an ad hoc network are major challenges that have to be further examined.

The rest of the chapter is organized as follows. In Section 10.2, ad hoc network topology construction and existing hardware capabilities are proposed. In Section 10.3, efficient pruning techniques are demonstrated. In Section 10.4, TL methodologies are introduced and analyzed. In Section 10.5, purely distributed FL methodologies are described. Section 10.6 introduces the main "bottlenecks" of TinyML and FL in peer-to-peer networks. Finally, Section 10.7 concludes this chapter.

10.2 Backbones and TinyML boards

10.2.1 Backbone construction

The communication reliability in edge distributed environments is of a great importance, since the research community focuses on the exploration of edge data processing and hence device independence. In *ad hoc* or *peer-to-peer* or *device-to-device* networks, all devices belonging to this network can communicate among each other without requiring a coordination of, e.g., a server node. However, these types of networks lack energy resources, since they consist of devices like sensors and hence vast amounts of information diffused to the network can cause resource depletion. So, this situation necessitates the construction of trustworthy ad hoc *backbones*, with the aim of reducing communication overhead by controlling information routing among the nodes of different local area networks (LANs) or subnetworks they need to interconnect. For instance, in [18], the combination of a *LoRa* low-power wide area network (LPWAN) single-hop star-like network topology and a mesh network architecture is used in order to implement cooperative learning in IoT networks. They take advantage of the LoRaWAN architecture protocol and the benefits that an LPWAN network topology offers, like long-range communication, low power consumption, or even low data rates, in conjunction with all-to-all communication that a mesh network provides to perform FL (see Section 10.5). However, the communication cost remains high, since every node can communicate with one another. Although there is a lot of existing literature [19], referring to smart routing protocols in distributed networks, there is little contribution regarding multi-layer resource-scarce networks. The main goal is to constrain routing of information, for instance as the distributed state-of-the art algorithms introduced in [20,21], by constructing backbones with small *cardinality*, meaning that every node is urged to transfer information between the minimum nodes, required in order for the message to be forwarded through the whole ad hoc network. By reducing unnecessary information exchange between the nodes, the energy consumption, required for every device to collaborate is reduced too, making these topologies "hospitable" to support collaborative training techniques (like FL methods) that will be further discussed in the next sections.

10.2.2 TinyML boards and embedded systems

Despite the endeavor for developing robust decentralized topology configurations, it is significant to take into account the nature of TinyML (see Section 10.3) edge devices that are part of the aforementioned networks. Specifically, TinyML embedded systems consist of MB or kB of flash memory, while they do not support energy-consuming applications, which provokes difficulty in deploying conventional approaches of ML/DL. TinyML devices, like the Raspberry Pi devices (*Raspberry Pi 3B+, Raspberry Pi 4*), based on Cortex A53 and A72 processors, respectively, belong to the family of low-power devices and have less than 2 GB of RAM [22,23].

Going further to the ultralow-power devices, like *Arduino Nano 33 BLE Sense*, the availability of flash memory is approximately 1 MB to 2 MB. This series of Arduino models features a Cortex-M4 processor (especially designed for ultralow-power devices and used for general purposes) [24,25], which reinforces DL/ML tasks. The series of STM32 microcontrollers, mentioned in [22,26,27], are also based on the family of Cortex processors. For example, the high-performance MCU *STM32F7* features a Cortex-M7 microprocessor with approximately 512 kB of SRAM and 2 MB of flash, while the general-purpose MCU *STM32F401* consists only of 96 kB of SRAM and 128 kB to 512 kB of flash memory. Similar performance is gained by the *STM NUCLEO L496ZG* (L4 – ultralow-power MCU), *STM DISCO F496NI* (F4 – over-balanced MCU), and *STM NUCLEO F767ZI* (F7 – high-performance MCU).

Furthermore, in [28], [29], PULP (Parallel Ultra-Low-Power Processing Platform) and Risc-V-based microprocessors, namely, *GAP8*, *VEGA*, and *Mr.Wolf*, are used for supporting parallel processing and multi-caching.

10.3 Neural network model reduction

ML/DL methods have high computational demands, since they need both sufficient data and much training execution time, in order to learn these data and converge. Optimization efforts like multicore parallel execution regarding both CPUs and GPUs, deep learning caching techniques [30], etc., try to bridge the gap between the efficient training/inference of a model and the enormous execution time/computational cost needed by such large models, e.g., GPT-x [31], having billions of connections among its neurons. However, talking about resource-constrained devices/environments, e.g., the Internet of Things (IoT), this is not feasible, taking into account that these devices lack computational power and memory (approximately MB or kB of RAM for example in sensors, embedded systems, etc., as described in Section 10.2.2) and the existence of a tremendous number of parameters, i.e., weights impact both inference time and training time. Conventional TinyML approaches support inference in these devices by compressing the ML/DL code and then deploying it in them. However, this technique does not enable adaptation to new raw data and thus it cannot provide satisfactory results in the future. In this work, we are going to demonstrate efficient methods (TinyML [32]) of minimizing computational cost and hence memory requirements of ML/DL algorithms during the training phase, so as to be capable of running on the aforementioned devices. A promising and extensively demonstrated concept of reducing computational demands of ML/DL methods is to reduce the size of the neural network we want to train by making it more "sparse," meaning that we can eliminate elements of the network (pruning) that do not or minimally contribute to its performance. We can classify pruning techniques into two main categories: one in which neurons are removed (e.g., dropout [33]) and one in which synapses (connections among neurons) are removed. Moreover, pruning is classified as static pruning, in which we prune the network in its initialization phase and fine-tune it in order to converge, or dynamic pruning, in which we prune and regrow the network iteratively in every epoch. The crucial part is to determine which pruning criteria work better in every case, regarding the nature of the network we use. It was observed in [34] and [6] that random pruning techniques are more efficient when they are applied for weight removal rather than neuron/filter removal in the network initialization phase. However, even in this case randomly eliminating information from the network may severely reduce network performance.

There exist straightforward techniques for pruning (removing unnecessary weights) after the training of a neural network and before inference, so as to accelerate the operation of the network. However, the real challenge is to do this during training. So how can neural model reduction be performed?

10.3.1 The lottery ticket hypothesis

Contemporary experience shows that there exists a small part of a dense neural network that can achieve (almost) the same test accuracy as the whole network. This observation was initially tested on fully connected multilayer perceptrons and convolutional neural networks (CNNs) in vision tasks and was summarized in the famous *lottery ticket hypothesis* described in [35] as follows:

A randomly-initialized, dense neural network contains a subnetwork (winning ticket) that is initialized such that – when trained in isolation – it can match the test accuracy of the original network after training for at most the same number of iterations.

In order to identify a winning ticket, an appropriate algorithm initializes and trains the whole network and after the completion of training removes the connections that correspond to the weights with the smallest values. After that pruning step, the algorithm restores the weights of the surviving connections to the values they got at the initialization step! The training and pruning could take place once at the end of the training (one-shot) or every time after a specific number of iterations, eliminating only a percentage of the surviving connections (iterative pruning).

A surge of works followed the initial article that generalized this technique to the *multi-prize lottery ticket hypothesis* [36] or applied the initial ideas to spiking neural networks [26,37,38], in few-shot learning [39], and so on.

10.3.2 Topology sparsification prospects derived from network science disciplines

The challenge in ML and the reason of introducing TinyML is that the learning process has shifted to the edge (IoT networks, sensors, etc.), meaning that every device has to be able to learn its own data, without being in need of, e.g., a server which reinforces it. Since the aforementioned devices (see Section 10.2.2) do not have enough computing power, the need to create more compact neural networks that still maintain their high accuracy was the main trigger.

The technique called *sparsification* is promising since it is aimed at reducing model size and hence memory requirements by dropping out unnecessary – regarding the training procedure – nodes/connections of the network. Additionally, there is great inspiration in the structure of the synapses in the human brain and hence the way the human brain processes incoming information. More specifically, the human brain creates thousands of synapses every day as it is exposed to new data (and consequently produces information), while it has the ability to make a decision of which of them to keep, since they are not all equally useful. At the same time, network science introduces the concepts of scale-free networks and based on them describes a variety of networks that we use in everyday life, such as the World Wide Web (web). More meticulously, in scale-free networks, the nodes follow a power law distribution, i.e., the nodes where most synapses end up are considered the most popular and consequently they are the nodes that transfer information between the nodes of the network.

So, inspired by the theory mentioned above, an idea was to remove the connections between the nodes of a bipartite fully connected feedforward *multilayer perceptron* (MLP) neuron network that do not actually contribute to the network (in other words, the weight values of these connections tend to be zero), while concurrently adding as many weights as we removed. The whole process of sparsification takes place during the training phase, in which we iteratively both remove and regrow (*dynamic sparsity during training*, according to [6]) a specific amount of connections (approximately 30%), with the aim of reorganizing the structure of the neural network. The work described in [40] (SET) was the first to correlate the functionality of scale-free networks with that of deep neural networks, but the initialization of synapses in thc first untrained network is done in a random way (Erdos–Renyi type network), while only the final network is structured, following a scale-free law. In [41], an optimized version of it was described. In [42], both the initial and the final network are reconstructed after every epoch of training not in a random way, but following network theories such as scale-free or small-world networks, with the aim of strengthening the nodes with the most attached connections (and thus those responsible for the guaranteed dissemination of information in the network) before and after every epoch. Furthermore, taking into account that the weight of a pruned connection is considered to be acquired knowledge [43], an experiment with keeping the weight value before connection removal and reassigning it to the new one was conducted. In other words, the weight values of the connection being restored are either random or the sum of the previous weights being deleted. Although the pretrained weights are some kind of preknowledge, after many connection modifications and hence weight replacements, the first task, for which weights were trained, tends to be "forgotten" (catastrophic forgetting [44] – one of the crucial problems in deep learning methodologies, attracting great scientific interest), and hence the methodology of keeping those weights tends to increase complexity of the algorithm rather than reinforcing model performance regarding accuracy levels in the long run (see the exact results in [42]).

So, practically, five algorithms were implemented, which differ in the way both the initial and the final network are redefined (either variants of scale-free [SF], like the one described in [45,46], or small world [SW], introduced by Watts and Strogatz in 1998 [47]), namely, *SF2SFrand*, *SF2SFba*, *SF2SW*, and *SW2SW*, respectively, with SF2SFrand being the one outperforming the other variants introduced. Regarding SW2SW implementations, we experimented with two values ($p = 0.02$ and $p = 0.075$), given to the rewiring probability of every node in the network, making the network configuration either denser or more random, respectively (see Table 10.2 for further details, in which specifically SF2SFrand seems to retain high orders of accuracy, while reducing training time approximately to 50% in some cases). It is worth highlighting that reinforcing the strongest nodes of every layer, the performance of the network is proportionately enhanced too. However, small world-based implementations fail to make the model generalize efficiently the new data to which it is exposed. The results are not satisfying not only with respect to accuracy levels ($\leq 70\%$), but also with respect to training time needed (hours of training). This occurs due to the fact that this type of network tends to be less strictly structured (increasing randomness, as probability p increases) as the probability of a node connecting to others in a layer increases and therefore the network is more similar to an Erdos–Renyi network. The results regarding the datasets mentioned in Table 10.1 are illustrated in Figs. 10.1 and 10.2 and Table 10.3. For the experimental evaluation, a three-layered MLP neural network is used, consisting of 1000 neurons each and the widely known ReLU activation function (defined as $ReLU(x) = (x > 0)?x : 0)$. In order to prevent misleading performance results in favor of a dataset, overfitting or L1 or L2 regularization is applied during the training phase.

There was further examination regarding the case where the percentage of connections, both pruned and restored, is not stable but varies (drawing inspiration from the ideas presented in [48]), linearly (the linear decreasing variation [*LDV*] method), exponentially (the exponential decay [*EXD*] method), or following the cosine rule (oscillating variation [*OSV*]) at each epoch, since from the beginning the pruned ones were insignificant for the process of training. The results [49] show that when the percentage of close-to-zero synapses is modified exponentially or oscillates following a cosine rule function, the accuracy of the model remains approximately 90%, as depicted in Fig. 10.3, and the time needed is also at the same level as for the baseline methods (MLP as bs1 and SET as bs2), since the technique used requires enough calculations, even if the number of connections removed is larger (see Fig. 10.4). The most impressive result, making the aforementioned methods congruous for TinyML devices, is that the memory footprint is further reduced (around 95%) (while tested on the fashion-mnist dataset and four different kinds of neural network topologies; see Table 10.1), as more superfluous connections were removed from the network (see Table 10.4).

10.4 Transfer learning

TL is a widely known technique, which aims at boosting ML model performance and learning ability by bequeathing already acquired knowledge from a source device to a target device [7,57–59]. Specifically, source devices like servers train ML algorithms asynchronously, without computational limitations, while they have plenty of data available, in order to accomplish successfully the training task. On the other hand, the target device (e.g., sensors, microcontrollers, etc.) lacks computational resources and storage capacity, making the deployment of ML algorithms unfeasible. So, according to the TL method, the model initially trained on a source device can be deployed to the target device by "freezing" some of its layers while retraining the remaining ones, using its weights as preknowledge. The specific approach is a promising technique,

Table 10.1 Characteristics of some of the datasets.

Dataset	Instances		Features	Classes
lung (.mat) [50]	203		3312	5
COIL20 (.mat) [51]	1440		1024	20
Fashion-Mnist [52]	train samples 60,000	test samples 10,000	28×28	10
CIFAR-10 [53]	train samples 50,000	test samples 10,000	32×32	10
CIFAR-100 [53]	train samples 50,000	test samples 10,000	32×32	100
ImageNet [54]	train samples 1,281,167	test samples 100,000	224×224	1000
Intel Classification Dataset [55]	train samples 14,000	test samples 3000	150×150	6
Kaggle [56] - temperature of five Chinese cities for a time period of five years (2010–2015).	52,585 measurements of real temperatures in degrees Celsius.			

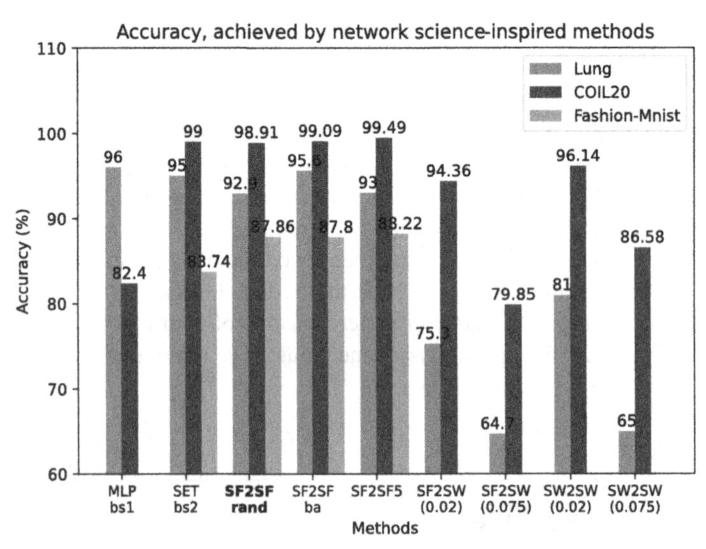

FIGURE 10.1

Accuracy gained during the training of network science-inspired algorithms with datasets mentioned in Table 10.1, described in Section 10.3.

as it combines not only predefined “knowledge,” but also on-device fine-tuning of the network, while simultaneously drastically reducing the time of online training, since it reduces the number of layers involved and therefore the total number of weights processed in the training phase.

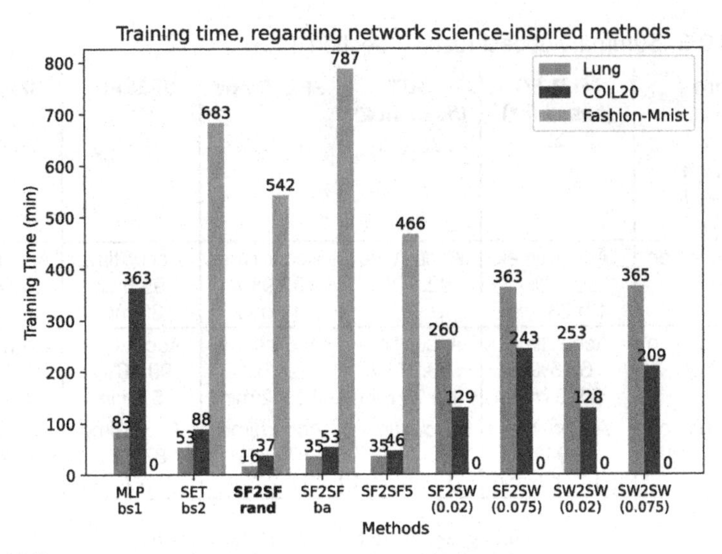

FIGURE 10.2

Training time (in minutes) of network science-inspired algorithms with datasets mentioned in Table 10.1, described in Section 10.3.

Table 10.2 Processes followed by the methods described in Section 10.3.

Method	Process followed
SF2SFbrand	We start from an initial non-arbitrarily connected node topology and conclude with a same type topology through training after linkage removal.
SF2SFba	We start from a Barabasi–Albert-based topology and conclude with a same type topology through training after linkage removal.
SF2SW (either $p = 0.02$ or $p = 0.075$)	We initialize a scale-free-like topology and reconfigure the final network topology into a small world-like network either with rewiring probability $p = 0.02$ (more structured topology) or $p = 0.075$ (more arbitrarily connected nodes) through training after linkage removal.
SW2SW (either $p = 0.02$ or $p = 0.075$)	We firstly initialize a small world-based network and redefine the final network topology into a small world-like network either with rewiring probability $p = 0.02$ (more structured topology) or $p = 0.075$ (more arbitrarily connected nodes) through training after linkage removal.

The most well-known case is the one in which only the last fully connected layer is fine-tuned, reinforcing the neural network model to adapt to the new data it is exposed to in the target device (raw data), as described in [57]. However, examining a classification task (which is one of the most frequently used tasks in ML/DL), we observe that fine-tuning only the last fully connected layer of a CNN, which is responsible for the final categorization of the training data, we can only modify the categories that the network recognizes, not somehow attain the ability of "adding" new categories (non-static training). In other words, in an image classification task,

Table 10.3 Summary of best results (Section 10.3).

Algorithm	MLP-FC (*baseline1*)	SET (*baseline2*)	SF2SFrand	SF2SFba	SF2SF(5)
Memory reduction (%) (*compared to MLP-FC*)	—	75.2%	76.31%	73.36%	74.00%
Performance on Lung.mat	Accur/time 99.52%/ 1 h 23 min	Accur/time 92.71%/ 38 min	Accur/time 92.9%/ 16 min	Accur/time 95.6%/ 32 min	Accur/time 92.8%/ 35 min
Performance on COIL20.mat	Accur/time 62.5%/ 6 h 3 min	Accur/time 99.37%/ 1 h 28 min	Accur/time 99.16%/ 1 h 42 min	Accur/time 99.16%/ 53 min	Accur/time 99.79%/ 1 h 13 min
Performance on Fashion-Mnist	Accur/time 49.94%/ 2 d 13 h 5 min	Accur/time 83.74%/ 11 h 22 min	Accur/time 87.86%/ 9 h 2 min	Accur/time 87.8%/ 13 h 7 min	Accur/time 88.22%/ 7 h 46 min

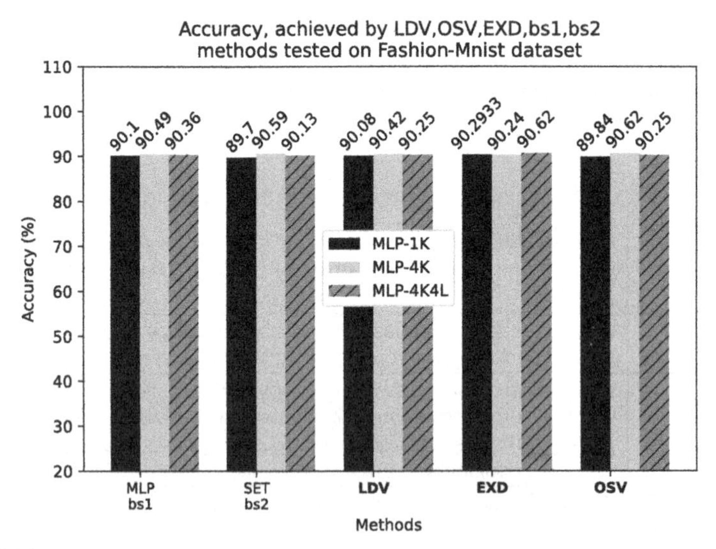

FIGURE 10.3

Accuracy gained during the training of different neural network topologies with LDV, EXD, and OSV algorithms and their baselines, described elaborately in Section 10.3.

the first layers of a CNN network (from whose functionality the main motivation of work [60] is derived) acquaint with the basic features of an input data image, and then this knowledge is transmitted among the next layers so as for the final layers to be able to "understand" the higher levels of the hierarchical structure of the image. This means that the final convolution layer is only responsible for categorizing the data processed by the network. So when the network is exposed to new data, it is of great importance to update the way that leads to the categorization of images and

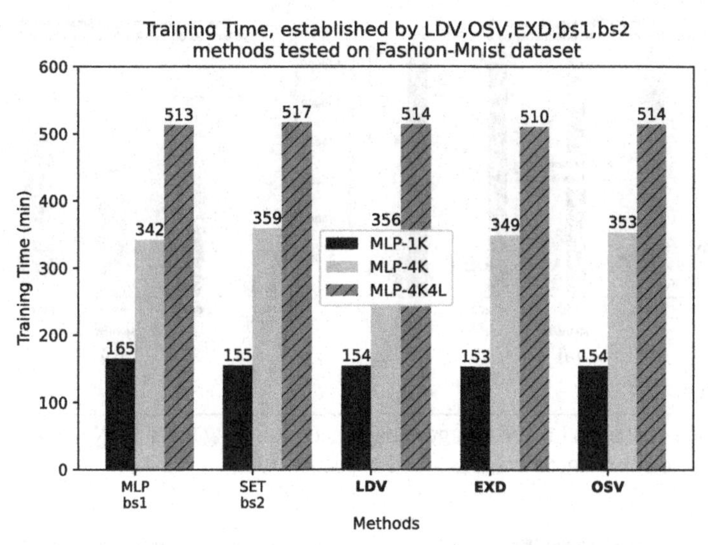

FIGURE 10.4

Training time (in minutes) of different neural network topologies using LDV, EXD, and OSV algorithms and their baselines, described elaborately in Section 10.3.

Table 10.4 Sparse MLP topologies, compared to dense ones, regarding memory footprint when using a non-stable number of pruned connections (Section 10.3).

Neural network architecture (# of hidden layers)	Size before sparsification	Size after sparsification	Reduction rate (%)
1000-1000-1000 (MLP-1K)	2.797.010	120.000	95.7%
4000-1000-4000 (MLP-4K)	11.185.010	350.000	96.8%
4000-2000-2000-1000 (MLP-4K4L)	17.155.010	380.000	97.7%

not merely the classification of the already formed categories. So, we delve into the existing research and try to tackle this problem by considering whether model performance can be improved if more than the last layers participate during the online training phase and more specifically how much its performance would improve if we retrain one or more of the previous (convolution) layers.

So, experiments were conducted [60] in which three different types of networks (Small CNN with 550,000 trainable parameters, EfficientNetB0 with 4,049,571 trainable parameters, and EfficientNetB2 with 7,768,569 trainable parameters) are used, in which more than the last fully connected layers are retrained. We call these approaches F_xC_y, where x and y denote the numbers of both fully connected and convolution layers retrained, respectively. The results, obtained by the experiments that are depicted in Figs. 10.5 to 10.8 and Table 10.5, show that the retraining of the last convolution layer can enhance model performance regarding accuracy, presenting a trade-off in the energy consumption required in order for the convolutional opera-

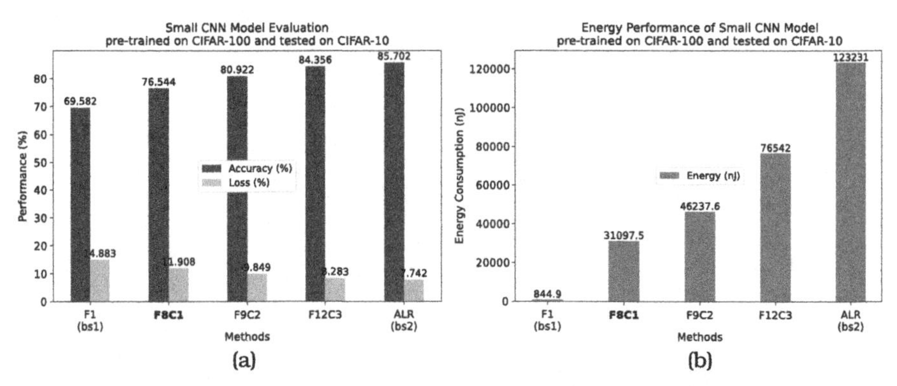

FIGURE 10.5

Performance evaluation (a) and energy consumption (b) using Small CNN neural network and transfer learning algorithms, described in Section 10.4.

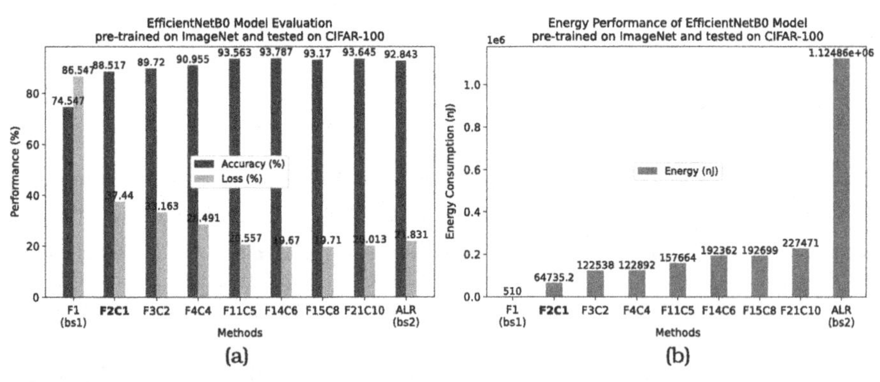

FIGURE 10.6

Performance evaluation (a) and energy consumption (b) using the EfficientNetB0 neural network and transfer learning algorithms, described in Section 10.4.

tion to be executed. In detail, the proposed method, namely, $F_x C_1$, outperforms both baseline methods (either reconfiguring the latter layer of a CNN network [$F1$-$bs1$] or including all the layers in the learning process[ALR-$bs2$]), attaining approximately 19% growth in model accuracy and about 61% faster convergence. To evaluate algorithms, accuracy, energy consumption in joules, and a categorical cross-entropy loss function were selected. Regarding energy consumption, we calculated the FLOPS an algorithm executes in every epoch of training; bearing in mind that a MAC operation in a 45 nm CMOS processor is tantamount to two FLOPS, we proportionately calculate the FLOPS consumed by our proposed approaches. The results are shown in Figs. 10.5 to 10.7.

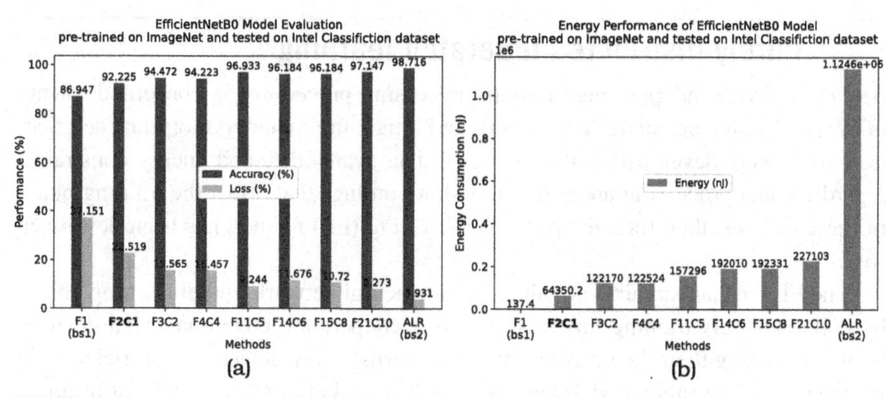

FIGURE 10.7

Performance evaluation (a) and energy consumption (b) using the EfficientNetB0 neural network and transfer learning algorithms, described in Section 10.4.

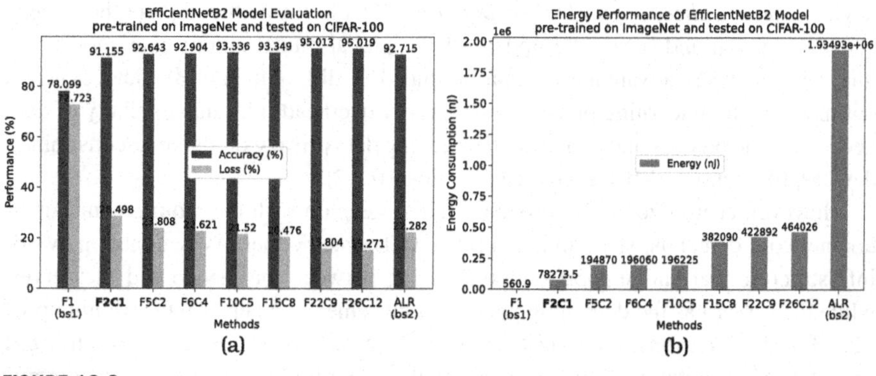

FIGURE 10.8

Performance evaluation (a) and energy consumption (b) using the EfficientNetB2 neural network and transfer learning algorithms, described in Section 10.4.

Table 10.5 Comparison of the proposed $F_x C_1$ method with its main competitor, the *baseline-1* method (Section 10.4).

Model used	Datasets for pretraining/testing	Increase in accuracy (%)	Increase in convergence rate (%)
Small CNN	CIFAR-100/CIFAR-10	10.01	19.98
EfficientNetB0	ImageNet/CIFAR-100	18.79	56.74
EfficientNetB0	ImageNet/Intel Classification	6.09	39.38
EfficientNetB2	ImageNet/CIFAR-100	16.79	60.81

10.5 Purely distributed federated learning

As far as device independence in data processing procedures is concerned, plenty of ways of reducing model parameters and hence the memory footprint (see Section 10.3) were developed, but the lack of data availability and energy constraints regarding data processing are still major shortcomings that affect the training phase of these devices; therefore, the *federated learning* (FL) method has been developed [61].

The FL mechanism aims at enhancing the neural network model learning ability by collaboratively training this model among cooperating edge connected devices, while preserving their data privacy. In other words, every device that participates in training trains the same model with its own data, and after some epochs of training, this device sends information like gradients produced during training (not the data themselves) either to a server, which then aggregates the sent model parameters, updates the model, and further broadcasts the global updated model to the rest of the participating devices (*centralized approach of FL*), or among the other participating devices (*decentralized – purely distributed FL* [PDFL]), which complete the parameter aggregation and hence the update of the global model among one another. In this way, FL takes advantage of data produced in different TinyML devices without incurring both learning process and memory overhead [62], due to plenty of data needed for the process and concurrently keeping data privacy of resource-constrained devices, like sensors, MCUs, etc. (see Section 10.2.2).

However, centralized FL approaches fail to keep up with the requirements an ad hoc network demands, since in a centralized FL scheme there is a suitable network infrastructure that can support communication between the device and the server, while the server has the demanding energy requirements to perform the training process. For these reasons, the transition from a centralized scheme to a decentralized one [9] has begun, taking into consideration the major problem that all-to-all communication costs result in. There are some efforts, mentioned in the existing literature, for example, [13–17,63,64], in which semi- or fully decentralized approaches are presented; however, they are not necessarily geared towards communication alleviating methods, as the following proposed methods. Since resource-scarce ad hoc environments needs decentralized FL solutions that can deal with the high communication cost, selecting which node participates is a critical issue and therefore the concept of *network clustering* is at the forefront of decentralized FL research [11,65].

10.5.1 Similarity-based selection of participating devices in PDFL

A practical way of mitigating information diffusion among the network of participating devices in FL is to fit them into groups (clusters) and only the representative node (device) of every cluster (clusterhead [CH]) is responsible for communicating with the other representative nodes. In this section, the configuration of the clusters is implemented using the MAX-MIN-D algorithm [66] and each CH is selected according to the well-known betweenness centrality algorithm [67], so as for the selected CH to be in the "center" of the cluster, or else to be equidistant from each node belonging to the cluster.

Moreover, an FL heuristic focusing on communication overhead reduction is demonstrated, in which the fact that some devices produce similar data is leveraged (for instance, connected devices, in 1-hop distance apart) and hence not all nodes are necessary in the training procedure. Taking into account the case of dealing with time-series data, the proposed way of seeking data similarity is based on the *discrete Fourier transform* (DFT). DFT mechanisms transform time-series data from the time domain to the frequency domain, easing even the comparison among time series, shifted through time. Furthermore, according to the Parseval's theorem, Euclidean distance can be utilized, even using only some of the first coefficients of the converted-to-frequency-space points, in order to calculate how similar the data of two different devices are. The first few coefficients are an adequate and representative sample, since the larger quantity of energy is gathered in them. The CH makes appropriate comparisons between all the nodes belonging to its cluster and hence eliminates a node or more which contain the same data as a previous node in the same cluster at the same epoch of training. It is worth highlighting that the DFT method for recognizing data similarity can be used not only for time-series data, but also for streaming data or image data, as described in [68,69].

We evaluated the aforementioned mechanism by conducting experiments using real time-series data produced by Kaggle (see Table 10.1). A part of the results is depicted in Figs. 10.9 and 10.10, in which either all nodes participate in the federation ($FEDLp2p\_ANP$) or a similarity-based criterion is used for selecting participating nodes ($FEDLp2p\_SNP$), respectively. In these figures, the performance of a representative node after the global model update phase is demonstrated, since after the global model update, all nodes have exactly the same model parameters and thus this node is an adequate sample in order to evaluate both the accuracy and the loss metric of the model. In the experiment in Fig. 10.10, either the first 5 or the first 10 coefficients of the DFT are leveraged in order to calculate the data similarity among two adjacent nodes. As we can observe, using a similarity-based criterion with 5 coefficients selected, the neural network converges faster than $FEDLp2p\_ANP$ and especially from the 15th round, the loss value remains stable and close to zero, while the accuracy attained by the model is approximately 99%. When running $FEDLp2p\_SNP$ with 10 coefficients, the results are more promising, since the accuracy values do not fluctuate, meaning that the model converges and the loss function is stably close to zero.

Finally, we can conclude that the participation of specific nodes in the federation can reduce the time complexity of algorithms, as the number of rounds of training is reduced due to faster convergence and the amount of data transmitted in the network is lower, reducing communication costs.

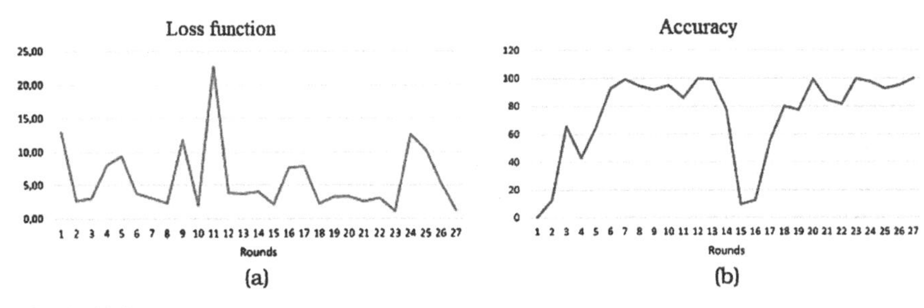

FIGURE 10.9

Demonstration of the performance (loss (a) and accuracy (b)) of $FEDLp2p\_ANP$ when running on a 20-node network, described in Section 10.5.

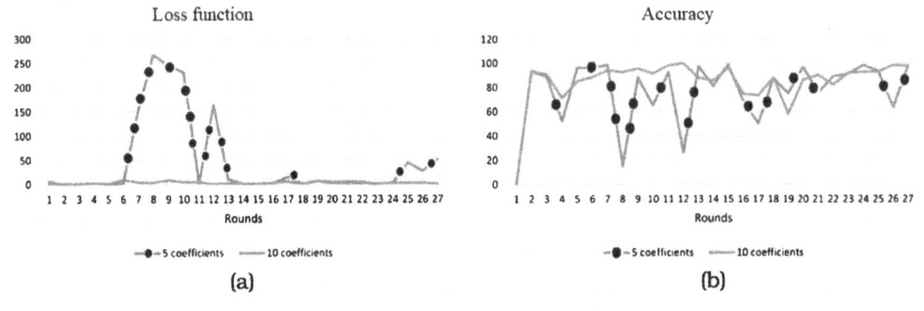

FIGURE 10.10

Demonstration of the performance (loss (a) and accuracy (b)) of $FEDLp2p\_SNP$ when running on a 20-node network, described in Section 10.5.

10.6 Challenges

- **Concept drift.** Taking into account that we are associated with IoT devices, like sensors or MCUs, after preinstalling an initially trained model on them, retraining it using data gathered by the device itself can make these devices converge faster. However, the problem that hinders the efficacy of this technique is that the streaming (raw) data of the target device may be of a different dimensionality (belonging to different feature spaces) compared to the ones inherited by the source device (heterogeneity of data); this phenomenon is called *concept drift*. Additionally, another form of concept drift is *class imbalance*, which is also a major problem in classification modeling approaches. In this case, the quantity of data every class consists of is not equal (or almost equal) between all the classes formed, since raw data preprocessing is not feasible in real-time TinyML algorithm deployment. So, incoming data are considered as biased or skewed incidental to this situation, leading to model performance degradation as the model is not adequately trained to make right predictions. Although there is a lot of work presented in the literature to prevent concept drift [70–72], there is still much room for improvement, since the complexity of tasks assigned to these kinds of devices increases.

- **Catastrophic forgetting**
  Making efforts to adjust incoming raw data to the existing knowledge, leading to the recognition of multiple tasks, in target devices can cause another problem, namely, *catastrophic forgetting*. Unlike humans' ability to learn more than one task and keep this knowledge in their memory, deep neural models execute some kind of "weight replacement," trying to fit to the new data they are exposed to. In [44], an approach to figure this problem out is demonstrated, based on recognizing which weights are the most significant ones from a probabilistic view. Furthermore, other approaches were also included, like in [73–77], specialized in either a specific network category, e.g., CNNs, or a type of learning scheme, like unsupervised learning. So, although the aforementioned annotated bibliography includes significant works for addressing this phenomenon, the actual challenge is to go further in finding ways for discovering efficient methods implementing few calculations to address this problem, since we are using resource-constrained devices, processing stream data in real-time scenarios.
- **Attacks.** ML/DL software methodologies suffer from attacks, aiming either at taking control over the neural network model and thus getting access to sensitive data or at *poisoning* the network – one of the most common attacks – by injecting deceitful data in it, in order for the model to export misleading results (white box attacks, backhaul attacks, etc.). Many approaches (adversarial ML) are already introduced in the existing literature, attempting to fortify DL algorithms, making the devices less susceptible to these attacks in [78]. From a hardware perspective, there are also vulnerabilities, according to [30], allowing Trojan viruses to invade the system. So, taking into consideration that the process of training takes place on the edge, there is still a lot of work that has to be done, so that DL models are secure and trustworthy, regarding the task they execute.
- **Benchmarking.** The lack of datasets, frameworks, and tasks adjusted to TinyML technology specifications in order to compare TinyML implementation feasibility for deployment is a serious deficiency that has to be mitigated, since this situation holds the scientific community back in releasing complete reliable TinyML systems.
- **Compatible-to-TinyML ad hoc decentralized topologies**
  As processing of data is shifting towards the edge, we anticipate the design of purely decentralized topologies, making the participating devices fully autonomous. Edge processing of data (either collaboratively among one another or with each device performing its own task) reduces the latency of data transmission (from device to server/cloud and vice versa), while data safety is ensured as data themselves do not circulate in the network. However, fully decentralized schemes demand secure, efficient, and trustworthy communication among the nodes of the network.
- **Communication cost in decentralized FL**
  In the literature, there is a plethora of FL algorithms, mainly in centralized environments, in which resource-limited devices are able to cooperatively train a global model, exchanging parameters, e.g., gradients, among each other. By shift-

ing to ad hoc peer-to-peer decentralized networks, there is not a central server coordinating the communication of devices, but devices themselves are responsible for keeping each other up-to-date, integrating the updated global model. So, the main "bottleneck" is the communication cost. Is it necessary to implement all-to-all communication? Is there an efficient way of electing which node can take place in global training? In Section 10.5, we briefly summarized some insights and demonstrated specific results regarding node participation protocols based on data similarity. However, the exploration of new efficient ways of solving this shortcoming in this field has yet to complete. Finally, is there a solution to reduce information diffused to the network, even more simple and less memory-expensive than model gradients? Purely decentralized learning schemes are an upsurging field, which the scientific community has to attach importance to.

10.7 Conclusions

All things considered, the necessity of developing new ML/DL techniques, applicable to edge resource-scarce devices, e.g., sensors, MCUs, etc., in networks like IoT, led to *TinyML* methodology, which reinforces device independence, regarding the processing of their own data (*decentralization of data*). In the existing literature, some static TinyML approaches have emerged, proposing inference-based methodologies, using ultracompact ML/DL models (see Section 10.3) so as to fit to these devices. However, the inference policy is not able to address the problem of heterogeneity of raw data, and this is the reason why *non-static TinyML* was introduced, which enables real-time training on the aforementioned devices. In order to perform on-device training, both the number of model parameters and the memory footprint of the neural network models have to be reduced during training, so *pruning* methods are used. The present chapter emphasizes the *sparsification* mechanism of a deep neural network, using network science concepts, like scale-free networks. Since the *transfer learning* (TL) mechanism bridges the gap between the random weight initialization and the all-layers training of a neural network, a new efficient TL method is also introduced (see Section 10.4).

Moreover, purely decentralized ML/DL methods have already appeared, in which node clustering protocols, *federated learning* (FL) schemes, or both are proposed in order to meet the resource specifications of this type of network. As extensively introduced in Section 10.5, an FL mechanism is proposed, accompanied by a promising node participation protocol in which nodes with similar data are discarded using DFT during the federation process. As a result, decentralized topologies necessitate the development of suitable reliable communication network infrastructures or else the construction of *backbones* as described in Section 10.2.

Finally, despite the fact that existing ML/DL or decentralized FL methodologies have paved the way for efficient training on edge devices, there are shortcomings that have yet to be solved. Some of the most crucial challenges in both TinyML and ad hoc network circumference are described in Section 10.6, which the scientific community has to cope with, so as to keep up with the requirements mandated by the new era of technological evolution.

References

[1] Armin Moin, Moharram Challenger, Atta Badii, Stephan Günnemann, Supporting AI engineering on the IoT edge through model-driven TinyML, in: IEEE 46th Annual Computers, Software, and Applications Conference (COMPSAC), 2022, pp. 884–893.

[2] Lachit Dutta, Swapna Bharali, TinyML meets IoT: A comprehensive survey, Internet of Things 16 (2021) 100461.

[3] Partha Pratim Ray, A review on TinyML: State-of-the-art and prospects, Journal of King Saud University: Computer and Information Sciences 34 (2021) 1595–1623.

[4] Martín Abadi, Ashish Agarwal, Paul Barham, Eugene Brevdo, Zhifeng Chen, Craig Citro, Greg S. Corrado, Andy Davis, Jeffrey Dean, Matthieu Devin, Sanjay Ghemawat, Ian Goodfellow, Andrew Harp, Geoffrey Irving, Michael Isard, Yangqing Jia, Rafal Jozefowicz, Lukasz Kaiser, Manjunath Kudlur, Josh Levenberg, Dandelion Mané, Rajat Monga, Sherry Moore, Derek Murray, Chris Olah, Mike Schuster, Jonathon Shlens, Benoit Steiner, Ilya Sutskever, Kunal Talwar, Paul Tucker, Vincent Vanhoucke, Vijay Vasudevan, Fernanda Viégas, Oriol Vinyals, Pete Warden, Martin Wattenberg, Martin Wicke, Yuan Yu, Xiaoqiang Zheng, TensorFlow: Large-scale machine learning on heterogeneous systems, Software available from tensorflow.org, 2015.

[5] Visal Rajapakse, Ishan Karunanayake, Nadeem Ahmed, Intelligence at the extreme edge: A survey on reformable TinyML, ACM Computing Surveys 55 (13s) (2022) 282.

[6] Torsten Hoefler, Dan Alistarh, Tal Ben-Nun, Nikoli Dryden, Alexandra Peste, Sparsity in deep learning: Pruning and growth for efficient inference and training in neural networks, Journal of Machine Learning Research 22 (2021) 241:1–241:124.

[7] Sinno Jialin Pan, Qiang Yang, A survey on transfer learning, IEEE Transactions on Knowledge and Data Engineering 22 (10) (2010) 1345–1359.

[8] Jason Yosinski, Jeff Clune, Yoshua Bengio, Hod Lipson, How transferable are features in deep neural networks?, in: NeurIPS, 2014.

[9] Sawsan Abdulrahman, Hanine Tout, Hakima Ould-Slimane, Azzam Mourad, Chamseddine Talhi, Mohsen Guizani, A survey on federated learning: The journey from centralized to distributed on-site learning and beyond, IEEE Internet of Things Journal 8 (2021) 5476–5497.

[10] Frank Po-Chen Lin, Seyyedali Hosseinalipour, Sheikh Shams Azam, Christopher G. Brinton, Nicolò Michelusi, Semi-decentralized federated learning with cooperative D2D local model aggregations, IEEE Journal on Selected Areas in Communications 39 (2021) 3851–3869.

[11] S. Hosseinalipour, S.-S. Azam, C.G. Brinton, V. Michelusi, N. abd Aggrawal, D.J. Love, H. Dai, Multi-stage hybrid federated learning over large-scale D2D-enabled fog networks, IEEE/ACM Transactions on Networking 30 (4) (2022) 1569–1584.

[12] Yuchang Sun, Jiawei Shao, Yuyi Mao, Jessie Hui Wang, Jun Zhang, Semi-decentralized federated edge learning with data and device heterogeneity, IEEE Transactions on Network and Service Management 20 (2) (2021) 1487–1501.

[13] Mohammad Mohammadi Amiri, Deniz Gündüz, Machine learning at the wireless edge: Distributed stochastic gradient descent over-the-air, in: IEEE International Symposium on Information Theory (ISIT), 2019, pp. 1432–1436.

[14] S. Sundhar Ram, Angelia Nedić, Venugopal V. Veeravalli, Distributed stochastic subgradient projection algorithms for convex optimization, Journal of Optimization Theory and Applications 147 (2008) 516–545.

[15] Hong Xing, Osvaldo Simeone, Suzhi Bi, Decentralized federated learning via SGD over wireless D2D networks, in: IEEE 21st International Workshop on Signal Processing Advances in Wireless Communications (SPAWC), 2020, pp. 1–5.

[16] Stefano Savazzi, Monica Nicoli, Vittorio Rampa, Federated learning with cooperating devices: A consensus approach for massive IoT networks, IEEE Internet of Things Journal 7 (5) (2020) 4641–4654.

[17] Chenghao Hu, Jingyan Jiang, Zhi Wang, Decentralized federated learning: A segmented gossip approach, ArXiv:1908.07782 [abs], 2019.

[18] Nil Llisterri Giménez, Joan Miquel Solé, Felix Freitag, Embedded federated learning over a LoRa mesh network, Pervasive and Mobile Computing 93 (C) (2023).

[19] Nico Saputro, Kemal Akkaya, Suleyman Uludag, A survey of routing protocols for smart grid communications, Computer Networks 56 (2012) 2742–2771.

[20] Dimitrios Papakostas, Theodoros Kasidakis, Evangelia Fragkou, Dimitrios Katsaros, Backbones for internet of battlefield things, in: IEEE/IFIP 16th Annual Conference on Wireless On-demand Network Systems and Services Conference (WONS), 2021, pp. 1–8.

[21] Evangelia Fragkou, Dimitrios Papakostas, Theodoros Kasidakis, Dimitrios Katsaros, Multilayer backbones for internet of battlefield things, Future Internet 14 (2002) 186.

[22] Simone Disabato, Manuel Roveri, Incremental on-device tiny machine learning, in: Proceedings of the 2nd International Workshop on Challenges in Artificial Intelligence and Machine Learning for Internet of Things, 2020.

[23] Ranya Aloufi, Hamed Haddadi, David Boyle, Emotion filtering at the edge, in: Proceedings of the 1st Workshop on Machine Learning on Edge in Sensor Systems, 2019.

[24] Lennart Heim, Andreas Biri, Zhongnan Qu, Lothar Thiele, Measuring what really matters: Optimizing neural networks for TinyML, ArXiv:2104.10645 [abs], 2021.

[25] Jingtao Li, Runcong Kuang, Split federated learning on micro-controllers: A keyword spotting showcase, ArXiv:2210.01961 [abs], 2022.

[26] Josen Daniel De Leon, Rowel Atienza, Depth pruning with auxiliary networks for TinyML, in: IEEE International Conference on Acoustics, Speech and Signal Processing (ICASSP), 2022, pp. 3963–3967.

[27] Colby R. Banbury, Chuteng Zhou, Igor Fedorov, Ramon Matas Navarro, Urmish Thakker, Dibakar Gope, Vijay Janapa Reddi, Matthew Mattina, Paul N. Whatmough, MicroNets: Neural network architectures for deploying TinyML applications on commodity microcontrollers, ArXiv:2010.11267 [abs], 2021.

[28] Miguel de Prado, Manuele Rusci, Alessandro Capotondi, Romain Donze, Luca Benini, Nuria Pazos, Robustifying the deployment of TinyML models for autonomous mini-vehicles, Sensors 21 (4) (2021).

[29] Antonio Pullini, Davide Rossi, Igor Loi, Giuseppe Tagliavini, Luca Benini, Mr.Wolf: An energy-precision scalable parallel ultra low power soc for IoT edge processing, IEEE Journal of Solid-State Circuits 54 (2019) 1970–1981.

[30] Taiwo Samuel Ajani, Agbotiname Lucky Imoize, Aderemi A. Atayero, An overview of machine learning within embedded and mobile devices–optimizations and applications, Sensors 21 (2021) 4412.

[31] OpenAI, et al., GPT-4 Technical Report, ArXiv:2303.08774 [abs], 2021.

[32] Youssef Abadade, Anas Temouden, Hatim Bamoumen, Nabil Benamar, Yousra Chtouki, Abdelhakim Senhaji Hafid, A comprehensive survey on TinyML, IEEE Access 11 (july 2023) 96922–96922.

[33] P. Baldi, P. Sadowski, The dropout learning algorithm, Artificial Intelligence 210 (2014) 78–122.

[34] Aidan N. Gomez, Ivan Zhang, Kevin Swersky, Yarin Gal, Geoffrey E. Hinton, Learning sparse networks using targeted dropout, ArXiv:1905.13678 [abs], 2019.

[35] J. Frankle, M. Carbin, The lottery ticker hypothesis: Finding sparse, trainable neural networks, in: Proceedings of the International Conference on Learning Representations (ICLR), 2019.

[36] J. Diffenderfer, B. Kailkhura, Multi-prize lottery ticket hypothesis: Finding accurate binary neural networks by pruning a randomly weighted networks, in: Proceedings of the International Conference on Learning Representations (ICLR), 2021.

[37] Y. Kim, Y. Li, H. Park, Y. Venkatesha, R. Yin, P. Panda, Lottery ticket hypothesis for spiking neural networks, in: Proceedings of the European Conference on Computer Vision (ECCV), 2022.

[38] M. Yao, Y. Chou, G. Zhao, X. Zheng, Y. Tian, B. Xu, G. Li, Probabilistic modeling: Proving the lottery ticket hypothesis in spiking neural network, https://arxiv.org/pdf/2305.12148, 2023.

[39] Yu Xie, Qiang Sun, Yanwei Fu, Exploring lottery ticket hypothesis in few-shot learning, Neurocomputing 550 (2023).

[40] Decebal Constantin Mocanu, Elena Mocanu, Peter Stone, Phuong H. Nguyen, Madeleine Gibescu, Antonio Liotta, Scalable training of artificial neural networks with adaptive sparse connectivity inspired by network science, Nature Communications 9 (1) (Jun 2018).

[41] Xiaolong Ma, Minghai Qin, Fei Sun, Zejiang Hou, Kun Yuan, Yi Xu, Yanzhi Wang, Yen kuang Chen, Rong Jin, Yuan Xie, Effective model sparsification by scheduled grow-and-prune methods, ArXiv:2106.09857 [abs], 2021.

[42] Evangelia Fragkou, Marianna Koultouki, Dimitrios Katsaros, Model reduction of feed forward neural networks for resource-constrained devices, Applied Intelligence 53 (2023) 14102–14127.

[43] Luciana Cavallaro, Ovidiu Bagdasar, Pasquale De Meo, Giacomo Fiumara, Antonio Liotta, Artificial neural networks training acceleration through network science strategies, Soft Computing 24 (2020) 17787–17795.

[44] James Kirkpatrick, Razvan Pascanu, Neil Rabinowitz, Joel Veness, Guillaume Desjardins, Andrei A. Rusu, Kieran Milan, John Quan, Tiago Ramalho, Agnieszka Grabska-Barwinska, Demis Hassabis, Claudia Clopath, Dharshan Kumaran, Raia Hadsell, Overcoming catastrophic forgetting in neural networks, Proceedings of the National Academy of Sciences 114 (13) (Mar 2017) 3521–3526.

[45] A.-L. Barabasi, R. Albert, Emergence of scaling in random networks, Science 286 (5439) (1999) 509–512.

[46] A.-L. Barabasi, Network Science, Cambridge University Press, 2016.

[47] Duncan J. Watts, Steven H. Strogatz, Collective dynamics of 'small-world' networks, Nature 393 (1998) 440–442.

[48] Q. Abbas, F. Ahmad, M. Imran, Variable learning rate based modification in backpropagation algorithm (MBPA) of artificial neural network for data classification, Science International 28 (3) (2016) 2369–2380.

[49] Andreas Chouliaras, Evangelia Fragkou, Dimitrios Katsaros, Feed forward neural network sparsification with dynamic pruning, in: Proceedings of the 25th Pan-Hellenic Conference on Informatics, 2021, pp. 12–17.
[50] Zi-Quan Hong, Jing-Yu Yang, Optimal discriminant plane for a small number of samples and design method of classifier on the plane, Pattern Recognition 24 (1991) 317–324.
[51] S.A. Nene, S.K. Nayar, H. Murase, Columbia Object Image Library (COIL-20), Technical Report CUCS-006-96, Columbia University, 1996.
[52] Han Xiao, Kashif Rasul, Roland Vollgraf, Fashion-MNIST: A novel image dataset for benchmarking machine learning algorithms, ArXiv:1708.07747 [abs], 2017.
[53] Alex Krizhevsky, Vinod Nair, Geoffrey Hinton, Learning multiple layers of features from tiny images, Technical Report, Computer Science Department, University of Toronto, 2009.
[54] Jia Deng, Wei Dong, Richard Socher, Li-Jia Li, Kai Li, Fei-Fei Li, ImageNet: A large-scale hierarchical image database, in: IEEE Conference on Computer Vision and Pattern Recognition, 2009, pp. 248–255.
[55] Puneet Bansal, Image classification, Intel, https://www.kaggle.com/datasets/puneet6060/intel-image-classification, 2019.
[56] Song Chen, Beijing PM2.5, UCI, Machine Learning Repository, https://doi.org/10.24432/C5JS49, 2017.
[57] Kavya Kopparapu, Eric Lin, TinyFedTL: Federated transfer learning on tiny devices, ArXiv:2110.01107 [abs], 2021.
[58] Fuzhen Zhuang, Zhiyuan Qi, Keyu Duan, Dongbo Xi, Yongchun Zhu, Hengshu Zhu, Hui Xiong, Qing He, A comprehensive survey on transfer learning, Proceedings of the IEEE 109 (1) (2021) 43–76.
[59] Oscar Day, Taghi M. Khoshgoftaar, A survey on heterogeneous transfer learning, Journal of Big Data 4 (2017) 1–42.
[60] Evangelia Fragkou, Vasileios Lygnos, Dimitrios Katsaros, Transfer learning for convolutional neural networks in tiny deep learning environments, in: Proceedings of the 26th Pan-Hellenic Conference on Informatics, 2022, pp. 145–150.
[61] H. Brendan McMahan, Eider Moore, Daniel Ramage, Seth Hampson, Blaise Aguera y Arcas, Communication-efficient learning of deep networks from decentralized data, in: Proceedings of the International Conference on Artificial Intelligence and Statistics (AISTATS), 2017, pp. 1273–1282.
[62] M. Ficco, A. Guerriero, E. Milite, F. Palmieri, R. Pietrantuono, S. Russo, Federated learning for IoT devices: Enhancing TinyML with on-board training, Information Fusion 104 (2023).
[63] J. Wang, A.K. Sahu, Z. Yang, G. Joshi, S. Kar, MATCHA: Speeding up decentralized SGD via matching decomposition sampling, in: Proceedings of the Indian Control Conference (ICC), 2019, pp. 299–300.
[64] A. Taya, T. Nishio, M. Morikura, K. Yamamoto, Decentralized and model-free federated learning: Consensus-based distillation in function space, IEEE Transactions on Signal and Information Processing over Networks 8 (2022) 799–814.
[65] F.P.-C. Lin, S. Hosseinalipour, S.S. Azam, Federated learning beyond the star: Local D2D model consensus with global cluster sampling, in: Proceedings of the IEEE Global Communications Conference (GLOBECOM), 2021.
[66] A.D. Amis, R. Prakash, T.H.P. Vuong, D.T. Huynh, Max-min d-cluster formation in wireless ad hoc networks, in: Proceedings of the IEEE Computer Communications Conference (INFOCOM), 2000, pp. 32–41.

[67] Meghana Nasre, Matteo Pontecorvi, Vijaya Ramachandran, Betweenness centrality - Incremental and faster, ArXiv:1311.2147 [abs], 2013.
[68] R. Agrawal, C. Faloutsos, A.N. Swami, Efficient similarity search in sequence databases, in: Proceedings of the International Conference on Foundations of Data Organization and Algorithms (FODO), 1993.
[69] C. Faloutsos, M. Ranganathan, Y. Manolopoulos, Fast subsequence matching in time-series databases, Technical Report CS-TR-3190, University of Maryland, 1993.
[70] Yujing Chen, Zheng Chai, Yue Cheng, Huzefa Rangwala, Asynchronous federated learning for sensor data with concept drift, in: IEEE International Conference on Big Data (Big Data), 2021, pp. 4822–4831.
[71] Jie Lu, Anjin Liu, Fan Dong, Feng Gu, João Gama, Guangquan Zhang, Learning under concept drift: A review, IEEE Transactions on Knowledge and Data Engineering 31 (12) (2019) 2346–2363.
[72] Simone Disabato, Manuel Roveri, Tiny machine learning for concept drift, ArXiv:2107.14759 [abs], 2021.
[73] Peter Jedlicka, Matús Tomko, Anthony V. Robins, Wickliffe C. Abraham, Contributions by metaplasticity to solving the catastrophic forgetting problem, Trends in Neurosciences 45 (2022) 656–666.
[74] Kibok Lee, Kimin Lee, Jinwoo Shin, Honglak Lee, Overcoming catastrophic forgetting with unlabeled data in the wild, in: IEEE/CVF International Conference on Computer Vision (ICCV), 2019, pp. 312–321.
[75] Abel S. Zacarias, Luís A. Alexandre, Overcoming catastrophic forgetting in convolutional neural networks by selective network augmentation, in: IAPR International Workshop on Artificial Neural Networks in Pattern Recognition, 2018.
[76] Sang-Woo Lee, Jin-Hwa Kim, Jaehyun Jun, Jung-Woo Ha, Byoung-Tak Zhang, Overcoming catastrophic forgetting by incremental moment matching, ArXiv:1703.08475 [abs], 2017.
[77] Irene Muñoz-Martín, Stefano Bianchi, Giacomo Pedretti, Octavian Melnic, Stefano Ambrogio, Daniele Ielmini, Unsupervised learning to overcome catastrophic forgetting in neural networks, IEEE Journal on Exploratory Solid-State Computational Devices and Circuits 5 (2019) 58–66.
[78] Anirban Chakraborty, Manaar Alam, Vishal Dey, Anupam Chattopadhyay, Debdeep Mukhopadhyay, Adversarial attacks and defences: A survey, ArXiv:1810.00069 [abs], 2018.

CHAPTER

Bayesian-driven optimizations of TinyML for efficient edge intelligence in LPWANs

11

Aristeidis Karras and Christos Karras
Computer Engineering and Informatics Department, University of Patras, Patras, Greece

11.1 Introduction

The integration of Bayesian optimization with tiny machine learning (TinyML) results in an inspiring synergy at the boundary of edge computing and machine learning. The aforementioned is the fundamental subject of our chapter. The chapter provides a comprehensive analysis of the integration of Bayesian optimization into TinyML. It thoroughly explores the theoretical and practical consequences of this inclusion. Our focus is on enhancing the efficiency of low-power wide area network (LPWAN) systems using Bayesian optimization techniques. Our aim here is to aid the academic community to further understand this rapidly expanding field of study by providing innovative methods.

A significant transformation in data processing has occurred as a consequence of the development of embedded machine learning: the replacement of centralized, high-performance processors with smaller, localized devices. This transition holds considerable importance in the domain of the Internet of Things (IoT), which encompasses the widespread utilization of interconnected sensors and devices. The significance of operating data processing algorithms on embedded devices is emphasized in [1], which highlights its substantial importance and the strengths of doing so. Concurrently, there is a growing scholarly interest in the domain of TinyML, as examined in the article [2]. TinyML is predicated on the utilization of machine learning algorithms to operate algorithms within IoT devices. Utilizing this methodology offers a number of benefits that warrant consideration. The benefits encompass reduced dependence on various data transfers, strengthened measures to ensure data security and protect privacy, diminished operational expenses, and expedited achievement of desired results. One notable advantage is that it facilitates the proper operation of IoT devices even in the absence of a continuous connection to cloud computing services. Moreover, the devices maintain the capacity to perform accurate machine learning operations. Therefore, TinyML is demonstrated to be a feasible and efficient option for a wide range of IoT implementations.

TinyML for Edge Intelligence in IoT and LPWAN Networks. https://doi.org/10.1016/B978-0-44-322202-3.00016-6

Bayesian optimization, an advanced statistical technique renowned for its ability to decipher complex black-box functions and traverse multivariate search spaces, is the subject of our initial investigation. LPWANs, the IoT, and TinyML are particularly noteworthy domains for the aforementioned optimization. The studies [3,4] have developed TinyML-based algorithms, variations of traditional AI, to address the unique requirements of devices that possess restricted processing power. Moreover, Bayesian optimization is an essential component of the TinyML enhancement procedure, which is the main scope of this chapter. By using Bayesian optimization techniques, we aim to refine the allocation of resources, guarantee accuracy in the selection of models and procedures, and improve the precision of hyperparameter tuning. By narrowly adhering to the TinyML objectives of minimizing energy consumption, environmental impact, and operational expenses linked to machine learning algorithms, this approach possesses the capacity to substantially enhance the performance of compact devices that have restricted resources. Analyzing the complex characteristics of Bayesian optimization brings attention to its efficacy in handling expensive or complex problems, an area where traditional optimization techniques often prove inadequate. This chapter presents a thorough examination of Bayesian optimization, including a discussion of its practical implementations and extent, merits and demerits, and possible obstacles. In addition to theoretical knowledge, this chapter explores a variety of practical scenarios and case studies, placing particular emphasis on the significant influence that Bayesian optimization has exerted in augmenting the functionality of TinyML within LPWAN settings. Comparative research has proven that such methods hold tangible potential.

Additionally, this chapter aims to provide insights into the integration of Bayesian optimization within LPWANs. Through our examination of the substantial gap in the current academic literature regarding Bayesian optimization and its implementation in TinyML, this chapter aims to make a substantial scholarly contribution to the ongoing discourse in this domain. We adeptly contribute to the existing corpus of knowledge and establish the foundation for groundbreaking insights by skillfully integrating theory and practice. Moreover, the practicality of the chapter extends beyond scholarly fields and incorporates various other industries. This chapter not only presents an equitable evaluation of the pragmatic obstacles that arise in LPWANs and IoT networks but also suggests innovative and robust approaches to overcome them. This chapter functions as a fundamental reference for academics and practitioners involved in the investigation of the most recent advancements in machine learning, including those working in the domains of IoT and LPWANs. It reduces intricate ideas to workable solutions. By implementing innovative approaches, the chapter broadens its reach and enhances the probability of producing a substantial contribution to the advancement of edge intelligence where devices act, compute, and produce results efficiently and effectively [5].

LPWANs are gaining recognition as a feasible substitute in the domain of machine-to-machine and IoT communication, which require long-lasting connectivity and low battery consumption [6], deliberately engineered to facilitate the long-distance transmission of compact data packets while simultaneously ensuring batteries' durability for years. Within the field of IoT applications, LPWANs are recognized as the most feasible solution for devices that must periodically transmit small data packets over vast distances while functioning for prolonged durations without necessitating battery replacement. Prominent examples of such applications encompass smart cities, remote monitoring, and industrial IoT. Furthermore, LPWANs offer an abundance of benefits, including comprehensive coverage, strong signals in both indoor and outdoor settings, and considerable device connectivity. The aforementioned attributes facilitate the integration of LPWANs into the ever-evolving infrastructure of the IoT.

However, several formidable challenges must be surmounted when traversing the terrain of LPWANs and IoT networks. The minimal power consumption demonstrated by IoT devices is the primary criterion for the critical need for energy-efficient algorithms. Inaccessible or remote areas pose a practical challenge to the routine replacement or recharging of batteries due to the operational effectiveness of these devices. In order to guarantee that these devices function optimally for prolonged durations without requiring regular maintenance, energy efficiency becomes imperative. Moreover, the immense scale of these networks poses a substantial barrier. The complexity of the optimization parameters and network architecture increases significantly when a solitary base station is capable of accommodating a substantial number of devices. The state space experiences exponential growth in proportion to the network size. This requires cutting-edge and sophisticated technologies to ensure uninterrupted operation despite the network's accelerated expansion. IoT devices frequently face constraints pertaining to computational capability, memory, and bandwidth [7,8]. To overcome these challenges, algorithms must be developed with optimal efficacy in mind while accounting for the resources at hand. To attain the requisite balance between optimizing resource utilization and preserving operational efficiency, it is critical to develop algorithms that are adaptable, lightweight, and capable of operating effectively notwithstanding these constraints. Therefore, it is imperative to maintain consistent effort in order to promote innovative solutions within the field of LPWANs and IoT networks to overcome these challenges.

By combining the concepts of Bayesian optimization, namely, the Markov Chain Monte Carlo (MCMC) approaches, with the operational mechanisms of TinyML, we highlight the significant obstacles and possible solutions in IoT and LPWANs [9]. Of particular interest are the problems of resource allocation and energy efficiency, where we highlight significant insights and cutting-edge solutions for reducing energy consumption and enhancing resource allocation efficiency [10,11]. In addition, we explore packet transmission – a crucial component of LPWANs. We demonstrate the instrumental role of Bayesian optimization in fine-tuning packet transmission strategies, leading to a decrease in data loss and an overall enhancement in network performance.

The primary aim of this study is to introduce Bayesian optimization into LPWANs, an innovative approach in a field where such methodologies have yet to be explored. This work represents a pioneering step, bridging a significant gap in the academic landscape and offering practical advancements for IoT and LPWANs.

Since there are no existing studies directly comparable to ours, this chapter not only contributes a novel perspective but also sets a foundational benchmark for future research. This development is in line with the progressive subjects addressed in TinyML for edge intelligence in IoT and LPWANs. Ultimately, the abbreviations utilized throughout this study serve to represent fundamental concepts and methodologies in the domains of TinyML and LPWANs, as detailed in Table 11.1.

Table 11.1 List of acronyms and their meanings.

Acronym	Meaning
TinyML	Tiny machine learning
IoT	Internet of Things
LPWAN	Low-power wide area network
MCMC	Markov Chain Monte Carlo
BOEOTN	Bayesian optimization for energy optimization in TinyML for LPWANs
BOTMS	Bayesian optimization for TinyML model selection and hyperparameter tuning
EBORA	Enhanced Bayesian optimization for resource allocation
BOEPT	Bayesian optimization for enhanced packet transmission
GP	Gaussian process
EI	Expected improvement
PI	Probability of improvement
UCB	Upper confidence bound
MLE	Maximum likelihood estimation

11.2 Background and related work

The process of improving machine learning models for use on LPWANs is related to the field of Bayesian-driven optimizations of TinyML for effective edge intelligence in LPWANs. LPWANs are wireless networks developed with extended range and minimal power consumption. LPWANs possess characteristics that render them exceptionally compatible with IoT devices. LPWANs are favored in IoT scenarios that necessitate devices to frequently transmit modest data loads across extensive geographic regions. This is because LPWANs possess the distinctive ability to facilitate data transmission over extensive distances while maintaining battery life for prolonged durations. Moreover, LPWANs exhibit remarkable adaptability as they function efficiently in both indoor and outdoor environments, thereby substantially broadening their coverage area. A significant portion of interconnected devices nowadays is LPWAN-based, demonstrating their indispensable status as it serves as a critical component in the exponentially growing IoT sector.

As a consequence of the proliferation of IoT devices, the significance of LPWANs as foundational components of the IoT architecture is becoming more and more obvious. The machine learning technique known as TinyML is implemented in the IoT ecosystem. Originally intended for microcontrollers, which are diminutive computational devices, this pioneering framework was developed for operating machine learning functions right on microcontrollers. The seamless integration of energy-efficient and compact devices is a critical element that is imperative for the successful execution of any IoT scenario. Nevertheless, the critical necessity for enhanced performance of machine learning models is emphasized by the growing intricacy of IoT systems. We methodically improve the models by implementing Bayesian optimization to fine-tune the hyperparameters using TinyML on microcontrollers. The increased efficacy of the model is a pivotal element that substantially contributes to its enhanced performance. This chapter aims to investigate an innovative methodology that highlights the significant influence of machine learning in our progressively advanced, data-centric, and interconnected society. Our methodology has been developed with the intention of mitigating the difficulties associated with enhancing machine learning models and IoT infrastructures. Bayesian optimization has exhibited considerable advantages in the realm of TinyML by improving the optimization process for models, particularly those intended for implementation on LPWANs. A highly effective method, Bayesian optimization facilitates the exploration of the hyperparameter space in an efficient manner. The primary objective of this research is to significantly decrease the number of iterations required to achieve the optimal solution by improving the identification of highly prospective search areas through the application of probabilistic models. Due to their exceptional precision and energy efficiency, the previously mentioned models demonstrate a notable degree of compatibility with low-power systems. The advent of machine learning systems that execute complex computations with remarkable precision and energy efficiency has been made possible by recent technological developments. In addition, this state-of-the-art methodology facilitates the smooth incorporation of numerous applications and accelerates the progress of energy-efficient IoT devices, offering remarkable flexibility to tackle the complexities present in various IoT scenarios. Bayesian optimization has become a prominent technique in the field of LPWAN implementations, offering significant potential for optimizing the use of TinyML functionalities. This chapter presents a novel approach to achieving edge intelligence efficacy in the context of LPWANs. In particular, we examine the potential of Bayesian optimization in LPWANs to augment the functionalities of TinyML. The fundamental issue under consideration pertains to the capability of performing machine learning operations remotely on peripheral devices. This eliminates the need to transfer data to a central server for additional analysis. Privacy protection and latency reduction are critical considerations that must be thoroughly incorporated into a comprehensive deployment of IoT devices. In order to effectively implement Bayesian optimization techniques over LPWANs, we will examine the importance of their integration. The utilization of these methods combined substantially improves the effectiveness and practicality of edge intelligence within the framework of the IoT and TinyML.

Nevertheless, the functionality of edge intelligence in IoT devices is constrained by a multitude of obstacles. A major problem of these devices is that they have restricted memory, storage capacity, and computational capabilities [12–17]. Despite the potential challenges posed by resource-intensive tasks and complex algorithms, these can be overcome by employing optimized algorithms and streamlined data-processing techniques. Moreover, the effectiveness of peripheral intelligence may be compromised as a result of its dependence on network connections that may be unreliable. The effective administration of these decentralized systems necessitates specialized expertise and resources owing to their intricate characteristics. The meticulous coordination and planning required to accommodate a diverse range of IoT devices is therefore a crucial element [18]. Another issue that arises is security, as peripheral devices might have inferior protective measures compared to cloud servers, making them more susceptible to possible attacks. This highlights the criticality of protecting both the devices and the data they process [19]. To summarize, the dynamic domain of edge intelligence encounters challenges regarding interoperability and compatibility on account of the absence of well-established protocols [20,21].

11.2.1 Bayesian algorithms in IoT scenarios

In the context of the IoT, Bayesian methodologies, encompassing naïve Bayes, Bayesian networks, and Bayesian regression, are deployed for a multitude of tasks like anomaly detection, predictive maintenance, and data fusion. These algorithms harness the principles of Bayesian inference to render predictions or reach decisions amidst uncertain conditions. This is especially pertinent in IoT contexts, where the data encountered are often characterized by noise or incompleteness, necessitating robust analytical approaches. Table 11.2 demonstrates some examples of Bayesian algorithms used in various IoT applications.

11.2.2 Bayesian-driven optimizations: applying the principles to LoRa in TinyML

We begin with a review of TinyML and LPWAN technologies, drawing on our prior work in [29]. Here we delve into the principles that underpin these technologies, their common applications, and the challenges they pose. The significance of energy optimization in these networks and the contribution of TinyML to this procedure are highlighted. In-depth analysis is conducted on the distinctive characteristics of LoRa, a key representative of LPWAN technologies. The symbol duration, symbol rate, and bit rate, which are critical in LPWAN environments for determining the energy efficiency and efficacy of TinyML applications, are given as follows:

$$T_{sym} = \frac{2^{SF}}{BW}, \tag{11.1}$$

$$R_{sym} = \frac{1}{T_{sym}}, \tag{11.2}$$

$$R_{bit} = \frac{R_{chip}}{SF}. \tag{11.3}$$

Table 11.2 Comparison of Bayesian algorithms in IoT applications.

Ref.	Algorithm	Application	Key features	Benefits	Limitations
[22]	Naïve Bayesian for cyberattack detection	Wireless sensor networks	Simple probabilistic model that applies Bayes' theorem	High accuracy in identifying attacks and low computational cost	Struggles with large or complex datasets
[23]	Bayesian clustering	Random partition models	Uses Bayesian inference for clustering and handles uncertainty effectively	Improved accuracy and efficiency in data clustering. Scalable to large datasets	Requires careful tuning of hyperparameters
[24]	Self-adaptive Bayesian for heart disease prediction	IoT healthcare systems	Adapts to varying data and is tailored for predicting heart diseases	Enhanced accuracy and reliability and suitable for real-time monitoring	Needs frequent retraining with new data
[25]	Bayesian convolution network for activity recognition	Wearable IoT devices	Combines Bayesian inference with CNNs and is optimized for wearable device data	Accurately recognizes human activities and is adaptable to different sensor types	Higher computational demand than simpler models
[26]	Bayesian optimization in machine learning	Urban tunneling prediction	Optimizes machine learning models for specific tasks with focus on surface settlement prediction	Increases prediction accuracy and reduces the risk of tunneling errors	Dependent on the quality of input data
[27]	Real-time probabilistic data fusion	Large-scale IoT applications	Two-layer architecture using Bayesian networks for event processing considering uncertainty	Effective in analyzing large-scale IoT data	Requires extensive computational resources
[28]	Bayesian hyperparameter optimization in intrusion detection	IoT intrusion detection system	Optimization of deep learning model hyperparameters	Significantly improved classification accuracy	Dependent on the initial configuration of the deep learning model

To integrate Bayesian-driven optimizations into this framework, it is necessary to modify these LoRa parameters in accordance with Bayesian principles. Consider, for example, the chip rate:

$$R_{chip} = SF \times R_{sym}. \tag{11.4}$$

The chip rate can be considered a random variable that adheres to a specific underlying probability distribution, considering the inherent uncertainty and disturbance present in IoT environments. By developing a chip rate that strikes a balance between energy consumption and data throughput, Bayesian methods enable one to deduce this distribution and make well-informed decisions.

The goal is to maximize the data rate R_{bit} within a specified energy budget E; this framework is applicable to optimization problems of this nature. A mathematical expression for this problem is as follows:

$$\begin{aligned} \max_{SF,BW} \quad & R_{bit} = SF \frac{BW}{2^{SF}}, \\ \text{subject to} \quad & E_{consumed}(SF, BW) \leq E. \end{aligned} \tag{11.5}$$

In this context, SF is the spreading factor, BW is the bandwidth, R_{bit} is the data rate, and $E_{consumed}$ is the energy expended in light of the selected SF and BW values. By representing the consumed energy as a black-box function and utilizing Bayesian optimization, it is possible to determine the optimal SF and BW in order to maximize the data rate while adhering to a predetermined budget for energy usage.

In light of the aforementioned problem statement, we shall now elaborate on the constituent elements that make up a Bayesian optimization framework. Bayesian optimization is a sequential design approach utilized to optimize black-box functions on a global scale while accounting for uncertainty. It requires the development of a probabilistic model for the objective function, which is then utilized to determine the optimal sampling location. In the present scenario, $E_{consumed}(SF, BW)$ may be regarded as the black-box function.

The function $E_{consumed}(SF, BW)$ is represented by a Gaussian process (GP) consisting of a mean function $m(x)$ and a covariance function $k(x, x')$ in our model. In order to move to the next step, an acquisition function $\alpha(x)$ is defined. This function aims to find a balance between exploration and exploitation by selecting places that have both a high function value and a high level of uncertainty.

A similar technique is applied [30,31], where two methods are presented to enhance the effectiveness of LPWAN by determining the optimal LoRa parameters, such as SF, BW, and the coding rate. This approach might potentially be used in future research to assess the performance limitations of TinyML applications over LPWANs.

11.2.3 Bayesian-driven optimization preliminaries for TinyML in LPWANs

LPWAN technologies, which are a subset of wireless communication technologies, are specifically designed to provide long-range communication while minimizing power consumption. Due to these characteristics, LPWAN is a suitable option for implementing widespread IoT applications [32]. However, the further optimization of TinyML networks may be particularly challenging because of the complex trade-offs between transmission range, data throughput, energy usage, and other network factors when integrating TinyML applications.

Let us consider a network comprising N devices, where each device i transmits data according to specific criteria, including a delivery range d_i, transmission power P_i, and data rate R_i. Let the goal be the maximization of the overall network performance while minimizing energy consumption. This problem can be formulated as

$$\max_{P_i, R_i, d_i} \sum_{i=1}^{N} Q(P_i, R_i, d_i), \quad \text{subject to} \quad \sum_{i=1}^{N} E_i(P_i, R_i, d_i) \leq E, \tag{11.6}$$

where $Q(P_i, R_i, d_i)$ represents a quality metric (such as data throughput, latency, or reliability) and $E_i(P_i, R_i, d_i)$ represents the energy consumed by device i for given values of P_i, R_i, and d_i. The function E_i is typically unknown and may be complex due to various network dynamics, so it can be modeled as a black-box function.

Bayesian optimization provides a probabilistic model to infer the distribution of these black-box functions. To illustrate this, we can redefine the function E_i as $f(x)$, where x refers to the parameters P_i, R_i, and d_i. Then, Bayesian optimization will help us find the optimal parameters that minimize the energy consumption, expressed as

$$x^* = \arg\min_{x \in X} f(x), \tag{11.7}$$

where X is the set of all possible parameters. The uncertainty about $f(x)$ is modeled using GP. The GP posterior predictive distribution is defined by a mean function $m(x)$ and a covariance function $k(x, x')$. After constructing the probabilistic model, we define an acquisition function that guides where to sample next. Common choices for the acquisition function include expected improvement (EI), probability of improvement (PI), and upper confidence bound (UCB):

$$EI(\mathbf{x}) = \mathbb{E}\left[\max(f(\mathbf{x}) - f(\mathbf{x}^+), 0)\right], \tag{11.8}$$

$$PI(\mathbf{x}) = P(f(\mathbf{x}) \geq f(\mathbf{x}^+) + \epsilon), \tag{11.9}$$

$$UCB(\mathbf{x}) = \mu(\mathbf{x}) + \kappa\sigma(\mathbf{x}), \tag{11.10}$$

where $\mathbf{x}^+$ is the current best solution, ϵ is a small constant, $\mu(\mathbf{x})$ and $\sigma(\mathbf{x})$ are the mean and standard deviation of the GP at $\mathbf{x}$, and κ is a tunable exploration parameter. Using these acquisition functions, we can explore and exploit the solution space to find optimal parameters that minimize energy consumption.

This Bayesian-driven optimization of TinyML for efficient edge intelligence in LPWANs represents a novel approach in the quest to enhance performance in IoT systems. The insights generated through this Bayesian optimization approach have the potential to guide the development of next-generation LPWAN technologies, balancing the intricate trade-off between energy efficiency, data rate, and transmission range.

In addition, we will explore the application of the GP regression method for simulating the objective function in Bayesian optimization. Furthermore, we will investigate the integration of maximum likelihood estimation (MLE) and MCMC techniques, as described in [33,34], in order to enhance the optimization procedure.

11.2.4 Energy optimization in TinyML

The proliferation of IoT devices and the increasing complexity of applications running on them necessitates an emphasis on energy optimization, especially in TinyML scenarios [35,36]. These situations frequently require limited-capacity devices to execute intricate computations, which inevitably results in substantial energy usage. Strategies including energy harvesting, sleep scheduling, and duty cycling have been extensively implemented to combat this. Nevertheless, the task of identifying an ideal configuration that minimizes energy consumption while maintaining optimal performance continues to be a formidable challenge.

To quantify this, consider an IoT device with a power profile $P(t)$ that is executing a TinyML application. The total energy consumption E during a specified time period T is defined as follows:

$$E = \int_0^T P(t)dt. \tag{11.11}$$

This energy consumption is complex to optimize due to its dependence on a variety of variables, including computation burden, wireless activity, and sensor readings, among others [37].

An intriguing strategy for tackling this optimization issue involves the application of Bayesian optimization methods. The aforementioned methods center on the optimization of an unidentified function $f(x)$, which represents the energy consumption for a specified set of control variables x (such as duty cycle length, sleep schedule, and data transmission rate). Consequently, the above can be reformulated as finding the optimal value of x^* that minimizes energy consumption, which can be denoted as

$$x^* = \arg\min_{x \in X} f(x). \tag{11.12}$$

Let X represent the set of all feasible control configurations. The solution to this optimization challenge necessitates achieving a delicate equilibrium between exploration, encompassing the discovery of novel configurations, and exploitation, entailing the utilization of configurations with established low energy consumption.

The integration of two statistical methodologies, namely, MLE and MCMC, within our Bayesian optimization framework has the potential to augment its robustness and efficacy. The MCMC technique is a widely employed method in computer science for efficiently exploring the space of control settings. This is achieved by generating samples from the posterior distribution of objective function parameters. By leveraging the principles of Markov chains, MCMC enables researchers to effectively navigate the vast solution space and obtain valuable insights for optimizing control systems. By incorporating intermittent energy surges, this approach guarantees a comprehensive examination of the entire spectrum of potential configurations, serving as a mechanism to circumvent entrapment in local minima. MLE is a widely used statistical method in the field of Bayesian optimization. It is employed to estimate the parameters of a given model [38]. The purpose of this method is to estimate the mean and covariance functions, which serve as essential parameters of the GP employed for modeling the objective function. The utilization of the TinyML framework facilitates the extraction of valuable insights pertaining to the energy consumption patterns exhibited by the device, through the analysis of observed data.

We now formulate all the above approaches for energy optimization in TinyML using Bayesian optimization, MCMC, and MLE. Utilizing a GP model, we assume that we have $GP(m(\mathbf{x}), k(\mathbf{x}, \mathbf{x}'))$ for the energy consumption function $f(\mathbf{x})$ with mean function $m(\mathbf{x})$ and covariance function $k(\mathbf{x}, \mathbf{x}')$. The function for determining the consumption of energy can be expressed as

$$f(\mathbf{x}) \sim GP(m(\mathbf{x}), k(\mathbf{x}, \mathbf{x}')). \tag{11.13}$$

Markov Chain Monte Carlo: MCMC utilizes Markov Chain Monte Carlo to sample from the posterior distribution of the GP parameters θ. GP refers to generalized linear models.

Given an initial parameter value θ_0, the MCMC sampler generates a sequence $\theta_0, \theta_1, ..., \theta_n$ such that as $n \to \infty$, $\theta_n \sim p(\theta|D)$, where D is the observed data:

$$\theta_n \xrightarrow{MCMC} p(\theta|D), \quad \text{as} \quad n \to \infty. \tag{11.14}$$

MLE is a statistical technique utilized to discern the parameters, denoted as θ, that maximize the likelihood function $L(\theta; D)$ based on the observed data D within the context of a GP model. Essentially, this strategy aims to identify the most likely combination of parameters, θ, that would produce the observed data based on the characteristics of the model. This approach is strong and effective for learning model parameters. The GP can be formulated as

$$\hat{\theta} = \arg\max_{\theta} L(\theta; D). \tag{11.15}$$

Bayesian optimization, a strategic approach that aims to find the optimal control configuration, can be represented as $\mathbf{x}$. The optimization goal here is to minimize energy consumption while balancing exploration and exploitation. The delicate equilibrium is preserved by the optimization of an acquisition function $a(\mathbf{x}; D, \theta)$, which may be exemplified by EI or UCB. We have

$$\mathbf{x} = \arg\max_{\mathbf{x}} a(\mathbf{x}; D, \theta). \tag{11.16}$$

Through the integration of these components, we may develop a customized iterative approach to optimize energy usage inside the TinyML framework. After configuring certain control parameters, data on energy use are collected. Subsequently, the GP model is revised using MLE and MCMC. During the execution of the methods, the most effective control configuration is determined by using Bayesian optimization. This repeated process of updating the model via modifications in the selection of control settings continues until a stable state of convergence is achieved. This technique serves as a thorough and robust solution to the energy optimization complexities inherent in TinyML. This enables an efficient method for optimizing models while also ensuring efficient use of resources, thereby improving the performance of TinyML deployments in a highly networked environment.

11.3 Methodology

11.3.1 Bayesian optimization techniques for model selection and hyperparameter tuning in TinyML

This section focuses on the utilization of Bayesian optimization techniques in TinyML, specifically for model selection and hyperparameter optimization. The process of hyperparameter tuning frequently poses a significant obstacle, manifesting as an optimization dilemma. As highlighted in [39], this can be resolved through the implementation of Bayesian optimization, which efficiently manages scenarios involving complex problems.

The primary aim is to achieve a balance between energy efficiency and the optimization of performance metrics, including accuracy and precision, for the TinyML models. The choice of TinyML models and their corresponding hyperparameters have a significant impact on both performance metrics and energy consumption, implying the process of identifying optimal models and hyperparameters could potentially be resource-intensive and time-consuming. Nevertheless, the employment of Bayesian optimization offers a robust and efficient strategy to overcome this challenge.

Imagine having a TinyML model, symbolized by a set of hyperparameters $\mathbf{x} = x_1, x_2, ..., x_p$. Let us denote the performance metric and energy consumption of this model by $f_1(\mathbf{x})$ and $f_2(\mathbf{x})$, respectively. The ultimate objective is to identify those hyperparameters $\mathbf{x}$ that can maximize $f_1(\mathbf{x})$ while minimizing $f_2(\mathbf{x})$. This essentially frames it as a multi-objective optimization problem, giving us a clear pathway to apply our robust Bayesian optimization techniques:

$$\max_{\mathbf{x}} \quad \& f_1(\mathbf{x}) \; \min_{\mathbf{x}} \quad \& f_2(\mathbf{x}). \tag{11.17}$$

To address this multi-objective problem, we can extend the Bayesian optimization framework introduced in Section 11.3.1.

Gaussian process model: We assume separate GP models for the performance metric and energy consumption:

$$f_1(\mathbf{x}) \sim GP(m_1(\mathbf{x}), k_1(\mathbf{x}, \mathbf{x}'))\ f_2(\mathbf{x}) \sim GP(m_2(\mathbf{x}), k_2(\mathbf{x}, \mathbf{x}')). \tag{11.18}$$

Bayesian optimization: We construct a multi-objective acquisition function $a(\mathbf{x}; D, \theta)$ that balances the trade-off between the two objectives. A popular choice based on previous works [40,41] is the hypervolume-based EI acquisition function:

$$a(\mathbf{x}; D, \theta) = \mathbb{E}[\max(HV(D \cup (\mathbf{x}, f(\mathbf{x}))) - HV(D), 0)], \tag{11.19}$$

where HV denotes the hypervolume (or the volume of the dominated region in the objective space), which provides a scalar measure of the quality of the Pareto frontier.

This Bayesian optimization framework for TinyML model selection and hyperparameter tuning offers an efficient and effective approach to navigate the complex trade-offs in TinyML applications, providing a pathway toward energy-efficient and high-performance TinyML models.

11.3.2 Bayesian optimization for resource allocation in TinyML-enabled LPWANs

In resource allocation problems [21,42,43], the primary goal is to efficiently utilize available resources to meet various objectives, such as maximizing network throughput, minimizing energy consumption, or reducing latency. In the context of TinyML-enabled LPWANs, resources can include things like transmission power, bandwidth, or even computational resources for executing machine learning models on devices.

Here, we use Bayesian optimization to address this problem. The primary advantage of this approach is its ability to handle complex, non-linear, and possibly unknown objective functions, which makes it suitable for resource allocation problems where the objective function can be difficult to model accurately.

Our approach involves formulating the resource allocation problem as a black-box optimization problem, where the goal is to minimize an unknown objective function $g(\mathbf{z})$ subject to some constraints. The objective function represents a performance metric we want to optimize (like energy efficiency), and $\mathbf{z}$ represents the resource allocation strategy, so we have

$$\min_{\mathbf{z} \in \mathcal{Z}} g(\mathbf{z}), \tag{11.20}$$

where $\mathcal{Z}$ is the search space of resource allocation strategies.

The Bayesian optimization procedure starts by placing a GP prior on $g(\mathbf{z})$. The GP is a flexible non-parametric model that can model complex, non-linear functions. In each iteration of the optimization process, we update our belief about $g(\mathbf{z})$ using the observations we have made so far and select the next point to evaluate by maximizing an acquisition function.

One popular choice based on previous works [44–47] for the acquisition function is the EI function:

$$EI(\mathbf{z}) = \mathbb{E}\left[\max(g(\mathbf{z}_{\text{best}}) - g(\mathbf{z}), 0)\right], \tag{11.21}$$

where $\mathbf{z}_{\text{best}}$ is the best point found so far.

The next resource allocation strategy to try is then given by

$$\mathbf{z}\text{next} = \arg\max \mathbf{z} \in \mathcal{Z} EI(\mathbf{z}). \tag{11.22}$$

In addressing the inherent uncertainty present within the GP model and the objective function, we apply MCMC methodologies, with the Metropolis–Hastings algorithm being one such efficient technique. These MCMC methods are utilized to generate samples derived from the GP posterior. Subsequently, these samples aid in approximating the EI function, thereby providing a solution for Eq. (11.22).

For the refinement of parameters intrinsic to the GP model, we employ MLE:

$$\theta\text{MLE} = \arg\max \theta \log p(\mathbf{y}|\mathbf{X}, \theta). \tag{11.23}$$

In the equation above, $\mathbf{y}$ signifies the observed function values, $\mathbf{X}$ represents the corresponding resource allocation strategies, and θ embodies the parameters inherent to the GP model. The MLE technique, in this case, serves as a powerful tool to maximize the likelihood of observing the empirical data $\mathbf{y}$, given the input $\mathbf{X}$ and the model parameters θ. Through this process, a highly accurate estimation of the GP model parameters is obtained, which subsequently influences the determination of the subsequent sampling point in our optimization dilemma. By integrating MCMC and MLE methodologies, a holistic strategy is established to effectively handle uncertainty and optimize parameters when Bayesian optimization is implemented on TinyML.

11.3.3 Improving packet transmission and reception in LPWANs with Bayesian optimization

For LPWAN networks to operate more efficiently, packet transmission must be enhanced. Our objective is to enhance the efficacy of packet transmission and reception through the careful consideration of various factors including transmission power, frequency band, data rate, and coding rate. The energy consumption of the network devices and the rate of transmission delivery can be significantly impacted by these variables. By optimizing these factors with Bayesian optimization, we can achieve a more dependable connection and extend the life of devices within the network. Each of these variables can have a significant impact on the energy efficiency and packet delivery rate of the network devices.

In order to conceptualize this as a Bayesian optimization issue, let us establish an objective function denoted as $h(\mathbf{w})$ which signifies the frequency at which packet transmission and reception are accomplished successfully. The vector of transmission parameters, denoted by **w**, may consist of elements such as transmission power, frequency band, data rate, coding rate, and more as shown below:

$$\max_{\mathbf{w}\in\mathcal{W}} h(\mathbf{w}), \tag{11.24}$$

where **w** represents the set of transmission parameters that are feasible. Similarly to the preceding as shown in Section 11.3.2, our assumption regarding the objective function is represented by a GP. To determine the subsequent point for evaluation, an acquisition function that strikes a balance between exploration and exploitation is maximized. The EI function is employed for this objective:

$$EI(\mathbf{w}) = \mathbb{E}\left[\max(h(\mathbf{w}_{\text{best}}) - h(\mathbf{w}), 0)\right]. \tag{11.25}$$

Then, the subsequent transmission parameters to attempt are denoted by

$$\text{wnext} = \arg\max \mathbf{w} \in \mathcal{W}EI(\mathbf{w}). \tag{11.26}$$

In order to mitigate the inherent unpredictability of the network caused by interference, fluctuating signal intensity, and other dynamic factors, we sample from the GP posterior using MCMC techniques. By providing approximations of the EI function, these examples direct the optimization procedure in the direction of more effective packet transmission strategies. GP model parameters are optimized through the utilization of MLE:

$$\theta\text{MLE} = \arg\max \theta \log p(\mathbf{y}|\mathbf{X}, \theta). \tag{11.27}$$

This iterative procedure continues until a predetermined condition is met, such as achieving the desired level of performance or completing a predetermined number of iterations.

By implementing this methodology, it is possible to optimize the parameters governing the transmission of data packets in LPWANs. Potential advantages encompass an increased rate of packet delivery success, improved energy efficiency, and a general enhancement in network performance. Significantly, this approach recognizes the fluctuating nature of the network, thereby facilitating the adaptive adjustment of transmission parameters. As a result, the confidence and resilience of communication within the LPWAN are progressively enhanced.

11.4 Proposed Bayesian optimization algorithms of TinyML in LPWANs

An overview of our TinyML-based Bayesian optimization algorithms for LPWANs is given in this section. The following algorithms are described below: Bayesian optimization for energy optimization in TinyML for LPWANs (BOEOTN) (Algorithm 11.1), Bayesian optimization for TinyML model selection (BOTMS) (Algorithm 11.2), enhanced Bayesian optimization for resource allocation (EBORA) (Algorithm 11.3), and Bayesian optimization for enhanced packet transmission (BOEPT) (Algorithm 11.4). Each algorithm specifically aims to handle important issues such as resource allocation, energy conservation, model selection, performance optimization, and so forth, that arise in LPWANs.

The BOEOTN algorithm, presented in Algorithm 11.1, is particularly designed to maximize TinyML applications' energy usage while they operate on LPWANs. Taking into account the unique properties of LPWANs, the technique uses Bayesian optimization to completely evaluate the most efficient control settings for data transmission rates and communication frequencies. Algorithm 11.2 presents the BOTMS technique, which tackles the model selection and hyperparameter optimization difficulties in TinyML applications. Through the efficient integration of exploration and exploitation components, an ideal model is generated, including hyperparameters that optimize efficiency while preserving accuracy and energy efficiency. Algorithm 11.3 provides a comprehensive description of EBORA, which is particularly made to optimize resource allocation for TinyML applications over LPWANs in an efficient manner. Because LPWANs have limited and shared resources, their main goal is to maximize network performance while using less energy. This strategy guarantees the network's long-term consistency. TinyML enhanced packet transmission optimization is the ultimate goal of the BOEPT algorithm, as shown in Algorithm 11.4. Using Bayesian optimization, the strategy repeatedly changes the control variables that affect TinyML applications' performance. The goal is to maximize packet transmission while minimizing energy use at normal levels.

Combining these techniques provides a broad framework to solve the particular difficulties presented by LPWANs and enhance the efficacy and efficiency of TinyML applications.

11.4.1 Bayesian optimization for energy optimization in TinyML for LPWANs (BOEOTN)

Energy efficiency is of great significance in LPWANs owing to the constraints of long-range communication and limited battery life. We present a novel Bayesian optimization for energy optimization in TinyML for LPWANs (as shown in Algorithm 11.1) within this context.

The energy optimization process in BOEOTN bears resemblance to that of TinyML, as elaborated in Section 11.2.4. However, BOEOTN distinguishes itself by catering specifically to the properties of low-power wide area network (LPWAN) connections. Its primary objective is to optimize the control settings associated with communication frequency and data transmission rate.

Regarding complexity, the primary computational cost in BOEOTN lies in fitting the GP model and maximizing the acquisition function. The computational complexity of fitting a GP is $O(n^3)$, where n is the number of observations, due to the

Algorithm 11.1 Bayesian optimization for energy optimization in TinyML for LP-WANs (BOEOTN).

1: **Output:** Optimal control settings $\mathbf{x}^*$
2: Randomly initialize control settings $\mathbf{x}^{(1)}, \mathbf{x}^{(2)}, ..., \mathbf{x}^{(n)}$
3: Measure energy consumption $f(\mathbf{x}^{(i)})$ for each $\mathbf{x}^{(i)}$, $i = 1, ..., n$, using Eq. (11.11)
4: Fit GP model on $\{(\mathbf{x}^{(i)}, f(\mathbf{x}^{(i)}))\}_{i=1}^{n}$ as per Eq. (11.13)
5: Initialize GP parameters θ_0
6: **while** stopping criterion not met **do**
7: Perform MCMC to generate a sequence $\theta_0, \theta_1, ..., \theta_n$ as per Eq. (11.14)
8: Estimate $\hat{\theta}$ by maximizing the likelihood $L(\theta; D)$ as per Eq. (11.15)
9: Define acquisition function $a(\mathbf{x}; D, \theta)$, such as EI or UCB
10: Determine $\mathbf{x}_{\text{next}}$ by maximizing $a(\mathbf{x}; D, \theta)$ as per Eq. (11.16)
11: Measure energy consumption $f(\mathbf{x}_{\text{next}})$
12: Update the GP with the new observation
13: **end while**
14: $\mathbf{x}^* \leftarrow \mathbf{x}_{\text{next}}$ that minimizes the energy consumption

inversion of the $n \times n$ covariance matrix. Maximizing the acquisition function involves an optimization procedure that can be done efficiently using gradient-based methods, assuming the function is differentiable. In practice, the number of iterations until convergence will depend on the complexity of the energy consumption function and the initial control settings.

11.4.2 Bayesian optimization for TinyML model selection and hyperparameter tuning (BOTMS)

BOTMS (Algorithm 11.2) uses the Bayesian optimization methodology described in Section 11.3.1 to navigate the complex trade-offs in TinyML applications. The BOTMS algorithm aims at selecting the optimal TinyML model and tuning its hyperparameters to simultaneously optimize the performance metrics and energy consumption.

BOTMS (Algorithm 11.2) provides an efficient and effective approach to navigate the complex trade-offs in TinyML applications, providing a pathway toward energy-efficient and high-performance TinyML models.

The complexity of the BOTMS algorithm (Algorithm 11.2) comes from three main steps: initial random initialization, GP regression, and the optimization of the acquisition function. The complexity of the initialization step is $\mathcal{O}(np)$, where n is the number of initial points and p is the number of hyperparameters. The GP regression step has a complexity of $\mathcal{O}(n^3)$ because it involves the inversion of an $n \times n$ matrix. The optimization of the acquisition function depends on the optimization algorithm used, typically proportional to the number of hyperparameters p and the number of iterations required for convergence.

Algorithm 11.2 Bayesian optimization for TinyML model selection and hyperparameter tuning (BOTMS).

1: **Output:** Optimal model and hyperparameters $\mathbf{x}^*$
2: Randomly initialize $\mathbf{x}^{(1)}, \mathbf{x}^{(2)}, ..., \mathbf{x}^{(n)}$
3: Evaluate $f_1(\mathbf{x}^{(i)})$ and $f_2(\mathbf{x}^{(i)})$ for $i = 1, ..., n$
4: Fit two GP models on $\{(\mathbf{x}^{(i)}, f_1(\mathbf{x}^{(i)}))\}_{i=1}^n$ and $\{(\mathbf{x}^{(i)}, f_2(\mathbf{x}^{(i)}))\}_{i=1}^n$, respectively, as per Eq. (11.18)
5: **while** stopping criterion not met **do**
6: Define the acquisition function $a(\mathbf{x}; D, \theta)$ based on current GPs as per Eq. (11.19)
7: Determine $\mathbf{x}_{\text{next}}$ by maximizing $a(\mathbf{x}; D, \theta)$ using a suitable optimization algorithm
8: Evaluate $f_1(\mathbf{x}_{\text{next}})$ and $f_2(\mathbf{x}_{\text{next}})$
9: Update the GPs with the new observations
10: **end while**
11: $\mathbf{x}^* \leftarrow \mathbf{x}_{\text{next}}$ that balances the trade-off between the performance metric and energy consumption

11.4.3 Bayesian optimization for resource allocation in TinyML-enabled LPWANs (EBORA)

The EBORA algorithm (Algorithm 11.3) is a principled strategy for optimizing resource allocation in TinyML-enabled LPWANs. The goal of EBORA is to find an optimal resource allocation strategy $\mathbf{z}^*$ that minimizes a performance measure $g(\mathbf{z})$.

The algorithm starts with an initial set of randomly generated resource allocation strategies. For each strategy, the performance measure $g(\mathbf{z})$ is evaluated. These initial points are used to fit a GP, which serves as a surrogate model for the performance measure. The GP is updated with each new observation as the algorithm progresses. An acquisition function is defined based on the current GP model, guiding the algorithm towards regions in the search space that are likely to yield improvements. The acquisition function is maximized to determine the next resource allocation strategy to be evaluated.

The process continues iteratively, updating the GP and optimizing the acquisition function, until a stopping criterion is met (such as a maximum number of iterations or a performance measure threshold). The resource allocation strategy that results in the minimum performance measure is chosen as the optimized solution.

- Initialization: Complexity is $\mathcal{O}(n)$, governed by the number of initial points n.
- GP regression: Overall complexity is $\mathcal{O}(m^2n^3)$, due to inversion of an expanding $n \times n$ matrix over m iterations.
- Optimization: Complexity depends on the optimization algorithm and can be proportional to the number of variables and iterations.

Algorithm 11.3 Enhanced Bayesian optimization for resource allocation in TinyML-enabled LPWANs (EBORA).

1: **Result:** Optimized resource allocation strategy $\mathbf{z}^{\text{opt}}$
2: Initialize $\mathbf{z}^{(1)}, \mathbf{z}^{(2)}, ..., \mathbf{z}^{(n)}$ randomly within $\mathcal{Z}$, the search space of resource allocation strategies (based on Eq. (11.20))
3: Evaluate performance metric $g(\mathbf{z}^{(i)})$ for each initial strategy $i = 1, ..., n$
4: Fit a Gaussian process $GP(\mu, K)$ on $(\mathbf{z}^{(i)}, g(\mathbf{z}^{(i)}))i = 1^n$ with parameters optimized using MLE (refer to Eq. (11.15))
5: **while** stopping criterion not met **do**
6: Define acquisition function $a(\mathbf{z}; GP)$ based on current GP and EI function (refer to Eq. (11.21))
7: Determine **znext** by maximizing $a(\mathbf{z}; GP)$ using a suitable optimization algorithm, aiming to explore the point with maximum EI (as per Eq. (11.22))
8: Evaluate $g(\mathbf{znext})$ to determine the performance of the new resource allocation strategy
9: Update the GP with the new observation $(\mathbf{znext}, g(\mathbf{z}_{\text{next}}))$
10: **end while**
11: $\mathbf{z}^{\leftarrow}$**znext** where $g(\mathbf{znext})$ is minimum, indicating the optimal resource allocation strategy

Therefore, the overall time complexity of the EBORA algorithm (Algorithm 11.3) is approximately $\mathcal{O}(m^2n^3)$, assuming that the complexity of the acquisition function optimization step is dominated by the complexity of the GP regression step. Here, m is the number of iterations of the algorithm and n is the number of initial points. This complexity can be a consideration when applying the EBORA algorithm to very large search spaces or when a very large number of initial points or iterations is needed.

11.4.4 Bayesian optimization algorithm for packet transmission in LPWANs (BOEPT)

The BOEPT algorithm (Algorithm 11.4) is designed to optimize the packet transmission parameters in LPWANs. The BOEPT algorithm utilizes Bayesian optimization, a powerful strategy for optimizing black-box functions, to enhance the network's performance metrics such as latency, throughput, and energy efficiency. The algorithm's performance and effectiveness will be evaluated in the following sections.

In the BOEPT algorithm (Algorithm 11.4), the most time-consuming parts are fitting the GP model and refining the acquisition function. If we have n data points, fitting the model has a time complexity of $O(n^3)$ because it involves inverting a big matrix – and this step happens every time we run the algorithm. The time it takes to refine the acquisition function depends on the method we use; for example, methods that use gradients typically have a time complexity of $O(n^2)$ per run. How many

Algorithm 11.4 Bayesian optimization for enhanced packet transmission in LPWANs (BOEPT).

1: **Result:** Optimized packet transmission parameters $\mathbf{w}^*$
2: Initialize $\mathbf{w}^{(1)}, \mathbf{w}^{(2)}, ..., \mathbf{w}^{(n)}$ randomly in $\mathcal{W}$
3: Evaluate $h(\mathbf{w}^{(i)})$ for $i = 1, ..., n$
4: **while** stopping criterion not met **do**
5: Fit a GP on $\{(\mathbf{w}^{(i)}, h(\mathbf{w}^{(i)}))\}_{i=1}^{n}$ with parameters optimized by MLE
6: Determine $\mathbf{w}_{\text{next}}$ by solving Eq. (11.26)
7: Evaluate $h(\mathbf{w}_{\text{next}})$
8: Update $\mathbf{w}^{(n+1)} \leftarrow \mathbf{w}_{\text{next}}$, $n \leftarrow n + 1$
9: **end while**
10: $\mathbf{w}^* \leftarrow \mathbf{w}_{\text{next}}$ where $h(\mathbf{w}_{\text{next}})$ is maximum

times we need to run these steps depends on how complex the acquisition function is and how many dimensions the problem has.

Consequently, as the iterations rise in number, the algorithm's complexity escalates considerably. Nonetheless, due to the sample efficiency of Bayesian optimization, it often demands significantly fewer iterations compared to other optimization methods. Furthermore, there are multiple strategies available to scale Bayesian optimization to larger problem sizes, such as utilizing sparse GP models or techniques for reducing dimensionality.

11.5 Experimental results

In this section, we present comprehensive simulations and evaluations of our proposed Bayesian optimization algorithms for TinyML in LPWANs, namely, BOEOTN (Algorithm 11.1), BOTMS (Algorithm 11.2), EBORA (Algorithm 11.3), and BOEPT (Algorithm 11.4).

In conducting our simulations, we dive into a new field by applying Bayesian optimization techniques to LPWANs. This new approach represents a significant differentiation from traditional methods in the field, placing our work at the forefront of research in this area. Consequently, there is a notable absence of similar studies or established baseline comparisons, as our methods pioneer the integration of Bayesian optimization within LPWAN contexts. This lack of precedent underscores the originality of our research and shows that our study presents a unique opportunity to set a foundational benchmark for future studies. Our simulations, therefore, focus on demonstrating the general applicability and effectiveness of our methods under diverse conditions, charting new paths for optimization in LPWANs. Further details about our simulation environment and the devices used are comprehensively outlined in Section 11.5.1.

11.5.1 Simulation interface and configuration parameters

The simulations were executed within our experimental configuration utilizing LoRa TTGO T-Beam devices, known for their efficient transmission of LoRa messages and minimal power consumption. This particular capability is of great importance in our specific scenario as it ensures compliance with the low-power demands of LPWANs. To create a practical and genuine testing configuration, we utilized 10 of the aforementioned devices. To create a network environment that required optimal resource allocation and efficient packet delivery, the devices were specifically configured. The methodologies presented here were developed through the open-source Arduino Software (IDE) along with the C programming language. The Arduino IDE environment provides a user-friendly and straightforward interface, which greatly simplifies the programming process for LoRa TTGO T-Beam devices. The mentioned capability enabled the accurate modification of operational attributes and facilitated the execution of simulation experiments. Furthermore, we consistently utilized the serial monitor feature to evaluate and monitor the results generated by the optimization algorithms. As a result of our experimentation, we have successfully assessed the efficiency of the modified LoRa TTGO T-Beam devices, and the acquired values were exported from the Arduino IDE and visually represented in the following figures. Under this configuration we managed to achieve a thorough assessment of the algorithms in practical scenarios, ensuring that the findings were based on a genuine implementation in a real-life case study.

Algorithm 11.1 is used to evaluate the energy consumption, efficiency, and expected energy efficiency of BOEOTN under various control setups and network circumstances. The results shown in Fig. 11.1 confirm that BOEOTN is useful in lowering the energy consumption of TinyML applications that operate on LPWANs. This demonstrates how BOEOTN may lengthen and strengthen these kinds of networks.

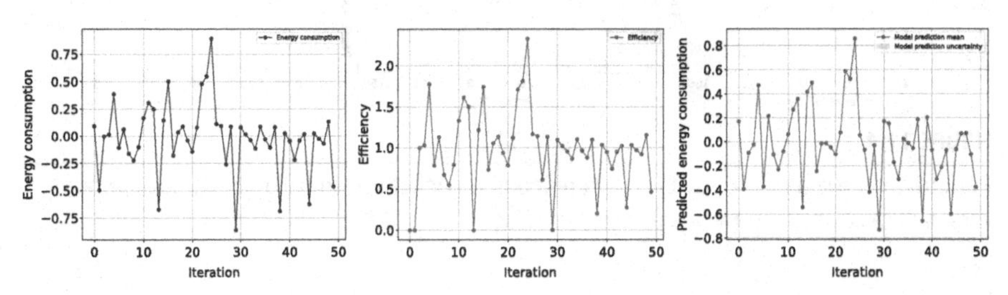

FIGURE 11.1

Performance evaluation of BOEOTN with respect to energy consumption, efficiency, and predicted energy efficiency.

The efficiency, expected energy consumption, and energy consumption of the chosen TinyML model, together with the hyperparameters, are used to assess BOTMS (Algorithm 11.2). BOTMS outperforms conventional model selection and hyperparameter-tuning techniques, due to its ability to efficiently manage the tradeoff between accuracy and energy consumption, as shown by the simulation results

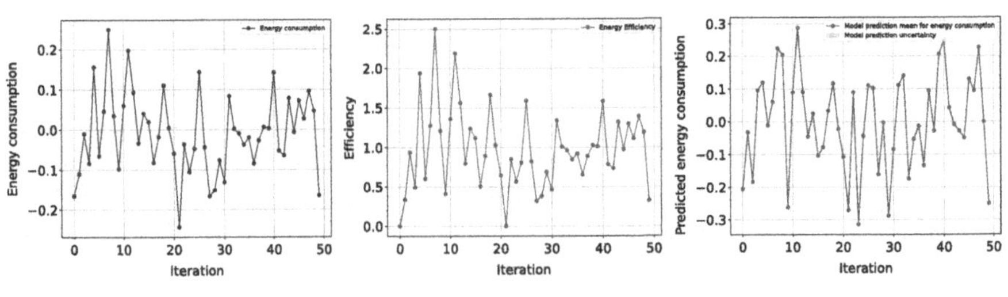

FIGURE 11.2

Performance evaluation of BOTMS with respect to energy consumption, efficiency, and predicted energy consumption.

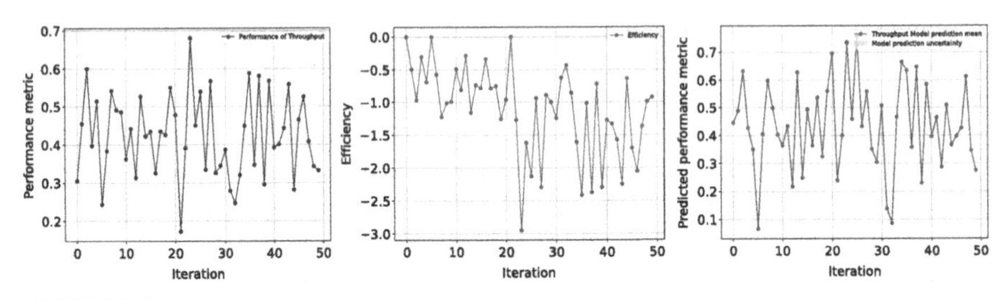

FIGURE 11.3

Performance evaluation of EBORA with respect to throughput, efficiency, and predicted throughput.

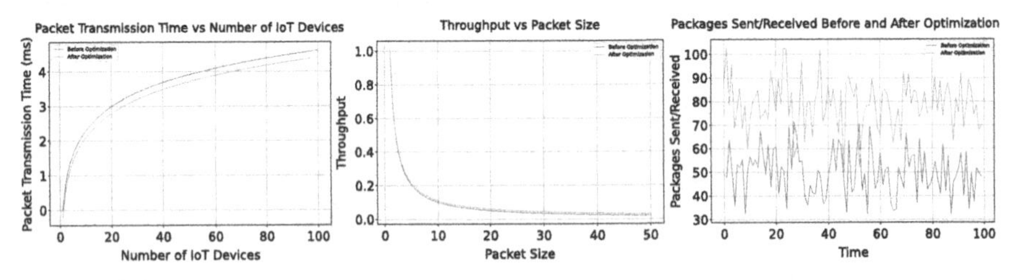

FIGURE 11.4

Performance evaluation of BOEPT with respect to packet transmission time, throughput versus packet size, and packages sent/received.

displayed in Fig. 11.2. EBORA (Algorithm 11.3) is assessed based on its capacity to allocate resources in an optimal manner to maximize network throughput while minimizing energy usage. Our simulations results shown in Fig. 11.3 reveal that EBORA significantly outperforms baseline resource allocation strategies, proving its efficiency in managing the limited and shared resources in LPWANs.

Lastly, the performance of BOEPT (Algorithm 11.4) is assessed by comparing the packet transmission time, throughput versus packet size, and performance of TinyML applications before and after tuning, especially packages sent/received. The results shown in Fig. 11.4 demonstrate that BOEPT is capable of achieving high performance while maintaining good throughput versus packet size, and performance is better after optimization, establishing it as a powerful tool for tuning TinyML applications. Overall, the simulation results validate the effectiveness and robustness of our proposed Bayesian optimization algorithms in addressing key challenges in TinyML-enabled LPWANs.

11.6 Conclusion and future directions

Throughout this chapter, we have conducted an exhaustive examination of the manner in which Bayesian optimization improves TinyML to facilitate edge intelligence in LPWANs. The effective integration of Bayesian optimization and TinyML showcases a remarkable capacity to tackle the distinct obstacles presented by the limited resources of low-power wide area network (LPWAN) infrastructures. The main challenges that we tried to overcome were energy consumption, computational workload, and model selection. Our study began with a comprehensive description of the foundational principles that form the basis of LPWANs, Bayesian optimization, and TinyML. A complete overview of the methodologies regarding the development of our four novel Bayesian optimization algorithms is given in Algorithms 11.1, 11.2, 11.3, and 11.4, where the inner workings of each method are presented.

Utilizing these algorithms, critical elements of TinyML in LPWANs were resolved, including resource allocation, prediction time reduction, energy optimization, model selection and hyperparameter adjustment, and resource allocation. Our Bayesian optimization techniques for prediction time reduction and energy optimization have demonstrated the capability to substantially prolong the longevity of IoT devices and decrease latency. In the interim, the utilization of an efficient resource allocation strategy in conjunction with the TinyML model selection and hyperparameter tuning methodology has proven successful in optimizing both model performance and resource allocation. As a result, the overall efficiency of the network has been enhanced. Despite the encouraging outcomes, several obstacles and constraints persist. For example, although Bayesian optimization demonstrates effectiveness in managing problems with a high number of dimensions, the computational expense may escalate exponentially as dimensionality increases. Furthermore, a critical component of Bayesian optimization, the exploration–exploitation trade-off, frequently necessitates meticulous adjustment and might not consistently result in optimal solutions.

Regarding future studies, a number of interesting possibilities are shown. Initially, the current work is devoted to improving Bayesian optimization techniques to handle higher-dimensional problems more effectively with less computing overhead. Furthermore, the integration of other optimization methods such as reinforcement learning and swarm intelligence into the TinyML ecosystem shows significant promise for improving its performance. To sum up, a wider variety of fields, including healthcare, industrial automation, and smart cities, may be investigated via the use of Bayesian optimization with TinyML. Different strategies can be needed in different industries to deal with specific challenges and limitations. Ultimately, this chapter concludes by demonstrating the remarkable power of Bayesian optimization strategies to improve the performance of TinyML inside LPWANs. We hope that our observations and algorithms will serve as a driving force behind further research and advancement in this quickly developing industry.

References

[1] P.P. Ray, A review on TinyML: state-of-the-art and prospects, Journal of King Saud University: Computer and Information Sciences 34 (4) (2022) 1595–1623.

[2] L. Dutta, S. Bharali, TinyML meets IoT: a comprehensive survey, Internet of Things 16 (2021) 100461.

[3] Y. Abadade, A. Temouden, H. Bamoumen, N. Benamar, Y. Chtouki, A.S. Hafid, A comprehensive survey on TinyML, IEEE Access (2023), https://doi.org/10.1109/ACCESS.2023.3294111.

[4] R. Kallimani, K. Pai, P. Raghuwanshi, S. Iyer, O.L. López, TinyML: tools, applications, challenges, and future research directions, arXiv preprint, arXiv:2303.13569, 2023.

[5] A. Karras, C. Karras, K.C. Giotopoulos, D. Tsolis, K. Oikonomou, S. Sioutas, Federated edge intelligence and edge caching mechanisms, Information 14 (7) (2023), https://doi.org/10.3390/info14070414, https://www.mdpi.com/2078-2489/14/7/414.

[6] B.S. Chaudhari, M. Zennaro, S. Borkar, LPWAN technologies: emerging application characteristics, requirements, and design considerations, Future Internet 12 (3) (2020), https://doi.org/10.3390/fi12030046, https://www.mdpi.com/1999-5903/12/3/46.

[7] Y.B. Zikria, H. Yu, M.K. Afzal, M.H. Rehmani, O. Hahm, Internet of Things (IoT): operating system, applications and protocols design, and validation techniques, https://doi.org/10.1016/j.future.2018.07.058, 2018.

[8] S. Zahoor, R.N. Mir, Resource management in pervasive Internet of things: a survey, Journal of King Saud University: Computer and Information Sciences 33 (8) (2021) 921–935.

[9] N. Michelusi, M. Levorato, Energy-based adaptive multiple access in LPWAN IoT systems with energy harvesting, in: 2017 IEEE International Symposium on Information Theory (ISIT), 2017, pp. 1112–1116, https://doi.org/10.1109/ISIT.2017.8006701.

[10] A. Karras, C. Karras, A. Giannaros, D. Tsolis, S. Sioutas, Mobility-aware workload distribution and task allocation for mobile edge computing networks, in: International Conference on Advances in Computing Research, Springer, 2023, pp. 395–407.

[11] A. Karras, C. Karras, I. Karydis, M. Avlonitis, S. Sioutas, An adaptive, energy-efficient DRL-based and MCMC-based caching strategy for IoT systems, in: I. Chatzigiannakis, I. Karydis (Eds.), Algorithmic Aspects of Cloud Computing, Springer Nature Switzerland, Cham, 2024, pp. 66–85.

[12] W. Yang, Z.Q. Liew, W.Y.B. Lim, Z. Xiong, D. Niyato, X. Chi, X. Cao, K.B. Letaief, Semantic communication meets edge intelligence, IEEE Wireless Communications 29 (5) (2022) 28–35.

[13] Y. Chen, F. Zhao, Y. Lu, X. Chen, Dynamic task offloading for mobile edge computing with hybrid energy supply, Tsinghua Science and Technology 28 (3) (2023) 421–432, https://doi.org/10.26599/TST.2021.9010050.

[14] C. Mwase, Y. Jin, T. Westerlund, H. Tenhunen, Z. Zou, Communication-efficient distributed AI strategies for the IoT edge, Future Generation Computer Systems 131 (2022) 292–308.
[15] T. Jain, Avaneesh, R. Verma, R. Shorey, Latency-memory optimized splitting of convolution neural networks for resource constrained edge devices, in: 2022 14th International Conference on COMmunication Systems & NETworkS (COMSNETS), 2022, pp. 531–539, https://doi.org/10.1109/COMSNETS53615.2022.9668356.
[16] S.S. Musa, M. Zennaro, M. Libsie, E. Pietrosemoli, Mobility-aware proactive edge caching optimization scheme in information-centric IoV networks, Sensors 22 (4) (2022) 1387.
[17] V. Rajapakse, I. Karunanayake, N. Ahmed, Intelligence at the extreme edge: a survey on reformable TinyML, ACM Computing Surveys 55 (13s) (2023) 1–30.
[18] V. Alvear-Puertas, P.D. Rosero-Montalvo, V. Félix-López, D.H. Peluffo-Ordóñez, Edge artificial intelligence for internet of things devices: open challenges, in: D.H. de la Iglesia, J.F. de Paz Santana, A.J. López Rivero (Eds.), New Trends in Disruptive Technologies, Tech Ethics and Artificial Intelligence, Springer Nature Switzerland, Cham, 2023, pp. 312–319.
[19] M. Jena, U. Das, M. Das, A pragmatic analysis of security concerns in cloud, fog, and edge environment, in: Predictive Data Security Using AI: Insights and Issues of Blockchain, IoT, and DevOps, Springer, 2022, pp. 45–59.
[20] K. Dolui, S.K. Datta, Comparison of edge computing implementations: fog computing, cloudlet and mobile edge computing, in: 2017 Global Internet of Things Summit (GIoTS), IEEE, 2017, pp. 1–6.
[21] B. Su, Z. Qin, Q. Ni, Energy efficient resource allocation for uplink LoRa networks, in: 2018 IEEE Global Communications Conference (GLOBECOM), IEEE, 2018, pp. 1–7.
[22] S. Ismail, H. Reza, Evaluation of naïve Bayesian algorithms for cyber-attacks detection in wireless sensor networks, in: 2022 IEEE World AI IoT Congress (AIIoT), 2022, pp. 283–289, https://doi.org/10.1109/AIIoT54504.2022.9817298.
[23] D.B. Dahl, D.J. Johnson, P. Müller, Search algorithms and loss functions for Bayesian clustering, Journal of Computational and Graphical Statistics 31 (4) (2022) 1189–1201.
[24] A.F. Subahi, O.I. Khalaf, Y. Alotaibi, R. Natarajan, N. Mahadev, T. Ramesh, Modified self-adaptive Bayesian algorithm for smart heart disease prediction in IoT system, Sustainability 14 (21) (2022) 14208.
[25] Z. Zhou, H. Yu, H. Shi, Human activity recognition based on improved Bayesian convolution network to analyze health care data using wearable IoT device, IEEE Access 8 (2020) 86411–86418, https://doi.org/10.1109/ACCESS.2020.2992584.
[26] D. Kim, K. Kwon, K. Pham, J.-Y. Oh, H. Choi, Surface settlement prediction for urban tunneling using machine learning algorithms with Bayesian optimization, Automation in Construction 140 (2022) 104331.
[27] A. Akbar, G. Kousiouris, H. Pervaiz, J. Sancho, P. Ta-Shma, F. Carrez, K. Moessner, Real-time probabilistic data fusion for large-scale IoT applications, IEEE Access 6 (2018) 10015–10027, https://doi.org/10.1109/ACCESS.2018.2804623.
[28] Y.N. Kunang, S. Nurmaini, D. Stiawan, B.Y. Suprapto, Improving classification attacks in IoT intrusion detection system using Bayesian hyperparameter optimization, in: 2020 3rd International Seminar on Research of Information Technology and Intelligent Systems (ISRITI), 2020, pp. 146–151, https://doi.org/10.1109/ISRITI51436.2020.9315360.
[29] N. Schizas, A. Karras, C. Karras, S. Sioutas, TinyML for ultra-low power AI and large scale IoT deployments: a systematic review, Future Internet 14 (12) (2022), https://doi.org/10.3390/fi14120363, https://www.mdpi.com/1999-5903/14/12/363.

[30] E. Sallum, N. Pereira, M. Alves, M. Santos, Performance optimization on LoRa networks through assigning radio parameters, in: 2020 IEEE International Conference on Industrial Technology (ICIT), 2020, pp. 304–309, https://doi.org/10.1109/ICIT45562.2020.9067310.
[31] S. Gupta, I. Snigdh, Applying Bayesian belief in LoRa: smart parking case study, Journal of Ambient Intelligence and Humanized Computing 14 (6) (2023) 7857–7870.
[32] K. Mekki, E. Bajic, F. Chaxel, F. Meyer, A comparative study of LPWAN technologies for large-scale IoT deployment, ICT Express 5 (1) (2019) 1–7.
[33] C. Karras, A. Karras, M. Avlonitis, S. Sioutas, An overview of MCMC methods: from theory to applications, in: I. Maglogiannis, L. Iliadis, J. Macintyre, P. Cortez (Eds.), Artificial Intelligence Applications and Innovations. AIAI 2022 IFIP WG 12.5 International Workshops, Springer International Publishing, Cham, 2022, pp. 319–332.
[34] C. Karras, A. Karras, M. Avlonitis, I. Giannoukou, S. Sioutas, Maximum likelihood estimators on MCMC sampling algorithms for decision making, in: I. Maglogiannis, L. Iliadis, J. Macintyre, P. Cortez (Eds.), Artificial Intelligence Applications and Innovations. AIAI 2022 IFIP WG 12.5 International Workshops, Springer International Publishing, Cham, 2022, pp. 345–356.
[35] A. Raha, S. Ghosh, D. Mohapatra, D.A. Mathaikutty, R. Sung, C. Brick, V. Raghunathan, Special session: approximate TinyML systems: full system approximations for extreme energy-efficiency in intelligent edge devices, in: 2021 IEEE 39th International Conference on Computer Design (ICCD), 2021, pp. 13–16, https://doi.org/10.1109/ICCD53106.2021.00015.
[36] Y. Zhang, D. Wijerathne, Z. Li, T. Mitra, Power-performance characterization of TinyML systems, in: 2022 IEEE 40th International Conference on Computer Design (ICCD), 2022, pp. 644–651, https://doi.org/10.1109/ICCD56317.2022.00099.
[37] A.S. Shah, H. Nasir, M. Fayaz, A. Lajis, I. Ullah, A. Shah, Dynamic user preference parameters selection and energy consumption optimization for smart homes using deep extreme learning machine and bat algorithm, IEEE Access 8 (2020) 204744–204762, https://doi.org/10.1109/ACCESS.2020.3037081.
[38] Y. Morita, S. Rezaeiravesh, N. Tabatabaei, R. Vinuesa, K. Fukagata, P. Schlatter, Applying Bayesian optimization with Gaussian process regression to computational fluid dynamics problems, Journal of Computational Physics 449 (2022) 110788.
[39] J. Wu, X.-Y. Chen, H. Zhang, L.-D. Xiong, H. Lei, S.-H. Deng, Hyperparameter optimization for machine learning models based on Bayesian optimization, Journal of Electronic Science and Technology 17 (1) (2019) 26–40.
[40] Z. Li, X. Wang, S. Ruan, Z. Li, C. Shen, Y. Zeng, A modified hypervolume based expected improvement for multi-objective efficient global optimization method, Structural and Multidisciplinary Optimization 58 (2018) 1961–1979.
[41] M. Abdolshah, A. Shilton, S. Rana, S. Gupta, S. Venkatesh, Expected hypervolume improvement with constraints, in: 2018 24th International Conference on Pattern Recognition (ICPR), IEEE, 2018, pp. 3238–3243.
[42] L. Zhao, J. Wang, J. Liu, N. Kato, Optimal edge resource allocation in IoT-based smart cities, IEEE Network 33 (2) (2019) 30–35.
[43] S.U. Minhaj, A. Mahmood, S.F. Abedin, S.A. Hassan, M.T. Bhatti, S.H. Ali, M. Gidlund, Intelligent resource allocation in LoRaWAN using machine learning techniques, IEEE Access 11 (2023) 10092–10106.
[44] Z. Xu, Y. Guo, J.H. Saleh, Efficient hybrid Bayesian optimization algorithm with adaptive expected improvement acquisition function, Engineering Optimization 53 (10) (2021) 1786–1804.

[45] H. Wang, B. van Stein, M. Emmerich, T. Back, A new acquisition function for Bayesian optimization based on the moment-generating function, in: 2017 IEEE International Conference on Systems, Man, and Cybernetics (SMC), IEEE, 2017, pp. 507–512.

[46] K. Yang, K. Chen, M. Affenzeller, B. Werth, A new acquisition function for multi-objective Bayesian optimization: correlated probability of improvement, in: Proceedings of the Companion Conference on Genetic and Evolutionary Computation, 2023, pp. 2308–2317.

[47] J. Wilson, F. Hutter, M. Deisenroth, Maximizing acquisition functions for Bayesian optimization, Advances in Neural Information Processing Systems 31 (2018).

CHAPTER

6TiSCH adaptive scheduling for Industrial Internet of Things 12

Nilam Pradhan[a], Bharat S. Chaudhari[a], and Marco Zennaro[b]

[a]*Department of Electrical and Electronics Engineering, Dr. Vishwanath Karad MIT World Peace University, Pune, India*

[b]*Science, Technology and Innovation Unit, Abdus Salam International Centre for Theoretical Physics (ICTP), Trieste, Italy*

12.1 Introduction

In the Internet of Things (IoT), everyday objects interact with each other over the Internet, addressing the needs of a specific application. Devices with built-in sensors collect and analyze the data, providing security, adaptation, and autonomous behavior. The IoT has a significant impact on several aspects of everyday life both in domestic and industrial fields [1], [2]. The Industrial Internet of Things (IIoT) has emerged with the integration of IoT technologies in industrial applications. The IIoT incorporates different industrial applications, including eHealth, transportation, agriculture, and the food industry. It integrates information technology (IT) with operational technology (OT). This integration has enabled the industry to achieve greater convergence in terms of automation and optimization with remote access and control. With the evolution of Industry 4.0, production processes are transformed with the help of real-time data analysis and decision-making. The IIoT has revolutionized the industry in terms of reliability and efficiency. In operational technology, the employment of machine-to-machine (M2M) communication, AI, and machine learning has allowed autonomous behavior in processing and manufacturing.

Tiny machine learning (TinyML) has emerged as an influential paradigm for deploying machine learning algorithms on resource-constrained devices. TinyML is powered by lightweight machine learning algorithms that bring intelligence to edge devices. The convergence of IIoT and TinyML has opened various opportunities for optimizing industrial processes through intelligent decision-making at the edge. However, deploying TinyML in IIoT introduces different challenges such as inadequate computing capability, constrained memory, and limited power. TinyML models can be trained to analyze sensor data and identify patterns that indicate potential equipment failures. These models can then be deployed on edge devices to continuously monitor equipment health and predict its maintenance requirements. In IIoT systems, energy efficiency is often a critical concern, especially for battery-

TinyML for Edge Intelligence in IoT and LPWAN Networks. https://doi.org/10.1016/B978-0-44-322202-3.00017-8

powered or energy-constrained devices. By leveraging TinyML, optimized machine learning models can be developed that are specifically designed to run efficiently on resource-constrained hardware. TinyML in IIoT offers immense potential for enhancing industrial processes. The applications, such as predictive maintenance, quality control, industrial manufacturing processes optimization, and software-defined processes leverage the power of machine learning to extract insights from sensor data gathered by edge devices. It has enabled proactive maintenance, enhanced operational efficiency, and reduced downtime in industrial environments.

IPv6 over the Time Slotted Channel Hopping (TSCH) mode of IEEE 802.15.4e (6TiSCH) technology [3], [4] is being developed as a promising scheduling solution for IIoT environments by providing determinism, reliability, and predictability. This chapter introduces 6TiSCH scheduling as a vital concept for addressing different challenges in IIoT. It explores the utilization of 6TiSCH scheduling in IIoT environments and discussing its benefits, challenges, and various scheduling strategies to enhance network performance.

12.2 Benefits of TinyML in IIoT

TinyML enables edge devices to function autonomously even when disconnected from the cloud. This is particularly valuable in industrial environments where network connectivity may be unreliable. Local machine learning models can continue to operate and make decisions based on locally collected data, ensuring continuous functionality. It offers various benefits in IIoT networks.

Real-time decision-making

With TinyML, edge devices can make real-time decisions based on the machine learning models running locally on the edge devices. This eliminates the need for round trips to the cloud for decision-making, resulting in faster response times. Time-sensitive applications in IIoT can benefit significantly from the immediate insights provided by TinyML. It enables real-time data analysis allowing for actionable intelligence at the edge, reducing the time needed for data transmission to the cloud, processing, and obtaining feedback.

Edge intelligence

With TinyML, machine learning models can be deployed directly on edge devices such as sensors, gateways, or controllers in IIoT systems. These models are optimized for running on resource-constrained devices with limited processing power, memory, and energy consumption. By running machine learning algorithms on edge devices they can perform data analysis and inference locally, enabling real-time decision-making. This brings intelligence to the edge and allows for localized decision-making without relying on constant connectivity to the cloud. In IIoT applications, where low-latency responses are crucial, the ability to process data and provide insights at the edge can significantly improve the system's overall responsiveness.

Adaptability and autonomy

TinyML allows edge devices to learn and adapt from incoming data over time. The machine learning models deployed on the devices can be continuously updated or retrained with new data, enabling adaptive behavior and improved performance. This adaptability empowers edge devices to make intelligent decisions based on changing conditions, without relying on frequent updates or cloud connectivity.

Privacy and security

By leveraging TinyML at the edge, sensitive data can remain local, minimizing the risks associated with transmitting sensitive information to the cloud. This not only enhances privacy but also mitigates security risks associated with transmitting sensitive information over the network. Edge-based processing reduces the attack surface and potential vulnerabilities of IIoT systems.

Reduced bandwidth requirements

TinyML in processing reduces the need for continuous data transmission from edge devices to the cloud. By running machine learning models on edge devices, only relevant insights or anomalies need to be communicated, minimizing the amount of data sent over the network. This reduces bandwidth requirements, eases network congestion, and lowers communication costs.

Offline functionality

TinyML allows edge devices to operate autonomously even when there is no network connectivity. Machine learning models deployed on the devices can continue to perform inference and make decisions without relying on the cloud. This offline functionality ensures continuous operation, even in environments with intermittent or unreliable network connectivity. By leveraging the computational capabilities of microcontrollers or low-power devices, TinyML can minimize the dependency on expensive cloud-based infrastructure for data processing and analysis.

Cost efficiency

By processing data and performing machine learning inference on edge devices, TinyML reduces the need for high-end, resource-intensive hardware or cloud infrastructure. This can result in cost savings in terms of hardware requirements, data storage, and cloud computing resources. Moreover, the reduced dependency on cloud services can lead to lower operational costs and subscription fees.

Energy efficiency

Edge intelligence powered by TinyML can contribute to energy-efficient IIoT systems. Instead of continuously sending data to the cloud for processing, which consumes energy and bandwidth, local processing on edge devices reduces the overall energy consumption. This is particularly beneficial in IIoT applications where devices are battery-powered or operate in remote locations with limited access to power sources.

12.3 IEEE 802.15.4e TSCH

The IEEE 802.15 Working Group (WG) has developed a standard to enable ultralow-cost, low-computational capability, ultralow-power consumption, and low-data rate communications. It achieves the objectives of a low-rate wireless personal area network (LR-WPAN) with ease of installation and reliable data transfer by using flexible protocols. This technology plays a vital role in IIoT applications allowing hop-to-hop communication. However, the basic IEEE 802.15.4 is prone to interference as it relies on a single channel frequency and also suffers from higher power consumption as devices are always in an ON-state. A single-channel approach for communication was a limiting factor for scalability and network performance because of its limited capability to operate a dense network in a contention-free manner. To address this limitation, the basic IEEE 802.15.4 medium access control (MAC) was revised, adding many new features, resulting in IEEE 802.15e TSCH [5]. It is a highly reliable and power-efficient technology for current industrial requirements. It enhances the MAC layer by keeping physical and security layers unchanged. It has various added MAC features, including the TSCH behavior.

IEEE 802.15.4e has enabled very low-duty cycle operation along with deterministic latency and scalability, which are essential conditions for time-critical applications. TSCH uses fixed-size time slots and multi-channel hopping. Time slots are grouped in a slotframe that repeats over time. During every time slot the device exchanges the data frame and an acknowledgment. The number of time slots S_n determines the slotframe size, and it can be made variable. Fig. 12.1 shows an illustration of the IEEE 802.15.4e TSCH schedule for an example network. Communication cells within a slotframe are used for interaction between a pair of devices whose coordinate is determined as $[timeOffset, channelOffset]$. Through time synchronization it minimizes the number of retransmissions and collision, whereas channel hopping helps in reducing the effects of multi-path fading and interference. Nodes B and D in Fig. 12.1 communicate with each other during time slots 1 and 2, and nodes A and C communicate during time slots 3 and 4. The total number of time slots that have elapsed since the start of the network is referred to as the Absolute Slot Number (ASN). Its value is global and used by all devices for channel frequency computation and counting frames. The slotframe is repeated after every S_n time slots, whereas ASN is continuously increasing.

The operational frequency f_{op} can be determined from the ASN and channel offset as

$$f_{op} = (ASN + co)\,\%C_n, \tag{12.1}$$

where co is the channel offset and C_n is the total number of channel offsets being used. ASN ensures the channel frequency alteration in every slotframe for the different channel offset values. Devices following a common schedule decide their role in every slot; to transmit data frames, to receive frames, or to sleep. This mechanism has enabled improved operational control and resource reservation across the network in a very efficient manner. TSCH became a key enabling technology for efficient processes in harsh industrial environments. The deterministic nature of TSCH ensures

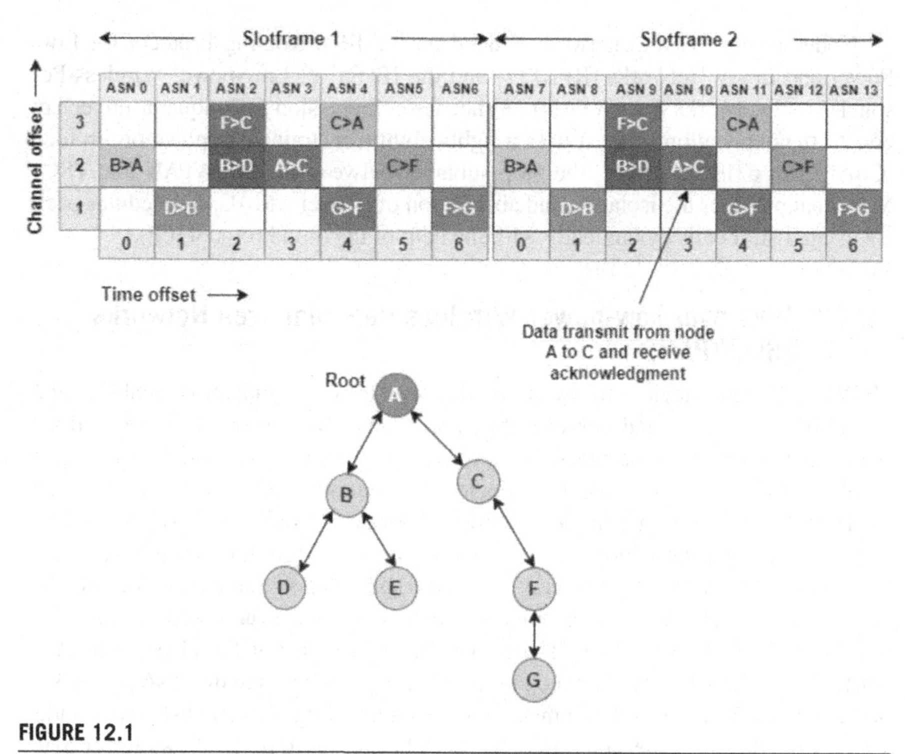

FIGURE 12.1

Basic TSCH scheduling structure.

predictable and interference-free communication, making it suitable for TinyML networks.

12.4 IPv6 over the TSCH mode of IEEE 802.15.4e (6TiSCH)

IEEE 802.15.4e executes the TSCH schedule but it does not specify rules to build it. To allow the use of IPv6 in the industrial-constrained network, a 6TiSCH WG was created [6]. It combines the IPv6 network and IEEE 802.15.4e TSCH, intending to build and manage the communication schedule. The 6TiSCH protocol is particularly suitable for industrial and smart city applications, where quality of service (QoS) support is important. It allows devices to prioritize different types of traffic according to their criticality level. The time-slotted communication technique used by the protocol helps to reduce energy consumption and extend battery life. Also, it helps to reduce packet loss and to improve operational efficiency, and it is designed to be scalable, enabling large-scale IIoT deployments. The protocol also provides mechanisms for devices to join or leave the network dynamically, without disrupting the ongoing operations.

Nodes in the 6TiSCH network depend on the IPv6 Routing Protocol for Low-Power and Lossy Networks (RPL) [7] and the IPv6 over Low-power Wireless Personal Area Networks (6LoWPAN) [8] header compression technique at upper layers. At the application layer, it uses a lightweight Constrained Application Protocol (CoAP) [9]. 6TiSCH defines the 6top sublayer between the 6LoWPAN and TSCH MAC that provides the isolation and abstraction of IP over a MAC. It schedules packets using TSCH cells dynamically with the help of the 6top Protocol (6P).

12.4.1 IPv6 over Low-power Wireless Personal Area Networks (6LoWPAN)

6LoWPAN allows devices to use IPv6 addresses for communication, enabling end-to-end IP connectivity and interoperability with other IPv6-enabled devices and networks. It is specifically designed for devices with limited resources. The main goal of 6LoWPAN is to allow the IIoT devices to communicate with IPv6-based networks without requiring complex gateway devices. By utilizing IPv6, 6LoWPAN allows the integration of low-power devices into the larger Internet infrastructure. It defines an adaptation layer that allows the fragmentation and reassembly of IPv6 packets into smaller units. It is suitable for transmission over low-power wireless links with limited packet size. The maximum packet size at the physical layer of IEEE 802.15.4e is 127 bytes, which is significantly smaller than the IPv6 packet size of 1280 bytes. That limits the frame size at the data link layer to 102 bytes with added security overhead. At the same time, the IPv6 header of 40 bytes allows fewer bytes for upper-layer protocols. To address this, 6LoWPAN implements header compression. It supports the network management for resource-constrained multi-hop mesh topology, with minimum configuration and self-convergence. Further, it specifies the address autoconfiguration procedures that generate IPv6 stateless addresses, reducing the configuration overhead on the devices.

12.4.2 IPv6 Routing Protocol for Low-Power And Lossy Networks (RPL)

Low-power and lossy networks (LLNs) consist of battery-operated resource-constrained devices. A dense network where thousands of devices are interconnected by unstable lossy links experiences a reduced packet delivery ratio (PDR). These characteristics invite many challenges in routing and scheduling problems. On that account, the Internet Engineering Task Force (IETF) ROLL WG has defined an application-specific RPL routing protocol for LLNs. An RPL defines an Objective Function (OF) as per the LLN application. It organizes the network as a destination-oriented directed acyclic graph (DODAG) to optimize routes to or from multiple roots. Along with point-to-point traffic, it supports the point-to-multipoint and multipoint-to-point patterns. Fig. 12.2 shows that the topology is divided into multiple DODAGs with roots acting as sinks joined by a common transit link at the backbone. A different set of DODAGs uses different OFs and applications. To build and manage DODAG,

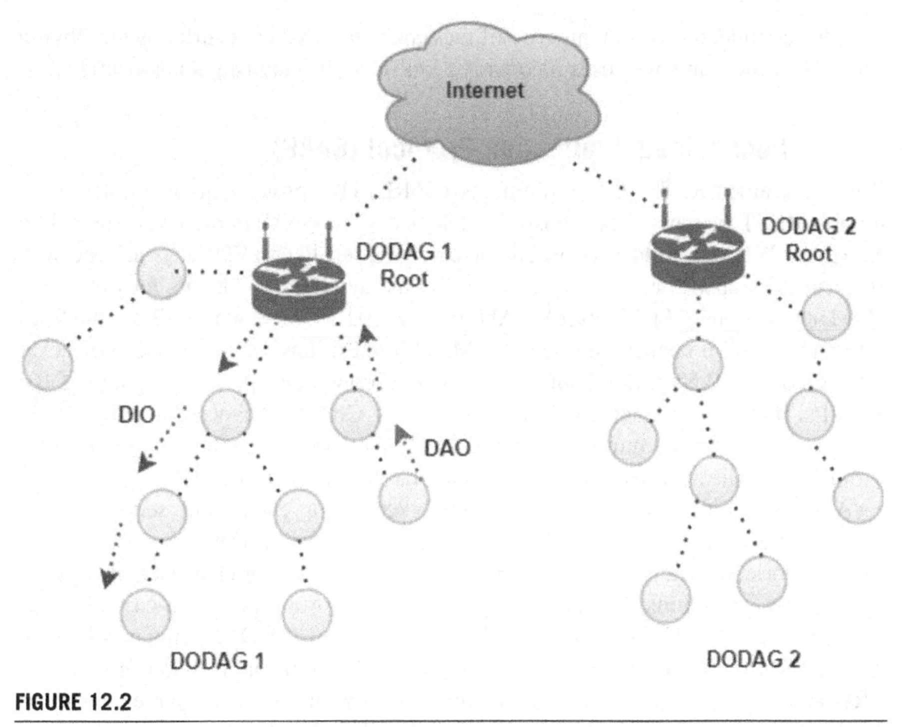

FIGURE 12.2

RPL DODAG framework.

the node uses DODAG information object (DIO) messages. The location of the node with respect to the root within a DODAG is represented by the rank value. The rank is incremented per hop as it progresses from root towards leaf as specified by the OF.

To construct the DODAG, some nodes are configured as a root. Nodes associated with DODAG advertise their affiliation and routing metrics by multi-casting DIO messages. After receiving DIO messages, neighbor nodes use their local information to join or maintain the DODAG and select a parent as per the defined OF. Nodes create a routing table for every destination by using the information in the DIO message via the parent. RPL optionally provisions downward routes by using destination advertisement object (DAO) messages. There are two approaches to facilitate the downward traffic: the storing and non-storing modes. In storing mode, the source sends a packet in an upward direction to the common ancestor and routed down to the destination, whereas in non-storing mode, the source sends a packet to the root and then down to the destination. The application can implement either storing or non-storing mode. RPL routing can be carried out in a centralized or distributed fashion. In a centralized routing model, routes are determined by the central controller. It is also responsible for the resource allocation and management of deterministic multi-hop networks. However, the distributed routing model is based on peer-to-peer information exchanges for routing and resource allocation. Furthermore, RPL supports the confidentiality and integrity of messages. It provides security by employing a link-layer mechanism or uses its own if a link-layer mechanism is not available.

12.4.3 Constrained Application Protocol (CoAP)

The Constrained RESTful Environments (CoRE) [10] group aims to implement a suitable REST framework for constrained devices. The CoAP interface developed by the CoRE WG is a light version client/server model, like HTTP. It is a basic web protocol developed for the special need of constrained environments for optimized M2M applications [11]. The 6LoWPAN in IEEE 802.15.4e networks allows the fragmentation of IPv6 packets into smaller MAC frames; however, it affects the packet delivery. 6LoWPANs suffer from higher packet drop with a typical throughput of 10s of kbit/s. Hence, to guarantee minimal packet loss, CoAP has evolved to reduce message overhead, thereby minimizing packet fragmentation. It can be easily interfaced with HTTP while achieving the specialized QoS requirements of LLNs.

A device with CoAP implementation executes a client and server model. A request/response message exchange is carried out asynchronously over a UDP datagram-oriented transport layer protocol. CoAP uses a shorter header of 4 bytes, followed by an options field and a payload. By supporting the multi-cast IP destination addresses, it allows multi-cast CoAP requests over UDP. It supports built-in service and resource discovery and also consists of significant Web notions such as URIs and Internet media. It allows for interaction with remote resources, enabling the deployment of TinyML IIoT devices in a distributed and interconnected manner.

12.5 6TiSCH architecture

The 6TiSCH architecture enables an IPv6 protocol suite over the IEEE Std 802.15.4e TSCH MAC, supporting the interoperability of IP networks with constrained wireless devices [12]. It depends on an IPv6 high-speed federating backbone with a complex routing structure to deliver larger scalability. An IPv6 subnet is represented as a 6TiSCH LLN mesh network. Nodes in the network depend on 6LoWPAN header compression and fragmentation to process IPv6 packets from upper layers. A node joining the 6TiSCH network acquires an IPv6 address and registers itself by employing 6LoWPAN neighbor discovery (ND), which is extended from the IPv6 ND. 6LoWPAN ND protects the address ownership within the subnet and its impressions. A particular 6TiSCH node acts as an RPL root referred to as the LLN border router (6LBR). Fig. 12.3 shows the 6TiSCH extended subnet configuration, consisting of multiple LLNs. It supports scalability by adding a high-speed backbone in the IPv6 network relying on RPL routing. It presents an IPv6 model that comprises multiple IEEE Std 802.15.4e TSCH subnets that are synchronized with the backbone. Any node acting as an RPL root is connected to the routing registrar that provides connectivity to the large factory plant network. The routing registrar called a backbone router (6BBR) is responsible for IPv6 ND operations such as registration redistribution in a

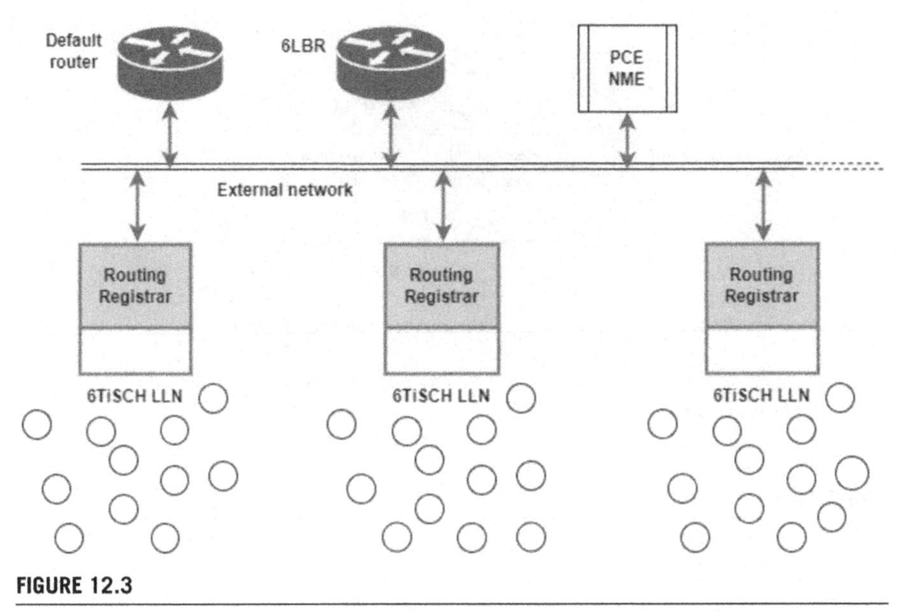

FIGURE 12.3

6TiSCH network configuration.

routing protocol for a mobile IPv6. The 6BBRs are located along the backbone and are synchronized with each other ensuring synchronization of several LLN subnets.

TDMA is a centralized scheduling approach where a central controller assigns fixed time slots to devices for communication. This strategy guarantees collision-free communication, efficient resource allocation, and optimal network performance. However, it may face scalability challenges when the number of devices increases significantly. 6TiSCH provides robust and reliable communication in the presence of interference and multi-path fading. The scheduling mechanisms enable devices to share the available bandwidth effectively, allowing for scalability in networks. The deterministic nature of 6TiSCH scheduling ensures that data transmission occurs at predefined times and channels. This predictability is crucial in TinyML environments where real-time data processing and decision-making are essential. Further, 6TiSCH scheduling optimizes energy consumption by minimizing idle listening and collisions. It addresses these challenges by providing robust and reliable communication.

12.6 6TiSCH protocol stack

At the IETF, a set of protocols and standards are being evolved to meet the growing need for IP devices in IIoT networks. The goal of IETF WGs is to integrate the various link layer technologies with the IP ecosystem. They manage the major constraints imposed by the underlying technologies in terms of payload, memory size, processing

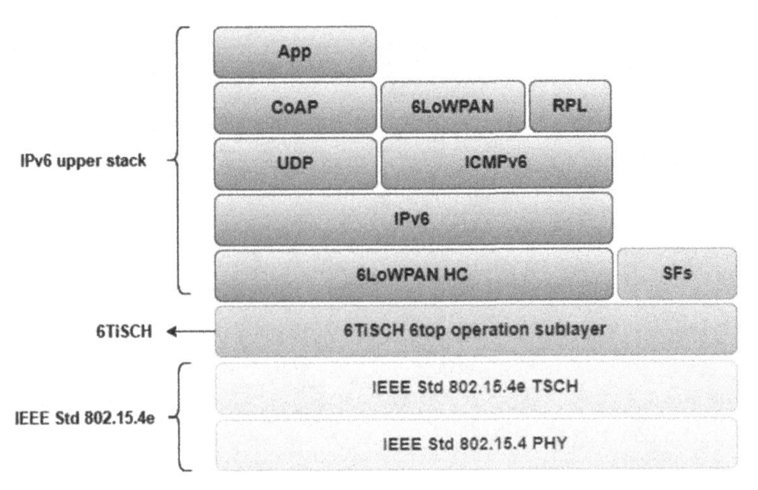

FIGURE 12.4

6TiSCH protocol stack.

power, and topology while ensuring IP compliance. 6TiSCH is one of the primary efforts to bring IPv6 to industrial low-power wireless networks, integrating TSCH with 6LoWPANs. It acts as the glue of different layers and enables improvements in header compression, IP-in-IP encapsulation, and 6LoWPAN ND. The 6TiSCH protocol stack shown in Fig. 12.4 is rooted in the IEEE 802.15.4e TSCH link layer and complements the IPv6 upper-layer protocol set. These protocols include a minimal configuration mechanism for network bootstrap, a security mechanism to support secure login, and a set of scheduling strategies. The 6TiSCH protocol stack is built upon IPv6 networking and TSCH technology, offering a robust communication solution for IIoT devices. By enabling precise scheduling of communication slots, 6TiSCH ensures predictable and interference-free data transmission in an industrial environment.

The physical and link layer leverages the IEEE 802.15.4e standard, which defines the physical characteristics and MAC protocol. The physical layer of the 6TiSCH protocol suite deals with the modulation schemes, data rates, and transmission power levels for wireless communication. It enables devices to establish a reliable link over the air interface. The MAC layer of 6TiSCH is based on the TSCH mechanism. It divides time into slots, allowing devices to access shared wireless media in a synchronized manner. The network layer of the 6TiSCH protocol suite is built upon the IPv6 protocol and enables the routing of packets in networks. The RPL provides adaptive and energy-efficient routing. The key components of this layer include the 6LoWPAN protocol compression that reduces overhead and efficiently utilizes the network protocol designed for IIoT networks. It establishes a mesh topology and enables reliable routing paths in the presence of intermittent connectivity and constrained resources. This is particularly important in resource-constrained environments where bandwidth and energy are limited.

The application layer of the 6TiSCH protocol suite is responsible for interfacing IIoT applications. The application layer encompasses various protocols and standards tailored for IIoT such as CoAP. CoAP provides tools to support the secure join process enabling low-overhead secure RESTful interactions. The 6TiSCH architecture limits the possible variations of the stack and recommends several base elements for IIoT applications. UDP and CoAP are used as the transport protocol of choice for applications and management as opposed to TCP and HTTP.

12.7 6TiSCH scheduling strategies

Scheduling strategies play a crucial role in optimizing the performance of IIoT networks. Different machine learning tasks in TinyML networks may have varying QoS requirements. It involves assigning different priority levels to various types of traffic, such as control messages, sensor data, or machine learning-related data. By prioritizing machine learning traffic, the network can allocate resources, ensuring that critical machine learning tasks meet their reliability requirements. QoS requirements can be achieved by carefully balancing resource allocation and assigning scheduling time slots for different traffic classes as per needs. Traffic-aware scheduling considers the traffic patterns and enhances the performance of networks by providing optimized scheduling while efficiently monitoring and managing network traffic.

The 6TiSCH architecture specifies rules to create and maintain the scheduling operation in a centralized, distributed, or autonomous manner as shown in Fig. 12.5. It is not necessary for every application to implement all methods of scheduling. However, minimal configuration must be executed by all applications, providing a foundation routing and time distribution.

12.7.1 Centralized scheduling

A centralized scheduling approach relies on a central PCE controller that determines and manages the schedule for each device in the network. It must learn the information related to any topological change to recompute the schedule and notify all devices regarding the change. The PCE minimizes the interface among the nodes by centrally monitoring and management of the schedule, enabling faster convergence. To manage resource allocation, it collaborates with the network management entity (NME) and determines the associated part of the schedule. It enables traffic engineering and determinism similar to the DetNet architecture. This approach requires the exchange of schedule information between the controller and the devices, ensuring synchronization and collision-free communication.

The state-of-the-art centralized scheme, a traffic-aware scheduling algorithm (TASA) [13], has been developed by using the matching and coloring mechanisms of graph theory. This technique enables a schedule for all devices across the entire network based on the topology and traffic conditions. In association with the RPL, the PCE uses the path information and related traffic to build the schedule. It allocates

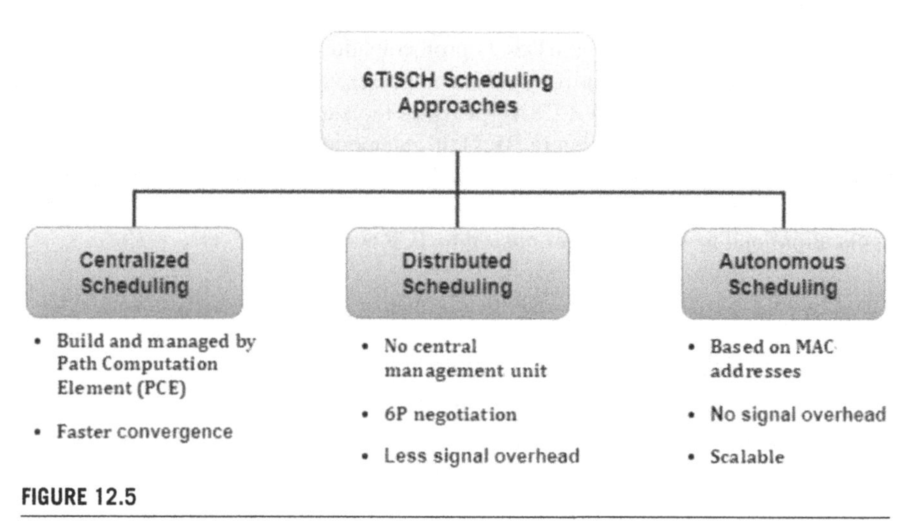

FIGURE 12.5

6TiSCH scheduling approaches.

a set of links in each slot based on queue levels and is updated in the next slot. The next slot will be chosen based on the traffic load at the device. It achieves the level of QoS in terms of a duty cycle and throughput. However, TASA uses the central unit to maintain the global network statistics, which produces high signal overhead. Further, the authors have optimized TASA by applying graph theory tools. They have determined the optimal threshold for the minimum number of time slots and duty cycle required for TASA to exhibit multi-channel, time-synchronized behavior. The work has demonstrated the immense benefits of TASA over traditional IEEE 802.15.4e approaches in terms of energy efficiency. It has laid the foundations of the IETF standardization 6TiSCH WG, intending to significantly improve IIoT data transmission over IEEE 802.15.4e TSCH and IETF 6LoWPAN/ROLL-enabled technologies.

Centralized scheduling techniques are more appropriate to static networks as they trigger the substantial signal overhead that affects network scalability. The main drawback is that it requires a central entity to generate and distribute the schedule, which can be a single point of failure and can cause network downtime if it fails. This approach is more appealing to an IIoT with constant real-time monitoring.

12.7.2 Distributed scheduling

In distributed scheduling, devices coordinate their communication slots without relying on a central controller. Devices exchange schedule information through periodic beacon frames, allowing them to align their time slots. The pair of nodes build and maintain their schedule by monitoring the local view of the network. It uses the 6P transaction to add, delete, or relocate cells dynamically, depending on the network topology and traffic conditions. It is more suitable for dynamic topology with time-varying traffic and to build a scalable network. Distributed scheduling enhances scalability, reduces control overhead, and provides fault tolerance in IIoT deployments. However, the 6P transaction control messages incur some signal overhead, resulting in a higher duty cycle and higher latency. It reduces the dependency on a central controller but may introduce additional synchronization overhead.

Several distributed scheduling algorithms are proposed in the literature. These techniques help in constructing a scalable network. Accettura et al. [14] have developed the Decentralized Traffic Aware Scheduling (DeTAS) algorithm as a modification of the earlier developed centralized scheduling technique TASA. It addresses the queue congestion problem of TASA, thus minimizing the packet drop caused by overflow. It is optimum collision-free scheduling that uses a small amount of information, hence maintaining very low signaling overhead. By leveraging the available 16 channels in IEEE 802.15.4 it enables several concurrent transmissions over a given time slot. It helps in reducing the network duty cycle by minimizing the active slots in a slotframe and improving the reliability of wireless links. To create a schedule, every node computes the amount of traffic towards its parent and reception traffic from children. Nodes forward this information to their parent, which in turn updates its local information. Each node in this technique assigns an alternate sequence of transmit and receive slots to avoid buffer overflow. The root node running DeTAS divides its children for even or odd time slots. When a node schedules an even time slot for transmission, it allocates an odd slot for reception, and vice versa. Furthermore, it keeps even-scheduled nodes and odd-scheduled nodes independent to build an optimal schedule. By allocating transmit and receive slots alternately it prevents queue overflow. Though DeTAS is distributed, nodes in a network update their traffic information and forward it to the root for schedule optimization, which involves significant signal overhead. Additionally, DeTAS initiates rescheduling procedures whenever a new node joins the network, alters its preferred parent, or experiences changes in traffic load.

Palattella et al. proposed on-the-fly (OTF) bandwidth reservation for 6TiSCH [15], which adapts the scheduling bandwidth to network requirements. The OTF module on top of the 6top sublayer optimizes the number of scheduled cells as per traffic flow, thus satisfying the QoS requirements. Cell over-provisioning is determined by the threshold value. Selecting the threshold value involves a trade-off; opting for a larger threshold reduces the number of computational operations needed but increases power consumption due to greater over-provisioning of cells.

12.7.3 Autonomous scheduling

Autonomous schedulers allocate cells by using a hash of the node's MAC addresses. They do not demand any information negotiation among devices, leading to minimized signal overhead. It neither uses a central management entity nor negotiation among neighbor nodes to build schedules. It builds a schedule with the help of simple scheduling rules and is more flexible. One of the benefits of the autonomous scheduling mechanism is its ability to adapt to changing network conditions. Devices can request changes to the schedule based on varying communication needs, and the network can dynamically adjust the schedule to accommodate device requests. This mechanism is highly scalable and adaptable, allowing the network to optimize communication performance in real time, even under variable conditions.

Several autonomous schedulers have been developed to reduce the overhead in cell negotiation. The 6TiSCH Minimal Scheduling Function (MSF) builds a schedule by using both autonomous and negotiated cells. The minimal communication requirement is enabled by the autonomous cell between the pair of nodes. As cells are uniformly distributed within a slotframe, nodes introduce more buffer time before forwarding a packet to their neighbors. The end-to-end delay increases linearly with an increasing number of hops and an increase in slotframe length. It contains higher jitter and lower reliability as an effect of packet retransmissions along the path. The Orchestra [16] node-based autonomous scheduler adapts its schedule by specified rules. Orchestra uses receiver-based shared (RBS) slots and sender-based shared (SBS) slots. SBS reserves cells based on the local information of the sender; RBS, on the other hand, assigns the cells based on the information of the receiver. Orchestra updates the schedule automatically as topology progresses without the additional signal overhead. It depends on existing local network information to manage the schedules, improving the robustness of TSCH. RBS uses single Rx cells for all neighbors, which leads to contention problems, whereas SBS uses Rx cells per neighbor, which leads to significant power consumption. It suffers from higher end-to-end packet drop as slotframe size is increased. Autonomous and traffic-aware scheduling for TSCH networks was proposed as an improvement of the basic Orchestra [17]. It introduces traffic diversity and handles the packets based on their criticality levels. By monitoring the traffic at every node, it reserves multiple cells successively, equal to the number of packets in a queue. The number of cells to be reserved is limited by the slotframe size. This technique minimizes the latency by emptying the queue in a minimum number of slotframes. However, a further rise in traffic may demand more consecutive slots that may collide with the preallocated cells. Cell collision results in packet drop and an increase in latency [18].

ALICE [19] allocates a unique cell for every link of a given direction between a pair of nodes by using the node's local information. It does not comprise negotiation overhead to collect neighbor information. ALICE outperforms Orchestra by improving throughput and reliability and lowering end-to-end latency. It addresses the contention and collision difficulties effectively. For a greater node density, it reduces the queue loss and takes the benefits of a shorter routing distance. However, ALICE uses one cell per slotframe irrespective of traffic in the network, providing insufficient transmission opportunities. For a longer slotframe, it may suffer from large queueing delays and packet drop.

12.7.4 Hybrid scheduling

Hybrid scheduling combines centralized, distributed, and/or autonomous approaches to achieve a balance between performance and scalability [20], [21]. The central controller approach offers flexibility and fault tolerance. The distributed technique supports dynamic networks with reduced control overhead to some extent, whereas an autonomous technique is scalable, has high performance, and is energy-efficient with negligible signal overhead. Hybrid scheduling combines the advantages of the above approaches to balance performance and scalability, offering flexibility and fault tolerance in TinyML environments. For example, it may involve a central controller, assigning fixed slots for critical and time-sensitive traffic while allowing devices to autonomously schedule non-critical data.

Despite its various benefits, the 6TiSCH protocol still faces several challenges that need to be addressed. With further research and development, the protocol has the potential to become the de facto standard for TinyML communication, enabling the development of innovative and transformative applications.

12.8 Deterministic scheduling for IIoT networks

Each node in a 6TiSCH network follows a predetermined schedule, ensuring communication at specific times and frequencies. This deterministic approach reduces interference and collisions, leading to more reliable and efficient information exchange among devices. In many applications, timely data acquisition and processing are critical. With 6TiSCH, devices can adhere to strict timing constraints, ensuring that the sensor data are collected and processed within specific time windows. Also, it makes efficient utilization of network resources, improving overall network capacity. To fully leverage the benefits of 6TiSCH deterministic scheduling, careful design and configuration are required. This includes determining the optimal slot allocation and channel hopping sequence, considering the network topology and traffic patterns. Additionally, synchronization among nodes is crucial to maintain the integrity of the schedule and ensure accurate communication timing.

The Low-latency Distributed Scheduling Function (LDSF) [22] has been developed. It divides the slotframe into several blocks for primary and secondary slot allocation and is repeated over time as shown in Fig. 12.6. The primary transmission has been scheduled in one block and retransmission in the subsequent block by using ghost slots. The transmitter uses ghost cells if it needs retransmission. The sender uses a fixed block within a slotframe based on its hop distance from the root. The technique provides high reliability and limits the buffering delay in retransmission by chaining the blocks. LDSF does not suffer from collisions even in a high-density network, enabling scalability. Also, it achieves a higher network lifetime as the sender and receiver do not have to wake up in every ghost cell.

The 6TiSCH Low Latency Autonomous (LLA) link-based scheduling scheme for deterministic 6TiSCH networks has been developed [23]. The technique is based on slotframe segmentation that achieves end-to-end transmission within a single slotframe. It is an autonomous scheduling scheme that utilizes the MAC addresses of the sender and receiver nodes to allocate cells. LLA requires neither cell negotiation with a neighbor nor a central controller, which helps in minimizing the signal overhead. For every link, the approach schedules a unique cell in a specified segment based on

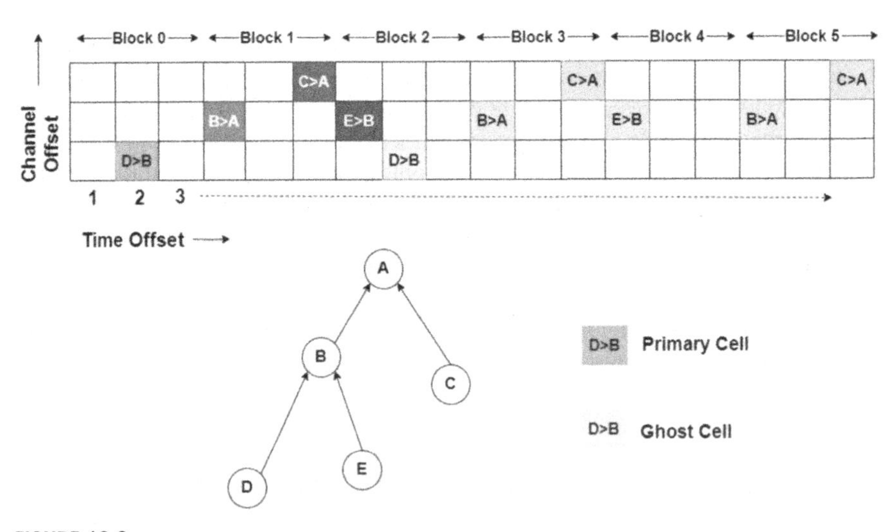

FIGURE 12.6

LDSF slotframe block organization.

the sender's hop distance from the root. All nodes at a given RPL level in the network will utilize the same segment for scheduling. LLA has shown an improvement in QoS in terms of end-to-end latency. It uses a slotframe of size 397 time slots for the EB frame, 31 for RPL control messages, and a variable for application data transmission. The TSCH EB frame uses channel offset 0, and the time offset for the EB frame is determined by using a hash of the address of the sender as

$$to_{EB}(i) = hash\,[ID(i)]\,\%SL_{EB}, \tag{12.2}$$

where $ID(i)$ is the MAC address of node i and SL_{EB} is the enhanced beacon slotframe size.

For a network with maximum hops H, this technique divides the slotframe into H segments, making each segment size $seg_{size} = SL_D/H$, where SL_D is the size of the application data slotframe. It determines a transmit time slot for application data in a specified segment as

$$to_D(ij) = hash\,[x \cdot ID(i) + ID(j)]\,\%seg_{size} + (H - d) \cdot seg_{size}, \tag{12.3}$$

where x is the direction of transmission. Channel offset is based on the sender's address as

$$co_D(i) = hash\,[ID(i)]\,\%C_n + 1. \tag{12.4}$$

An RPL message uses one fixed time slot offset 0 and channel offset 1. The LLA slotframe design for the example network in Fig. 12.7 is shown in Fig. 12.8. The sender i is at a hop distance of d, receives the packet in a segment $(H - d)$, and forwards it to its parent in the immediately next segment.

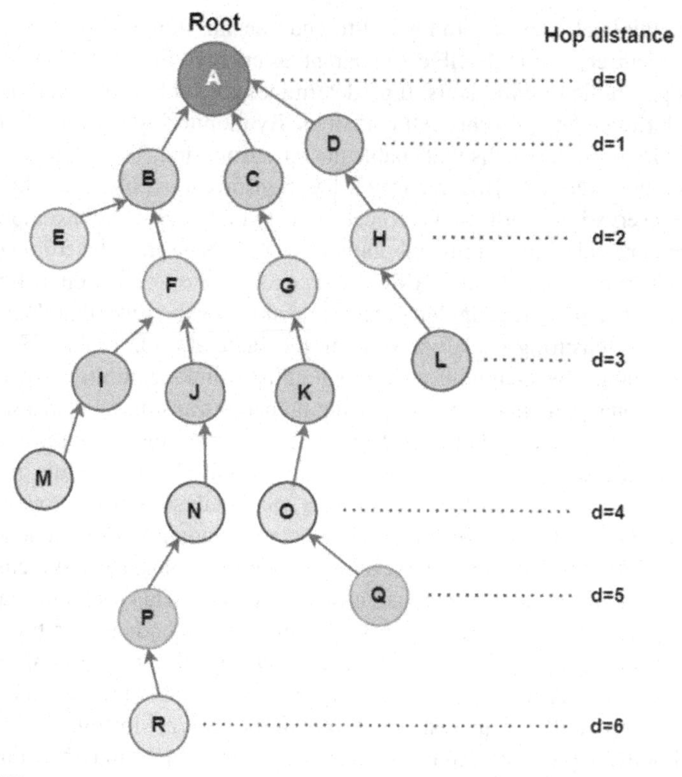

FIGURE 12.7

Example DODAG network for LLA.

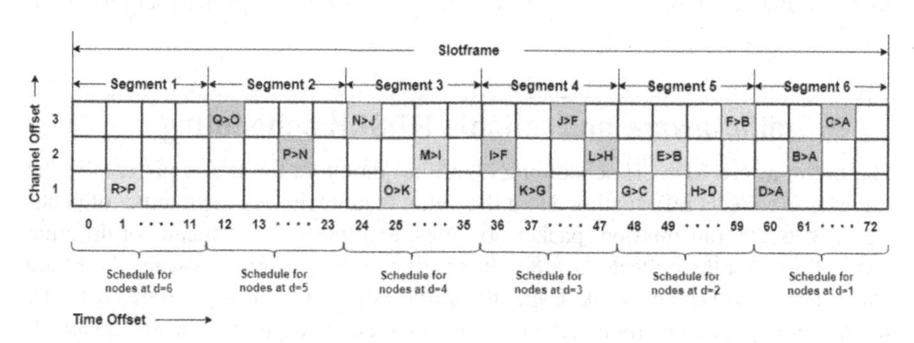

FIGURE 12.8

LLA slotframe design.

The Recurrent Low-Latency Scheduling Function (ReSF) [24] addresses the traffic condition with recurrent transmissions; data transmission repeats after a specific period. This method models such traffic problems as an integer linear program and maintains a minimal latency path for end-to-end transmission. It activates this path to handle the recurrent traffic. ReSF also guarantees energy efficiency by activating resources as per traffic requirements. It predetermines packet loss by reserving backup cells in a slotframe and prevents cell collisions. By monitoring the link quality, it reserves the supplementary cells that enable the retransmission. The main goal of ReSF is to reduce the latency in IIoT recurrent data transmissions; however, this is at the cost of the reservation overhead. Overhead leads to higher power consumption.

For time-critical applications, the authors in [25] have introduced a centralized Adaptive MUlti-hop Scheduling (AMUS) that enables progressive multi-hop allocation. This method allocates supplementary time slots for the links that demand more retransmissions, ensuring packet delivery in the same slotframe. AMUS optimizes resource utilization by assigning provisional cells to links exhibiting lower performance. Additionally, it allocates multiple simultaneous transmissions in a single time slot, preventing interference between them. It helps in limiting interference and collisions, which reduces the end-to-end latency. Also, a centralized schedule gives rise to fast network convergence. Further, it addresses energy consumption through the End-of-Queue (EoQ) notification mechanism. It uses redundant cells in a schedule for further reliability improvement. However, the node may consume more energy as it wakes up frequently without any transmit/receive activity. This problem is addressed by the EoQ notification mechanism, which minimizes redundant data transmission. In this mechanism, the sender notifies the receiver after the transmission of the last packet in its queue. After receiving an EoQ message a node turns off its radio and switches to sleep mode for the rest of the slots of the current slotframe.

In a TinyML deployment, multiple machine learning tasks may be running concurrently, each with its own timing requirements. By providing a predictable and synchronized communication framework, 6TiSCH can ensure the operation of different tasks simultaneously, avoiding interference and contention for network resources.

12.9 Traffic-aware and reliable 6TiSCH scheduling

The traffic-aware 6TiSCH scheduling algorithm monitors the traffic patterns in the network. It collects information about the traffic load, including the number of pending packets for transmission, packet priorities, and the traffic demands of different nodes. Based on the collected traffic information, the algorithm estimates the future traffic demands of the network. Using the traffic estimation, the algorithm adapts the slot allocation dynamically to accommodate the predicted traffic demands. It can adjust the slot allocation for different nodes based on their traffic priorities, ensuring that nodes with higher data transmission requirements utilize more slots and vice versa.

A traffic-aware link-based autonomous scheduling technique [26] has been developed for the 6TiSCH IIoT applications. Authors have extended the work in [23] by considering the traffic condition of the nodes. It is based on the segmentation principle with consecutive segments being assigned to the nodes along the path from

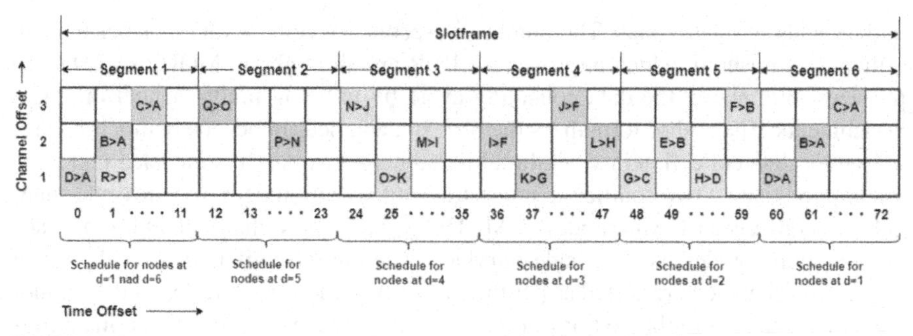

FIGURE 12.9

Traffic-aware slotframe scheduling.

source to destination. In the advancement, it analyzes the traffic at nodes which are close to the root. These nods are the bottleneck for the overall traffic. They act as forwarders for the packets from their descendants as well as their own packets. The node schedule is built by considering traffic conditions at a distance of one hop. It adds the supplementary cells for nodes with heavy traffic load as shown in Fig. 12.9 for the network in Fig. 12.7. The technique improves QoS in terms of PDR and latency [27]. Latency improvement is achieved by using slotframe segmentation. There are fewer nodes close to the root handling higher traffic as compared to other nodes in the network.

A Distributed Broadcast-based Scheduling (DeBraS) [28] algorithm was proposed for dense deployments. Every node but the root maintains a queue towards its parent. By employing multiple radios every node can exchange packets over multiple channels in the given time slot. The optimization solution assigns pairs of cells to receive maximum packets at the root. It reduces collisions while improving throughput, without increasing the network overhead in collecting real-time information from neighbors. It requires less frequent information exchange with other nodes and experiences fewer retransmissions due to a reduced number of collisions. The algorithm reduces latency and improves the throughput in a highly dense network with complex traffic patterns but at the expense of increased power consumption. Muraoka et al. [29] suggested Simple Distributed Scheduling with Collision Detection housekeeping techniques. Tx-housekeeping detects collisions by comparing the PDR of different cells communicating with the same neighbor, then to mitigate collisions it relocate cells to alternative positions within the schedule. Rx-housekeeping detects collisions when overhearing packets from a node which is not the neighbor. The technique reduces collisions and improves network scalability.

Adaptive Autonomous Allocation (A^3) [30] for TSCH networks is an improvement of ALICE. The lack of adaptability in ALICE for dynamic traffic, network density, and topology degrades the performance in terms of latency, PDR, throughput, and network lifetime. A^3 is an adaptive scheduler that estimates the number of transmit and receive cells to be scheduled for a link based on traffic over that link. It allocates one primary cell in a specified zone and for retransmission multiple secondary cells in other zones. The number of secondary cells is varied as per the cell utilization threshold, which improves the PDR and throughput. MODESA [31] emphasizes minimizing the data collecting cycle by utilizing multi-channel time slot assignments. The author formalizes the problem and determines the optimal time required for data collection in several multi-channel networks. It is modeled under the assumption that no two conflicting nodes transmit at the same time on the same channel in a convergecast. Sink nodes in MODESA have more than one radio interface and all source nodes have a single interface. The algorithm iterates over the set of source nodes, which are sorted as per their priorities. In every iteration, it determines the nodes that are eligible to transmit in a current time slot. It prioritizes critical traffic and provides mechanisms to meet latency or reliability constraints for specific applications.

By considering the traffic patterns and adapting the slot allocation based on predicted demands, the traffic-aware 6TiSCH scheduling algorithm [32] can optimize the network's resource utilization, minimize delays, and improve the overall performance of a TinyML network. This algorithm aims to optimize the usage of network resources to improve the network's reliability and robustness.

12.10 Energy efficiency and reliability considerations

Energy efficiency in IIoT scheduling refers to the optimization of energy consumption to maximize the overall system performance. This is particularly important in IIoT deployments where devices are battery-operated or have limited access to power sources [33]. Allocation of computationally intensive tasks from resource-constrained edge devices to more powerful cloud or fog nodes can reduce the energy consumption of the devices. Utilizing sleep modes or low-power states during idle periods can significantly reduce energy consumption. Devices can be programmed to enter sleep modes when not actively involved in data acquisition, processing, or communication. Optimizing the routing and scheduling protocols and communication patterns in networks can minimize energy usage. Reliability refers to the ability of the system to perform the task consistently without failures or disruptions. This is highly important in industrial applications where downtime or system failures can lead to significant losses. The QoS requirements ensure that critical tasks are given priority, reducing the likelihood of system failures or delays. Several techniques have been developed considering the energy efficiency and reliability in IIoT scheduling that can reduce costs and improve industrial processes.

Opportunistic anycast scheduling [34] is proposed by leveraging a MAC layer and routing behavior which can be implemented in PCE. After receiving a packet from a group of nodes, one of the nodes opportunistically forwards a packet. By monitoring channel quality and selecting a next-hop node it creates an optimal schedule, improving reliability while reducing energy consumption. The authors have addressed the problem by carrying out a dynamic and stochastic formulation. A child sends a packet to a set of parents at the beginning of a dedicated cell. Parents receiving it correctly should respond with an acknowledgment. The child ranks all parents; when any parent does not listen to acknowledgment for a specified time, it forwards the packet. All other parents stop complying after receiving an acknowledgment from any parent.

Adaptive static scheduling [35] is an energy-efficient schedule for TSCH networks. It avoids idle listening overhead by allowing a pair of nodes to activate a subset of their allocated slots. Further, to support bursts of data packets, the nodes can dynamically allocate additional cells without redistributing new schedules. The scheduler allows the nodes to use a cell as needed by overprovisioning them. It schedules sufficient dedicated cells within a slotframe to handle the high traffic load. The pair of nodes adaptively agree upon the number of slots needed by them. Initially, some of the slots are activated and a few of them remain inactive. It selects the number of active slots from the exponentially weighted moving average (EWMA) of the utilization level of activated cells. After successful packet delivery, the pair of nodes replicate the schedule in the next slotframe with the same slot value. Adaptive static scheduling is less appropriate for busy TSCH networks with very high traffic. Certainly, in heavy traffic conditions cells have a sparser distribution; consequently, maintaining inactive cells would be inefficient.

Domingo-Prieto et al. proposed the Distributed PID-Based Scheduling mechanism [36], which implements an industrial model referred to as proportional, integral, and derivative (PID) control. It calculates the number of cells for each node at the start of every slotframe and optimizes the queue size according to the traffic request. The objective of the PID controller is to keep a lower packet count in the queue and to reduce the number of schedule cells to minimize power consumption. By introducing an integral constant, efforts are made to reduce the instability of the network. This technique is very reactive in bursty traffic but in case of lower traffic demands, this approach has more signal overhead and a higher duty cycle. This is due to cell computation operations which are carried out at the start of every slotframe.

Immense research work is being carried out to improve QoS in IIoT applications. Though scheduling algorithms offer certain benefits in terms of different performance parameters, they still face some challenges. Handling the topologies where each node can direct traffic to multiple parents is difficult. There may be packet collision when two pairs of nearby nodes choose the same cells for information exchange. Some schedulers deal with the best-effort track, which corresponds to the class of traffic with the lowest priority. They are unable to provide a similar solution in another critical task. The real challenge is to implement an algorithm which builds schedules depending on the topology and data traffic requirements and satisfies QoS requirements.

12.11 6TiSCH simulation software

Simulator software plays a crucial role in 6TiSCH scheduling algorithm implementations by providing a realistic and controlled environment for testing, modification, and optimization. Simulators allow for the validation of systems before their deployment in real-world environments [37]. They help identify and rectify issues, evaluate system performance, and verify the effectiveness of different network configurations without disrupting actual production processes. 6TiSCH simulation software is being developed that enables flexible implementation, switching at different layers of the 6TiSCH protocol stack, and rapid evaluation under different network scenarios. It allows researchers and developers to create virtual network topologies, configure various parameters, and evaluate different scheduling algorithms and protocols. Simulators offer support for centralized, distributed, and autonomous scheduling approaches. Users can monitor various metrics such as packet delivery ratio, latency, throughput, energy consumption, scalability, and others.

These tools allow researchers to gain insights into the behavior of the network under different scheduling conditions and make informed decisions to optimize 6TiSCH scheduling algorithms. This capability is essential for understanding the performance characteristics and complexity of 6TiSCH networks as they grow. This flexibility allows users to compare and analyze the advantages and limitations of different scheduling approaches in 6TiSCH networks. Moreover, they support the integration of external tools and frameworks. Researchers can integrate additional analysis tools, network protocols, or scheduling algorithms into the simulation environment to enhance its capabilities. This extensibility enables users to explore new ideas and evaluate novel approaches to 6TiSCH scheduling. Different simulator programs are discussed below.

Contiki-NG Cooja: Contiki-NG is an open-source operating system designed for IoT devices. It is a lightweight and highly flexible platform that provides a variety of features and protocols to enable efficient communication and resource management in IoT networks [38], [39]. Contiki provides a comprehensive implementation of 6TiSCH scheduling, offering a range of scheduling mechanisms and algorithms to meet diverse requirements. 6TiSCH scheduling framework in Contiki includes several algorithms, such as Orchestra, MSF, SF0, and distributed hopping sequences. Contiki-NG support for 6TiSCH scheduling is not limited to the software stack, but it also provides integration with hardware platforms that are compatible with the IEEE 802.15.4e standard. By leveraging the Contiki-NG lightweight and flexible architecture, developers can build highly efficient and scalable IoT networks. It offers a rich set of features, such as low power consumption, support for IPv6, and a modular architecture for easy customization.

Cooja is a powerful network simulator and is an integral part of the Contiki-NG operating system [40], providing developers with a versatile tool for testing, debugging, and evaluating their IIoT deployments. It has been developed in Java and is user-extendable through different plugins; also, it can execute native C programs. It uses MSPSim for emulation of MSP430 hardware platforms. The Contiki operating system can be executed on physical hardware as well as on the Cooja simulator. Cooja plays a crucial role in simulating and analyzing the performance of 6TiSCH scheduling. It allows developers to create virtual network topologies, emulate various IoT devices, and simulate the behavior of 6TiSCH networks in a controlled and repeatable manner. It provides an accurate representation of the timing and synchronization aspects of 6TiSCH networks, allowing developers to observe and analyze the behavior of the network under different conditions.

OpenWSN: OpenWSN is an open-source software and hardware platform designed for wireless sensor networks (WSNs) [41]. It provides a range of tools, protocols, and frameworks to enable the development of reliable and energy-efficient IoT applications. OpenWSN plays a significant role in implementing 6TiSCH scheduling and contributing to the overall efficiency of IoT networks. OpenWSN offers a complete set of tools and protocols to support 6TiSCH scheduling. It provides a modular software stack that includes the OpenWSN firmware, which runs on resource-constrained IoT devices, and the OpenWSN operating system, which provides higher-level abstractions and interfaces for application development. The software stack is specifically designed to optimize resource usage and energy efficiency, making it well suited for 6TiSCH networks. It uses an OpenSim simulator that runs the OpenWSN firmware on devices in a simulated network. OpenWSN also offers hardware platforms that allow the deployment of 6TiSCH applications on a variety of IoT devices.

6TiSCH Simulator: The 6TiSCH Simulator [42] is another powerful tool designed for simulating and evaluating the performance of 6TiSCH networks. 6TiSCH is an extension of the IPv6 protocol that enables time-synchronized communication in low-power wireless networks. It provides a comprehensive simulation environment that accurately models the behavior of 6TiSCH networks. This Python-based simulator strictly simulates time according to individual TSCH time slots and offers web visualization. After every run, the simulator gathers and stores the statistics. It is a PDF file generated by Python scripts included in the source code release. It is faster than Contiki Cooja; however, it is highly abstract and specifically designed for adherence to the 6TiSCH proposed standards. Lacking non-standard schedulers and routing protocols limits the extension evolutions.

12.12 Challenges and limitations

The primary objective of 6TiSCH is to enhance QoS in the IIoT; however, it faces certain challenges in scheduling.

- Synchronization among devices poses a significant challenge in 6TiSCH scheduling to avoid collisions. However, maintaining precise synchronization in dynamic IIoT environments with high mobility, frequent topology changes, and variable link qualities is a non-trivial task. Devices may experience clock jitter, interference, or communication delays, leading to synchronization errors. Creating resilient synchronization mechanisms capable of addressing such challenges and delivering precise time synchronization is essential for efficient scheduling.
- Allocation of resources, such as time slots, channels, and transmit power, is a critical aspect of 6TiSCH scheduling. Decisions are based on factors like network

load, traffic patterns, interference, and energy constraints. Optimizing the allocation of these resources to ensure efficient network operation, minimize collisions, and avoid interference is a complex task, especially when dealing with a large number of devices.
- Network scalability is another important challenge in 6TiSCH scheduling. As the number of devices in a network grows, the scheduling overhead increases. The coordination and management of a large number of devices become more challenging, potentially leading to increased latency and reduced network performance. Designing scalable scheduling algorithms and protocols capable of efficiently managing devices in a densely populated network is crucial to meet the increasing demands of IIoT deployments.
- To satisfy determinism, many IIoT applications require real-time communication with strict latency constraints. Scheduling real-time traffic alongside best-effort traffic in a shared network introduces additional challenges. Scheduling mechanisms must prioritize and allocate resources to meet the timing requirements of real-time flows while not adversely affecting other non-real-time traffic.
- Energy consumption is a critical concern in IIoT deployments, involving battery-operated devices. Scheduling algorithms need to balance the need for seamless communication and maintaining synchronization while lowering energy consumption. Energy-efficient scheduling algorithms and power management strategies are required to prolong the battery life of devices and enable sustainable operation in energy-constrained IIoT environments.

Massive research work is being carried out to mitigate these scheduling challenges. The goal is to enhance performance in terms of latency, PDR, reliability, and energy efficiency.

12.13 Conclusion and future directions

6TiSCH is an evolving technology that aims to provide deterministic, reliable, and energy-efficient communication for IIoT networks. The 6TiSCH architecture uses the TSCH mode of IEEE 802.15.4e to provide robust and predictable communication while leveraging IPv6 to enable end-to-end communication and integration with other IP-based networks. The 6TiSCH protocol uses a different scheduling mechanism that allows devices to transmit and receive data on predefined time slots, thereby avoiding collisions. In a centralized approach, the scheduling function is performed by a centralized entity called the PCE, which maintains a schedule for each device in the network based on their communication requirements and energy constraints. In the distributed approach, the device negotiates a required number of cells with neighbors using 6P messages, whereas the autonomous approach uses device MAC addresses to build a schedule. The scheduling algorithm ensures that each device receives an equal share of the available bandwidth and that the network remains stable and resilient to changes in the topology and traffic patterns. In the future, TinyML-enabled IIoT can leverage the benefits of 6TiSCH technology in real-time decision-making because of its determinism and robustness.

References

[1] L. Atzori, A. Iera, G. Morabito, The internet of things: a survey, Computer Networks 54 (15) (Oct. 2010) 2787–2805, https://doi.org/10.1016/j.comnet.2010.05.010.

[2] S.N. Ghorpade, M. Zennaro, B.S. Chaudhari, Introduction to internet of things, in: SpringerBriefs in Applied Sciences and Technology, 2022, https://doi.org/10.1007/978-3-030-88095-8_1.

[3] M.R. Palattella, P. Thubert, X. Vilajosana, T. Watteyne, Q. Wang, T. Engel, 6TiSCH wireless industrial networks: determinism meets IPv6, in: S.C. Mukhopadhyay (Ed.), Internet of Things: Challenges and Opportunities, Springer, 2014, https://doi.org/10.1007/978-3-319-04223-7_5.

[4] P. Thubert, T. Watteyne, M.R. Palattella, X. Vilajosana, Q. Wang, IETF 6TSCH: combining IPv6 connectivity with industrial performance, in: Proceedings - 7th International Conference on Innovative Mobile and Internet Services in Ubiquitous Computing, IMIS 2013, 2013, pp. 541–546, https://doi.org/10.1109/IMIS.2013.96.

[5] M. Standards Committee of the IEEE Computer Society, IEEE Std 802.15.4TM-2015, IEEE Standard for Low-Rate Wireless Networks, http://www.ieee.org/web/aboutus/whatis/policies/p9-26.html, 2016.

[6] X. Vilajosana, T. Watteyne, T. Chang, M. Vucinic, S. Duquennoy, P. Thubert, IETF 6TiSCH: a tutorial, IEEE Communications Surveys and Tutorials 22 (1) (2020), https://doi.org/10.1109/COMST.2019.2939407.

[7] T. Winter, et al., RPL: IPv6 routing protocol for low power and lossy networks, IETF RFC 6550, https://datatracker.ietf.org/doc/html/rfc6550. (Accessed 12 June 2023), 2012.

[8] N. Kushalnagar, et al., IPv6 over low-power wireless personal area networks (6LoWPANs): overview, assumptions, problem statement, and goals status, IETF RFC 4919, https://datatracker.ietf.org/doc/html/rfc4919. (Accessed 12 June 2023), 2007.

[9] Z. Shelby, K. Hartke, C. Bormann, CoAP: the constrained application protocol, IETF, RFC 7252, https://datatracker.ietf.org/doc/html/rfc7252. (Accessed 12 June 2023), 2014.

[10] C. Amsüss, Z. Shelby, M. Koster, C. Bormann, P. van der Stok, Constrained RESTful Environments (CoRE) resource directory, IETF RFC 9176, https://www.rfc-editor.org/info/rfc9176, 2022.

[11] B.S. Chaudhari, M. Zennaro, LPWAN Technologies for IoT and M2M Applications, Academic Press, 2020, https://doi.org/10.1016/B978-0-12-818880-4.00020-X.

[12] P. Thubert, RFC 9030: an architecture for IPv6 over the time-slotted channel hopping mode of IEEE 802.15.4 (6TiSCH), IETF RFC 9030, https://www.rfc-editor.org/info/rfc9030, 2021.

[13] M.R. Palattella, N. Accettura, M. Dohler, L.A. Grieco, G. Boggia, Traffic aware scheduling algorithm for reliable low-power multi-hop IEEE 802.15.4e networks, in: IEEE International Symposium on Personal, Indoor and Mobile Radio Communications, PIMRC, 2012, https://doi.org/10.1109/PIMRC.2012.6362805.

[14] N. Accettura, E. Vogli, M.R. Palattella, L.A. Grieco, G. Boggia, M. Dohler, Decentralized traffic aware scheduling in 6TiSCH networks: design and experimental evaluation, IEEE Internet of Things Journal 2 (6) (Dec. 2015) 455–470, https://doi.org/10.1109/JIOT.2015.2476915.

[15] M.R. Palattella, et al., On-the-fly bandwidth reservation for 6TiSCH wireless industrial networks, IEEE Sensors Journal 16 (2) (Jan. 2016) 550–560, https://doi.org/10.1109/JSEN.2015.2480886.

[16] S. Duquennoy, B. Al Nahas, O. Landsiedel, T. Watteyne, Orchestra: robust mesh networks through autonomously scheduled TSCH, in: SenSys 2015 - Proceedings of the 13th ACM Conference on Embedded Networked Sensor Systems, Association for Computing Machinery, Inc, Nov. 2015, pp. 337–350, https://doi.org/10.1145/2809695.2809714.

[17] S. Rekik, N. Baccour, M. Jmaiel, K. Drira, L. Grieco, L. Alfredo Grieco, Autonomous and traffic-aware scheduling for TSCH networks, Computer Networks 135 (2018) 201–212, https://doi.org/10.1016/j.comnet.2018.02.023.

[18] S. Rekik, N. Baccour, M. Jmaiel, K. Drira, A performance analysis of Orchestra scheduling for time-slotted channel hopping networks, Internet Technology Letters 1 (3) (May 2018), https://doi.org/10.1002/itl2.4, John Wiley and Sons Inc.

[19] S. Kim, H.S. Kim, C. Kim, Alice: autonomous link-based cell scheduling for TSCH, in: IPSN 2019 - Proceedings of the 2019 Information Processing in Sensor Networks, Association for Computing Machinery, Inc, Apr. 2019, pp. 121–132, https://doi.org/10.1145/3302506.3310394.

[20] A. Karaagac, I. Moerman, J. Hoebeke, Hybrid schedule management in 6TiSCH networks: the coexistence of determinism and flexibility, IEEE Access 6 (Jun. 2018) 33941–33952, https://doi.org/10.1109/ACCESS.2018.2849090.

[21] J.N. Tsitsiklis, K. Xu, On the power of (even a little) centralization in distributed processing, in: Performance Evaluation Review, 2011, pp. 161–172, https://doi.org/10.1145/1993744.1993759.

[22] V. Kotsiou, G.Z. Papadopoulos, P. Chatzimisios, F. Theoleyre, LDSF: low-latency distributed scheduling function for industrial internet of things, IEEE Internet of Things Journal 7 (9) (Sep. 2020) 8688–8699, https://doi.org/10.1109/JIOT.2020.2995499.

[23] N.M. Pradhan, B.S. Chaudhari, M. Zennaro, 6TiSCH low latency autonomous scheduling for industrial internet of things, IEEE Access 10 (2022) 71566–71575, https://doi.org/10.1109/ACCESS.2022.3188862.

[24] G. Daneels, B. Spinnewyn, S. Latré, J. Famaey, ReSF: recurrent low-latency scheduling in IEEE 802.15.4e TSCH networks, Ad Hoc Networks 69 (Feb. 2018) 100–114, https://doi.org/10.1016/j.adhoc.2017.11.002.

[25] Y. Jin, P. Kulkarni, J. Wilcox, M. Sooriyabandara, A centralized scheduling algorithm for IEEE 802.15.4e TSCH based industrial low power wireless networks, in: IEEE Wireless Communications and Networking Conference, WCNC, Institute of Electrical and Electronics Engineers Inc., 2016, https://doi.org/10.1109/WCNC.2016.7565002.

[26] N. Pradhan, B.S. Chaudhari, Traffic-aware autonomous scheduling for 6TiSCH networks, International Journal of Computers & Applications 44 (11) (2022) 1039–1046, https://doi.org/10.1080/1206212X.2022.2103889.

[27] N. Pradhan, B.S. Chaudhari, 6TiSCH scheduling for IIoT by slotframe fragmentation, in: 2022 IEEE 2nd International Conference on Mobile Networks and Wireless Communications, ICMNWC 2022, Institute of Electrical and Electronics Engineers Inc., 2022, https://doi.org/10.1109/ICMNWC56175.2022.10031737.

[28] E. Municio, K. Spaey, S. Latré, A distributed density optimized scheduling function for IEEE 802.15.4e TSCH networks, Transactions on Emerging Telecommunications Technologies 29 (7) (Jul. 2018), https://doi.org/10.1002/ett.3420.

[29] K. Muraoka, T. Watteyne, N. Accettura, X. Vilajosana, K.S.J. Pister, Simple distributed scheduling with collision detection in TSCH networks, IEEE Sensors Journal 16 (15) (Aug. 2016) 5848–5849, https://doi.org/10.1109/JSEN.2016.2572961.
[30] S. Kim, H.-S. Kim, C. Kim, A3: adaptive autonomous allocation of TSCH slots, in: IPSN' 21: The 20th International Conference on Information Processing in Sensor Networks, May 2021, Nashville, TN, USA, 2021.
[31] R. Soua, P. Minet, E. Livolant, MODESA: an optimized multichannel slot assignment for raw data convergecast in wireless sensor networks, in: 2012 IEEE 31st International Performance Computing and Communications Conference, IPCCC 2012, 2012, https://doi.org/10.1109/PCCC.2012.6407742.
[32] N. Pradhan, B.S. Chaudhari, P.D. Khandekar, Enhanced time slotted channel scheduling for constrained networks, in: 2023 Fifth International Conference on Electrical, Computer and Communication Technologies (ICECCT), 2023, pp. 1–5, https://doi.org/10.1109/ICECCT56650.2023.10179825.
[33] S. Ghorpade, M. Zennaro, B.S. Chaudhari, Towards green computing: intelligent bio-inspired agent for IoT-enabled wireless sensor networks, International Journal of Sensor Networks 35 (2) (2021), https://doi.org/10.1504/IJSNET.2021.113632.
[34] T. Huynh, F. Theoleyre, W.J. Hwang, On the interest of opportunistic anycast scheduling for wireless low power lossy networks, Computer Communications 104 (May 2017) 55–66, https://doi.org/10.1016/j.comcom.2016.06.001.
[35] X. Fafoutis, A. Elsts, G. Oikonomou, R. Piechocki, I. Craddock, Adaptive static scheduling in IEEE 802.15.4 TSCH networks, in: IEEE World Forum on Internet of Things, WF-IoT 2018 - Proceedings, 2018, https://doi.org/10.1109/WF-IoT.2018.8355114.
[36] M. Domingo-Prieto, T. Chang, X. Vilajosana, T. Watteyne, Distributed PID-based scheduling for 6TiSCH networks, IEEE Communications Letters 20 (5) (May 2016) 1006–1009, https://doi.org/10.1109/LCOMM.2016.2546880.
[37] A. Elsts, TSCH-Sim: scaling up simulations of TSCH and 6TiSCH networks, Sensors (Switzerland) 20 (19) (Oct. 2020) 1–17, https://doi.org/10.3390/s20195663.
[38] A. Kurniawan, Practical Contiki-NG, Apress, 2018, https://doi.org/10.1007/978-1-4842-3408-2.
[39] S. Duquennoy, A. Elsts, B. Al Nahas, G. Oikonomo, TSCH and 6TiSCH for Contiki: challenges, design and evaluation, in: Proceedings - 2017 13th International Conference on Distributed Computing in Sensor Systems, DCOSS 2017, Institute of Electrical and Electronics Engineers Inc., Jan. 2018, pp. 11–18, https://doi.org/10.1109/DCOSS.2017.29.
[40] A. Velinov, A. Mileva, Running and testing applications for Contiki OS using Cooja simulator, in: International Conference on Information Technology and Development of Education, August 2016.
[41] T. Watteyne, et al., OpenWSN: a standards-based low-power wireless development environment, European Transactions on Telecommunications 23 (5) (Aug. 2012) 480–493, https://doi.org/10.1002/ett.2558.
[42] E. Municio, et al., Simulating 6TiSCH networks, Transactions on Emerging Telecommunications Technologies 30 (3) (Mar. 2019), https://doi.org/10.1002/ett.3494.

CHAPTER

Securing TinyML in a connected world

13

Rachana Yogesh Patil[a], Mamta Bhamare[b], Yogesh H. Patil[c], and Aparna Bannore[d]

[a]*Pimpri Chinchwad College of Engineering, Pune, India*

[b]*Department of Computer Engineering and Technology, Dr. Vishwanath Karad MIT World Peace University, Pune, India*

[c]*D. Y. Patil College of Engineering, Pune, India*

[d]*SIES Graduate School of Technology, Navi Mumbai, India*

13.1 Introduction

TinyML has experienced explosive growth because it enables the deployment of machine learning models in applications like IoT gadgets, wearables, and embedded systems, all of which suffer from resource constraints at the network's edge [1]. TinyML gives these gadgets the ability to think for themselves and make decisions in real time, which has far-reaching implications. Concerns about the safety and reliability of these systems are exacerbated by TinyML's expanding role in the Internet of Things (IoT).

Due to the sensitivity of processed data and the importance of the decisions made by these models, security is of paramount importance in TinyML [2]. Critical aspects of TinyML security include the protection of sensitive data, the reliability of predictions, the mitigation of adversarial attacks, the protection of intellectual property, the prevention of device compromise, and the observance of regulations. By putting security first, businesses can create reliable TinyML systems that gain users' trust and pave the way for TinyML's easy adoption across a wide range of industries [3].

The article provides an in-depth examination of the difficulties, dangers, and potential solutions involved in protecting TinyML in the modern, interconnected world [4,5]. Our goal is to protect TinyML deployments and enable their secure and reliable operation by illuminating the research and industry landscape regarding the unique security considerations and implications.

We start by looking at how adversaries could potentially steal TinyML devices' machine learning models. In order to steal sensitive data such as model parameters, architecture, or other intellectual property, adversaries may use methods like reverse engineering firmware or analyzing device memory. Knowing the nature of these model extraction risks is essential for creating defenses and preserving TinyML model privacy [6].

Model tampering is another major issue, which occurs when an attacker changes the deployed machine learning models in order to make inaccurate predictions or take

TinyML for Edge Intelligence in IoT and LPWAN Networks. https://doi.org/10.1016/B978-0-44-322202-3.00018-X

advantage of security flaws. This could lead to malicious activity, data corruption, or unauthorized access. We investigate potential vectors for model tampering and the results for TinyML applications' safety and trustworthiness. Protecting the accuracy and reliability of TinyML models requires investigating possible countermeasures against tampering.

TinyML systems place a premium on privacy, especially when handling sensitive information. TinyML devices are vulnerable to eavesdropping and interception because adversaries can potentially gain access to or infer private information. Strong security mechanisms and privacy-preserving techniques adapted to low-resource settings are necessary for protecting user privacy and preventing unauthorized access to sensitive data [7].

Furthermore, exhaustion of resources is a significant difficulty in TinyML security. Since these devices are often underpowered, attackers may attempt denial-of-service attacks by flooding them with data or demanding computations. To protect TinyML systems from malicious actions, it is essential to lessen the impact of resource exhaustion attacks.

Another major risk faced by TinyML devices is the possibility of physical assault. Intruders could modify the hardware, the power supply, or the firmware in an effort to steal information or compromise the system's security. Robust physical security measures, such as tamper-resistant designs, secure bootstrapping, and secure hardware implementations, are necessary to protect against physical attacks [8].

Future research in TinyML security should concentrate on a few main areas to address these issues. Among these are the investigation of privacy-preserving techniques like secure multi-party computation and differential privacy, the development of robust defenses against adversarial attacks like adversarial training, and guaranteeing the trustworthiness and explainability of TinyML models.

13.1.1 Evolution of TinyML

TinyML's transition from cloud machine learning to mobile machine learning/edge machine learning, as shown in Fig. 13.1, is a major step forward for the machine learning industry. At first, most machine learning work was done in the cloud, thanks to its powerful servers and ample resources that made complex computations and the storage of large datasets relatively painless. This cloud-first strategy posed latency and privacy issues because it restricted machine learning model deployment on devices with internet access [9].

On-device machine learning became necessary as mobile and edge computing became more popular. The term "TinyML" is used to describe the practice of deploying machine learning models on low-powered edge devices. This shift has completely altered the landscape of intelligent computing by allowing devices like smartphones, wearables, and IoT gadgets to make decisions in real time.

TinyML's development over time has allowed for several significant improvements [10]. Because it processes data locally, without relying on cloud connectivity, it has significantly reduced latency. Time-sensitive applications, like autonomous vehicles or real-time monitoring systems, can benefit greatly from this.

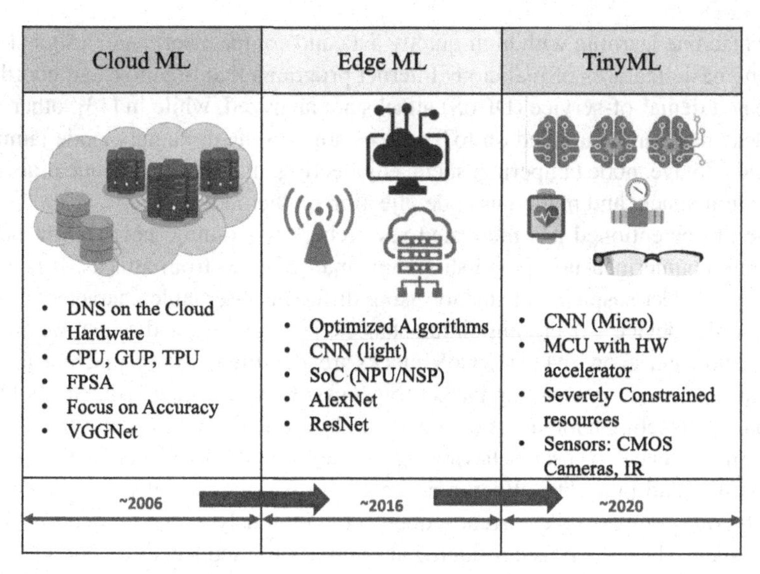

FIGURE 13.1

Evolution of TinyML.

By storing private information locally on the device, TinyML eliminates the potential for data exposure during transmission to the cloud. This is absolutely essential for apps that handle sensitive user information like medical records, bank records, and so on.

Reliability and robustness have also been enhanced by deploying machine learning models at the edge. A reliable internet connection is not necessary for TinyML devices to function, making them ideal for use in off-the-grid settings.

TinyML has improved its power efficiency as it has developed. Battery life can be increased and devices can run continuously if computations are performed locally, thus reducing energy consumption.

Combining TinyML with edge computing technologies like edge servers and gateways, a new distributed computing paradigm has emerged. As a result, multiple edge devices can work together and share information to make more informed decisions.

13.2 Related work

A growing field of research, TinyML security, examines the security aspects of developing machine learning models for devices with limited resources. IoT gadgets are typically constructed using a framework and architecture that reflects the security of the underlying framework and architecture. In TinyML, co-design between hardware and software is fundamental. Such systems should overlap the orientations of optimized machine learning with high-quality data and compact software design [11]. In [12], the basic features of malicious Internet programs that organize and coordinate distributed denial-of-service (DDoS) attacks are analyzed, while in [13], other types of attacks that can be applied on IoT devices are presented, namely, node jamming, physical damage, node tampering, social engineering, malicious node injection, sleep deprivation attack, and malicious code injection on the node.

The aforementioned has motivated researchers to examine new approaches for detecting, countermeasuring, and shielding smart devices from attacks. It is a typical way to detect suspicious behavior using different observable characteristics that compose the device's operating circumstances, such as power dissipation, ambient temperature, and so on. In [14], DDoS attacks are identified through machine learning techniques, e.g., for controlling packet transmission rate, packet size, etc. In [15], a method for detecting anomalous operations of house IoT devices is presented, which can learn sequences of user behaviors according to conditions such as time of day, temperature, and humidity. When a command to perform an operation is received, the technique compares the current sequence to previously learned sequences for the present state. The surgery is considered abnormal if the sequences do not match.

Another approach that uses machine learning-based feature-group-clustering techniques, nodding, and parameter use for proper education systems is found in [16]. In [17], two approaches were proposed that include deep automated encoder models for analyzing time series collected by gravitational wave detectors and provide a classification tag (noise or real signal).

Modern machine learning techniques, such as deep neural networks (DNNs), have made substantial progress in recent years in achieving very high accuracy for a specific set of tasks, such as image classification, object recognition, natural language processing, and medical data analysis. However, these DNNs have high processing, memory, and energy requirements, and they are subject to a variety of security vulnerabilities.

The author of [18] presented challenges and cross-layer frameworks for building highly energy-efficient and robust machine learning systems for TinyML and edge AI applications, which jointly leverage optimizations at different software and hardware layers, e.g., neural acceleration, memory access optimizations, approximations, hardware-aware NAS, and network compression.

TinyML must contend with extra security vulnerabilities that can impact tiny devices, which often rely on less assistance from the hardware and the operating system to achieve security and, if deployed in the field, might be vulnerable to physical attacks.

The MPAI-AIF framework and also the IEEE P3301 standard produced by the MPAI community and the IEEE P3301 Artificial Intelligence Framework Working Group provide some answers and support for easy implementation of TinyML on microcontroller units (MCUs) [19].

Finally, Myridakis et al. [20–23] studied physical features, such as uninterrupted power supply, and extracted information that in combination with thresholds or ranges of values can be used to detect irregularities in IoT devices. Efforts have been made to design an autonomous system that will provide greater security at a minimal price.

13.3 Issues and challenges in securing TinyML

TinyML security presents various distinct problems and challenges. When a machine learning model is deployed on a device with limited resources, such as a microcontroller or a low-power edge device, the term "TinyML" is used. TinyML has a lot of potential for different applications, but it is important to make sure these gadgets are secure [24]. Several issues and challenges in securing TinyML are discussed in this section.

13.3.1 Resource constraints in securing TinyML

TinyML systems can be difficult to secure because of the resource limitations that are frequently present in such hardware. Here are some important resource limitations to take into account when protecting TinyML:

- *Limited processing capacity*

TinyML devices frequently contain low-power microcontrollers or specialized hardware accelerators that have limited processing capabilities. Due to these limitations, it is difficult to build complicated security methods that demand a lot of computational power.

- *Memory restrictions*

TinyML devices often have constrained RAM and storage capacity. The quantity of data that can be processed or saved is constrained, making it difficult to implement resource-intensive security algorithms or store substantial datasets related to security.

- *Energy efficiency*

Because TinyML devices frequently run on batteries or in energy-constrained settings, power consumption is a major challenge [25]. Security features that need constant connection, encryption, or decryption can drain a device's battery quickly and negatively affect overall performance.

- *Constraints on bandwidth*

TinyML devices are frequently connected to spotty or low-bandwidth wireless networks. This can make it difficult to deploy secure communication methods like encryption and authentication since they might need to transmit and process more data.

- *Limited sensor input*

Due to their compact form or particular use cases, TinyML devices may have limited input capabilities. This may have an impact on the accessibility and caliber of sensor data utilized for security-related operations like anomaly or intrusion detection.

13.3.2 Data privacy issues in TinyML

- *Data collection and storage*

TinyML devices frequently acquire and interpret private information from their environment, such as sound, photos, or sensor readings. When these data are gathered and retained, privacy concerns surface, especially if they contain sensitive or personally identifiable information [26]. Data must be carefully obtained, securely stored, and subject to the necessary privacy protections.

- *Data transmission*

Data transmission to external servers or other devices by TinyML devices may be necessary for processing or analysis. This transmission may put your privacy in danger, especially if the data are not adequately secured or if the communication protocols are weak. To avoid unauthorized access or interception, it is essential to protect data while they are being transmitted.

- *Inference privacy*

TinyML models are often installed directly on the hardware, enabling inference and decision-making on the fly. On the other hand, the inference procedure itself may disclose confidential facts about the data being processed. With access to the gadget or its output, adversaries could be able to deduce personal or confidential information. To mitigate these concerns, strategies like differential privacy or secure multi-party computation can be used.

- *Model inversion attacks*

TinyML models may be hacked or subjected to reverse engineering in order to obtain private data used for training. Model inversion attacks use the model's outputs to infer properties from the training data, possibly disclosing sensitive or private data. Careful consideration of the model architecture, training data privacy, and potential adversarial scenarios is necessary to defend against such attacks.

- *Unauthorized access and tampering*

Given the resource-constrained nature of TinyML devices, they may be more vulnerable to unauthorized access or physical tampering. Adversaries can try to take over the gadget or change its operation, which could jeopardize data privacy. These hazards can be reduced by putting in place robust access controls, secure boot processes, and tamper-resistant hardware.

- *Data sharing and third-party involvement*

In certain scenarios, TinyML devices may interact with third-party services or share data with external entities. When data are shared with these parties, privacy concerns arise because they might be used in ways that go beyond what the user intended or in ways that are inconsistent with their expectations [27]. For the sake of privacy, it is crucial to have open data sharing procedures and secure user consent for such sharing.

13.3.3 Adversarial attacks on TinyML

- *Adversarial perturbations*

Adversarial perturbations entail modifying or adding undetectable noise to input data in order to trick the TinyML model. These alterations may result in inaccurate classifications or outputs from the model. Techniques like the Fast Gradient Sign Method (FGSM), Projected Gradient Descent (PGD), or the Carlini and Wagner assault can be used to create adversarial perturbations.

- *Evasion attacks*

Through the creation of inputs that are purposefully made to avoid detection or categorization, evasion attacks try to deceive the TinyML model. Attackers create inputs that take advantage of flaws in the decision boundaries of the model, causing the model to misclassify or miss some patterns. Attacks that evade detection can be carried out using gradient-based techniques or optimization algorithms.

- *Model poisoning*

In model poisoning assaults, the TinyML model's training data are manipulated by the adversary. Attackers try to influence the behavior of the model or introduce flaws by introducing malicious or well-constructed samples into the training dataset. Attacks using model poisoning, which take place during the training phase of the model, might be particularly difficult to identify.

- *Model inversion*

Model inversion attacks aim to extract sensitive information or characteristics of the training data used to develop the TinyML model. Attackers use the model's outputs to deduce details about specific training samples, potentially invading privacy or disclosing sensitive information.

13.3.4 Firmware updates issues in TinyML

- *Limited resources*

TinyML devices often have constrained memory, processor, and storage capacities. It can be difficult to incorporate upgrades within the device's limitations because firmware updates may call for a lot of storage or computing power.

- *Connectivity constraints*

TinyML devices may have limited or intermittent connectivity to the internet or update servers. This can make it difficult to establish reliable and timely connections for downloading firmware updates. Limited connectivity can result in delays or failed attempts to update the firmware.

- *Over-the-air update mechanisms*

Remote firmware updates without manual involvement are made possible through over-the-air (OTA) update techniques. However, due to memory constraints, security considerations, and potential compatibility difficulties with the device's firmware and software stack, integrating OTA update functionality in TinyML devices can be challenging.

- *Compatibility and versioning*

It is essential that firmware updates work with any software or models already installed on TinyML devices. Updates to software libraries, model formats, or interfaces may be necessary to accommodate changes in firmware versions, creating compatibility issues. It can be difficult to ensure backward compatibility and manage several firmware versions.

13.3.5 Lack of standards while using TinyML

TinyML can face a number of difficulties and downsides due to the absence of industry standards. The following are some major concerns with utilizing TinyML without standards:

- *Interoperability*

Interoperability between various TinyML systems, frameworks, and devices is lacking in the absence of standardized formats, protocols, and interfaces. Because of this, creating and deploying TinyML models in different hardware and software systems may be challenging [28]. Interoperability is necessary for the seamless integration and exchange of TinyML solutions, which limits their wide acceptance and makes it more difficult for the TinyML community to work together.

- *Reproducibility*

TinyML tests and results are difficult to repeat and replicate across several platforms due to a lack of standards. TinyML models and methodologies are difficult to compare and test due to inconsistent implementations, differences in hardware requirements, and non-standard evaluation procedures. Scientific progress, knowledge exchange, and the ability to build upon prior work are all hampered by reproducibility problems.

- *Development efficiency*

When constructing TinyML solutions, developers must work harder and with more complexity in the lack of standards. Fragmentation and duplication of effort in creating standardized tools, libraries, and frameworks result from the absence of these elements. As a result, the TinyML ecosystem experiences longer development times, reduced code reuse, and slower innovation.

13.3.6 Power consumption issues in TinyML

TinyML system design and deployment must take power consumption into account. These systems are frequently used on devices with restricted resources, including microcontrollers or edge devices with tight power budgets. The following power usage considerations should be considered when using TinyML:

- *Energy efficiency*

To maximize battery life and permit continued operation in energy-constrained conditions, TinyML systems must be highly energy-efficient. To reduce the computational and memory requirements of TinyML models, power-efficient algorithms, model structures, and optimization methods are essential.

- *Hardware selection*

The proper selection of hardware elements has a big impact on how much power is used. Compared to general-purpose processors, microcontrollers and specialized low-power processors with built-in hardware accelerators can offer more energy-efficient alternatives. Overall energy efficiency is also improved by choosing peripherals and sensors with minimal power requirements.

- *Model optimization*

Model optimization techniques can significantly reduce power consumption in TinyML systems. Techniques like model compression, quantization, and pruning reduce the size and complexity of the model, leading to lower computational requirements and decreased power consumption during inference.

- *Data transmission*

Particularly in wireless communication contexts, data transmission can be a power-intensive process. Energy might be used up significantly when sending data to remote servers or devices for processing or analysis. In order to reduce power consumption during data transmission, communication protocols can be improved, data transfer sizes can be decreased, and effective encryption techniques can be used [29].

- *Sensor power management*

Sensor data are a common source of inference for TinyML systems. It is essential to effectively manage sensor power usage. By activating sensors only when necessary and tailoring the sampling frequency based on the needs of the application, methods including duty cycling, sensor fusion, and intelligent sampling can reduce power usage.

13.4 Existing solutions and approaches for TinyML security

The integrity and security of TinyML devices must be maintained by securing the boot process. Although small devices have limited resources, it is possible to develop safe boot procedures using existing technologies. Here are a few typical methods for secure boot in TinyML:

- *Cryptographic signatures*

A popular method for ensuring the legitimacy and integrity of firmware during the boot process is the use of cryptographic signatures. A private key is used to sign the firmware image, and the device uses the associated public key to verify the signature. This makes sure that on boot, only trusted and signed firmware images are loaded.

- *Secure boot loader*

The safe boot procedure cannot be implemented without a secure bootloader. Only verified and reliable firmware images are loaded and run, thanks to the bootloader. Before enabling the firmware image to execute, it checks its cryptographic signatures to guard against unauthorized or modified firmware.

- *Read-only memory (ROM) protection*

To stop unauthorized modifications, it is essential to protect the read-only memory where the basic boot loader is stored. The integrity of the boot loader is maintained during the boot process thanks to techniques like read-only memory locking or write protection methods that stop unauthorized writes or modifications.

- *Chain of trust*

By establishing a chain of trust, it can be made sure that each part of the boot process is tested before being executed. In order to do this, it is necessary to securely validate the legitimacy and integrity of each step in the boot process, from the bootloader to the operating system and applications.

- *Hardware security features*

A safe boot procedure can be improved by hardware security capabilities that some microcontrollers and specialized TinyML devices have built in. These capabilities could include hardware-based encryption and decryption, secure key storage, and secure boot modes that require firmware image verification.

- *Firmware update authentication*

To stop unauthorized or malicious firmware upgrades, secure boot methods must be extended to the firmware update procedure. Only authorized firmware updates are uploaded to the device by using secure channels or cryptographic signatures to authenticate and check the integrity of firmware upgrades.

13.5 Threat models in TinyML

Several factors should be taken into account when evaluating the threat model in the context of TinyML, wherein machine learning models are deployed on low-power devices. TinyML's data usage is a major source of potential vulnerabilities. However, machine learning comes with a laundry list of risks that could compromise systems and dampen the usefulness of machine learning models [30].

TinyML's threat model as shown in Fig. 13.2 accounts for all possible dangers that could arise from using machine learning models on low-power gadgets. Model extraction, in which an adversary attempts to retrieve the machine learning model from the device by, for example, reverse engineering the firmware or analyzing memory, is one of the many factors taken into account by this model. Another issue is model tampering, in which bad actors change the model's settings or inputs to make it make inaccurate predictions or act in unexpected ways. Furthermore, adversaries may employ adversarial attacks, which are aimed at fooling the model, by manipulating input data to bring about unexpected behavior or to take advantage of vulnerabilities. Data transmitted between the TinyML device and other systems need to be protected from eavesdropping and modification, which is why communication security is so crucial. Concerns about privacy arise when handling sensitive data, and denial-of-service attacks can lead to a system's resources being depleted [31]. The threat model for TinyML should also account for physical attacks, supply chain risks, and adversarial machine learning.

The following are some possible threats that should be considered.

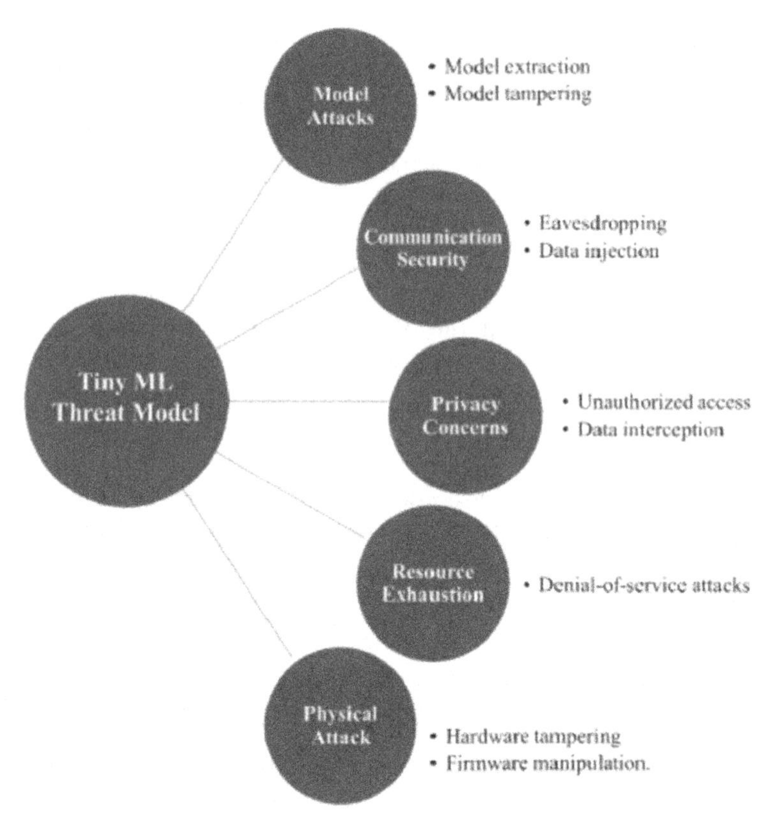

FIGURE 13.2

TinyML threat model.

13.5.1 Model attack

Model attacks represent potential threats related to model security, such as model extraction, tampering, or poisoning attacks.

As adversaries may try to stealthily retrieve the machine learning model installed on the device, model extraction is a major issue for TinyML systems. In order to get at the private data connected to the model, the extraction process typically involves exploiting vulnerabilities or using reverse engineering methods.

Reverse engineering the firmware is one tactic that attackers may use. Intruders can learn sensitive information about the model's parameters, architecture, and implementation details through reverse engineering the firmware. To learn how the model works and extract its essential parts, they may debug it, examine its binary code, or disassemble its firmware.

The device's memory can also be analyzed as an alternative method. The TinyML device's memory can be inspected by adversaries, who can then retrieve the model and any data it contains. The model's parameters, architecture, and other crucial details can be extracted by an attacker using memory analysis techniques like memory dumping or debugging to reveal the representation of the model stored in the device's memory.

Model extraction could be pursued for various reasons. Potential threats include appropriation of intellectual property, unauthorized use of the model, and the use of model vulnerabilities in subsequent attacks [32]. The extracted models can be used for malicious purposes or reverse-engineered to learn about proprietary algorithms.

Model tampering in TinyML systems threatens machine learning model deployment. Attackers may tamper with the deployed model to manipulate its behavior or make incorrect predictions, compromising system integrity and reliability. Model tampering has many methods. Attackers can alter model decisions and predictions by changing model parameters like weights and biases. During model training or inference, malicious data can be injected; adversaries can manipulate the learning process by providing carefully crafted inputs.

Attackers may also manipulate the model using TinyML implementation vulnerabilities to gain unauthorized access to the model or execution environment, exploit software vulnerabilities, corrupt memory, or perform buffer overflow attacks. Attackers can control the model's behavior, functionality, or the entire system once inside. Attackers can also poison model training data, manipulating the model by injecting biased or malicious data into the training dataset.

13.5.2 Communication security

Communication security is an important part of the risk model for TinyML systems. Communication between the TinyML device and external systems is vulnerable to interception and tampering by hostile actors, who pose a significant threat to the system's safety and reliability.

Eavesdropping is a major security risk because it allows attackers to steal private information sent between the TinyML device and other systems [33]. This can be anything from private data to a proprietary model or your name and address. Adversaries can intercept private information and use it for their own ends if they listen in on the conversation.

In addition, the communication channel is vulnerable to the injection of malicious commands or data. If an attacker is able to compromise the TinyML device's security or disrupt its normal functioning, they may attempt to do so by altering the commands sent to it or by injecting malicious data. The integrity of the machine learning model may be compromised as a result of unauthorized access, altered device behavior, or the introduction of false data.

Multiple options exist for enhancing communication security and reducing associated risks. Protecting data in transit between your TinyML device and other networks requires the implementation of secure communication protocols like Transport Layer Security (TLS). By using encryption, information sent over a network is safe from snooping and manipulation.

On top of that, though, there must be robust authentication and access control mechanisms in place to ensure that only authorized users have access. Methods like mutual authentication, in which both the TinyML device and the external system verify each other's identities before establishing a connection, fall into this category.

13.5.3 Privacy concerns

When a TinyML system is handling sensitive data, concerns about privacy play a major role in the threat model. The privacy of the user's information can be breached in a number of ways: either by an attacker specifically targeting the device or by an adversary intercepting data in transit.

The security of personal data stored on the TinyML device is a major cause for concern. Personal information, financial data, and medical records are all examples of sensitive information that could be compromised if an attacker were able to exploit a security flaw or gain physical access to the device. Identity theft, fraud, and other invasions of privacy are all possible outcomes should such data be compromised [34].

There is also the risk that inference attacks will be used to steal private data from the device by an adversary. They may attempt to infer hidden information about the TinyML model by examining its actions or results. In a healthcare setting, for instance, an attacker may try to extrapolate information about a patient's condition from the model's predictions even if that information was not made public.

Threat actors may also attempt to intercept data in transit between the TinyML device and other systems, in addition to gaining unauthorized access or making inferences from it. They might be able to access private information by intercepting it as it travels over wired or wireless channels. Eavesdropping, man-in-the-middle attacks, and network sniffing are just a few examples of how this can happen.

There are a variety of approaches that can be taken to deal with privacy issues in TinyML systems. Sensitive information should be encrypted using a strong technique before being transmitted or stored. This makes it so that no one, not even the bad guys, can read the data without the right decryption keys.

13.5.4 Resource exhaustion

Considering that adversaries may try to launch denial-of-service attacks, which consume all of the device's processing power, it is important to keep this in mind. Because of their limited resources, TinyML devices are particularly vulnerable to attacks that deplete those resources.

There are a number of methods that can be used by an adversary to drain a TinyML device of its power. The device may be overwhelmed with data or operations that place heavy demands on its limited resources. Attackers can make a device unresponsive, slow, or even crash by taxing its processing power to its limit.

Exploiting loopholes in the system's method of managing its resources is yet another strategy. To deplete system resources, attackers may cause memory leaks, buffer overflows, or high CPU utilization. These kinds of attacks can starve a device of memory, processing power, or other essentials, making it less effective at its primary tasks [35].

Attacks that deplete a system's resources can have devastating effects. TinyML devices are vulnerable to these attacks because they can cause disruptions in service, system instability, or even complete inoperability. Resource exhaustion attacks can have devastating effects on vital applications like healthcare and industrial control systems, as well as substantial financial losses.

13.5.5 Physical attacks

Physical attacks can come in the form of tampering with the hardware, the power supply, or the software. Data breaches, compromised device integrity, and inoperable devices are all possible outcomes of these types of attacks.

Physical attacks can take many forms, including tampering with the TinyML device's hardware components. In order to tamper with the device's internals, an attacker may attempt to disassemble it. Soldering, jumpering, and adding new components are all viable options. To extract sensitive data, inject malicious functionality, or circumvent security measures, attackers may try to tamper with the hardware [36].

Tampering with the TinyML device's power supply is another potential entry point for physical attacks. An assault on the power grid could take many forms, including the introduction of voltage spikes, power surges, or supply noise. If the device's power source is compromised, it becomes vulnerable to instability, malfunctions, and even physical damage from its adversaries.

In addition, the TinyML device's firmware could be a target of an attack. The firmware, which governs the device's operation and behavior, can be altered by adversaries if they gain physical access to the device. Attackers can compromise a device in a number of ways by messing with its firmware, including by inserting malicious code, stealing data, or changing how it works.

13.6 Applications of TinyML

Industry, consumer, and military innovation will be enabled by machine learning/AI algorithms on embedded devices. Healthcare, industrial monitoring, and consumer electronics all greatly benefit from adding a layer of intelligence to embedded devices. TinyML applications as shown in Fig. 13.3 are expected to grow rapidly in some of the following domains in the near future.

13.6.1 Surveillance

In the current phase of development, TinyML primarily works on creating neural networks for computer vision and audio processing. Running TinyML-based code on MCU-powered devices allows high accuracy in detection of objects, persons, faces, simple activities, and voice activities. TinyML will allow for the creation of new kinds of pervasive, lightweight, and low-cost surveillance, monitoring, and identification applications. It is best suited for applications that require a short, categorized

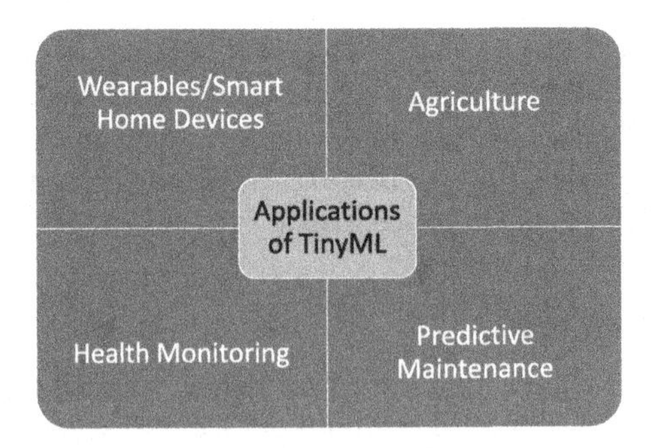

FIGURE 13.3

Applications of TinyML.

answer (such as person identified or not) rather than supplying complete contextual information [13].

13.6.2 Healthcare

There are a number of monitoring and health solutions that can be significantly improved with TinyML. It is imperative that wearable technologies are connected to the internet and protected from outside intrusion since they continuously sense and transmit physiological and activity data. Inference models based on TinyML for signal denoising, temporal analysis, and classification may be used to assess all gathered data in real time, thereby reducing the need to continuously stream data. Several companies are looking towards TinyML-style frameworks to improve personal health products like hearing aids [14].

13.6.3 Agriculture

The IoT may play an important role in automated and smart agriculture. Data transport and privacy concerns are among the challenges associated with this technology. TinyML is a new technology that delivers low-cost, high-efficiency hardware capable of locally running complex machine learning models and neural networks, hence addressing many of the aforementioned IoT issues. TinyML-based systems are capable of monitoring ambient conditions and soil moisture as a complete solution for watering plants autonomously. The goal of the author of [15] was to investigate the feasibility of building a system based on TinyML technology that could be taught and adapted to varied climatic conditions and landscapes, as well as climate change.

13.6.4 Autonomous systems

Industrial IoT practices are rapidly expanding as a consequence of advancements in the automated systems of conventional production divisions, increased networking, and sophisticated sensing technologies. TinyML capabilities can help with further enhancements to these operations. Predictive data maintenance can be utilized on a variety of industrial equipment running on embedded technology. TinyML can assist a variety of industrial monitoring systems with time-sensitive requirements.

These are just a few examples of the many possible applications of TinyML. The ability to deploy machine learning models on resource-constrained devices opens up a wide range of opportunities across industries, enabling the utilization of intelligent and efficient systems in various domains.

13.7 Implications for future research in TinyML security

The field of TinyML security will benefit greatly from ongoing and future research aimed at resolving existing and new threats. As TinyML becomes more widely used in areas like healthcare, smart homes, industrial automation, and wearable devices, it is becoming increasingly important to find novel approaches to protecting user data [37–39]. Some important directions for the future of TinyML security research include the following.

Increased deployment in edge devices: Research should address the unique security challenges associated with these different environments as TinyML continues to be deployed in a wide range of edge devices like IoT devices and wearables. This involves making threat models tailored to edge deployments and developing lightweight security mechanisms that can function effectively with limited computational resources.

Improved accuracy of models: TinyML models have made great strides in accuracy, but more work is needed to improve their robustness and reliability. This requires investigating methods for protecting against adversarial attacks, enhancing model-training techniques, and creating tools for detecting and dealing with data drift in low-resource settings.

Integration with blockchain technology: Combining TinyML with blockchain technology can improve the safety and reliability of applications that handle sensitive information. Secure model sharing and collaborative learning in blockchain networks, as well as the potential for secure and verifiable model inference using smart contracts, are all areas that could benefit from further study.

Greater focus on explainability: There is a growing need for model explainability and transparency due to the increasing use of TinyML models in mission-critical applications like healthcare and autonomous systems. More study is needed on how to interpret and explain TinyML models to promote transparency, confidence, and moral judgment in real-time software.

Integration with augmented and virtual reality: TinyML's compatibility with augmented reality (AR) and virtual reality (VR) technologies paves the way for novel studies of privacy and security. There is room for further research into privacy-preserving techniques for user data in mixed-reality settings, model integrity verification for immersive experiences, and secure data exchange between TinyML-powered edge devices and AR/VR platforms.

Overall, the changing deployment landscape, the need for robust and trustworthy solutions, and the investigation of emerging technologies to further strengthen security, privacy, and dependability should all guide future TinyML security research. Researchers can pave the way for TinyML's widespread adoption and secure integration by focusing on these specific areas.

References

[1] P.P. Ray, A review on TinyML: state-of-the-art and prospects, J. King Saud Univ, Comput. Inf. Sci. 34 (4) (2022) 1595–1623.

[2] L. Dutta, S. Bharali, TinyML meets IoT: a comprehensive survey, IEEE Int. Things J. 16 (2021) 100461.

[3] C.R. Banbury, V.J. Reddi, M. Lam, W. Fu, A. Fazel, J. Holleman, X. Huang, R. Hurtado, D. Kanter, A. Lokhmotov, D. Patterson, Benchmarking TinyML systems: challenges and direction, arXiv preprint, arXiv:2003.04821, 2020.

[4] V. Tsoukas, A. Gkogkidis, A. Kampa, G. Spathoulas, A. Kakarountas, Enhancing food supply chain security through the use of blockchain and TinyML, Information 13 (5) (2022) 213.

[5] A. Dutta, S. Kant, Implementation of cyber threat intelligence platform on Internet of Things (IoT) using TinyML approach for deceiving cyber invasion, in: 2021 International Conference on Electrical, Computer, Communications and Mechatronics Engineering (ICECCME), IEEE, October 2021, pp. 1–6.

[6] V.J. Reddi, B. Plancher, S. Kennedy, L. Moroney, P. Warden, A. Agarwal, C. Banbury, M. Banzi, M. Bennett, B. Brown, S. Chitlangia, Widening access to applied machine learning with TinyML, arXiv preprint, arXiv:2106.04008, 2021.

[7] H. Han, J. Siebert, TinyML: a systematic review and synthesis of existing research, in: 2022 International Conference on Artificial Intelligence in Information and Communication (ICAIIC), IEEE, February 2022, pp. 269–274.

[8] S.A.R. Zaidi, A.M. Hayajneh, M. Hafeez, Q.Z. Ahmed, Unlocking edge intelligence through tiny machine learning (TinyML), IEEE Access 10 (2022) 100867–100877.

[9] M. Shafique, T. Theocharides, V.J. Reddy, B. Murmann, TinyML: current progress, research challenges, and future roadmap, in: 2021 58th ACM/IEEE Design Automation Conference (DAC), IEEE, December 2021, pp. 1303–1306.

[10] P. Andrade, I. Silva, M. Diniz, T. Flores, D.G. Costa, E. Soares, Online processing of vehicular data on the edge through an unsupervised TinyML regression technique, ACM Trans. Embed. Comput. Syst. (2023), https://doi.org/10.1145/3591356.

[11] P.P. Ray, A review on TinyML: state-of-the-art and prospects, J. King Saud Univ, Comput. Inf. Sci. 34 (2022) 1595–1623, https://doi.org/10.1016/j.jksuci.2021.11.019.

[12] K. Angrishi, Turning internet of things (IoT) into internet of vulnerabilities (IoV): IoT botnets, arXiv:1702.03681, 2017. [Google Scholar].

[13] D. Sopori, T. Pawar, M. Patil, R. Ravindran, Internet of things: security threats, Int. J. Adv. Res. Comput. Eng. Technol. 6 (2017) 263–267. [Google Scholar].

[14] R. Doshi, N. Apthorpe, N. Feamster, Machine learning DDoS detection for consumer internet of things devices, in: Proceedings of the 2018 IEEE Security and Privacy Workshops (SPW), San Francisco, CA, USA, 24 May 2018, pp. 29–35. [Google Scholar].

[15] M. Yamauchi, Y. Ohsita, M. Murata, K. Ueda, Y. Kato, Anomaly detection in smart home operation from user behaviors and home conditions, IEEE Trans. Consum. Electron. 66 (2020) 183–192. [Google Scholar] [CrossRef].

[16] S. Muller, J. Lancrenon, C. Harpes, Y. Le Traon, S. Gombault, J.M. Bonnin, A training-resistant anomaly detection system, Comput. Secur. 76 (2018) 1–11. [Google Scholar] [CrossRef][Green Version].

[17] R. Corizzo, M. Ceci, E. Zdravevski, N. Japkowicz, Scalable auto-encoders for gravitational waves detection from time series data, Expert Syst. Appl. 151 (2020) 113378. [Google Scholar] [CrossRef].

[18] https://forums.tinyml.org/t/tinyml-talks-on-february-1-2022-energy-efficiency-and-security-for-tinyml-and-edgeai-a-cross-layer-approach-by-muhammad-shafique/752/1.

[19] https://forums.tinyml.org/t/tinyml-talks-on-june-20-2023-standardized-ai-architectures-for-secure-tinyml-by-andrea-basso/1178.

[20] D. Myridakis, G. Spathoulas, A. Kakarountas, Supply current monitoring for anomaly detection on IoT devices, in: Proceedings of the 21st Pan-Hellenic Conference on Informatics, Larissa, Greece, 28–30 September 2017, pp. 1–2. [Google Scholar].

[21] D. Myridakis, G. Spathoulas, A. Kakarountas, D. Schoinianakisy, J. Lueken, Anomaly detection in IoT devices via monitoring of supply current, in: Proceedings of the 2018 IEEE 8th International Conference on Consumer Electronics-Berlin (ICCE-Berlin), Berlin, Germany, 2–5 September 2018, pp. 1–4. [Google Scholar].

[22] D. Myridakis, G. Spathoulas, A. Kakarountas, D. Schinianakis, J. Lueken, Monitoring Supply Current Thresholds for Smart Device's Security Enhancement, in: Proceedings of the 2019 15th International Conference on Distributed Computing in Sensor Systems (DCOSS), Santorini Island, Greece, 29–31 May 2019, pp. 224–227. [Google Scholar].

[23] D. Myridakis, G. Spathoulas, A. Kakarountas, D. Schinianakis, Smart devices security enhancement via power supply monitoring, Future Internet 12 (2020) 48. [Google Scholar] [CrossRef][Green Version].

[24] K. Prabhu, B. Jun, P. Hu, Z. Asgar, S. Katti, P. Warden, Privacy-preserving inference on the edge: mitigating a new threat model, in: Research Symposium on Tiny Machine Learning, 2021.

[25] B. Sudharsan, J.G. Breslin, M. Tahir, M.I. Ali, O. Rana, S. Dustdar, R. Ranjan, OTA-TinyML: over the air deployment of TinyML models and execution on IoT devices, IEEE Internet Comput. 26 (3) (2022) 69–78.

[26] S. Soro, TinyML for ubiquitous edge AI, arXiv preprint, arXiv:2102.01255, 2021.

[27] S. Bhasin, D. Jap, S. Picek, On (in)security of edge-based machine learning against electromagnetic side-channels, in: 2022 IEEE International Symposium on Electromagnetic Compatibility & Signal/Power Integrity (EMCSI), IEEE, August 2022, pp. 262–267.

[28] N. Quadar, A. Chehri, G. Jeon, M.M. Hassan, G. Fortino, Cybersecurity issues of IoT in Ambient Intelligence (AmI) environment, IEEE Int. Things Mag. 5 (3) (2022) 140–145.

[29] K.A. Nagaty, IoT commercial and industrial applications and AI-powered IoT, in: Frontiers of Quality Electronic Design (QED). AI, IoT and Hardware Security, Springer International Publishing, Cham, 2023, pp. 465–500.

[30] F. Parastar, M. Sepahi, M. Moudi, On-device ML For the Current and the Emerging Networks: A Survey on Current Approaches and Challenges, https://doi.org/10.1145/3503823.3503836.

[31] V. Tsoukas, E. Boumpa, G. Giannakas, A. Kakarountas, A review of machine learning and TinyML in healthcare, in: 25th Pan-Hellenic Conference on Informatics, November 2021, pp. 69–73.

[32] H. Doyu, R. Morabito, M. Brachmann, A TinyMLaaS ecosystem for machine learning in IoT: overview and research challenges, in: 2021 International Symposium on VLSI Design, Automation and Test (VLSI-DAT), IEEE, April 2021, pp. 1–5.

[33] D. Piątkowski, K. Walkowiak, TinyML-based concept system used to analyze whether the face mask is worn properly in battery-operated conditions, Appl. Sci. 12 (1) (2022) 484.

[34] N. Tekin, A. Acar, A. Aris, A.S. Uluagac, V.C. Gungor, Energy consumption of on-device machine learning models for IoT intrusion detection, IEEE Int. Things J. 21 (2023) 100670.

[35] A. Hussain, N. Abughanam, J. Qadir, A. Mohamed, Jamming detection in IoT wireless networks: an edge-AI based approach, in: Proceedings of the 12th International Conference on the Internet of Things, November 2022, pp. 57–64.

[36] M. Lord, TinyML, Anomaly Detection, Doctoral dissertation, California State University, Northridge, 2021.

[37] W. Raza, A. Osman, F. Ferrini, F.D. Natale, Energy-efficient inference on the edge exploiting TinyML capabilities for UAVs, Drones 5 (4) (2021) 127.

[38] S.K. Priya, N. Balaganesh, K.P. Karthika, Integration of AI, blockchain, and IoT technologies for sustainable and secured Indian public distribution system, in: AI Models for Blockchain-Based Intelligent Networks in IoT Systems: Concepts, Methodologies, Tools, and Applications, Springer International Publishing, Cham, 2023, pp. 347–371.

[39] X. Zhu, R. Li, J. Hu, H. Zhang, Q. Luo, Q. Duan, K. Jiang, A guided task and obstacle alert robot system based on TinyML and augmented reality, in: Proceedings of the 2022 3rd International Conference on Control, Robotics and Intelligent System, August 2022, pp. 225–229.

CHAPTER

TinyML applications and use cases for healthcare 14

Mamta Bhamare, Pradnya V. Kulkarni, Rashmi Rane, Sarika Bobde, and Ruhi Patankar

Department of Computer Engineering and Technology, Dr. Vishwanath Karad MIT World Peace University, Pune, India

14.1 Introduction

TinyML, an emerging technology, is a distinct technique the scientific community proposes to construct secure autonomous devices capable of gathering, analyzing, and notifying without providing information to other parties. This study examines how TinyML, an emerging technology that requires the integration of machine learning algorithms, contributes to healthcare applications at the edge and the solutions it may provide, particularly for wearable technologies. It also discusses how TinyML might improve neural networks to offer devices in sectors such as healthcare intelligence and autonomy. Apart from the several possible applications of TinyML in healthcare, these devices have raised significant concerns regarding the privacy of personal data that these services collect and store.

Microcontrollers with limited computational resources and memory make TinyML feasible for implementing machine learning models. TinyML can be used in healthcare to improve accuracy, reduce latency, and guarantee security and privacy. TinyML has several potential applications in healthcare [1][2][3][4], including the following.

1. **Wearable devices:** TinyML can power devices that measure numerous health metrics, such as heart rate, blood pressure, and oxygen level, such as fitness trackers, smartwatches, and biosensors. In an emergency, these devices can trigger real-time alerts based on the data collected through these devices.
2. **Remote patient monitoring:** A healthcare provider can monitor the health status of patients remotely through TinyML. Patients with diabetes, hypertension, and heart disease will benefit the most from this medication.
3. **Diagnosis and treatment:** Patients can benefit from TinyML when it comes to diagnosis and treatment. With TinyML algorithms, healthcare providers can diagnose patients more accurately and recommend personalized treatment plans based on medical history, symptoms, and lab results.
4. **Drug discovery:** A comprehensive evaluation of drug interactions, molecule structures, and cellular pathways can also be performed using TinyML to speed up

TinyML for Edge Intelligence in IoT and LPWAN Networks. https://doi.org/10.1016/B978-0-44-322202-3.00019-1

the drug discovery process. In addition to identifying new drug targets, researchers can develop more effective treatments based on this information.

14.2 Wearable devices

Scientists suggest a new means of building autonomous, safe, ecofriendly gadgets that can gather, analyze, and alarm without communicating data to the outside world using TinyML technology. Medical devices can be made intelligent and autonomous using neural networks optimized by TinyML [5]. Many communication and AI techniques are examined in [6], suitable for the next generation of wearable devices. With the help of TinyML, the author of [7] has developed a device, BandX, that determines human activities based on wearable sensors.

14.2.1 Wearable healthcare device examples

The popularity of wearable healthcare gadgets has grown in recent years as they can measure various health metrics. The categorization of different wearable healthcare devices [3] is shown in Fig. 14.1. A few wearable devices discussed in this chapter are among those types. The different types of wearable devices are listed below and illustrated in Fig. 14.2.

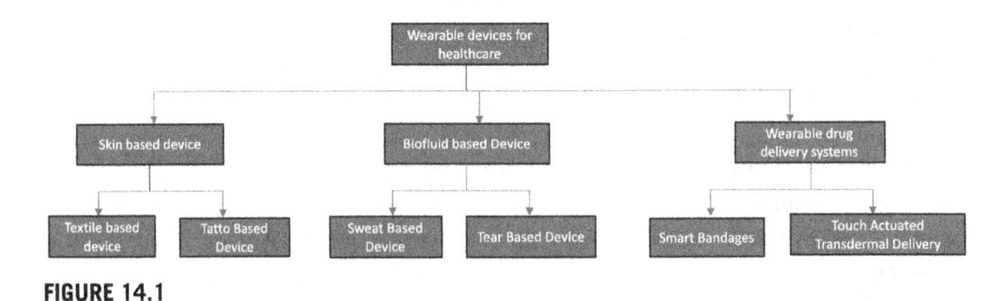

FIGURE 14.1

Classification of wearable healthcare devices.

FIGURE 14.2

Wearable devices.

1. Wearable fitness trackers: Fitbit, Garmin, and Apple Watch measure exercise activities such as steps taken, kilometers traveled, calories burned, and heart rate. They frequently have functions like GPS, sleep tracking, and exercise tracking [4].
2. Smartwatches: Smartwatches combine smartphone notification, texting, and music-playing capabilities with fitness monitoring functions. Additionally, they may monitor heart rate, sleep habits, and physical activity.
3. Heart rate monitors: These tools measure and monitor heart rate continually. During activity, at rest, or even under stress, they might offer helpful insights into heart health.
4. Blood pressure monitors: With wearable blood pressure monitors, blood pressure may be conveniently and continuously monitored throughout the day. They can assist people in tracking changes in their blood pressure and offer information to healthcare specialists.
5. Glucose monitors: People with diabetes may continually check their blood glucose levels thanks to wearable glucose monitors. These gadgets generally employ sensors to determine the blood or interstitial fluid glucose levels and send the information to a smartphone or specialized device.
6. ECG monitors: The heart's electrical activity is captured and recorded by electrocardiogram (ECG) monitors. Wearable ECG monitors, like the Apple Watch Series 4 and subsequent models, can help users understand abnormal heartbeats and perhaps identify diseases like atrial fibrillation.
7. Sleep trackers: Wearable bands and smartwatches measure sleep patterns and keep track of sleep duration, quality, and phases. They can provide information on sleep disruption, sleep efficiency, and probable sleep disorders.
8. Posture correctors: Wearable posture correctors employ sensors to identify the user's body position and deliver feedback, prompting them to keep their posture correct. They can aid in the prevention or redress of bad postural habits.
9. UV exposure correctors: These wearable technologies assess the amount of ultraviolet (UV) radiation exposure and deliver real-time information or alerts to assist users in preventing skin cancer.
10. Smart clothing: Sensors and electronics are integrated into clothing to make "smart" clothing capable of tracking vital indications such as heart rate, breathing rate, body temperature, and activity levels. They can be used for remote patient monitoring (RPM), healthcare applications, and tracking athletic performance.

The number of users who use wearable devices during 2019–2023 has drastically increased due to the COVID-19 pandemic. Fig. 14.3 shows the trend in using wearable devices [1].

The wearable healthcare technology market is surging, and its maturation will put more wearable technology in the hands of consumers and businesses. According to an Insider Intelligence study, the number of people who use health and fitness apps will increase from 88.5 million in 2022 to 91.3 million by 2023.

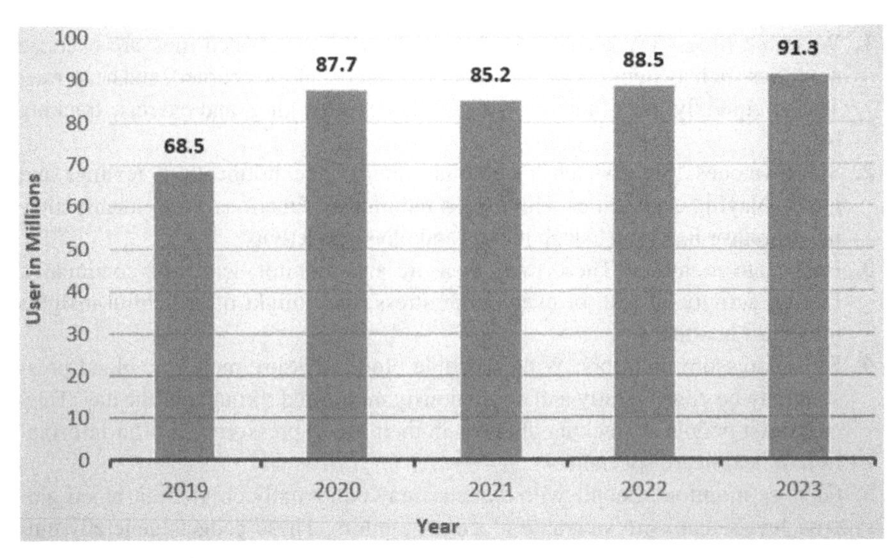

FIGURE 14.3

Health and fitness app users.

The user of wearable devices would always want to have easy access [2]. However, different factors affect healthcare device use in adults. These factors are:

1. user-friendly device,
2. keeping privacy confidential,
3. security of information and data,
4. personalized care,
5. easy maintenance,
6. cost of healthcare,
7. coordination with healthcare providers,
8. health coach access.

14.2.2 TinyML in healthcare

TinyML uses machine learning models in limited-resource devices, such as microcontrollers, which are frequently used in wearable technology [8]. TinyML allows machine learning algorithms to operate in-house on these devices, providing real-time data processing and inference without the need for cloud-based services. With emerging TinyML technology, the author of [9] designed and developed a wearable device to prevent infectious disease transmission. TinyML is widely used in healthcare, as shown in Fig. 14.4, with applications including the following:

1. COVID-19: a wrist gadget for monitoring vital signs;
2. hearing aid: neural augmentation of speech;

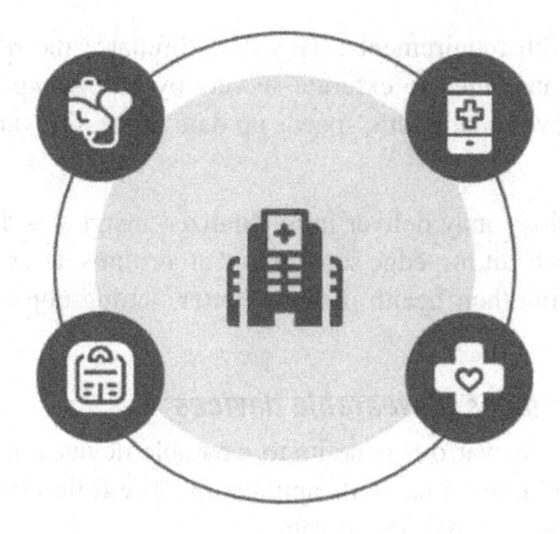

FIGURE 14.4

TinyML in healthcare.

3. home healthcare: assistance from patients and caregivers;
4. framework: a framework for eHealth sectors.

14.2.3 Wearable gadget advantages and disadvantages

14.2.3.1 Wearable gadget advantages

TinyML integration into wearable devices has several benefits, including greater energy economy, enhanced privacy, real-time analysis, and a more responsive and customized user experience. Here are some key advantages:

1. Device usage: Wearable technology offers a high level of convenience since it is made to be portable and readily worn on the body. Wearable devices do not significantly hurt or interfere with regular activities and may be worn comfortably all day.
2. Continuous monitoring and feedback: The capacity to track numerous health measurements and activities in real time is provided by wearable technology. Instantaneous feedback and insights into a user's health and fitness may be obtained by monitoring their heart rate, steps, sleep habits, and other data.
3. Recognition of gestures: TinyML models' lightweight nature helps to minimize inference latency, resulting in a more responsive user experience. It is helpful in applications such as gesture recognition or voice commands when prompt and precise replies are required.
4. Scalability: TinyML models are lightweight and can operate on various wearable devices, from smartwatches to fitness trackers. This scalability enhances TinyML apps' adaptability across different wearables.

5. Reduced bandwidth requirements: TinyML eliminates the requirement for vast raw data to be transmitted to external servers by processing information on the device itself. It saves bandwidth, speeds up data processing, and reduces reaction times.

Wearable technology may deliver individualized insights and suggestions based on individual data with cutting-edge sensors and algorithms. Users may enhance their well-being by knowing their health patterns better, setting objectives, and selecting wisely.

14.2.3.2 Disadvantages of wearable devices

While TinyML provides various benefits to wearable devices, it also has some potential drawbacks and limitations in its application. The following aspects should be considered when using TinyML in wearables:

1. Limited model complexity: TinyML models are meant to be lightweight, which implies they may have complexity and capacity restrictions. This can affect the capacity to manage highly complicated jobs or recognize complex patterns.
2. Overfitting and generalization: TinyML models may be prone to overfitting because of their limited model capacity, especially if the training dataset does not include all conceivable situations. It might result in poor generalization performance.
3. Resource constraints: Wearable electronics are frequently constrained in processing power, memory, and battery life. Running TinyML models on such resource-constrained hardware may result in model complexity and real-time performance sacrifices.
4. Security concerns: TinyML models deployed on wearable devices may expose them to security issues such as adversarial assaults. TinyML models may be more vulnerable to certain forms of assault due to their lightweight nature.
5. Data privacy issues: While on-device processing improves privacy, it also means that sensitive health data are kept on the device. It raises questions regarding the security of data saved on the device, particularly if it is lost or stolen.

14.3 Remote patient monitoring

14.3.1 Remote patient monitoring examples in healthcare

An RPM system is a healthcare solution that uses TinyML technology to monitor patients' health remotely. TinyML is a branch of machine learning that uses machine learning models on limited-resource gadgets like microcontrollers, which are miniature, low-power computer gadgets. Fig. 14.5 depicts RPM, which is widely used [10]. The existing system uses signal processing, a methodology for remote diagnosis and treatment of patients.

The components involved in the TinyML-based systems are the following.

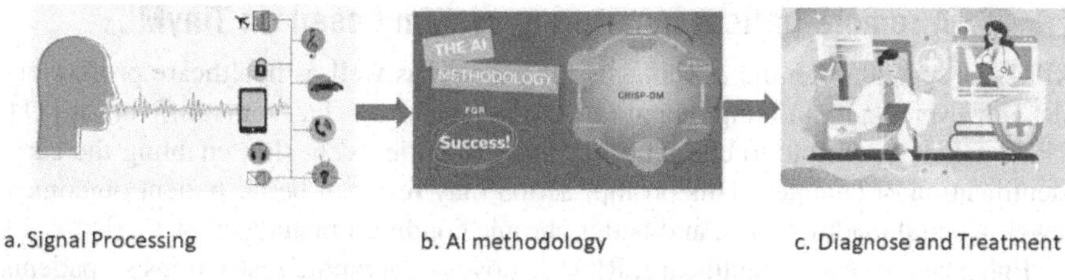

FIGURE 14.5

Remote patient monitoring architecture.

- **Sensor devices**

These are little, wearable gadgets with numerous sensors to gather vital health information. Some examples include accelerometers, temperature sensors, blood pressure monitors, and heart rate monitors. These sensors collect the patient's live physiological data.

1. Data preprocessing: The sensor devices gather unprocessed data, which are later formatted and processed for additional analysis. In this step, the raw sensor data are cleaned, the appropriate transformations are made, and pertinent features are extracted.
2. TinyML models: This step uses machine learning models created especially for deployment on devices with limited resources, including microcontrollers. These models have been trained to examine the preprocessed sensor data and identify trends or categories that pertain to the patient's health. The TinyML models have been enhanced to operate effectively on the sensor devices' constrained processing power.
3. Inference and decision-making: The preprocessed data are processed in real time by the TinyML models installed on the sensor devices to produce insights or forecasts about the patient's health. Vital signs, activity levels, anomaly detection, and other pertinent health measures are examples of these data. If an irregular or critical situation is found based on the study, the system can send alerts or messages to carers or medical professionals.
4. Data transmission and storage: The sensor devices securely send the analyzed data and any alerts or notifications created to a central server or cloud platform. It enables carers or medical professionals to access the patient's health information remotely. The data can be kept for further study, trend tracking, or report generation.
5. User interface and utilization: Continuous patient monitoring, early health issue diagnosis, decreased hospital visits, and improved patient outcomes are all advantages of an RPM system built on TinyML. By utilizing TinyML technology, healthcare practitioners can effectively monitor patients with chronic diseases or those in need of long-term care and expand the accessibility of healthcare services to rural locations.

14.3.2 A remote patient monitoring system based on TinyML

RPM has several potential advantages for patients as well as healthcare professionals. However, using and implementing RPM systems also has some difficulties [11]. RPM makes it possible to continuously monitor patients' health, enabling the early identification of changes. This prompt action may result in better patient outcomes, fewer hospital readmissions, and better chronic condition management.

Enhanced access to healthcare: RPM removes geographic restrictions so patients can access high-quality medical treatment wherever they are. Patients in remote or underserved areas with trouble accessing medical services can benefit most. By reducing unnecessary hospital stays, ER visits, and outpatient appointments, RPM can reduce healthcare costs. It promotes proactive care management, preventing issues and requiring expensive procedures.

Enhanced patient engagement: It provides access to real-time health data and insights; RPM empowers patients to take an active role in their healthcare. Self-care activities boost adherence to treatment regimens and lifestyle modifications by increasing patient engagement.

Early detection and intervention: RPM's continuous monitoring enables the early identification of abnormal patterns or solemn events. Healthcare professionals can take quick action to avert adverse outcomes and give fast interventions or make changes to treatment regimens.

14.3.3 Remote patient monitoring: challenges and potential benefits

14.3.3.1 Challenges in remote patient monitoring

1. Technical infrastructure: RPM system implementation calls for a solid technical foundation, which includes dependable internet connectivity, secure data transfer, and interoperability with current healthcare systems. It can be challenging to overcome infrastructure constraints, particularly in distant or resource-limited places.
2. Data security and privacy: RPM entails the gathering, transferring, and storing of private patient health information. Although important, ensuring data security, privacy, and compliance with legal requirements like the Health Insurance Portability and Accountability Act (HIPAA) may be challenging and expensive.
3. User adoption and engagement: The ability of patients to embrace and interact with the technology is essential for a successful RPM implementation. Some patients, especially the elderly or those with low digital literacy, may find it challenging to use and comprehend the RPM devices and interfaces.
4. Workflow integration: It can be challenging to integrate RPM into the healthcare workflows and processes that are currently used. To include RPM data in clinical decision-making, care coordination, and follow-up processes, healthcare professionals must modify their current procedures.
5. Reimbursement and regulatory barriers: RPM's varied healthcare systems and geographical areas have various reimbursement schemes and regulatory frame-

works. Healthcare providers would have trouble getting paid for RPM services, which could hinder the acceptance and sustainability of such programmers.

Collaboration between healthcare providers, technology suppliers, policymakers, and regulatory agencies is necessary to address these issues. Continuous technological advancements and focusing on overcoming these constraints can revolutionize and improve healthcare delivery, patient outcomes, and access to high-quality care through RPM.

14.3.3.2 Benefits of remote patient monitoring

RPM that uses TinyML can improve healthcare delivery and patient outcomes. Here are some of the primary benefits:

1. Real-time monitoring: With TinyML, health parameters such as blood pressure, blood glucose, and heart rate can be continuously monitored in real time. As a result, healthcare providers can detect abnormalities or changes quickly in a patient's condition, enabling them to intervene quickly and prevent potential problems.
2. Reduced burden on healthcare systems: Health issues can be detected and addressed early using RPM with TinyML, reducing the burden on healthcare systems and preventing problems from escalating.
3. Remote accessibility: RPM with TinyML is beneficial for patients living in remote areas, as it enables medical professionals to monitor and manage their patients remotely without requiring them to travel long distances for regular checkups.
4. Increased patient engagement: With real-time data, patients may actively engage in their health management. It empowers patients and motivates them to adopt lifestyle adjustments or follow treatment regimens more closely.
5. Data-driven personalized care: TinyML offers patient data processing at the edge (on the device), enabling personalized and context-aware healthcare. It can lead to more tailored and successful treatment programs depending on the needs of each patient.

Ultimately, RPM combined with TinyML holds tremendous promise for transforming healthcare by providing personalized, real-time, and cost-effective monitoring and management of patient health.

14.4 Role in diagnosis and treatment by healthcare providers

With the introduction of TinyML, healthcare has been recognized as one of the sectors with the highest growth potential [12]. TinyML technology has the potential to revolutionize healthcare diagnosis and treatment. TinyML refers to using machine learning algorithms on devices with limited resources, such as microcontrollers and low-power circuits. TinyML brings the capability of machine learning closer to the

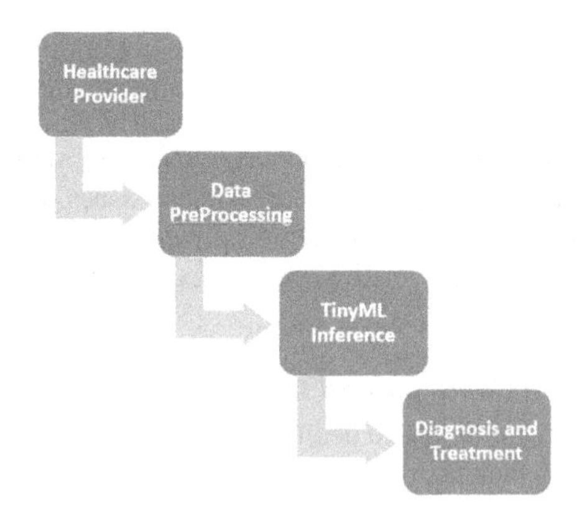

FIGURE 14.6

Diagnosis and treatment by healthcare providers.

point of care through the ability to enable local processing and inference on these devices, providing various benefits in healthcare settings. TinyML, in general, helps healthcare providers by delivering machine learning capabilities at their fingertips, allowing for faster and more accurate diagnosis and personalized treatment suggestions. TinyML integration with healthcare systems can improve patient outcomes and increase healthcare delivery efficiency. Some ways in which TinyML can help are covered in this section.

Fig. 14.6 shows the process of diagnosis and treatment provided by healthcare providers. The healthcare provider represents the healthcare professional or service provider responsible for diagnosing and treating patients. They play a crucial role in the complete process. Medical images, sensor readings, physiological signals, and other data are received from patients by healthcare providers. These data must be preprocessed to extract essential features and ensure compatibility with the TinyML model. The preprocessed data are fed into a TinyML model trained to perform a specific task, such as disease diagnosis or giving treatment recommendations. The TinyML model runs on a resource-constrained device such as a microcontroller, allowing real-time reasoning directly at the point of care. Based on the results of the TinyML inference, the healthcare provider makes a diagnosis and develops a treatment plan suitable for the patient. A TinyML model helps in decision-making by providing valuable insights and predictions.

TinyML can monitor the patient's blood pressure from their home. Ahmed and Hassan [13] built a complete end-to-end prototype system that performs machine learning interpretation on various machine learning techniques on peripherals powered by a microcontroller unit (MCU) to predict important blood pressure parameters such as systolic blood pressure (SBP), diastolic blood pressure (DBP), and mean arterial blood pressure (MAP) with ECG and photoplethysmography (PPG) sensors. The system uses TinyML technologies to enable continuous health monitoring on the extreme edge, close to the sensors attached to the patient's body, regardless of network connection or cloud infrastructure. The proposed solution is trained and tested using over 500 hours of 12,000 intense treatment unit data cases. Despite having finite memory, computation, and power budgets, the proposed solution achieved results comparable to state-of-the-art server-based solutions.

To improve the tolerability of COVID-19 in the emergency department, Fyntanidou et al. [14] created a wrist gadget that monitors vital signals such as body temperature, breathing rhythm, and blood oxygen saturation. The neural network responsible for respiration evaluation runs locally on the wearable device, eliminating data transmission to the cloud, utilizing TinyML technology, and safeguarding sensitive patient data.

According to one study [15], the revolutionary TinyML framework for healthcare can (1) consolidate machine learning methods, (2) select or even adapt machine learning models, (3) optimize for better decision-making, and (4) learn and adapt to improve performance. Additionally, the system can handle various eHealth sectors, including mental health, body scanning, hygiene monitoring, and symptom tracking.

In [16], a combination of TinyML and TensorFlow Lite is proposed for personalized home healthcare, to assist patients with rehabilitation, patients with chronic and acute conditions, and caregivers' physical and mental well-being in high-stress situations such as the COVID-19 pandemic.

Tiny RespNet, a convolutional neural network-aware model, was developed by Rashid et al. [17]. The approach is scalable and can be used in multimodal environments. Xilinx Artix-7 100T FPGA is used to implement consisting of parallel processing capability. It has a modest power consumption of 245 mW, and its energy efficiency has improved 4.3 times compared to the existing options. NVIDIA Jetson TX2 SoC is used to implement the model collated to the TX2 CPU's processing power. The Tiny RespNet system can perform categorization using audio recordings, patient speech, and demographic data. The respiratory symptoms associated with cough detection are identified using three datasets: ESC-50, FSDKaggle2018, and CoughVid. The framework is capable of detecting dyspnea.

14.4.1 Advantages and challenges

When it comes to treatment and diagnosis, the integration of TinyML can offer several advantages.

- Real-time and continuous monitoring: Data from patient devices can be monitored in real time and continuously using TinyML, which enables machine learning models to be deployed directly on edge devices [18]. This can be particularly useful when an immediate reaction or intervention is required.
- Personalized and adaptive therapy: TinyML enables personalized and adaptive therapies because machine learning models can be applied directly to the device. Models can learn from individual patient data, adapting to specific characteristics and optimizing treatment plans tailored to each patient's needs.
- Cost effectiveness: TinyML's ability to process on-device eliminates the need for expensive cloud infrastructure and external computing resources. It can lead to

cost savings in data storage, transmission, and computing resources, making treatment and diagnosis more affordable.

- Lower latency and better response time: By performing data processing and analysis on the device, TinyML minimizes the need for data transmission to external servers or cloud platforms. It significantly reduces latency and improves response times, ensuring faster diagnoses and treatment decisions.
- Improved privacy and data security: With TinyML, sensitive patient data can be processed and analyzed locally on the device, reducing the need to transfer data to and store data on external servers. It can help solve data protection issues and improve data security because data remain on the device or local network.

TinyML has the potential to transform healthcare by allowing machine learning models to be deployed on resource-constrained devices such as wearables, implantable devices, and edge devices. However, various problems must be overcome before TinyML may be successfully implemented in healthcare. Here are a few of the significant challenges.

- Limited computing resources: Medical equipment frequently has limited processing power, memory, and energy. Developing and deploying machine learning models that can run efficiently on such devices is difficult. TinyML methods must be optimized to function within these limits while maintaining acceptable accuracy.
- Limited power: A major research challenge for TinyML is the limited power of peripherals. The TinyML paradigm should handle the energy efficiency function. TinyML models require a certain amount of energy to maintain the maximum accuracy of the algorithms. Describing a global hardware platform-wide power management module is impossible because of the differences in their designs and preprocessing processes. Power-aware mismanagement in TinyML systems is possible as edge devices are typically linked to sensors and other peripherals. As a result, designing an energy-efficient TinyML system remains a crucial problem [19].
- Cost: The cost of large-scale deployment can be significant, even if the cost per device is low. Thus, the success of low-cost platforms requires overcoming financial challenges [20].
- Limited memory: TinyML's expansion is hampered by another layer of resistance: limited memory. As the size of SRAM is in kilobytes and flash memory is less than 1 MB, it becomes challenging to exploit on edge devices. Traditional machine learning inference can use larger memory peaks (e.g., GB in size), which are unavailable on edge systems. TinyML-aware devices cannot dedicate their cores to system-under-test (SUT) procedures. As a result of the tiny memory footprint of edge technology, TinyML systems have developed [19].
- Data security and privacy: Healthcare data are extremely sensitive and subject to stringent privacy restrictions. Deploying TinyML models on edge devices may cause data security and privacy concerns. Patient data security and compliance

with regulatory criteria are critical concerns in implementing TinyML in healthcare.

- Model interpretability: Interpretability of decisions made using TinyML models is critical, especially in medical environments where transparency and explainability are crucial. The deep learning models commonly used in TinyML are notorious for being black, which makes it challenging to understand the reasons behind their predictions. Ensuring the interpretability of TinyML models is an ongoing area of research.
- Data limitations and diversity: Machine learning models require large and diverse datasets for training to ensure generalizability and robustness. However, obtaining large and varied datasets in some medical fields can be difficult due to data protection regulations, limited data availability, or imbalances. These limitations can affect the efficiency and effectiveness of TinyML models.

Solving these problems will require algorithm development, hardware improvements, specialized optimization techniques, data management strategies, and collaboration between healthcare professionals, machine learning experts, and regulatory agencies. Overcoming these challenges will help unlock the full potential of TinyML for therapeutic and diagnostic applications.

14.5 Drug discovery

TinyML has a lot of potential for drug discovery. Machine learning and AI are primarily used in drug discovery to predict bioactivity and physical properties. Molecular machine learning for drug discovery is a field where academia and industry are both playing important roles [21][22][23]. It makes it possible to analyze enormous amounts of molecular data effectively and to identify prospective therapeutic candidates more quickly. TinyML could transform the drug discovery process by speeding up chemical screening and optimization, ultimately developing new and effective drugs.

TinyML can be used in a variety of ways to speed up the drug development process. Following is one of the applications where TinyML is used.

Predictive modeling: TinyML facilitates the development of predictive models that analyze vast molecular datasets. These models predict the effectiveness and safety of potential drug candidates, minimizing the need for time-consuming and expensive lab experiments. An iterative cycle of distinct processes can summarize the process of realizing actionable prediction models in drug discovery [24].

- Data collection: The first step involves collecting relevant data, such as molecular information, compound properties, and biological assay results. This dataset serves as the foundation for training the predictive model.
- Feature extraction: The collected data are processed to extract informative features representing molecular and compound characteristics. These features could

include molecular descriptors, chemical fingerprints, or other representations that capture essential properties of the compounds.

- Model training: Using the extracted features and the labeled dataset containing known outcomes (e.g., activity or toxicity), TinyML algorithms, such as neural networks or decision trees, are trained to learn patterns and correlations between the features and the desired prediction.
- Model optimization: Once the initial TinyML model is trained, it may undergo optimization with techniques like pruning or quantization to reduce its size and make it suitable for deployment on low-power devices without compromising its predictive performance.
- Model deployment: The optimized TinyML model is deployed on low-power, resource-constrained devices, such as microcontrollers or embedded systems. It allows for on-device processing, making real-time or near real-time predictive modeling possible.
- Prediction process: New, unseen compound data are fed into the deployed TinyML model. The model applies its learned algorithms to analyze the input features and generate predictions regarding the compound's effectiveness or safety.
- Prediction results: The prediction results are generated by the TinyML model and can be used for various purposes, such as prioritizing compounds for further testing, identifying potential lead candidates, or screening out compounds with undesirable properties.

Fig. 14.7 represents the process of predictive modeling. In Fig. 14.7, each step of the predictive modeling process is described, from data collection to the final prediction results. The TinyML model is deployed on a low-power device, enabling on-device processing for real-time predictions.

Specific implementation and architecture of TinyML in predictive modeling can vary depending on the algorithms, models, and hardware used. Fig. 14.7 gives a broad overview of the components and methodology for using TinyML for predictive modeling in the drug development process.

The following are some examples of where TinyML is used in drug discovery.

- Virtual screening: Virtual screening becomes feasible by deploying TinyML algorithms on portable devices. This computational approach enables the rapid screening and prioritization of large compound libraries, swiftly identifying promising molecules.
- Toxicity prediction: Early detection of potential toxicity issues is crucial in drug discovery. TinyML algorithms can be trained on toxicology databases to predict compound toxicity, enabling researchers to filter out hazardous candidates and focus on safer alternatives.
- Drug repurposing: TinyML contributes to exploring new therapeutic applications for existing drugs. Through analysis of extensive datasets on drug properties and molecular interactions, TinyML models identify opportunities for repurposing, accelerating the search for new treatments across various diseases.

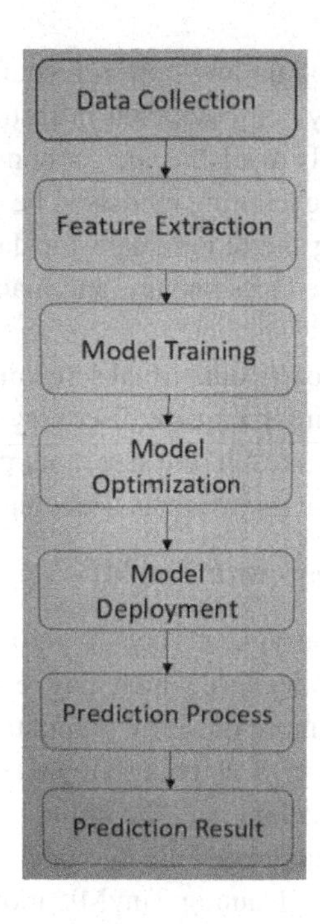

FIGURE 14.7

Flow of predictive modeling.

- Personalized medicine: TinyML algorithms can be employed in personalized medicine approaches. By analyzing individual patient data, including genetic information and health records, TinyML helps identify tailored drug treatments that align with specific patient needs, resulting in more effective and targeted therapies.

Overall, integrating TinyML into drug discovery holds significant potential to enhance efficiency, reduce costs, and expedite the identification of new drugs and treatments.

14.5.1 Advantages and challenges

Drug discovery with TinyML offers several advantages and also presents specific challenges. Here we list some of them.

- **Advantages of drug discovery with TinyML**
  1. TinyML makes it possible to analyze enormous datasets and complex molecular data in real time or very close to real time, which helps researchers accelerate drug discovery.

2. TinyML's implementation on low-power, resource-constrained devices eliminates the need for pricey computational infrastructure, making the discovery of new drugs more readily available and economical.
3. TinyML enables machine learning models to be deployed immediately on mobile devices, eliminating the requirement for data transfer to remote servers. This improves privacy, reduces latency, and makes real-time decision-making possible.
4. TinyML algorithms can carry out virtual screening and predictive modeling on mobile devices, facilitating the quick discovery and prioritization of prospective drug candidates. As a result, the screening procedure is expedited and the number of labor-intensive experiments is decreased.

- **Challenges of drug discovery with TinyML**

1. Limited computational resources: TinyML algorithms must operate within the limitations of low-power devices, which may restrict the complexity and depth of analysis compared to more powerful computational resources.
2. Model size and complexity: Designing TinyML models that balance accuracy and model size can be challenging. The limited memory and processing power of tiny devices may require trade-offs in model complexity and performance.
3. Dataset size and quality: Training TinyML models often requires extensive, high-quality datasets. Obtaining and curating such datasets, particularly in the context of drug discovery, can be time-consuming and resource-intensive.
4. Interpretability and explainability: TinyML models, especially deep learning models, can be black-box models that lack interpretability. Interpreting and explaining the decisions made by these models in drug discovery may pose challenges, especially in regulatory and ethical contexts.
5. Generalization and robustness: Ensuring the generalization and robustness of TinyML models across different datasets, drug targets, and biological contexts remains challenging. Models trained on limited data may struggle to generalize well to diverse scenarios.

Despite these challenges, ongoing research and advancements in TinyML continue to address these limitations, making it a promising technology for accelerating drug discovery and revolutionizing pharmaceutical research.

14.6 Ethics and TinyML

TinyML [25][26] applies machine learning models to resource-constrained devices such as microcontrollers and Internet of Things (IoT) devices. While TinyML offers exciting possibilities for edge computing and real-time reasoning, it raises several ethical questions and challenges. Here we list some of them.

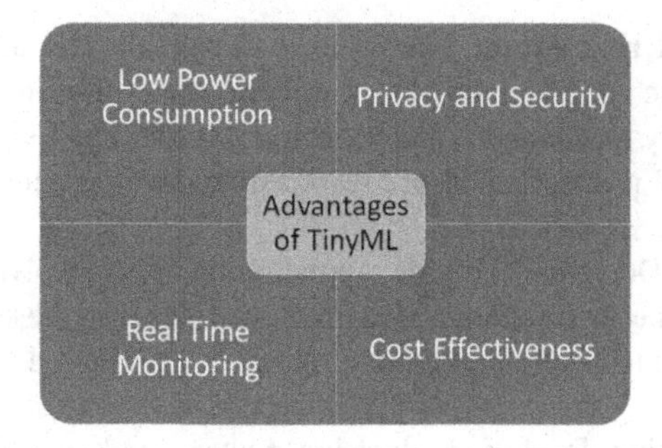

FIGURE 14.8

TinyML: issues and challenges.

Privacy: TinyML devices often use sensitive data collected from sensors or user interactions. Privacy is critical to protecting individuals' personal information, mainly when these devices have limited security measures or data encryption capabilities.
Algorithmic bias: Machine learning models used in TinyML applications can be susceptible to bias in the training data. Biased designs can lead to discriminatory results or reinforce existing social inequalities when used in devices. Addressing and mitigating algorithmic bias is critical to ensuring fairness and equal treatment.
Informed consent: The deployment of TinyML devices in different contexts, such as healthcare or smart homes, raises questions about informed consent. Users must be fully aware of the information collected, the purpose of its use, and the potential risks associated with implementing machine learning algorithms on their devices.
Security risks: TinyML devices often have limited computing power and memory, making them vulnerable to security threats. Ethical issues arise when these devices are compromised, leading to unauthorized use, data breaches, or malicious activity.
Transparency and explainability: Machine learning models used in TinyML tools are typically complex, so understanding their decision-making processes is difficult. Ensuring the transparency and explainability of these models becomes essential for accountability, building trust and identifying potential biases or errors.
Energy efficiency and environmental impact: Since TinyML devices are often battery-operated or have limited energy resources, optimizing their energy efficiency is crucial. Ethical aspects include, among other things, reducing the environmental impact of devices, considering their life cycle and considering electronic waste.
Ownership and control: The deployment of TinyML devices raises questions about the ownership and control of the data collected and the resulting insights. To prevent misuse or abuse, it is essential to ensure that users have control over their data and a say in how they are used. Fig. 14.8 shows the issues and challenges related to TinyML.
Access and equity: Implementation of TinyML in different domains should consider issues of access and equity. Ensuring that the benefits of TinyML technology are available to all individuals and communities regardless of socioeconomic status or geographic location becomes an ethical imperative.

Accountability and regulation: The rapid development and adoption of TinyML devices highlights the need for ethical frameworks, standards, and regulations. Establishing accountability mechanisms and regulatory guidelines can help resolve ethical issues, protect users' rights, and ensure responsible development and deployment of TinyML technology.
Ethical trade-offs: Designing TinyML applications often involves ethical trade-offs between accuracy, energy efficiency, and privacy. Balancing these trade-offs requires careful consideration to ensure that moral principles are adhered to and potential risks are mitigated.

It is imperative that developers, decision-makers, and stakeholders actively engage in discussions about these ethical issues and challenges to promote responsible and ethical deployment of machine learning in resource-constrained devices.

In this section, we first examine how TinyML's features can present particular privacy challenges. In more detail, we go over how TinyML's small size and stealthy nature create issues for informed consent and how the potential for its use in various scenarios forces us to reconsider how we view privacy in terms of the public and private worlds. It introduces Nissenbaum's defense of privacy as a right to contextual integrity as a framework for addressing these difficulties. According to Nissenbaum, every setting is linked to norms that control the proper flow of information and, as a result, our expectations of privacy. Using this analytical framework, students consider how novel practices or gadgets could subvert context-related informational norms.

14.6.1 TinyML's potential challenges in healthcare

TinyML has significant potential in healthcare applications as it offers real-time monitoring, predictive analytics, and personalized care opportunities. However, several challenges and ethical issues arise regarding implementing TinyML in healthcare. Here we list some of them.

Data protection and security: Healthcare data are extremely sensitive and require strict protection measures. To protect patient data from unauthorized access, breaches, and misuse, TinyML needs to be implemented with robust data protection and security measures.
Informed consent: The use of TinyML in healthcare may involve collecting and analyzing personal health information. Obtaining informed consent from patients becomes critical so that they understand how their data will be used, the potential benefits, and any associated risks or limitations.
Algorithmic bias and fairness: TinyML models trained on health data can inherit data biases, leading to potential disparities in care and outcomes. It is essential to intervene in the algorithms, ensure fairness of the model predictions, and minimize the possibility of discrimination between different patient groups.
Transparency and explainability: To establish confidence and responsibility, healthcare practitioners and patients must comprehend how TinyML models produce predictions. Techniques must be created that enable health professionals to analyze and validate the outcomes while also offering explanations or insights into the think-

ing behind the model's choices. The use of TinyML in healthcare should be viewed as a decision assistance instead of a replacement for medical knowledge. Another ethical consideration is ensuring healthcare personnel have the expertise and abilities to analyze and apply the insights offered by TinyML models correctly.

TinyML models must be validated and their effectiveness assessed in healthcare contexts to guarantee their dependability and safety. The need for these models to go through thorough validation procedures and, if required, clinical trials before deployment, particularly for important or high-risk health applications, is raised by ethical concerns.

It might be challenging to establish accountability and culpability when TinyML models are used to make medical decisions. Clear frameworks and norms are required to share responsibility, especially when decisions are made independently using models. When applying TinyML in healthcare, equity and access issues must be considered. Making sure that the advantages of TinyML technology reach various patient populations, including underserved communities, and avoiding escalating already existing healthcare disparities are important ethical considerations.

TinyML models used in healthcare applications must be continually monitored and updated to maintain their efficacy and dependability. Ethical considerations include establishing mechanisms for ongoing monitoring, maintenance, and revising models to reflect patient population changes or medical guidelines.

Regulatory compliance: Healthcare is a highly regulated industry and healthcare settings using TinyML models must comply with relevant regulations such as privacy laws (e.g., HIPAA in the US) and medical devices. Compliance with these regulations is critical to protecting patients' rights and ensuring the responsible use of TinyML in healthcare.

For TinyML to be used effectively in healthcare, collaboration between practitioners, technologists, policymakers, and ethics experts is essential. While utilizing the potential advantages of TinyML to enhance health outcomes, it is critical to prioritize patients' security, privacy, and equity.

14.6.2 TinyML – privacy concerns

Since TinyML requires data processing and analysis on devices with limited resources, privacy is crucial in its implementation. Here we list a few difficulties and things to consider when using TinyML to address privacy concerns.

Data collection and storage: Personal health information and user behavior data are often collected by TinyML devices. Regarding privacy protection, it is essential to collect data only for intended purposes and with the user's consent.

Information minimization: Applications using TinyML must adhere to the information minimization principle by gathering only the data required to complete a task. This helps safeguard user privacy by lowering the possibility of unauthorized access or data breaches.

Secure data transmission: Secure protocols and encryption are required for data transmission from TinyML devices to remote servers or cloud platforms to avoid eavesdropping or unauthorized access while transmission occurs.
Device processing: Data processing and inference can be done on the device to increase privacy, reducing the need to communicate raw data or sensitive information to other servers. Through the provision of valuable insights, on-device processing can aid in protecting user privacy.
User consent and control: Gaining users' informed consent is critical before deploying TinyML. Users must control their data and understand how they are collected, used, and shared. Transparent opt-in and opt-out mechanisms and granular control options empower users to make informed privacy choices.
Anonymization and aggregation: An individual's privacy can be protected by methods such as anonymization and aggregation. Aggregating data from multiple users can provide insight without revealing specific individual data, reducing the risk of reidentification.
Edge computing and local processing: With edge computing infrastructure, data can be processed and analyzed locally on a device or at the network's edge. It reduces the need to send data to remote servers and minimizes data protection risks associated with data transmission and storage.
Privacy by design: It is essential to integrate privacy considerations into TinyML applications from the beginning of their design and development. The system's privacy can be ensured by implementing privacy-enhancing technologies, conducting privacy impact assessments, and adhering to privacy frameworks and regulations.
Secure firmware and software updates: Ensuring the security and privacy of TinyML devices requires regular firmware and software updates to eliminate vulnerabilities and security risks. Timely and secure updates are essential to protect user privacy and combat emerging threats.
Compliance with data protection regulations: TinyML applications must comply with relevant data protection regulations and data protection laws, such as the General Data Protection Regulation (GDPR) in the European Union or the California Consumer Privacy Act (CCPA) in the US. Adherence to these policies protects user privacy and creates a framework for responsible data processing.

Addressing TinyML's privacy concerns requires a privacy-first approach, strong safeguards, a user-centered design, and compliance with privacy regulations. By prioritizing privacy considerations, organizations can build user trust and ensure responsible and ethical adoption of TinyML technology.

14.7 Conclusion

TinyML is still a new field, but its use in healthcare shows great promise for improving patient care. TinyML is used in healthcare for various applications, ranging from RPM to disease detection and personalized medicine.

In addition to reducing hospital visits and improving patient safety, TinyML could enable wearable devices to monitor vital signs and continuously detect anomalies in real time [11]. It is possible to optimize treatment recommendations and drug doses using TinyML models to analyze patient-specific data [8]. TinyML algorithms can analyze medical images, such as X-ray images, CT scans, and MRIs, to assist healthcare providers in making accurate and timely diagnoses [27]. The use of TinyML-powered devices in chronic disease management can improve treatment adherence and health outcomes by monitoring patients with diabetes or asthma and providing feedback [10]. In addition to detecting critical events, TinyML allows immediate intervention, adding value to patient care [28].

TinyML-based assistive technologies can also provide real-time audio feedback to individuals with disabilities, which helps them navigate their surroundings more effectively [29]. In addition to optimizing medicine delivery systems, TinyML can improve medication adherence and reduce side effects by modifying doses and timing based on individual needs [30]. To ensure patient safety and trust in the healthcare system, TinyML must address data privacy, security, and ethical concerns. The full potential of TinyML in healthcare can only be realized by ongoing research and collaboration between technology developers, healthcare providers, and regulators [31].

References

[1] Maha Diab, Esther Rodriguez-Villegas, Embedded machine learning using microcontrollers in wearable and ambulatory systems for health and care applications: a review, IEEE Access (2022) 1–1 https://doi.org/10.1109/ACCESS.2022.3206782.

[2] M. Zennaro, B. Plancher, V. Janapa Reddi, TinyML: Applied AI for Development, https://sdgs.un.org/sites/default/files/2022-05/2.1.3-9-Zennaro-TinyML.pdf, May 2022.

[3] F. Sabry, T. Eltaras, W. Labda, K. Alzoubi, Q. Malluhi, Machine learning for healthcare wearable devices: the big picture, Journal of Healthcare Engineering 2022 (2022) 4653923, https://doi.org/10.1155/2022/4653923. PMID: 35480146; PMCID: PMC9038375.

[4] S. Beniczky, P. Karoly, E. Nurse, P. Ryvlin, M. Cook, Machine learning and wearable devices of the future, Epilepsia 62 (Suppl 2) (2021) S116–S124, https://doi.org/10.1111/epi.16555. Epub 2020 Jul 26. PMID: 32712958.

[5] Vasileios Tsoukas, Eleni Boumpa, Georgios Giannakas, Athanasios Kakarountas, A review of machine learning and TinyML in healthcare, in: 25th Pan-Hellenic Conference on Informatics (PCI 2021), November 26–28, 2021, Volos, Greece, ACM, New York, NY, USA, 2021, 8 pages, https://doi.org/10.1145/3503823.3503836.

[6] R. Sanchez-Iborra, LPWAN and embedded machine learning as enablers for the next generation of wearable devices, Sensors 21 (15) (Jul. 2021) 5218, https://doi.org/10.3390/s21155218.

[7] B. Saha, R. Samanta, S. Ghosh, R.B. Roy, BandX: an intelligent IoT-band for human activity recognition based on TinyML, in: Proceedings of the 24th International Conference on Distributed Computing and Networking, Jan. 2023, pp. 284–285, https://doi.org/10.1145/3571306.3571415.

[8] M.S. Islam, D. Kwak, Edge intelligence in healthcare: tiny machine learning for improving the efficiency of healthcare services, Electronics 10 (6) (2021) 676.
[9] R.M. Umutoni, et al., Integration of TinyML-based proximity and couch sensing in wearable devices for monitoring infectious disease's social distance compliance, in: Proceedings of the 2023 12th International Conference on Software and Computer Applications, Feb. 2023, pp. 349–355, https://doi.org/10.1145/3587828.3587880.
[10] T. Bansal, F. De La Torre, M.M. Ghassemi, S.Z. Masood, Machine learning in COVID-19 ICU risk prediction: a comparative study, IEEE Journal of Biomedical and Health Informatics 24 (10) (2020) 2820–2829.
[11] J. Chiang, Edge A.I.: TinyML as a disruptive healthcare technology, Electronics 10 (4) (2021) 424.
[12] R. Sanchez-Iborra, A.F. Skarmeta, TinyML-enabled frugal smart objects: challenges and opportunities, IEEE Circuits and Systems Magazine 20 (3) (2020) 4–18.
[13] K. Ahmed, M. Hassan, tinyCare: a tinyML-based low-cost continuous blood pressure estimation on the extreme edge, in: 2022 IEEE 10th International Conference on Healthcare Informatics (ICHI), Rochester, MN, USA, 2022, pp. 264–275, https://doi.org/10.1109/ICHI54592.2022.00047.
[14] Barbara Fyntanidou, Maria Zouka, Aikaterini Apostolopoulou, Panagiotis D. Bamidis, Antonis Billis, Konstantinos Mitsopoulos, Pantelis Angelidis, Alexis Fourlis, IoT-based smart triage of Covid-19 suspicious cases in the Emergency Department, in: 2020 IEEE Globecom Workshops (GCWkshps), IEEE, 2020, pp. 1–6.
[15] Prafulla Kumar Padhi, Feranando Charrua-Santos, 6G enabled tactile Internet and cognitive internet of healthcare everything: towards a theoretical framework, Applied System Innovation 4 (2021) 66.
[16] Srihari Yamanoor, Narasimha Sai Yamanoor, Position paper: low-cost solutions for home-based healthcare, in: International Conference on COMmunication Systems & NETworkS (COMSNETS), IEEE, 2021, pp. 709–714.
[17] H.A. Rashid, H. Ren, A.N. Mazumder, T. Mohsenin, Tiny RespNet: A Scalable Multimodal TinyCNN Processor for Automatic Detection of Respiratory Symptoms, https://api.semanticscholar.org/CorpusID:231400757, 2020.
[18] H. Han, J. Siebert, TinyML: a systematic review and synthesis of existing research, in: 2022 International Conference on Artificial Intelligence in Information and Communication (ICAIIC), Jeju Island, Korea, Republic of, 2022, pp. 269–274, https://doi.org/10.1109/ICAIIC54071.2022.9722636.
[19] P.P. Ray, A review on TinyML: state-of-the-art and prospects, Journal of King Saud University: Computer and Information Sciences 34 (4) (2022) 1595–1623.
[20] D. Situnayake, MLOps for TinyML, https://sites.google.com/g.harvard.edu/tinyml/lectures?authuser=0#h.m9uxfxjs8d5u.
[21] C. Tyrchan, E. Nittinger, D. Gogishvili, A. Patronov, T. Kogej, Chapter 4—Approaches using A.I. in medicinal chemistry, in: T. Akitsu (Ed.), Computational and Data-Driven Chemistry Using Artificial Intelligence, Elsevier, 2022, pp. 111–159, https://doi.org/10.1016/B978-0-12-822249-2.00002-5.
[22] D.V.S. Green, Using machine learning to inform decisions in drug discovery: an industry perspective, in: Machine Learning in Chemistry: Data-Driven Algorithms, Learning Systems, and Predictions, vol. 1326, American Chemical Society, 2019, pp. 81–101, https://doi.org/10.1021/bk-2019-1326.ch005.
[23] N. Stephenson, E. Shane, J. Chase, J. Rowland, D. Ries, N. Justice, J. Zhang, L. Chan, R. Cao, Survey of machine learning techniques in drug discovery, Current Drug Metabolism 20 (2019) 185–193.

[24] Andrea Volkamer, Sereina Riniker, Eva Nittinger, Jessica Lanini, Francesca Grisoni, Emma Evertsson, Raquel Rodríguez-Pérez, Nadine Schneider, Machine learning for small molecule drug discovery in academia and industry, Artificial Intelligence in the Life Sciences (ISSN 2667-3185) 3 (2023) 100056.

[25] M. Shafique, T. Theocharides, V.J. Reddy, B. Murmann, TinyML: current progress, research challenges, and future roadmap, in: 2021 58th ACM/IEEE Design Automation Conference (DAC), San Francisco, CA, USA, 2021, pp. 1303–1306, https://doi.org/10.1109/DAC18074.2021.9586232.

[26] A. Dutta, S. Kant, Implementation of cyber threat intelligence platform on Internet of Things (IoT) using TinyML approach for deceiving cyber invasion, in: 2021 International Conference on Electrical, Computer, Communications and Mechatronics Engineering (ICECCME), Mauritius, Mauritius, 2021, pp. 1–6, https://doi.org/10.1109/ICECCME52200.2021.9590959.

[27] M.S. Mahmud, M.S. Kaiser, A. Hussain, Applications of deep learning and reinforcement learning to biological data, IEEE Transactions on Neural Networks and Learning Systems 32 (3) (2020) 865–882.

[28] P. Bhatt, G. Srivastava, Internet of things-based remote health monitoring systems: a review, Journal of Healthcare Engineering 2020 (2020) 1–14.

[29] R. Banerjee, S. Mitra, Artificial Intelligence (A.I.) for assistive healthcare: applications, benefits, and challenges, ACM Transactions on Accessible Computing 14 (1) (2021) 1–29.

[30] Y. Zhang, Z. Luo, Z. Hou, A review of artificial intelligence and machine learning applications in smart drug delivery systems, Pharmaceutical Research 38 (5) (2021) 983–1000.

[31] J.H. Chen, S.M. Asch, T. Delbanco, Unlocking the power of digital health data for better care, The New England Journal of Medicine 381 (19) (2019) 1798–1801.

CHAPTER

15

Machine learning techniques for indoor localization on edge devices

Integrating AI with embedded devices for indoor localization purposes

Diego Méndez, Daniel Crovo, and Diego Avellaneda
Department of Electronics Engineering, Pontificia Universidad Javeriana, Bogotá, Colombia

15.1 Introduction

The Internet of Things (IoT) has become a pervasive technological paradigm, widely applied to many different verticals, such as precision agriculture, retail applications, smart cities, and Industry 4.0, just to name a few examples [1]. As part of the IoT, location-based services play a paramount role in order to support context-aware applications, providing services that react and interact with the users according to the deployed scenario, and how this environment evolves or dynamically changes over time [2].

As stated before, these systems require context awareness, which ultimately implies a need for sensors to characterize and measure the environment where they are deployed. As part of this environment characterization, localization is the most important variable to be estimated. Outdoor localization has been widely studied and developed, and many solutions are available, mostly based on the Global Positioning System (GPS), which provides an acceptable spatial resolution for most applications. However, indoor localization imposes several technical restrictions that cannot be solved following the same GPS-based approach, as will be presented in more detail later [3].

This chapter offers an overview of the technical challenges that need to be overcome in the development of indoor positioning systems and presents different techniques and mechanisms that have been proposed to deal with these issues for indoor localization of objects and people. Most of these implementations rely on a centralized implementation on a *computationally powerful* device, such as personal computers or on-premise servers, as well as cloud-based deployments. However, relying

TinyML for Edge Intelligence in IoT and LPWAN Networks. https://doi.org/10.1016/B978-0-44-322202-3.00020-8

on cloud resources is not always possible as some applications are latency-sensitive (going to the cloud and back would not satisfy such a temporal restriction), do not have enough bandwidth to support this bidirectional communication (as is the case for low-power wide area networks [LPWANs]), or simply do not have an internet connection at all. In addition to this, some solutions require devices to be installed in the object or on the person being tracked, which also imposes specifications in terms of form (light and small), cost (cheap), and power supply (battery-powered).

It is in this complex technical scenario where TinyML, machine learning on tiny devices, is born. TinyML is a new technological paradigm that studies the challenges related to supporting and running these ever-increasingly complex machine learning models on resource-restrained embedded devices, such as small microcontrollers with very limited computational power and memory. This chapter also presents very recent efforts to support machine learning-based indoor localization techniques on these embedded devices [4].

The rest of the chapter is organized as follows. Section 15.2 presents the problem and challenges of indoor localization. Section 15.3 provides a general review of the related work in the area of indoor localization techniques. Section 15.4 details an interesting example of a TinyML-based solution for indoor localization based on fingerprinting of Bluetooth Low Energy (BLE) signals. Finally, Section 15.5 presents the conclusions and general trends in this area.

15.2 The challenges of indoor localization

The outdoor localization problem has been widely studied, and solutions are mainly based on global navigation satellite systems (GNSSs), e.g., GPS, operated by the United States. In the GPS case, a constellation of satellites provides localization (and time) information to a GPS receiver on Earth that utilizes this information to estimate its own location. In some cases, the localization precision can be improved by integrating a known static location that also receives these GPS signals and then transmits correction information to the moving object being tracked, as presented in Fig. 15.1a.

However, these GPS-based solutions suffer from the line of sight (LoS) problem, reducing its effectiveness when it is not possible for the receiver to *see* at least four satellites to calculate a precise location. This situation worsens when the application requires to track an object in an indoor location. In most cases, the strength of the GPS signals is too low to penetrate building structures; hence, the receiver detects even fewer satellites, if any (see Fig. 15.1b). With this in mind, for an indoor localization scenario, a new positioning infrastructure is needed as devices will not be traceable using a typical GNSS approach.

This new infrastructure for indoor localization application requires the deployment of devices (beacons, access points, etc.) that would either advertise themselves or collect advertising information from other devices. With this advertising information, there are two approaches to solve the indoor localization problem: range-based or range-free.

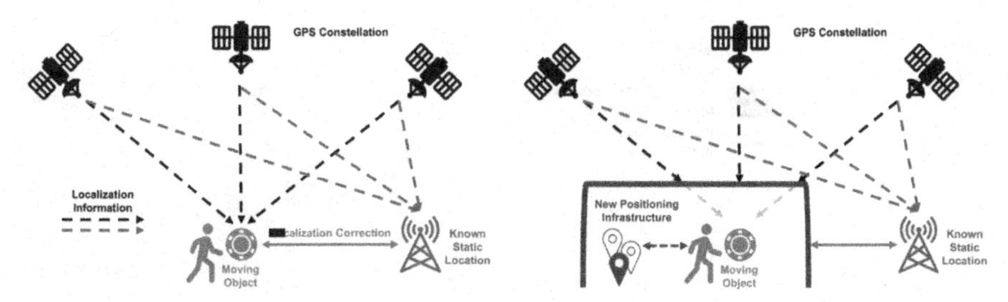

(a) The outdoor localization scenario using the Global Positioning System (GPS).

(b) The indoor localization problem where GPS signals are not received.

FIGURE 15.1

The outdoor vs. the indoor localization problem.

In a range-based approach, the advertised information is used to determine the distance between the transmitter(s) and the receiver(s), but due to the many different considerations (obstructions, multi-path packets, etc.), this approach can suffer from low precision. In range-free systems, the localization is estimated by characterizing the environment where the devices or objects need to be tracked [5], which also implies an enormous effort. Next, one solution for each of these approaches will be presented: trilateration and fingerprinting.

15.2.1 Trilateration

Trilateration is a geometric-based method to compute the unknown relative or absolute position of an object. It requires at least three reference points with known coordinates and the distances from these to the target. Using the distances as the radius, it is possible to form circles and the unknown coordinates will be the overlapping point between them. For indoor positioning systems, anchor nodes, also called beacons, are deployed at fixed locations and the mobile node or tag needs to be tracked. Generally, these devices communicate through a variety of wireless protocols and there are different techniques to measure or infer the distances between them such as angle of arrival, time of arrival, time difference of arrival, and received signal strength indicator (RSSI). RSSI-based approaches based are widely utilized due to their computational simplicity, availability in different wireless technologies, and low cost of deployment.

To illustrate how this method works, consider the standard trilateration setup shown in Fig. 15.2. In this particular implementation, the mobile node continuously broadcasts packets (dashed gray arrows) and the anchor nodes $[ST_1, ST_2, ST_3]$ are receiving them. The anchor nodes utilize the received RSSI value to compute the distance to the mobile node, by solving Eq. (15.1), which is derived from the log path loss model and denotes the power law relationship between the distance and the

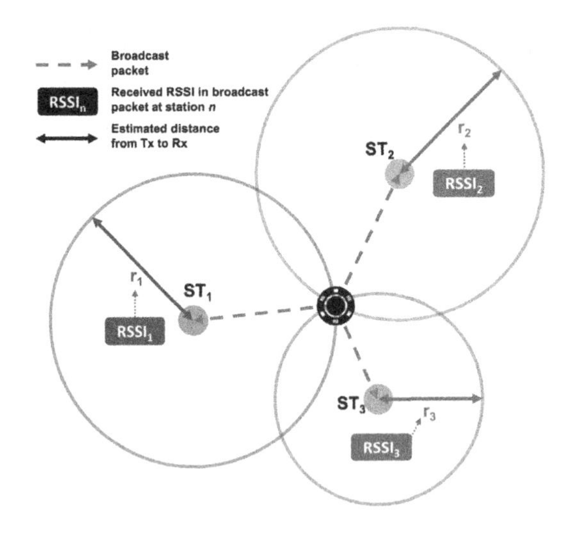

FIGURE 15.2

A simple representation of the trilateration technique.

received power:

$$RSSI_{dB}(d) = RSSI_{dB}(d_0) - n * 10 * \log_{10}\left(\frac{d}{d_0}\right), \tag{15.1}$$

where n is the path loss exponent, d_0 is the distance from the referenced point to the common transmitter, and $RSSI_{dB}(d)$ is the RSSI value at d_0 distance to the transmitter [6].

Once distances r_1, r_2, and r_3 are determined, the location of the mobile node is found by calculating the coordinates (x, y) of the overlapping point of the three circles formed from the radii $[r_1, r_2, r_3]$. Denoting the anchor node coordinates as (x_1, y_1), (x_2, y_2), and (x_3, y_3), the mobile node position can be computed by solving the following set of equations:

$$(x - x_1)^2 + (y - y_1)^2 = r_1^2, \tag{15.2}$$

$$(x - x_2)^2 + (y - y_2)^2 = r_2^2, \tag{15.3}$$

$$(x - x_3)^2 + (y - y_3)^2 = r_3^2. \tag{15.4}$$

Regardless of its simplicity and low cost of implementation, the trilateration method using RSSI has shown limitations. When in the presence of objects and environmental interference, the RSSI measurements are not stable and have high variance, leading to inaccurate RSSI values, which in turn generate events where there is not a single overlapping point but an enclosed region where the circles or spheres intersect or they might not intersect at all.

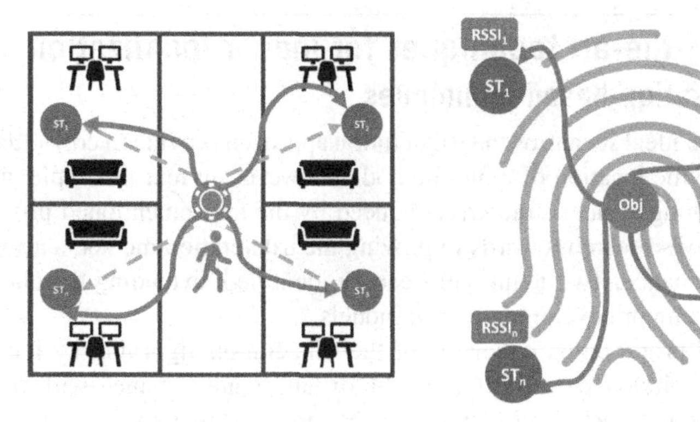

FIGURE 15.3

The general concept of radio fingerprinting.

15.2.2 Fingerprinting

Since estimating distance based on the log path loss model has been proven not to be very precise, a different approach is followed in the fingerprinting technique. A mobile node continuously broadcasts packets (dashed gray arrows), but since there are many obstacles (walls, furniture, humans, etc.) the LoS is not possible and the received RSSI will not help much in estimating the distance between transmitter and receiver.

Moreover, the received RSSI values may vary a lot with small variations in the location of the mobile node. However, this inner problem of the space and high variance of the received RSSI values becomes an advantage under the fingerprinting approach. Different locations generate very different sets of received RSSI values, almost unique (a *fingerprint*), for each location (as represented in Fig. 15.3).

With that in mind, it is possible to associate (label) each set of received RSSI values $[RSSI_1, RSSI_2, RSSI_3...RSSI_n]$ for that particular location, in order for an intelligent system to learn each of these fingerprints. After learning each of these fingerprints, a model of the space is created. Now, by using this model, a new collected RSSI can be *classified* to a particular location (label class).

As will be presented next, many indoor localization techniques have been designed, implemented, and tested, but most of them rely on the computational capabilities of the cloud, an on-premise data center, or a desktop. However, relying on these devices might sometimes not be possible because of a poor internet connection (if any), low-latency requirements, or power-limited applications. With this in mind, TinyML becomes a feasible solution to implement these intelligent systems on the edge.

15.3 State-of-the-art techniques for indoor localization

15.3.1 Trilateration-based techniques

As stated before, in ideal scenarios the trilateration approach provides a cost-effective means to estimate the location of a mobile node. However, in real-life deployments for indoor positioning systems, the error induced by the abovementioned problems has driven the active research towards improving the trilateration method's accuracy and overall performance. Two main paths can be identified: extending the classical trilateration algorithm or developing hybrid models.

Ibwe et al. [7] proposed an extension of the trilateration algorithm by incorporating an adaptive circle expansion stage. The distances are obtained with the log path loss model from the RSSI measurements for various wireless protocols. Subsequently, the circle expansion stage verifies if the intersection of the circles exists; if not, the smallest radius is multiplied by a factor m until the circles have one overlapping point. Experimental evaluations are conducted in two different environments and with different wireless technologies. The average mean squared error values are 0.719 m for BLE, 0.517 m for WiFi, 0.793 m for LoRaWAN, and 0.741 m for ZigBee, reducing the positioning error by 4%, 22%, 33%, and 17%, respectively, compared to the classical trilateration approach.

Implementing hybrid models is an alternative to improve the performance of the trilateration method. The two principal methodologies are applying optimization techniques or using machine learning models. Yang et al. [8] applied a Gaussian filter to reduce the measured noise of the raw RSSI values. To take into account specific environmental factors, the path loss exponent and transmit power of the log path loss model are obtained by a least square curve fitting algorithm which usually is assumed as a fixed value based on empirical models. Afterwards, a non-linear error function is formulated from the standard trilateration so that the minimum value is the target node coordinates. Finally, the error function is minimized using a Taylor series approximation. The experimental setup consisted in four anchor nodes using the CC2530 module as a transmitter and one target node acting as the receiver with the same module and performing the calculations to estimate its position.

An extensive study by Kia et al. [9] proposed a three-phase approach for indoor localization by fusing range measurements from WiFi and UWB devices. In the first phase, position-dependent errors in WiFi RTT range measurements are investigated using an anechoic chamber. The second phase focuses on observing the environment and extracting Gaussian distributions to model device behavior. Finally, in the third phase, range fusion is achieved by combining calibrated WiFi RTT and UWB two-way ranging (TWR) ranges using a fusion algorithm based on Gaussian process model parameters and RSSI values. The proposed approach proved to improve positioning accuracy by leveraging the strengths of both technologies was tested on commercially available Pozyx Creator tags.

Luo et al. [10] proposed a hybrid model which groups the anchor nodes into N sets of three nodes and calculates the distances from the mobile node to the anchor nodes based on the RSSI value and the log path loss model. Subsequently, it computes the coordinates of the unknown node using trilateration, generating N sets of positions that are fed into a K-means clustering model to yield k disjoint clusters and remove position results with significant errors. Finally, considering that locations with high error values will warp the cluster center during clustering, the coordinates with the least error correspond to the maximum member number. Simulation results demonstrated that the proposed model outperformed the classical trilateration method, reducing the average position error from 2.05 m to 1.47 m and improving the accuracy by 27.9%.

In general, users carry the target nodes in their pockets, bags or hands, inducing high variability in the measured wireless signals characteristics used to calculate distances from anchor nodes to target nodes, therefore increasing the error in the estimated position. To overcome this problem, Sung et al. [11] developed a UWB-based system. UWB uses TWR technology to calculate distances between transmitters and receivers by measuring the time of flight. The whole system comprises three fundamental components: a neural network, a Kalman filter, and a standard trilateration coordinate estimation algorithm. The artificial neural network (ANN) classifier uses the channel impulse response (CIR) and the calculated distances as input features, to classify between the no-LoS scenarios (pocket, bags, hands) and LoS. The Kalman filter-based distance error correction receives the probability information from the ANN and distance data to adjust the Kalman filter parameters. The proposed system was implemented in a NUCLEO F429ZI development board with a DWM3000 UWB module. It was tested with three anchor nodes and one mobile node, and the experimental results showed an accuracy improvement of 20.84% to 27.22% with an error tolerance of ± 25 cm compared to the traditional trilateration method.

The work in [12] presents a machine learning approach to estimate distances from anchor nodes to the target node by using an ANN in which the inputs are the RSSI values, and then the 3D coordinates of the target are calculated with trilateration. The whole system was implemented on the edge using an ESP32. The proposed system's maximum error does not exceed 0.6 m and the execution time of the calculations is 40 ms.

15.3.2 Fingerprinting-based techniques

Fingerprinting-based techniques can be implemented with multiple wireless technologies, such as WiFi, UWB, RFID, ZigBee, and Bluetooth. Besides, RSSI and channel state information (CSI) are the prevalent signal characteristics for generating fingerprints. Optimization algorithms such as particle swarm optimization (PSO) can be used to determine the positions of mobile nodes; for instance, a WiFi RSSI-based system has been proposed by Zheng et al. [13] in which a two-panel fingerprint homogeneity model evaluates fingerprint similarity of the cosine distance and the Euclidean distance, constraining the bias in fingerprint similarity characterization. Based on the previous stage, an objective function is formulated and optimized with PSO to find the optimal coordinate for the mobile node. The experimental validation was conducted with 10 WiFi APs and 187 reference points separated 1 m apart from each other which were measured to generate the fingerprints. The proposed method obtained a root mean squared error (RMSE) of 2.45, outperforming KNN, SVM, linear regression, and random forest by 11.25%, 16.25%, 33.56%, and 36.76%, respectively. Although the experimental results showed a substantial improvement compared to machine learning models, the positions were calculated offline on a computer with 16 GB of RAM and an Intel Core i7 CPU.

Generally speaking, fingerprinting indoor positioning systems (IPSs) via optimization are more computationally complex and require more time to calculate positions compared to machine learning approaches. To this end, machine learning algorithms are being extensively studied for IPSs because of the resource constraints of edge devices.

Cross-device fingerprint compatibility is one of the challenges that IPSs have to overcome, considering that the offline phase of a fingerprinting-based technique is usually performed with a specific device and that real-life scenarios comprise a variety of target devices with different communication modules, introducing random variances in RSSI measurements due to the heterogeneity of IoT and edge devices.

Zhou et al. [14] argued that most of the IPS-related works focused on improving localization accuracy and did not take into account the diversity of WiFi RSSI data; therefore, they designed a hybrid system performing a series of statistical tests on the WiFi signal distributions to account for every AP contribution degree on the target node location, and then they used a KNN classifier which uses the previous contribution degree as a weight during calculation to find the matching reference point. The system was implemented as an Android application and tested with four Samsung S7568 smartphones. The proposed approach considering AP contribution degrees achieved a mean error of 1.85 m and the baseline method obtained a mean error of 2.16 m.

Ye et al. [15] proposed EdgeLoc, a real-time and high-accuracy system which aimed to overcome the abovementioned challenge. EdgeLoc data flow consists of two computation components: edge server and edge devices. The edge server stores the RSSI fingerprint dataset and processes the data into an image-like matrix to train the CapsNet module, a convolutional neural network model. Edge devices collect RSSI data and perform lightweight data preprocessing and location calculations. Besides, the edge devices in the localization step download the trained weights of the model from the server to perform real-time position estimation. A Raspberry Pi and three Android phones were employed as edge devices, and a laptop was used as the edge server. The authors performed several experiments; using their own database, the system achieved a localization accuracy of 96%, and when tested on the UJIIndoorLoc dataset, which contains samples collected from 25 different Android devices, the proposed system achieved an average localization error of 7.93 m, compared with 8.10 m for the KNN method.

Altaf et al. [16] developed a system introducing a bag-of-features (BoF) model for data preprocessing, which generates a vocabulary of features by clustering with a K-means clustering model from the raw WiFi RSSI data collected, which allows to achieve noise robustness and improves the distinction between fingerprints. The vocabulary is used to create feature vectors for training and testing the KNN classification model. A Huawei KIW-L21 smartphone ran an application with the implemented system, and the experimental results showed that the proposed system achieved a mean error of 1.58 m, whereas the baseline KNN method yielded a mean error of 1.77.

Although WiFi is the most researched technology for developing fingerprint-based IPSs because of the availability of already deployed APs and because it is present in almost every edge device, Bluetooth-based IPSs are a reliable approach to implement IPSs. For instance, Aranda et al. [17] conducted a performance evaluation of different machine learning algorithms for fingerprint-based systems utilizing iBKs Plus BLE beacons and five different smartphones as target nodes in the fingerprint collection process. Moreover, the authors evaluated eight different models against three available BLE datasets from previous works [18–20]. The model with the least average normalized error was a neural network classifier (0.68), followed by SVM, which obtained an average normalized error of 0.75.

BLE is substantially superior to WiFi regarding power consumption and is cheaper to implement. Besides, it can be available in smaller devices with less computational power. Polak et al. [21] developed an RSSI fingerprint BLE-based IPS with STM32 MCUs and BL652 modules for the anchor nodes and target nodes. A Raspberry Pi communicates through ZigBee with the nodes and acts as a gateway. The authors implemented four machine learning classifier models: KNN, SVM, random forest, and MLP. Although the best model was the random forest, with an accuracy in the test set of 0.9995 and a mean error of 0.0019 m, all models obtained a mean error of less than 0.022 m.

As a summary, Table 15.1 presents and compares the most relevant related work in the area of indoor localization.

15.4 A tiny indoor localization implementation

As presented before, the work devoted to indoor localization has proposed and implemented different techniques, which mostly depend on the cloud or desktop computers. However, depending on these computationally powerful devices is not always feasible; hence, two different implementations of fingerprinting indoor localization will be presented now, both based on the TinyML paradigm.

As a new infrastructure is normally necessary to support indoor localization systems, there are two main network architectures to implement this: (1) intelligence at the edge infrastructure and (2) intelligence at the mobile device. In the former, the edge infrastructure is in charge of the collection, centralization, and estimation processes in order to locate of the mobile device. In the latter, the mobile device is in charge of collecting the messages coming from the infrastructure and estimating its own location. Fig. 15.4 is a graphical representation of these architectures.

Table 15.1 Summary of the most relevant related work in the area of indoor localization.

Work	Technique	Wireless Technology	Deployment	Comments
Ibwe et al. [7]	Trilateration-RSSI	WiFi, BLE, ZigBee, LoRaWAN	Offline computation	The circle expansion algorithm improves the accuracy of the classical trilateration algorithm.
Yang et al. [8]	Trilateration-RSSI, Bayesian filtering	ZigBee	Edge	The parameters of the log path loss model are obtained via least square curve fitting. The trilateration technique is improved based on the extreme value theory and Bayesian filtering.
Kia et al. [9]	Trilateration, RSSI, GP, RTT, TWR, and ML	WiFi, UWB	Edge	WiFi and UWB range fusion based on machine learning and Gaussian process for range correction, the position is estimated with trilateration.
Luo et al. [10]	Trilateration, RSSI, K-means	Not specified	Computer simulation	Hybrid trilateration approach with K-means clustering. Computationally expensive, long calculation time.
Sung et al [11]	Trilateration, TWR, machine learning, CIR, Kalman filtering	UWB	Edge	A hybrid approach consisting of an ANN classifier to discern between LoS and no-LoS scenarios, Kalman filtering for distance correction and trilateration to calculate location.
Zheng et al. [13]	Fingerprint, RSSI, PSO	WiFi	Cloud	An optimization-based approach.
Zhou et al. [14]	Fingerprint, RSSI, machine learning	WiFi	Edge smartphone	Aims to reduce the variability induced by the heterogeneity of IoT devices using statistical tests on WiFi signal distributions. A KNN classifier is used to estimate localization.
Ye et al. [15]	Fingerprint, RSSI, ML	WiFi	Edge smartphone	Converts the RSSI data into an image-like matrix to use convolutional neural networks.
Altaf et al. [16]	Fingerprint, RSSI, machine learning	WiFi	Edge smartphone	Bag-of-features for data preprocessing and creating feature vectors for the KNN classifier.
Aranda et al. [17]	Fingerprint, RSSI, machine learning	BLE	Edge smartphone	A performance evaluation of eight different machine learning models.
Polak et al. [21]	Fingerprint, RSSI, machine learning	BLE	Edge	An STM32-based implementation.

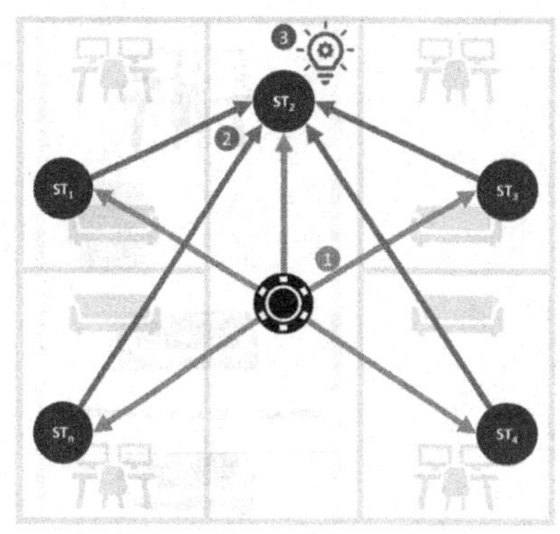

Intelligence at the Edge Infrastructure

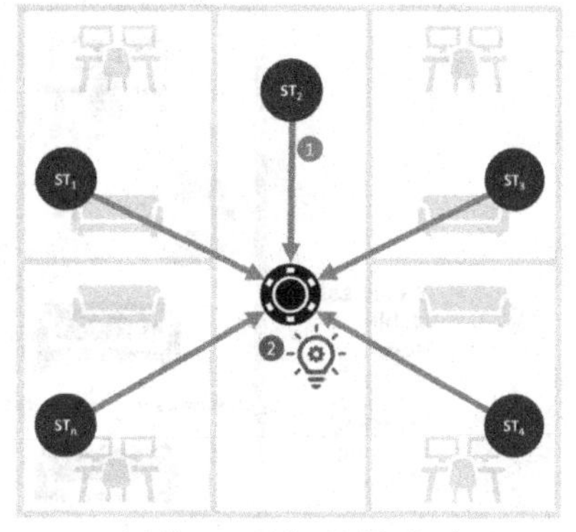

Intelligence at the Mobile Device

FIGURE 15.4

Two different network architectures for fingerprinting on the edge.

The process of fingerprinting when the intelligence is deployed at the edge infrastructure can be summarized as follows (see Fig. 15.4):

1. The mobile device sends broadcast packets (e.g., BLE or WiFi). As part of these messages, all scanning stations (the infrastructure) receive a corresponding RSSI. (first step – green arrows; light gray in print version)
2. All scanning stations centralize the received RSSIs to a master scanning station (marked with the orange bulb; dark gray in print version). (second step – blue arrows; mid gray in print version)
3. The master scanning station consolidates the RSSI values received by all the scanning stations and estimates the location of the mobile device, applying the previously trained model based on fingerprinting. (third step – orange bulb; dark gray in print version)

On the other hand, the process of fingerprinting when the intelligence is deployed at the mobile device can be summarized as follows (see Fig. 15.4):

1. The scanning stations are constantly sending broadcast packets, and the mobile device is now in charge of collecting all these messages and the corresponding RSSI values. (first step – green arrows; light gray in print version)
2. As the mobile device now has all the RSSI values from all the incoming broadcast messages, it is now possible to apply the previously trained model based on fingerprinting. (second step – orange bulb; dark gray in print version)

   - *Although this approach seems better (fewer steps), nonetheless it imposes stronger computational and memory requirements on the mobile device, which is normally battery-powered.*

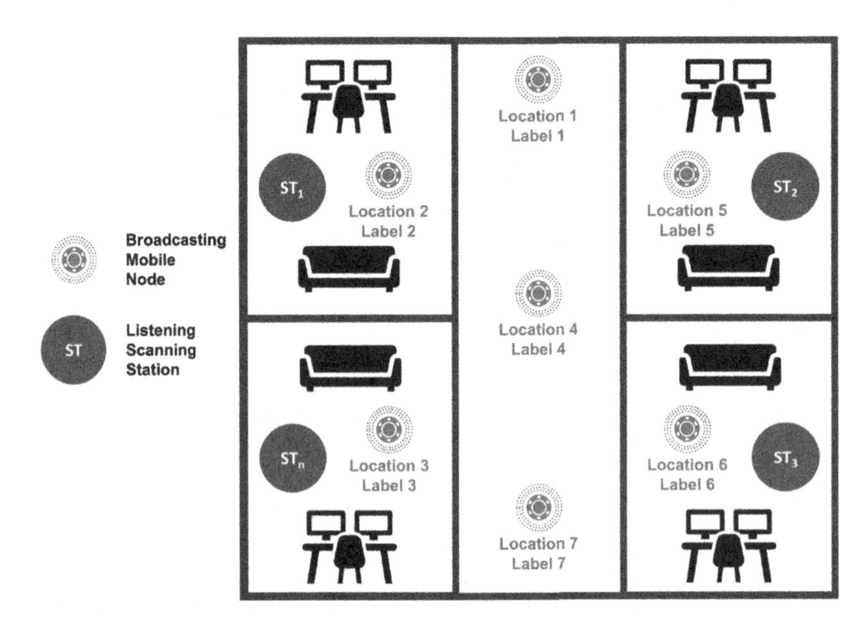

FIGURE 15.5

A graphical representation of the fingerprinting technique utilizing a classification approach. For each location, training data must be collected and properly labeled.

15.4.1 Intelligence at the edge infrastructure – classification approach

For the first case of intelligence at the edge infrastructure, a classification approach has been proposed and implemented. As for any machine learning-based system, a first data collection stage is required, in which the mobile device is located in different locations (while continuously broadcasting) and the received data are labeled accordingly at each scanning station, as presented in Fig. 15.5. As explained before, the scanning stations centralize this information in order to train the model at the mast scanning station. As this is a classification approach, each location will correspond to a label and to an output class [22].

For this system, Edge Impulse[1] has been selected as the framework to support all the development stages: data collection, model selection, training, testing, and deployment. Edge Impulse is a cloud-based machine learning operations (MLOps) platform for edge intelligence, easily allowing the collection of data utilizing the embedded devices, and later also supporting the deployment process [23].

[1] Edge Impulse website: https://edgeimpulse.com.

15.4.1.1 The proposed classification solution

The proposed system has been implemented utilizing an HM-10 module for the BLE-based tag device that will be carried by the object to be tracked. The scanning stations are based on two modules, an ESP32 and an ESP8266, to simultaneously support BLE (collect RSSI values) and WiFi (centralize packets to the mast scanning station) communication. The scanning stations, which is in charge of collecting all the RSSI values and infer the object's location, has been implemented on a Raspberry Pi Zero W.

In total, five scanning stations were implemented and deployed in a two-bedroom apartment (approx. 80 m^2). The exact location of the scanning stations is irrelevant as the system does not estimate distance with the RSSI values; nonetheless, it is ideal to uniformly distribute them in the area under analysis, in order to increase the probability that each location is being covered by more than one scanning station. Data were collected and labeled in 26 different locations in a small apartment. While collecting data from each location, the tag device was manipulated in different ways, in order to introduce noise to the measurements, modeling a more realistic application scenario.

15.4.1.2 The ML model on Edge Impulse

After being collected, all the information was loaded to Edge Impulse in JSON format. The RSSI measurements coming from each scanning stations were considered as a time series, a moving average filter was implemented at the device to soften the signals, and a window size of eight measurements was defined as an input for Edge Impulse. Fig. 15.6a shows an RSSI sample loaded to Edge Impulse, where the behavior of the signals is analyzed for each corresponding scanning station.

A flatten preprocessing stage is defined in Edge Impulse in order to calculate different features for each sample. The flatten module calculates the following features for a time series: root mean square, standard deviation, skewness, average, minimum, maximum, and kurtosis. A particularly useful functionality available in Edge Impulse is the feature explorer, where the user can easily visualize the behavior of each of the calculated features.

Fig. 15.6b presents the behavior of the average (one of the seven features calculated by the flatten module) of the RSSI value from three scanning stations (only three because of visualization limitations) and for a subset of four different locations ([3, 1], [3, 2], [4, 1], [4, 2]). Even though all the selected locations belong to the same small space (approx. 2.38 m × 2.18 m), the feature point cloud show a clear clustering of points, and therefore separability of classes. This visualization option allows the designer to easily test different preprocessing techniques and estimate its performance before training the model.

Fig. 15.7 presents the final version of the machine learning designed in Edge Impulse. As a classification approach has been selected, a deep neural network was implemented with the following structure:

- an input layer with 35 neurons: 5 times series (one per scanning station) and a flatten module that generates 7 features for each time series;

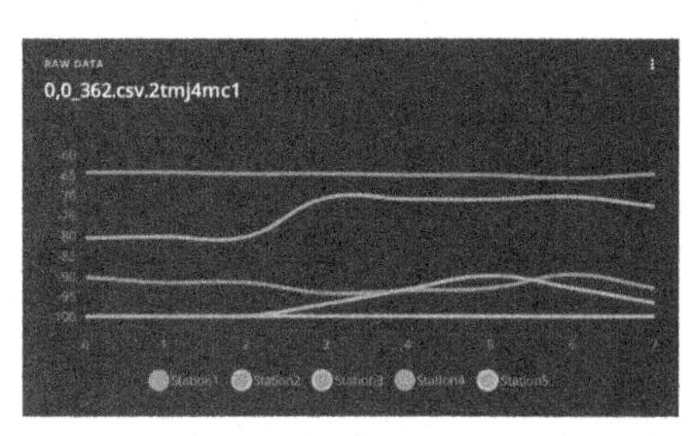

(a) A sample loaded to Edge Impulse. Each line is a time series of eight RSSI values acquired, and the system integrates five scanning stations.

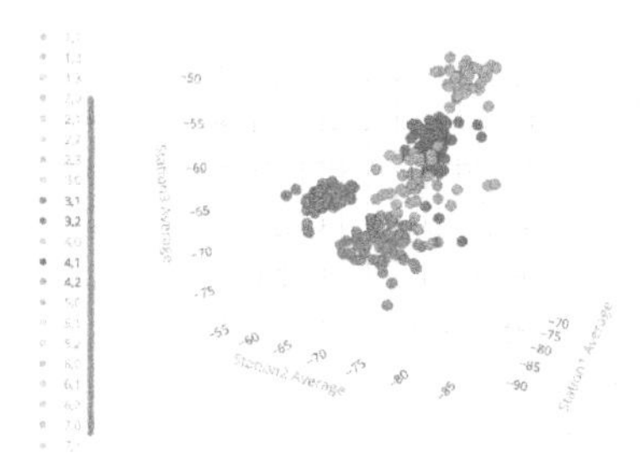

(b) The behavior of the average of the RSSI values received at three different scanning stations for four different locations.

FIGURE 15.6

A sample loaded to Edge Impulse and the feature explorer for four locations.

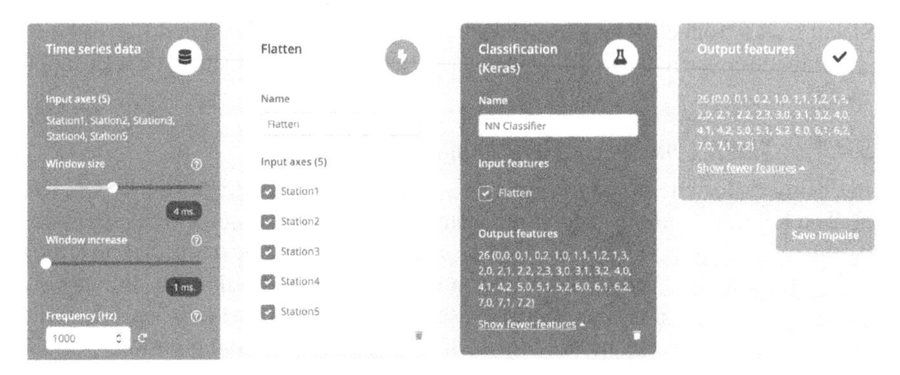

FIGURE 15.7

The impulse (machine learning model) designed on the Edge Impulse platform.

- two hidden layers: one dense hidden layer with 10 neurons and a second dense hidden layer with 20 neurons;
- an output layer with 26 neurons: one output neuron for each of the 26 labeled locations.

15.4.1.3 Results and discussion

After being trained, the machine learning model achieves an accuracy of 94% for all 26 classes with the validation set on the Edge Impulse platform. Fig. 15.8 shows the feature explorer on the validation set, where items in green (light gray in print version) are classified correctly, while items in red (dark gray in print version) are misclassified.

To really measure the final performance of the system, the model is exported and deployed to the master scanning station. Moreover, during the deployment, new samples were acquired by the system in order to utilize completely new data under real operational conditions. During this data collection process, the tag was again

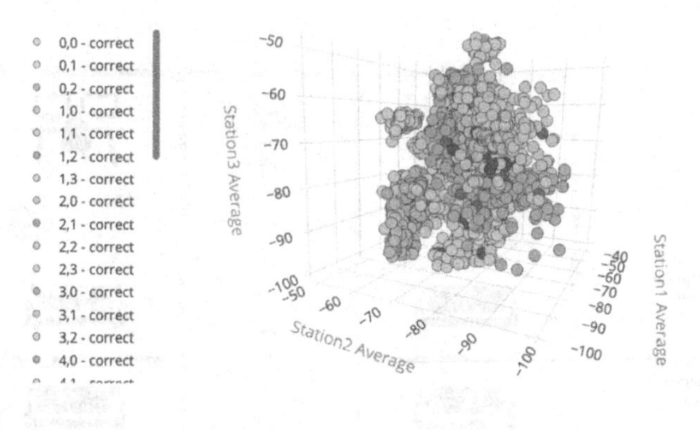

FIGURE 15.8

Feature explorer for the subset of samples on the validation set. A final accuracy of 94% was achieved on the Edge Impulse platform.

manipulated, simulating a more realistic scenario. A final accuracy of 88% has been achieved with this system running on the master scanning station.

Considering that for most indoor applications the traceable objects will not move very fast in the area under analysis, e.g., a person walking, the errors made by the system were analyzed. After realizing that the errors made by the system appear in the middle of many properly classified locations, while the tag is not moving, a postprocessing stage has been proposed. A simple median filter was implemented in the last five classification outcomes. With this simple filter, the system's accuracy increases to 94%.

Although the final implementation achieves a very high accuracy, it comes with the burden of a completely new infrastructure deployed in the area of interest. In addition to this, the infrastructure has to implement a sophisticated communication mechanism to allow the collection of the RSSI values and their later centralization for inference purposes at the master scanning station. Nonetheless, this cost at the infrastructure reduces the requirements on the tag device, as it only needs to send broadcast packets in order to be located, which is paramount as these mobile devices are normally battery-powered.

15.4.2 Intelligence at the mobile device – regression approach

The classification approach presented before achieves a very high accuracy during validation and deployment. However, it is impossible to estimate the location of the device in locations outside the previously defined set of labeled locations. With that in mind, a regression approach is now proposed and implemented. Fig. 15.9 shows a graphical representation of the regression-based mobile-centric system for fingerprinting indoor localization, where it is now the stations that work as beacons and the mobile device is in charge of listening for broadcast packets to create the fingerprint.

In this case, the goal is to train a model capable of generating a continuous output for the (X, Y) coordinates of the mobile devices. Although it is still necessary to label

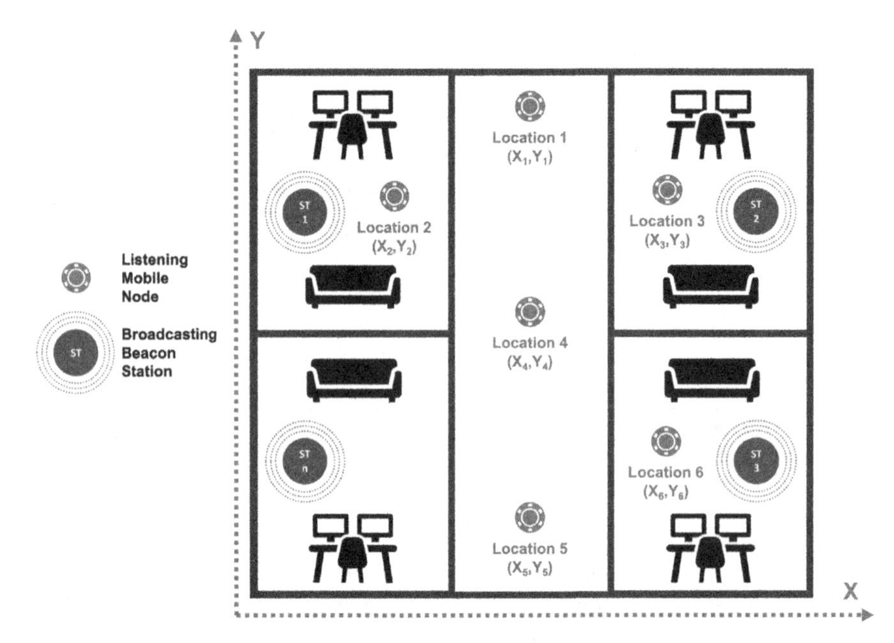

FIGURE 15.9

A graphical representation of the fingerprinting technique utilizing a regression approach.

the data collected by the mobile device, the label is now composed of the (X, Y) coordinates for each labeled location. In contrast to the classification-based systems, this regression approach would allow to estimate the location of the mobile device in locations different to the ones used for training.

15.4.2.1 The proposed regression solution

The proposed system was implemented utilizing five Arduino Tiny Machine Learning Kits[2] (integrating an Arduino Nano 33 BLE board) for both the beacons and the mobile device. Four boards were set to operate as beacons with fixed locations, which are continuously broadcasting packets. The fifth board operates as the mobile device to be tracked and constantly scans for the IDs of each beacon node.

By collecting the available broadcast packets, the mobile device constructs a set of RSSI values, coming from the beacon devices. During the data collection stage, for each labeled location, the mobile node stores the measurements (RSSI values and beacon IDs) in a plain text file on the available flash memory, every 100 ms for 3 minutes and 30 seconds.

[2] Arduino Tiny Machine Learning Kit website: https://store.arduino.cc/products/arduino-tiny-machine-learning-kit.

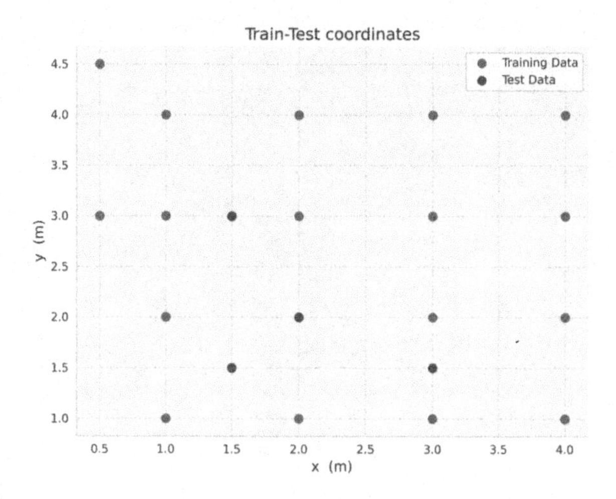

FIGURE 15.10

Location coordinates where the mobile device performed measurements. Blue dots (mid gray in print version) indicate the locations and measurements used for training and testing, while the red locations (dark gray in print version) were used for validation purposes.

The whole system was deployed in a small 35-m^2 studio apartment, where the beacons were uniformly distributed and installed on walls. The mobile device collected the RSSI values from various locations in a 2D 1 m × 1 m grid. Fig. 15.10 is a graphical representation of the location of the devices, in which blue dots (mid gray in print version) indicate the points used for training and the red ones (dark gray in print version) were used for validation purposes. A moving average filter with a window size of 1 s was applied to soften the RSSI signals and to reduce the induced error from outliers. An example of the resulting waveforms is depicted in Fig. 15.11.

15.4.2.2 Machine learning regression model analysis

In order to estimate the location of the mobile device as a continuous value, it is required to predict at least two target variables, i.e., the (X, Y) coordinates. This problem is commonly known as multivariate regression. Considering that the Edge Impulse platform has limited availability of regression models and does not include a multivariate regression model, an analysis of different algorithms was conducted utilizing scikit-learn.[3]

The preprocessed data were used to train the following basic models: multivariate linear regression, decision tree regressor, random forest regressor, XGBoost regressor, and an ANN. The training set had 30,631 data samples and the test set had 7463 samples (80/20 split). Various experiments were performed, with different numbers of hidden layers and neurons, and the best-performance architecture was as follows:

[3] scikit-learn website: https://scikit-learn.org/stable/.

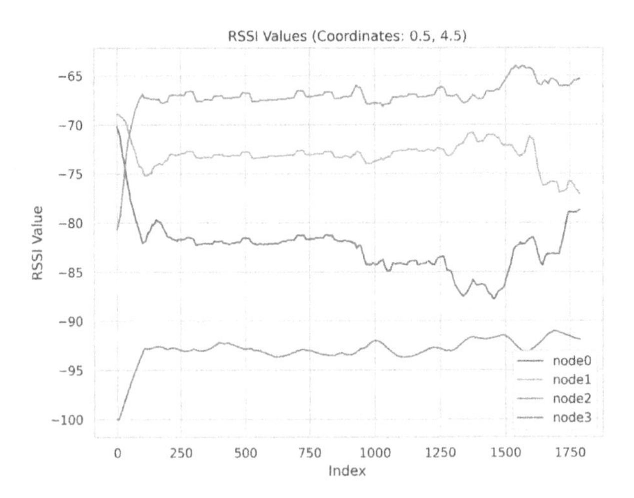

FIGURE 15.11

An example of the softened RSSI values received from the available four beacon nodes, after applying a moving average filter.

- an input layer with four neurons for the RSSI values of the four beacons,
- four dense hidden layers with 128, 64, 32, and 16 neurons,
- an output layer with two neurons for the (X, Y) coordinates,
- the loss function was the mean squared error,
- Adam optimizer was selected.

Since the data type that the system produced was in the form of a time series, a gated recurrent unit recurrent neural network (GRU) was implemented. GRUs are specific neural networks designed to work on time-series data, which are similar to long short-term memory (LSTMs) but with fewer parameters, making them a good fit to be implemented in resource-constrained edge devices.

To be able to use the collected data in a GRU model, sequences of a determined size need to be generated. For this particular case, the best result was obtained with a sequence length of 20 samples, that is, 2 s of RSSI measurements. Moreover, various iterations were done to evaluate which architecture of the GRU performed better, and the following configuration was found to be the best:

- Two GRU units with 64 neurons each. One acts as the input layer and receives the sequence vector with the RSSI values. The second is a hidden layer to improve the performance of the whole model.
- Two dropout layers with probability 0.2. This reduces overfitting of the model and in general reduces error in data that the model has not yet seen.
- A dense output layer with two neurons to predict (X, Y) coordinates.

15.4.2.3 Results and discussion

The models were trained and afterwards evaluated with the validation set, as depicted in Fig. 15.10. Considering that the goal is to be able to predict the location as a continuous value, the coordinates of the test set had values in intermediate locations of the 2D grid, e.g., at (1.5, 1.5). The RMSE for each model was:

- linear regression: 0.63 m.
- decision tree: 0.81 m.
- random forest: 0.73 m.
- XGBoost: 0.95 m,
- ANN: 0.62 m,
- GRU: 0.58 m.

As expected, neural networks yielded better results than the basic machine learning models. The GRU model outperformed the ANN because of the nature of the data being a time series. However, the linear regression algorithm obtained a close RMSE value, and it is simpler and lightweight and has low inference time, making it a good alternative model to deploy in resource-constrained devices. In contrast, ANN models are in general more computationally expensive, require more storage space, and have longer inference times.

15.5 Conclusions

Indoor localization is a paramount component of many IoT-based applications, such as home automation, retail, warehouse logistics, and human tracking, to name a few. However, GNSS-based solutions are not feasible as satellite signals are not available in these cases. For such scenarios, with newly deployed infrastructure, trilateration and fingerprinting appear as possible solutions.

However, distance estimation in indoor environments has proven to have low precision, limiting the application of the trilateration technique. Nonetheless, fingerprinting-based solutions offer a better performance, as they do not depend on the estimation of distance but simply on the association of a set of measurements (mostly RSSI) with a certain location.

For fingerprinting to work, the consolidation of these RSSI values is necessary, which can happen on the edge infrastructure or the mobile device being tracked. Two TinyML-based solutions have been presented: (1) one based on intelligence at the edge infrastructure with a deep neural network classification technique and (ii) another based on intelligence at the mobile device, but implementing a regression mechanism on the embedded device. Although the best performance is achieved with the classification-based solution, the location estimation is restricted to the places where the data were collected and labeled. On the other hand, the regression approach offers a lower performance, but it is possible to estimate the location of the mobile device in any position, as long as it is under the coverage of the deployed infrastructure.

Some open challenges remain in the area of indoor localization based on the TinyML paradigm:

- Although the classification approach offers a better performance, it requires capturing a lot of data in each location to be estimated. How many samples per location are necessary to properly characterize the environment under analysis?
- In the case of the regression approach, how many samples and how many locations are necessary to properly characterize the environment under analysis and to achieve a certain estimation error?
- Some other variables can be used to estimate location. For instance, WiFi CSI has been widely used for human activity recognition (HAR) in indoor environments. However, if HAR has been proved to be possible, to what extent could CSI be applied for indoor localization purposes?

References

[1] L.B. Furstenau, Y.P.R. Rodrigues, M.K. Sott, P. Leivas, M.S. Dohan, J.R. López-Robles, M.J. Cobo, N.L. Bragazzi, K.-K.R. Choo, Internet of things: conceptual network structure, main challenges and future directions, Digital Communications and Networks 9 (3) (2023) 677–687, https://doi.org/10.1016/j.dcan.2022.04.027, https://www.sciencedirect.com/science/article/pii/S2352864822000827.

[2] P.S. Farahsari, A. Farahzadi, J. Rezazadeh, A. Bagheri, A survey on indoor positioning systems for IoT-based applications, IEEE Internet of Things Journal 9 (10) (2022) 7680–7699, https://doi.org/10.1109/JIOT.2022.3149048.

[3] H. Obeidat, W. Shuaieb, O. Obeidat, R. Abd-Alhameed, A review of indoor localization techniques and wireless technologies, Wireless Personal Communications 119 (1) (2021) 289–327, https://doi.org/10.1007/s11277-021-08209-5.

[4] D.L. Dutta, S. Bharali, TinyML meets IoT: a comprehensive survey, Internet of Things 16 (2021) 100461, https://doi.org/10.1016/j.iot.2021.100461, https://www.sciencedirect.com/science/article/pii/S2542660521001025.

[5] P. Singh, N. Mittal, R. Salgotra, Comparison of range-based versus range-free WSNs localization using adaptive SSA algorithm, Wireless Networks 28 (4) (2022) 1625–1647, https://doi.org/10.1007/s11276-022-02908-y.

[6] T. Yang, A. Cabani, H. Chafouk, A survey of recent indoor localization scenarios and methodologies, Sensors 21 (23) (2021) 8086, https://doi.org/10.3390/s21238086, https://www.mdpi.com/1424-8220/21/23/8086.

[7] K. Ibwe, S. Pande, A.T. Abdalla, G.M. Gadiel, Indoor positioning using circle expansion-based adaptive trilateration algorithm, Journal of Electrical Systems and Information Technology 10 (1) (2023) 10.

[8] B. Yang, L. Guo, R. Guo, M. Zhao, T. Zhao, A novel trilateration algorithm for RSSI-based indoor localization, IEEE Sensors Journal 20 (14) (2020) 8164–8172, https://doi.org/10.1109/JSEN.2020.2980966.

[9] G. Kia, L. Ruotsalainen, J. Talvitie, Toward accurate indoor positioning: an RSS-based fusion of UWB and machine-learning-enhanced WiFi, Sensors 22 (9) (2022), https://doi.org/10.3390/s22093204, https://www.mdpi.com/1424-8220/22/9/3204.

[10] Q. Luo, K. Yang, X. Yan, J. Li, C. Wang, Z. Zhou, An improved trilateration positioning algorithm with anchor node combination and K-means clustering, Sensors 22 (16) (2022), https://doi.org/10.3390/s22166085, https://www.mdpi.com/1424-8220/22/16/6085.
[11] S. Sung, H. Kim, J.-I. Jung, Accurate indoor positioning for UWB-based personal devices using deep learning, IEEE Access 11 (2023) 20095–20113, https://doi.org/10.1109/ACCESS.2023.3250180.
[12] B. Bengherbia, M. el amine Bougheriza, M. Bouhedda, A. Tobbal, A. Toubal, H. Hentabeli, Implementation of a 3D indoor positioning system using MLP-based distance estimation for IoT applications, in: 2022 International Conference on Electrical Engineering, Computer and Information Technology (ICEECIT), 2022, pp. 209–214, https://doi.org/10.1109/ICEECIT55908.2022.10030667.
[13] J. Zheng, K. Li, X. Zhang, Wi-Fi fingerprint-based indoor localization method via standard particle swarm optimization, Sensors 22 (13) (2022), https://doi.org/10.3390/s22135051, https://www.mdpi.com/1424-8220/22/13/5051.
[14] M. Zhou, Y. Li, M.J. Tahir, X. Geng, Y. Wang, W. He, Integrated statistical test of signal distributions and access point contributions for Wi-Fi indoor localization, IEEE Transactions on Vehicular Technology 70 (5) (2021) 5057–5070, https://doi.org/10.1109/TVT.2021.3076269.
[15] Q. Ye, H. Bie, K.-C. Li, X. Fan, L. Gong, X. He, G. Fang, EdgeLoc: a robust and real-time localization system toward heterogeneous IoT devices, IEEE Internet of Things Journal 9 (5) (2022) 3865–3876, https://doi.org/10.1109/JIOT.2021.3101368.
[16] S.B. Altaf Khattak, Fawad, M.M. Nasralla, M.A. Esmail, H. Mostafa, M. Jia, WLAN RSS-based fingerprinting for indoor localization: a machine learning inspired bag-of-features approach, Sensors 22 (14) (2022), https://doi.org/10.3390/s22145236, https://www.mdpi.com/1424-8220/22/14/5236.
[17] F.J. Aranda, F. Parralejo, F.J. Álvarez, J.A. Paredes, Performance analysis of fingerprinting indoor positioning methods with BLE, Expert Systems with Applications 202 (2022) 117095, https://doi.org/10.1016/j.eswa.2022.117095, https://www.sciencedirect.com/science/article/pii/S0957417422005012.
[18] P. Baronti, P. Barsocchi, S. Chessa, F. Mavilia, F. Palumbo, Indoor bluetooth low energy dataset for localization, tracking, occupancy, and social interaction, Sensors 18 (12) (2018), https://doi.org/10.3390/s18124462, https://www.mdpi.com/1424-8220/18/12/4462.
[19] F.J. Aranda, F. Parralejo, F.J. Álvarez, J. Torres-Sospedra, Multi-slot BLE raw database for accurate positioning in mixed indoor/outdoor environments, Data 5 (3) (2020), https://doi.org/10.3390/data5030067, https://www.mdpi.com/2306-5729/5/3/67.
[20] G.M. Mendoza-Silva, M. Matey-Sanz, J. Torres-Sospedra, J. Huerta, BLE RSS measurements dataset for research on accurate indoor positioning, Data 4 (1) (2019), https://doi.org/10.3390/data4010012, https://www.mdpi.com/2306-5729/4/1/12.
[21] L. Polak, S. Rozum, M. Slanina, T. Bravenec, T. Fryza, A. Pikrakis, Received signal strength fingerprinting-based indoor location estimation employing machine learning, Sensors 21 (13) (2021), https://doi.org/10.3390/s21134605, https://www.mdpi.com/1424-8220/21/13/4605.
[22] D. Avellaneda, D. Mendez, G. Fortino, A TinyML deep learning approach for indoor tracking of assets, Sensors 23 (3) (2023), https://doi.org/10.3390/s23031542, https://www.mdpi.com/1424-8220/23/3/1542.
[23] S. Hymel, C. Banbury, D. Situnayake, A. Elium, C. Ward, M. Kelcey, M. Baaijens, M. Majchrzycki, J. Plunkett, D. Tischler, A. Grande, L. Moreau, D. Maslov, A. Beavis,

J. Jongboom, V.J. Reddi, Edge impulse: an MLOps platform for tiny machine learning, arXiv:2212.03332, 2023.

CHAPTER

Embedded intelligence in Internet of Things scenarios: TinyML meets eBPF

16

Irene Bru-Santa, Jorge Gallego-Madrid, Ramon Sanchez-Iborra, and Antonio Skarmeta
Department of Information and Communication Engineering, University of Murcia, Murcia, Spain

16.1 Introduction

In recent years, the exponential growth of data traffic in communication networks has led to the increasing demand for new technologies capable of managing this traffic efficiently and securely. One of the most promising solutions for this problem is the Extended Berkeley Packet Filter (eBPF), an innovative technology integrated into the Linux kernel that provides a secure and efficient mechanism for extending kernel functionalities at runtime without requiring kernel source code changes. eBPF enables sandboxed programs to run in a privileged context without compromising the security or performance of the Linux kernel. This approach provides the flexibility to add new capabilities to the operating system dynamically, like natively compiled code. When combined with other tools, such as the Linux eXpress Data Path (XDP), eBPF can be used effectively in network management. XDP provides a secure execution environment for custom packet processing applications that are executed in the device driver context. Custom XDP applications are written in higher-level languages, like C, and compiled into custom bytecodes. XDP defines a limited execution environment in the form of a virtual machine that executes eBPF code. This environment executes programs directly in the kernel context, allowing for custom processing at the earliest possible point after the packet is received from the hardware. By combining eBPF with XDP, it is possible to achieve efficient traffic packet processing with the flexibility to add new features at runtime [1].

One of the areas where these two technologies have enormous potential is the Internet of Things (IoT). IoT devices are notably constrained, typically having limited processing power, memory, and energy resources. These constraints present a challenge for some traditional resource-intensive solutions. Although machine learning solutions have shown remarkable capabilities in recent years, their use in IoT devices remains a challenge. To address this issue, tiny machine learning (TinyML) has emerged as a promising field of research that aims to bring machine learning capabil-

TinyML for Edge Intelligence in IoT and LPWAN Networks. https://doi.org/10.1016/B978-0-44-322202-3.00021-X

ities to small, low-power devices such as microcontrollers, wearables, or IoT devices [2]. As intelligent edge devices become increasingly in demand, there is a growing need for machine learning models that can run efficiently on such devices, without requiring large computing or memory resources. Hence, the TinyML research community focuses on developing algorithms, techniques, and hardware architectures that enable the deployment of machine learning models on resource-constrained devices. In particular, the combination of TinyML with eBPF and XDP can provide a powerful set of tools for managing network traffic on resource-constrained devices with intelligent decision-making capabilities.

One of the main advantages of using machine learning in eBPF is the ability to improve the accuracy and efficiency of packet filtering and processing. By training machine learning models on large datasets of network traffic, it is possible to develop algorithms that can classify packets based on complex criteria, such as application-level protocols, content types, or security threats [3]. With the implementation of these models into the device's kernel, lightweight solutions can be created that provide fast response times, low resource consumption, and scalability. Moreover, machine learning-enhanced eBPF programs can adapt to changing network environments. Machine learning models can be continuously updated and refined based on new data, allowing them to learn from past experiences and adjust their behavior accordingly. This is particularly important in the context of IoT, where network conditions and security risks can vary depending on the device type, location, and usage scenario.

However, while machine learning-based eBPF offers promising capabilities, there are still some challenges that need to be addressed [4]. In particular, eBPF programs loaded into the Linux kernel are subject to some restrictions to ensure the system's stability and security. For example, eBPF programs have limited access to C language libraries and do not support operations with external libraries. They also do not support non-static global variables, floating-point operations, and loops, as the eBPF verifier does not allow them to ensure that all programs loaded into the kernel finish. In addition, eBPF follows a passive event-based model, where programs are called for every incoming packet, so it is difficult to perform asynchronous tasks. All these constraints pose limitations for the practical implementation of most machine learning algorithms, and most of them have not been successfully implemented yet.

Therefore, this chapter aims at presenting real advances in the implementation of complex machine learning models, such as neural networks, within the Linux kernel by exploiting all the capabilities offered by eBPF. Thus, the main objective of this work is to explore the potential of machine learning and eBPF in enhancing the performance of IoT devices without excessively augmenting their computational workload. The study will involve identifying an appropriate use case that considers the constraints imposed by the limited resources of IoT devices. Given the limitations of traditional security measures for IoT devices, cybersecurity is a promising area of application. As each end node in an IoT deployment is a potential entry point to the entire network infrastructure, improving their resilience against attacks is crucial to the overall system security. However, it is worth noting that this study will not only concentrate on the chosen use case (IoT devices' cybersecurity) but also investigate the process of implementing machine learning models within the eBPF framework given its complexity. In this line, the goal is to identify potential limitations of eBPF and explore possible solutions to overcome them, not only for the specific use case but also for other applications in the future. Consequently, this work aims at contributing to the progress of the field of machine learning-based eBPF development while providing insights into the potential benefits of using machine learning in IoT systems.

The rest of this chapter is organized as follows. Section 16.2 provides a literature review considering the latest related work. The design and implementation of our solution is presented in Section 16.3. The results obtained from experimental tests are shown and discussed in Section 16.3. Finally, Section 16.4 closes the chapter by highlighting the main findings and drawing future research lines.

16.2 State-of-the-art

The study and understanding of eBPF, XDP, and TinyML concepts is essential to address today's performance challenges within computing and networking environments. The primary objective of this section is to provide a comprehensive overview of the current state-of-the-art in these areas and review recent advances, challenges, and limitations.

16.2.1 eBPF and XDP

16.2.1.1 eBPF system

The eBPF system consists of interconnected components that handle the compilation, verification, and execution of source code for developed applications. The workflow of the eBPF system (Fig. 16.1) starts with an eBPF program in a high-level language, primarily restricted C, although other options are also accepted, such as P4. The program is then converted into an Executable and Linkable Format (ELF) object with executable eBPF code, using the compilation technologies provided by Low-Level Virtual Machine (LLVM). Subsequently, a specialized eBPF loader facilitates the insertion of the ELF object into the kernel. To ensure the integrity and security of the operating system, the kernel employs a verifier that performs a static program analysis on the eBPF instructions. The verifier examines various aspects of the program, including its size (limited to 4096 assembly instructions until kernel version 5.2, which increased that number to 1 million for privileged users), program termination, memory address validity, and execution path depth. This verification occurs after the code has been compiled and during the program's loading into the kernel. For a comprehensive understanding of the purpose and functioning of the eBPF verifier, the reader is referred to [5]. Once the verification is completed, the kernel proceeds with a dynamic translation, also known as just-in-time (JIT) compilation. An insightful study on the performance of the JIT compiler can be found in [6].

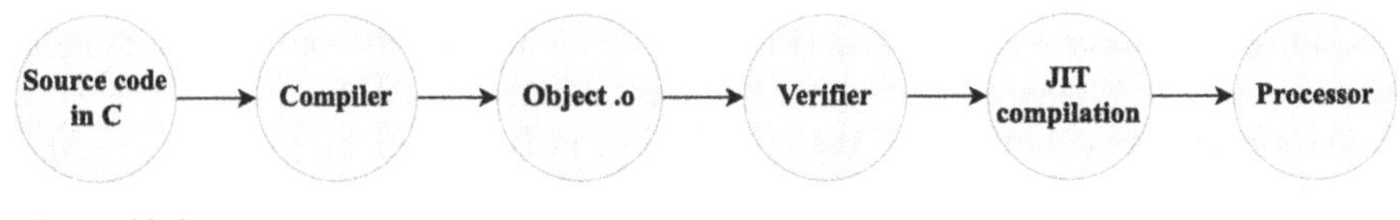

FIGURE 16.1

Workflow of the eBPF system.

In recent times, significant efforts have been dedicated to the creation of effective tools for the development and debugging of eBPF programs. For example, a noteworthy collaboration between VMWare and the University of British Columbia (US) resulted in the implementation of a P4 compiler specifically designed for eBPF [7]. Furthermore, it is worth highlighting the influential open-source project known as the BPF Compiler Collection (BCC),[1] which serves the purpose of streamlining the development process of eBPF programs by offering a comprehensive set of frontends. These frontends enable developers to interact with the eBPF system using high-level languages like Python and Go, thus facilitating the adoption of eBPF within various software ecosystems. BCC is one of the most important initiatives of IOVisor,[2] which is an important open-source project that aims to provide a framework for building and deploying dynamic and programmable networking and security applications using eBPF technology. In addition, libbpf[3] is also a remarkable open-source library that greatly simplifies the development of eBPF programs within the Linux kernel. It offers an intuitive API for interacting with eBPF code, providing functions for program compilation, shared memory map management, and kernel program loading, abstracting eBPF code complexity.

16.2.1.2 XDP

To facilitate the execution of an eBPF program, it is necessary to establish its attachment to a hook, which functions as a customized programming interface. Hooks are used for intercepting packets before the call or during execution in the operating system. The Linux kernel exposes several hooks to which eBPF programs can be attached, so they can be triggered every time an event occurs, such as the sending or receiving of a packet. In this discussion, our focus will be on the eXpress Data Path (XDP) hook.

XDP is the lowest layer of the Linux kernel network stack. It is present only on the RX path, inside a device's network driver, allowing packet processing at the earliest point in the network stack, even before memory allocation is done by the OS. Within this hook, programs pose the capability to promptly make decisions regarding incoming packets and execute arbitrary modifications on them, effectively circumventing additional overhead imposed by processing within the kernel as packets should cross

[1] https://github.com/iovisor/bcc.
[2] https://www.iovisor.org/.
[3] https://github.com/libbpf/libbpf.

the whole Linux network stack. XDP has different modes of operation that encompass various approaches for optimizing packet processing performance:

- XDP Native Mode: This mode is used when programs are executed directly by the driver, bypassing the operating system. Only some high-speed device drivers, such as i40e, nfp, mlx*, and the ixgbe family, already have this functionality.
- XDP Generic: The execution is performed by the operating system when devices lack native driver-level support. However, it still achieves efficiency by avoiding additional steps required for socket buffer allocation and emulation.
- XDP Offload: The eBPF program is offloaded to compatible programmable NICs. This offers the best performance.

Currently, although the XDP layer is well suited for many applications, as aforementioned it can only process ingress traffic. To process egress traffic, the closest layer to the NIC that has access to the entire Ethernet frame is the traffic control (TC) layer. A comparison between XDP and TC is made in [1]. For a more comprehensive understanding of eBPF and XDP, we also recommend referring to [5].

16.2.1.3 eBPF and machine learning

As previously stated, the integration of machine learning models with eBPF and XDP technologies offers a powerful toolkit for managing network data traffic, incorporating intelligent decision-making capabilities. This combination enables fast response times, minimal resource utilization, scalability, and adaptability. Nonetheless, there exist certain challenges that obstruct the implementation of machine learning algorithms within the Linux kernel. One of the primary limitations is the constrained access to C language libraries, which restricts the availability of essential resources for machine learning tasks. Furthermore, the inherent limitations on floating-point operations within the kernel environment or the absence of loops in kernel programs complicate the implementation of machine learning algorithms. Consequently, the pool of solutions integrating machine learning algorithms directly into the Linux kernel using eBPF remains relatively limited.

One example in this context is the work in [8], which successfully implements an intrusion detection system (IDS) using decision trees (implemented as concatenated if/else instructions) within the kernel, exclusively relying on eBPF. This IDS can determine the maliciousness of each packet and achieves a notable 20% performance increase compared to the user-space implementation. However, it is noteworthy that the predominant approach in this field involves the utilization of eBPF for data collection, while machine learning techniques are implemented in user space to perform analysis and decision-making tasks. For example, in the study presented in [9], the adoption of eBPF for network monitoring is proposed, accompanied by machine learning methodologies in user space for the analysis and diagnosis of network performance. This approach aims to identify network bottlenecks and address related issues. The authors of [10] proposed a solution for identifying and classifying microservices. Identification is achieved by employing an eBPF module to trace system calls. With these data, a combination of Bayesian learning and long short-term memory (LSTM) autoencoders is capable of accurately identifying numerous real-world microservices with 99% precision, while incurring only an additional 1–2% CPU usage. Furthermore, in [11] an intelligent load balancer designed for the Linux kernel was introduced, which leverages eBPF for system data collection and implements a multilayer perceptron (MLP) model into the Linux kernel. However, it is important to highlight that the integration of the forward pass mechanism from the MLP is made as an optional kernel configuration, which intriguingly does not involve the utilization of eBPF. It is also interesting to remark that, to overcome the constraints posed by the limited support for floating-point operations within the kernel, the proposal suggests the adoption of fixed-point arithmetic.

Despite an extensive review of the literature, no existing proposal implementing a neural network within the Linux kernel using eBPF, as proposed in this work, has been found.

16.2.2 TinyML

TinyML, an emerging field at the intersection of machine learning and embedded systems, holds immense promise for enabling AI capabilities on resource-constrained devices. As the demand for intelligent applications grows in domains such as IoT, wearable devices, and edge computing, the traditional reliance on cloud-based processing becomes impractical due to latency, privacy concerns, and network bandwidth limitations. TinyML addresses these challenges by bringing machine learning models directly onto edge devices, allowing for real-time, low-power, and privacy-preserving AI inference. In this section, we explore the principles, applications, and advancements in TinyML, highlighting its potential to revolutionize various industries and empower a new era of intelligent edge computing.

16.2.2.1 Applications

Numerous IoT applications have already permeated our everyday personal and professional lives. However, a significant proportion of IoT devices heavily rely on network-related operations, which may not always be the optimal approach. That happens, for example, with wearables such as "smart" watches, which heavily depend on a smartphone connected via Bluetooth to carry out most processing tasks. By integrating machine learning capabilities within resource-efficient objects, their reliance on a central device can be mitigated, thereby substantially enhancing the current landscape of IoT applications.

- eHealth: Wearable technology is a rapidly growing sector with diverse applications such as health monitoring, visual assistance, hearing aids, personal sensing, and activity recognition. Furthermore, wearable tech advancements are anticipated to bring about augmented reality, real-time voice recognition, language translation, context-aware support systems, and precise indoor/outdoor positioning services. These developments have the potential to revolutionize various domains and open up new possibilities for users [14].

- Smart spaces: Machine learning-enabled embedded devices will revolutionize various scenarios, such as smart cities and cognitive buildings [15]. They will transform existing IoT-based monitoring and surveillance systems into intelligent and autonomous entities capable of decentralized decision-making. These devices can be easily deployed in remote and rural areas, creating interconnected smart spaces that seamlessly interact with each other. Their low cost will enable their adoption in disadvantaged areas, stimulating the local economy and business activities. This advancement holds immense potential for enhancing collective intelligence and improving the overall quality of life in diverse environments [16].
- Vehicular services: TinyML's expansion in the vehicular market [17] opens new opportunities for cooperative intelligent transportation systems (C-ITSs). Integration of microcontroller-based onboard units (OBUs) enables the seamless inclusion of personal mobility devices like shared bikes, electric mopeds, and scooters into existing vehicular networks. This integration enhances their capabilities within the transportation system, allowing access to important vehicular services such as driving safety, device monitoring, and route planning. This advancement contributes to the development of C-ITSs and smart city ecosystems.
- Industry 4.0: TinyML has the potential to revolutionize the industrial and manufacturing sectors as part of the Industry 4.0 revolution [12]. By integrating machine learning-based decision support systems into MCUs, these systems can intelligently determine whether to handle specific computation tasks themselves or offload them to higher processing layers like edge or cloud computing. This integration of intelligence throughout the production chain will enhance production processes, improve decision-making, and optimize asset tracking in the industry.
- Smart agriculture and farming: The digital transformation in sectors like agriculture [18] will be enhanced by machine learning-based smart monitoring and control systems, improving crop and animal well-being and product quality. The integration of machine learning-enabled devices will elevate everyday activities and home appliances, with applications such as biometric e-locks, context-aware home automation, and well-being apps. TinyML's revolution will enable intelligent pet trackers, smart garden irrigation control, and advanced safety and security systems. This advancement has the potential to transform industries and enhance the quality of life.

For more information about TinyML applications, the reader is referred to [2].

16.2.2.2 Libraries and frameworks

Various libraries and tools have been developed to facilitate the implementation of machine learning algorithms on resource-constrained platforms in the context of TinyML. These frameworks can be classified into three main categories. The first category involves converting pretrained models from popular machine learning libraries like TensorFlow,[4] scikit-learn,[5] or PyTorch[6] into formats suitable for execution on limited-resource MCUs. This approach allows the deployment of machine learning models on devices with restricted capabilities. The second category focuses on integrating machine learning libraries directly into MCUs, enabling local training and analysis capabilities. This empowers the device to generate models from its own data, potentially improving model accuracy and supporting unsupervised learning algorithms. The third category involves the use of dedicated co-processors that augment the main computing unit with specialized machine learning processing capabilities. Although this approach enhances computational performance, it is less commonly used due to the increased cost and complexity of the processing platforms.

TensorFlow Lite[7] is a widely used library in the field of TinyML, developed by Google. It provides a comprehensive set of tools for adapting TensorFlow models to run efficiently on mobile and embedded devices. The library consists of two primary components: the converter and the interpreter. The converter is responsible for transforming TensorFlow models into optimized code that can be executed on resource-constrained platforms. The interpreter is responsible for running the code generated by the converter. TensorFlow Lite supports a variety of optimized models, including those based on neural networks, and enables their execution on platforms such as smartphones, embedded Linux systems, and MCUs. This versatility allows for the deployment of efficient machine learning algorithms across a range of devices in the TinyML ecosystem. Microsoft has also made its own contribution to the TinyML field with the release of the Embedded Learning Library (ELL), an open-source framework.[8] This powerful tool enables the design and deployment of pretrained machine learning models on resource-constrained platforms like Arduino, Raspberry Pi, and micro:bit.

Several frameworks in the field of TinyML primarily focus on neural networks. However, there are notable initiatives that extend their support to encompass other machine learning techniques, including the naive Bayes classifier, decision trees, and k-nearest neighbors (k-NN), among others. MicroML,[9] for instance, is a noteworthy project that facilitates the adaptation of support vector machine (SVM) and relevance vector machine (RVM) algorithms into C code, ensuring compatibility with diverse microcontrollers such as Arduino, ESP8266, and ESP32. Another prominent framework, emlearn,[10] plays a crucial role in the TinyML landscape by providing the means to generate portable C99 code. This code is derived from popular Python libraries like scikit-learn and Keras. The emlearn framework supports a wide array of machine learning algorithms, including random forest (RF), decision trees, Naive

[4] https://www.tensorflow.org/.
[5] https://scikit-learn.org/.
[6] https://pytorch.org/.
[7] https://www.tensorflow.org/lite.
[8] https://microsoft.github.io/ELL/.
[9] https://github.com/eloquentarduino/micromlgen.
[10] https://github.com/emlearn/emlearn.

bayes, and linear models. The following sections describe how emlearn can be applied to this work to port a trained MLP neural network in Python to restricted C.

A more comprehensive study on TinyML frameworks can be found at [2].

16.2.2.3 Challenges

Despite the promising potential demonstrated by numerous applications of TinyML, some of the key issues to investigate in this domain are the following:

- Inconsistent power usage: Power consumption is a significant issue in TinyML, as there are variations in power requirements among different MCUs and data processing and machine learning services demand higher power compared to simple data gathering tasks. It is crucial to make fast advancements in power management to ensure the reliability of always-on battery-powered applications. Additionally, while processor performance in MCUs has been quickly increasing, battery capacity and technology have been improving at a much slower rate.
- Memory constraints: The integration of novel machine learning models in MCUs faces two main challenges, namely, the size complexity of the models and the need for innovative optimization techniques for each algorithm. Aggressive compression is required to shrink the models, but it is crucial to maintain performance parameters and balance memory footprint and system accuracy. Currently, IoT devices rely heavily on cloud resources for complex analysis, but researchers are actively working towards a breakthrough that would enable large models to run smoothly on tiny MCUs.
- Cost: As research in TinyML progresses, there is a growing need for improved performance and reliability in MCUs. However, balancing cost reduction with meeting the performance and reliability goals of MCUs can be challenging. This is particularly important in large-scale IoT applications where hundreds or thousands of sensors are deployed, making cost reduction a significant concern in the field of TinyML.

The issues presented here offer an important step forward in the future research direction. A more comprehensive study about the challenges that TinyML has to face is reported in [13].

16.3 Design and implementation

16.3.1 Machine learning model development

As discussed before, the primary objective of this work is to introduce an advanced machine learning-based traffic processing solution into the Linux kernel, utilizing the capabilities of eBPF. Among various machine learning algorithms, we have specifically opted for MLP neural network models for our purpose. MLP neural networks have gained considerable prominence in network monitoring and management [19] due to their capacity to handle non-linear relationships between input vectors and output labels and their exceptional pattern recognition capabilities, enabling the detection of subtle anomalous behaviors that may not be immediately evident [20]. Furthermore, MLP neural networks are renowned for their high accuracy and speed, making them an optimal choice for systems with low-latency requirements. While previous research has successfully implemented decision tree-based solutions in eBPF [8], our work goes a step further by implementing a more complex machine learning algorithm like MLP.

For the implementation of such neural network, we utilized an MLP based on the Python library scikit-learn.[11] Each perceptron employs an activation function to determine whether the neuron activates based on the weighted sum of inputs. Among various options, we chose to use the Rectified Linear Unit (ReLU) activation function, defined as follows:

$$\phi : \mathbb{R} \rightarrow (0, +\infty), \tag{16.1}$$

$$\phi(x) = \max(0, x). \tag{16.2}$$

The primary reason behind this decision is that the ReLU activation function, unlike many others, does not rely on exponential operations, making it feasible for implementation within the Linux kernel (further details on this are explained in the next section). Moreover, ReLU offers simplicity and computational efficiency by avoiding resource-intensive calculations, making it well suited for real-time processing applications, including those developed in the eBPF framework. Additionally, ReLU is a versatile activation function suitable for various use cases. Particularly, it has proven valuable in binary classification tasks, such as intrusion or malicious traffic detection, and in identifying connectivity and performance issues within networks.

In the context of MLPs, the choice of activation function in the final layer significantly influences the network's output behavior. For networking tasks, classification MLPs are highly relevant, and they commonly utilize the sigmoid activation function on the output layer to effectively address them. The sigmoid or logistic function is especially suitable for binary or multi-label classification, generating an interpretable output between 0 and 1 for each label. This enables the model to assign probabilities to each class and make predictions accordingly.

Once the machine learning model is trained, the next step involves porting the MLP obtained from scikit-learn to an eBPF program. This process includes an intermediate step based on the TinyML paradigm [2]. Specifically, the emlearn[12] library is utilized to transform the Python model into ANSI C code, significantly simplifying the integration of the neural network within an eBPF program. The emlearn library offers conversion utilities that generate floating-point arrays containing the weights and biases of each layer, along with a complete structure representing the trained neural network model.

---

[11] https://scikit-learn.org/stable/.
[12] https://github.com/emlearn/emlearn.

16.3.2 Implementation in eBPF

Integrating the MLP model into an eBPF program is a challenging task due to the strictness of the eBPF verifier, which rigorously checks the program before allowing it to be loaded into the kernel. To accomplish this integration, the emlearn library has been employed, providing headers with functions that implement the neural network's logic. As previously explained, the neural network's layers involve performing a linear combination of the previous layer's outputs with the current layer's weights and biases. This is followed by the application of a non-linear activation function. Despite the apparent simplicity of the C code generated by emlearn, directly integrating the MLP model into eBPF presents challenges due to constraints within this environment. To overcome this issue, certain modifications have been introduced into the emlearn library. These changes adapt the generated code, ensuring it meets the requirements of the eBPF verifier and makes it acceptable for integration within the eBPF environment.

eBPF programs are unable to handle floating-point operations. This poses a significant challenge when working with neural networks obtained from scikit-learn, as their weights and biases often involve fractional numbers. To overcome this obstacle, a solution based on fixed-point representation has been devised. This approach utilizes a 32-bit format, with 16 bits dedicated to the integer part, 15 bits for the fractional part, and 1 bit for the sign. To address this, additional functions have been developed to convert all floating-point values into the fixed-point representation. Consequently, the arithmetic operations had to be adjusted to accept it. Specifically, multiplication operations need support of 64-bit integers to ensure the necessary precision. Furthermore, the ReLU activation function has been adapted to work with this integer type and verify if the values are positive.

However, the fixed-point representation was not enough to implement the sigmoid function, as it contains divisions and exponential operations (Eq. (16.3) shows its mathematical representation), which do not have support in eBPF. As explained before, this function is utilized in the final layer of the MLP to generate the ultimate output. The sigmoid function computes a value ranging from 0 to 1, after which classification is performed. Specifically, if x denotes the output of the last perceptron and $\phi(x)$ represents the value of the sigmoid function (16.3), the model outputs 1 when $\phi(x) > 0.5$ (and vice versa). Consequently, an alternative approach can be made (16.4):

$$\phi(x) : \mathbb{R} \rightarrow (0, 1), \ \phi(x) = \frac{1}{1 + e^{-x}}, \tag{16.3}$$

$$\forall a, b \in \mathbb{R}, \ a < b \Leftrightarrow \phi(a) < \phi(b). \tag{16.4}$$

Considering that $\phi(0) = 0.5$,

$$\phi(x) > 0.5 = \phi(0) \Leftrightarrow x > 0. \tag{16.5}$$

In summary, the sigmoid function in the last layer of the trained model can be replaced by a simple comparison. By examining the fixed-point output of the last

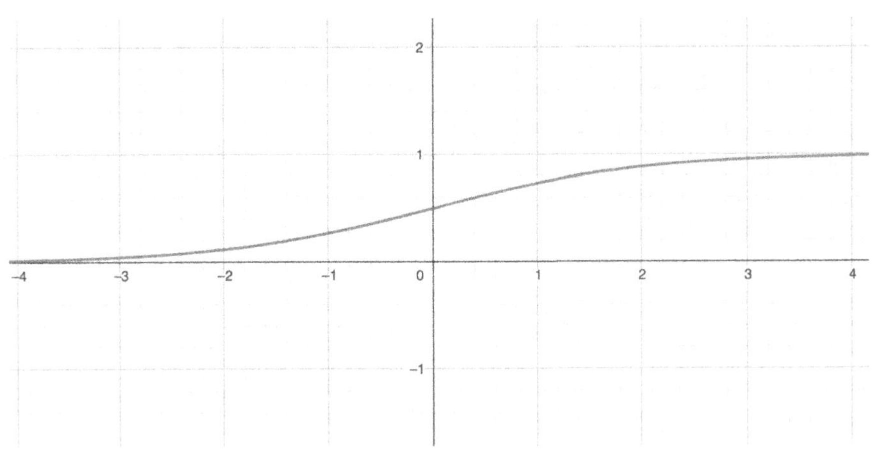

FIGURE 16.2

Sigmoid function.

layer before applying the activation function, if it is greater than 0, the MLP returns the value 1, and vice versa. These concepts can be better understood by referring to the graphical representation of the function (Fig. 16.2).

To comply with the eBPF verifier's restrictions on unbounded loops, necessary adjustments have been made in the code. All loops have been modified to have a known number of iterations at compile time. This involved replacing variables representing the number of layers, neural network outputs, and auxiliary buffer lengths with MACROS that have fixed global values. To enable the linear combination of weights and biases in each layer, constant static arrays have been employed. These arrays store the number of inputs and outputs for each layer and, utilizing the `pragma unroll` directive, all loops are loaded into the kernel unrolled. So, the implementation supports neural networks with varying numbers of layers and neurons, ensuring a high degree of generalization. However, it is important to acknowledge that eBPF has additional limitations to safeguard the kernel itself, such as restricting the maximum number of instructions a program can contain and the depth of nested calls. These limitations can impact the number of layers and neurons that can be used in the eBPF program. Despite this, the adjustments made allow for effective integration of neural networks into eBPF programs, providing a degree of flexibility and generality.

To facilitate the analysis of network traffic, the implementation of a header parsing mechanism is essential. In the eBPF context, this process requires vigilant monitoring of memory access boundaries to guarantee that all accesses occur within the appropriate reserved space. While certain libraries exist to aid in parsing widely used headers like Ethernet, IP, or TCP, additional headers classified as "less common," such as 6LoWPAN, GTP, and others, necessitate manual parsing. Consequently, specialized handling is required to properly interpret these less prevalent headers within the eBPF program. By adopting such mechanisms, eBPF can efficiently process various types of network traffic, enabling a more comprehensive analysis of network data.

In scenarios where aggregated traffic analysis is preferred over per-packet analysis, the collected data from parsed packet headers must be persistently stored between eBPF program executions. Since eBPF programs are event-driven, such as triggered by the reception of a packet, there is no inherent state maintained between executions. To address this requirement, eBPF maps come into play. These maps enable the storage of information, allowing the application of the implemented MLP model at specific points in time, such as when a predefined sampling time window has elapsed. By leveraging eBPF maps, the collected data can be accumulated and organized efficiently. As each time window concludes, the MLP model can analyze the aggregated data, leveraging the conclusions or decisions it has been trained for. This approach enables more comprehensive traffic analysis, facilitating the extraction of meaningful insights from the accumulated data over time.

In conclusion, by taking these considerations into account, we have laid a robust groundwork for the integration of a trained MLP model within the Linux kernel. With careful attention to the aspects discussed earlier, the subsequent chapter will showcase the results attained from an extensive validation and evaluation test of the proposed solution. This comprehensive validation will shed light on the effectiveness and efficiency of our approach, demonstrating its potential benefits in real-world scenarios.

16.4 Case of study

16.4.1 Context

IoT security is a relevant field of application for eBPF-based solutions. As discussed earlier, IoT devices typically lack sufficient processing power, memory, and energy resources, making it challenging to implement traditional security solutions, especially those based on machine learning algorithms, on these devices. IoT devices often collaborate in wireless sensor networks (WSNs), where multiple nodes (typically constrained devices) work together to sense and control the surrounding environment using wireless communication technologies. In consequence, securing IoT deployments is of utmost importance since each IoT device can serve as a potential entry point to the entire network infrastructure. To address this, it is crucial to develop defense mechanisms tailored specifically to IoT devices, considering their limitations and the unique threats they face in their operational environments. By harnessing eBPF, it can be created lightweight but sophisticated security solutions that run directly in the kernel. These solutions empower IoT devices to autonomously detect and respond to security threats promptly and efficiently. Additionally, eBPF enables the enforcement of security policies at the kernel level, adding an extra layer of self-protection against potential attacks.

To enable such networks, a widely used technology is IPv6 over low-power wireless personal area networks (6LoWPAN). This open standard, defined by the IETF, allows IPv6 packets to be transmitted over low-power wireless networks, such as those based on IEEE 802.15.4. The main purpose of 6LoWPAN is to make IPv6 usable by highly constrained devices, enabling them to directly connect to the Internet. Another related technology is Routing Protocol for Low-Power and Lossy Networks (RPL), which is a proactive routing protocol designed for wireless networks with low power consumption and susceptibility to packet loss. RPL operates on top of IEEE 802.15.4 and 6LoWPAN, allowing quick creation of network routes, efficient sharing of routing information, and adaptation to topology changes, making it highly suitable for WSNs with constrained devices. As a result, RPL is one of the most used routing protocols in these types of systems.

Given these considerations, the objective of this use case is to demonstrate and validate the potential of eBPF in enhancing the self-protection capabilities of constrained IoT devices. This involves executing attack detection algorithms powered by MLP models directly within the Linux kernel. In the following sections, we will outline the steps taken to implement this use case, aiming to improve experiment reproducibility.

16.4.2 Dataset generation

To integrate machine learning models into the Linux kernel of the end devices, a dataset is required for training the models. To gather reliable data while maintaining the security of operational WSNs, we utilized the Cooja simulator to create a controlled environment. This simulation environment allowed the emulation of networks of various sizes, comprising interconnected devices running on the Contiki open-source operating system, which is specifically designed for IoT devices and supports both 6LoWPAN and RPL protocols.

The focus of the investigation is on detecting the "Hello Flood" attack that targets the RPL protocol. To conduct this attack, the RPL-Attack-Framework[13] has been employed. This framework includes different modules responsible for emulating root, malicious, and benign nodes within a WSN managed by the RPL protocol. In our use case, the "Hello Flood" attack involves one or more malicious nodes frequently generating routing information requests in the form of RPL's Hello messages, forcing other nodes to respond to these messages. The main objective of this attack is overwhelming the network with control messages, causing the depletion of benign nodes' resources, such as their batteries.

In Fig. 16.3, we can see the simulated scenario comprising a single root node (#1), 10 sensor nodes (#2–#11), and one malicious node (#12) within the communication range of the root node. The Cooja simulator's Radio Messages Tool was used to capture the traffic generated during a 10-minute simulation, producing a .pcap file for dataset generation. The focus of our experiments was to enhance the self-protective capabilities of the root node, which plays a crucial role in the RPL network architecture.

[13] https://rpl-attacks.readthedocs.io/en/latest.

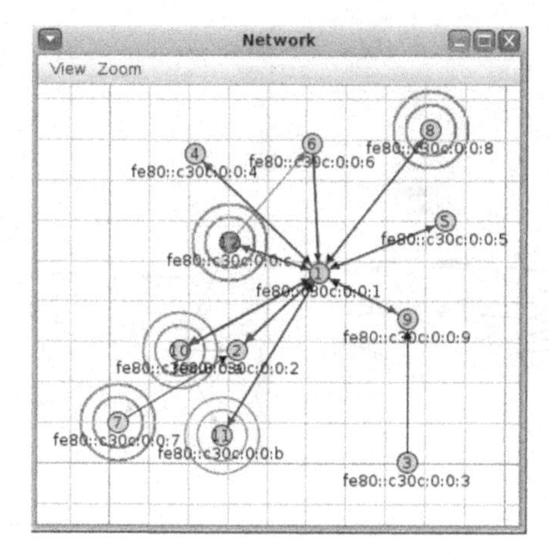

FIGURE 16.3

Deployed scenario in Cooja simulator.

To process the dataset, we extracted RPL packets received by the root node from the unprocessed .pcap file. Each row in the dataset contains packet counts for four distinct RPL message types: DIS, DIO, DAO, and others (as specified in IETF's RFC 6550[14]). These counts represent the number of packets originated from the same source and received by the root node within specific time frames. We conducted evaluations over multiple time frames (1 s, 5 s, and 10 s) to determine the most suitable data aggregation, ensuring the highest accuracy for machine learning-based attack detection models. Furthermore, each row includes the IP address of the corresponding source node. Although this attribute has not been used to train the machine learning models, it could potentially aid in identifying a detected attacker and triggering appropriate countermeasures (which is beyond the scope of this work). Additionally, each row is assigned a binary label indicating whether the source node is classified as malicious or benign. It is important to note that the dataset we used contains data that can be easily collected in real time by any node during operational scenarios. Specifically, these data include the counts of DIS, DIO, DAO, and other messages received from each neighboring node within a defined time frame, serving as input features for the machine learning model embedded in eBPF. Consequently, both the training dataset and resulting models possess a high degree of realism and are applicable to real-world WSN scenarios.

16.4.3 Model training

During this stage, we made use of the generated dataset, consisting of the RPL messages received by the root node from neighboring motes, to train various models.

[14] https://datatracker.ietf.org/doc/html/rfc6550.

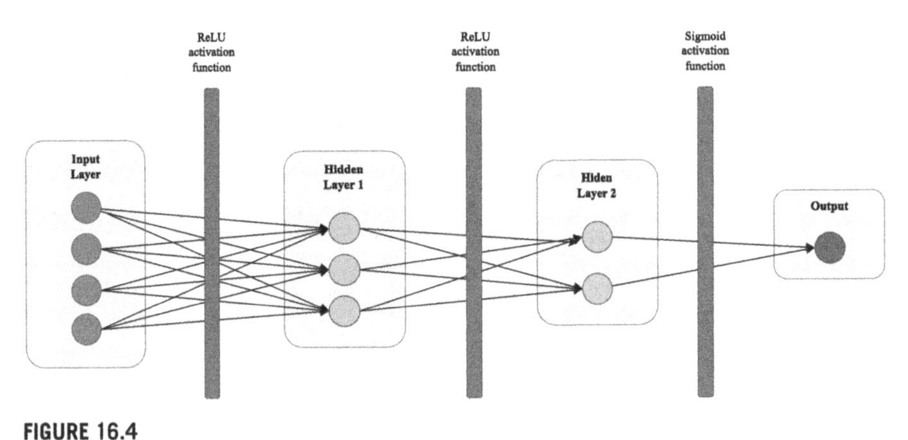

FIGURE 16.4

MLP final model.

To this end, we implemented MLP models tailored for classification tasks using the Python library scikit-learn. These models were trained on datasets associated with time frames of 1 s, 5 s, and 10 s, appropriately aggregating the information in the dataset.

By employing the activation functions mentioned before and conducting multiple experiments with various MLP configurations, such as adjusting the number of hidden layers and neurons per layer, we observed that a neural network architecture with a reduced number of layers and neurons yields satisfactory accuracy results, as discussed later. It is crucial to avoid an excessive number of layers or neurons to prevent overfitting, which could compromise the model's generalization capabilities. However, it is worth mentioning that the employed methodology has potential for implementing solutions involving more complex neural networks. Specifically, for our tests, we have settled on an MLP model comprising two hidden layers, composed of three and two perceptrons. In each layer, we use the ReLU activation function, while the sigmoid function is employed in the final layer as explained above. Fig. 16.4 provides a graphical representation of this neural network architecture. With these models, we achieve remarkable prediction metrics, as shown in Table 16.1.

16.4.4 eBPF adaptation

Once the neural networks (MLP models) are trained, they should be adapted to eBPF, as explained in Section 16.3. To accomplish this, we made use of the emlearn library to extract the MLP structure, including the weights and biases associated with each layer. Additionally, fixed-point conversion tools facilitate the integration of MLPs within the Linux kernel.

To detect the "Hello Flood" attack in the RPL protocol using the MLP model, we implement an eBPF-based header parsing procedure to analyze incoming network packets. As mentioned earlier, this parsing process requires careful boundary checks to ensure memory access remains within expected limits. The parsing procedure con-

Table 16.1 MLP metrics with different aggregation time frames.

Aggregat. time frame	Accuracy	Confusion matrix			Precision		Recall	
			Predicted negative	Predicted positive	Negative	Positive	Negative	Positive
1 s	0.99	**Real negative**	302	0	0.99	1	1	0.98
		Real positive	1	81				
5 s	1	**Real negative**	189	0	1	1	1	1
		Real positive	0	21				
10 s	1	**Real negative**	104	0	1	1	1	1
		Real positive	0	11				

siders that the link-level header conforms to IEEE 802.15.4 standards, followed by the 6LoWPAN header, which contains the compressed IPv6 address. The RPL message type can be identified based on a byte value of 0 for DIS messages, 1 for DIO messages, and 2 for DAO messages.

To persistently store the collected data across eBPF program executions, we have utilized pinned eBPF maps. This mechanism allows us to track messages transmitted by each IPv6 address to the root node across various time intervals (1 s, 5 s, and 10 s). Once a time window is completed, the MLP model is applied to analyze the aggregated data and determine the presence of an attack. This iterative process ensures continuous device protection. It is essential to highlight that the implementation of advanced countermeasures against the attack is beyond the scope of this research. Thus, when an attacker is detected, the root node promptly discards packets originating from that specific node (identified by its IPv6 address) to prevent unnecessary resource consumption, although more advanced techniques may be implemented.

16.4.5 Testbench

The experiments involved two devices: A regular computer responsible for generating traffic and an IoT device running the eBPF program to process the received packets and execute the embedded neural network. The regular computer was a desktop powered by an Intel i5-3470 CPU with four cores and 8 GB of RAM. In turn, the IoT device used was a Raspberry Pi 3 Model B V1.2, equipped with a four-core CPU and 1 GB of RAM. Both devices were connected via an Ethernet cable, with a maximum theoretical throughput speed of 100 Mbps, as the network interface of the Raspberry Pi was limited to that rate. To generate the traffic for the experiments, we used the tcpreplay[15] tool. This tool allowed us to replicate the captured traffic from the raw dataset mentioned earlier. By replaying the traffic extracted from the original .pcap file (see Section 16.4.2), it was sent from the desktop computer at the desired speed and received at the Ethernet port of the Raspberry Pi, where the eBPF program was attached.

In addition to the eBPF implementation of the neural network, we have also developed another version of our program where the neural network is decoupled from the parsing mechanism and executed in Linux's user space. By doing so, we can compare the performance of the neural network operating in both the kernel and user space, as depicted in Fig. 16.5.

Therefore, we prepared two scenarios: one where the parsing of packets and neural network execution were performed in a privileged context inside the kernel and another where the neural network calculations were moved to the user space. As mentioned earlier, we have chosen three different time frames (1 s, 5 s, and 10 s) to assess the best data aggregation in terms of the machine learning model's accuracy and to inspect the impact on end device performance. To ensure reliable results, each experiment was executed 10 times over a duration of 60 seconds. This approach helped

[15] https://tcpreplay.appneta.com/.

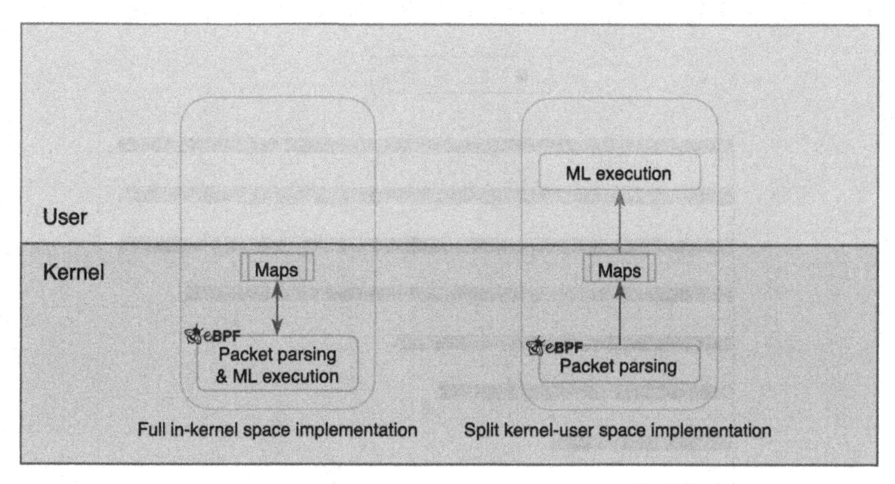

FIGURE 16.5

Conceptual diagram of the two implementations.

avoid singularities in the obtained outcomes and provided statistical significance to the results.

16.4.5.1 Results

The first batch of tests were aimed at measuring the maximum achievable data rate between both devices with no eBPF program attached to the NIC, obtaining a baseline performance of the setup. To do so, the maximum bandwidth achieved with diverse packet sizes was measured by using the iPerf3 tool. Fig. 16.6 shows the results of these tests. Note how, as expected, with lower packet sizes, the data rate decreases. This is because the capabilities of the IoT device's interface are exceeded with the high number of packets per second to process. With big packets, the maximum bandwidth achieved was above 90 Mbps, with a decrement down to 15 Mbps with 30-B packets. Although in IoT networks it is not usual to see such data rates, we injected this kind of traffic to push the limits of the eBPF program and the IoT device to evaluate their performance in stringent conditions.

Once the baseline performance of the setup is known, the eBPF program is loaded and the rest of the experiments are conducted. Fig. 16.7 presents the maximum data rate achieved without packet loss per configuration. Two variables were considered in the tests: (1) whether the machine learning model is executed in user or kernel space and (2) the duration of the time window to aggregate the gathered input and execute the MLP. The average packet length employed in the experiment being 90 B, there is no notable decrement in the obtained bandwidth values when compared with the baseline scenario shown in Fig. 16.6. These results demonstrate that the eBPF program does not reduce the performance of the IoT device and it is not acting as a bottleneck. When comparing both scenarios, the highest data rates are obtained when the machine learning model is executed in user space and the parsing of the

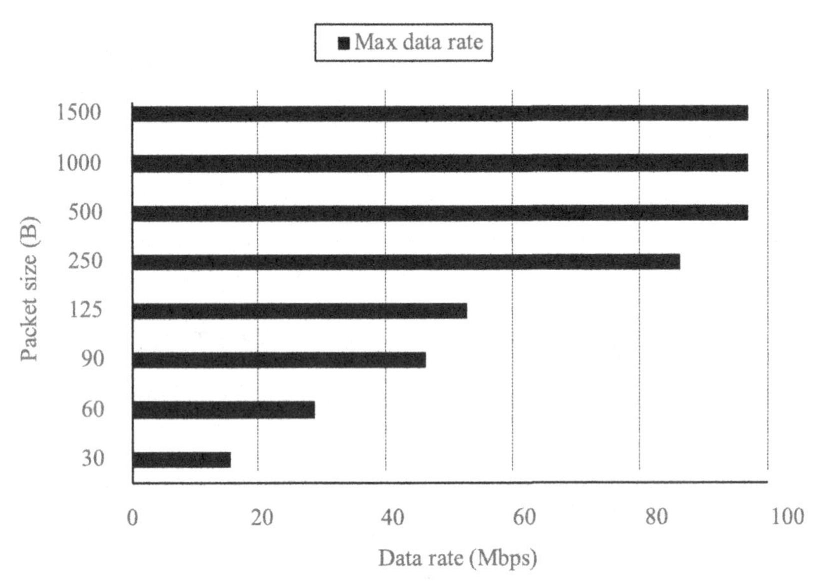

FIGURE 16.6

Maximum bandwidth supported by the physical link.

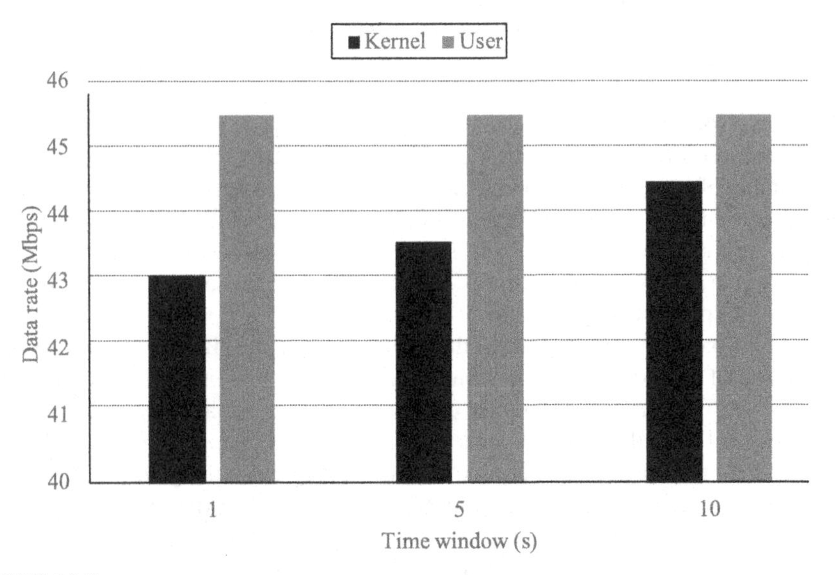

FIGURE 16.7

Maximum bandwidth per configuration.

packets is executed by the eBPF program in kernel space. When everything is run at kernel level, the data rate decreases by 5% in the worst case (1 s time window) when compared to the other configuration. The decrement in the performance is due to the execution of the machine learning model each time the time window expires; while this happens, the program cannot process any additional packets coming to the NIC. This is also the reason why the bandwidth increases when the time window gets longer in the kernel space scenarios. These results are expected because the machine learning model is executed with lower frequency with greater time frames, and hence the eBPF program can process more packets as it is not being blocked by this task. This behavior can be considered as a drawback of the implementation in kernel space, but the decrement in the performance of the IoT device is just 5% with high-bandwidth conditions, an infrequent situation in IoT environments.

Regarding resource consumption, Fig. 16.8 presents a real-time graph of the CPU consumption in terms of software interruptions obtained from the execution of an experiment using an aggregation time window of 5 s. Although both graphs oscillate on average around the same CPU usage range (60%–80%), the CPU usage of the in-kernel implementation is lower than that of the user space one. In this way, running the neural network in the Linux kernel reduces CPU consumption by 4% on average (dashed lines). The reduction of CPU usage is of utmost importance in constrained scenarios such as the ones found in TinyML architectures, given the computational limitations of end nodes. Note that these graphs show stable behavior of the CPU just before starting to drop packets due to CPU saturation (100% usage). This is the reason why the experiments were performed at the maximum possible data rate with no packet drops, and a 100% usage of the CPU is not reached in the graphs.

Finally, the execution time of the machine learning model in both scenarios is depicted in Fig. 16.9, depending on the time window to aggregate the gathered data. The 97% decrease obtained in the execution time of the neural network when it is executed in kernel space is remarkable, when compared with its execution at user level. This notable difference is because eBPF programs are run in a very efficient way and, in consequence, the execution of the machine learning model is faster in the privileged context. Besides, the runtime is not affected by the length of the time window, as it is not impaired by the data aggregation frequency. We consider these results of significant importance, as they demonstrate the advantages of executing complex algorithms within a privileged context to achieve notable improvements in the performance of the solution when compared with its execution at user level.

The obtained outcomes evidence the advantages of the eBPF ecosystem to execute software in kernel space to improve the performance of computing resource usage and reduce the packet processing time. Therefore, these kinds of solutions are interesting choices to be installed in fog or edge scenarios where executing any type of machine learning algorithm-based software is necessary, as in the considered use case of cyberattack self-protection. Furthermore, given its more than acceptable performance in commodity hardware, it serves as a cost-effective alternative to process traffic flows in an intelligent way.

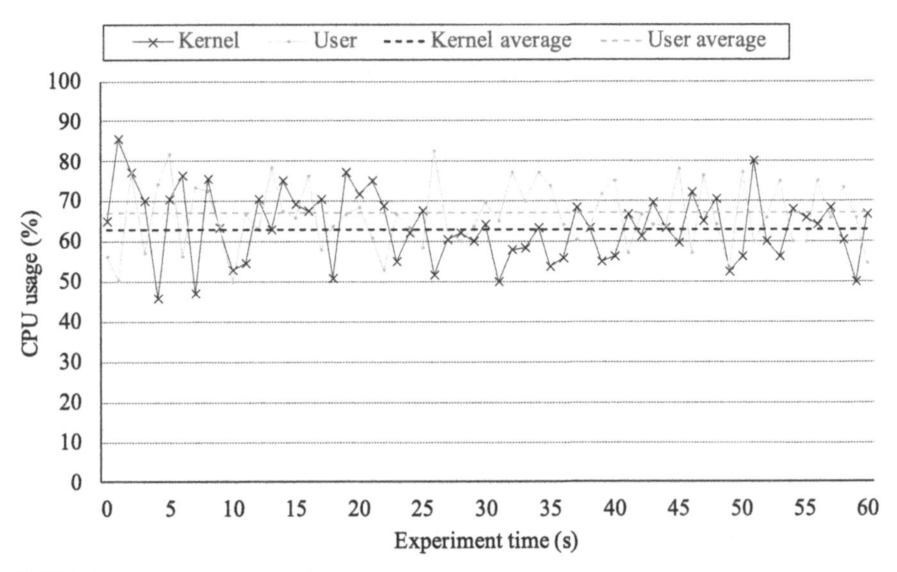

FIGURE 16.8

CPU usage.

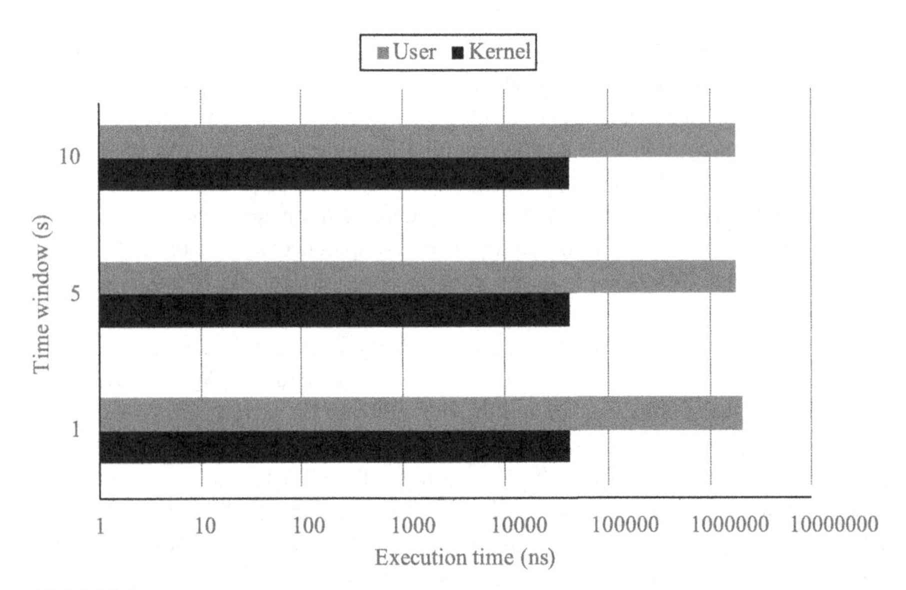

FIGURE 16.9

Machine learning model execution time.

16.4.6 Conclusion

This chapter focused on exploring the latest advancements in eBPF and TinyML, leveraging their combined potential to enhance packet filtering and processing. This integration results in lightweight solutions that exhibit rapid response times, minimal resource utilization, and the ability to scale effectively. Using the acquired knowledge, we effectively implemented an MLP within the eBPF framework into the Linux kernel. To achieve this, we thoroughly examined its implementation in Python, comprehending the fundamental concept of perceptrons, their layered organization, and the mathematical operations involved in classification tasks. Subsequently, we leveraged the emlearn library to facilitate the transition to eBPF, enabling the generation of a portable ANSI C structure.

Nonetheless, the limitations of eBPF presented a substantial obstacle. Primarily, the absence of floating-point operations in the Linux kernel necessitated the adoption of fixed-point arithmetic, utilizing 32-bit integers and adjusting arithmetic procedures accordingly. Moreover, we tackled the implementation of the sigmoid function from a mathematical standpoint to overcome the constraints posed by exponential and division operations. Additionally, we successfully mitigated the limitations imposed by unbounded loops within the Linux kernel by leveraging MACROS and static variables.

Once these limitations were successfully addressed, we conducted a study using a real-world use case. Specifically, we focused on cybersecurity in IoT, as constrained devices face challenges in implementing traditional security solutions. We trained various MLPs and implemented them into the Linux kernel using eBPF. Additionally, we developed a parsing mechanism to store the packet types obtained from the neighboring nodes, allowing persistent storage across different executions using pinned maps. To validate our approach, experiments were conducted on a Raspberry Pi 3 to compare the performance between the neural network in the kernel using eBPF and in user space. The obtained results demonstrated a substantial improvement when compared to a conventional implementation approach. Specifically, we achieved a significant reduction of over 95% in the execution time of the machine learning model, when operating within the Linux kernel as opposed to user space. Besides, a decrease in the CPU usage was evidenced, too. These findings have demonstrated the feasibility of implementing MLPs in eBPF programs, offering several key insights.

Looking ahead, several directions can be explored to extend and enhance this work. Firstly, the integration of some other complex neural network architectures, such as convolutional neural networks, in eBPF can provide more advanced classification capabilities for WSNs. Additionally, investigating optimization techniques specific to fixed-point arithmetic in eBPF can further improve the efficiency and accuracy of the implemented MLPs. Exploring approaches such as quantization-aware training and adaptive precision scaling can mitigate the potential loss of information caused by fixed-point representation, ensuring reliable and robust performance. Moreover, exploring the potential of hardware acceleration for eBPF-based neural networks is an interesting path for future research. Leveraging specialized hardware, such as SmartNICs or FPGAs, can potentially boost the processing capabilities of eBPF-based classifiers, enabling even more demanding functionality.

Acknowledgments

This work has been supported by the Spanish Ministry of Science and Innovation (MCIN) and the Spanish Research Agency (AEI), under the project ONOFRE-3 (Grant PID2020-112675RB-C44 funded by MCIN/AEI/10.13039/501100011033); by the Fundación Séneca-Agencia de Ciencia y Tecnología de la Región de Murcia, under the FPI Grant 21429/FPI/20, and co-funded by Odin Solutions S.L., Spain; by the Fulbright Commission in Spain under the Fulbright grant 00003/FLB/21; by the European Commission under the 5GASP (Grant No. 101016448) and NANCY (Grant No. 101096456) projects; and by the Spanish Ministry of Economy and Digital Transformation, under the project CERBERUS-HADES (Grant No. TSI-063000-2021-62).

References

[1] M.A.M. Vieira, M.S. Castanho, R.D.G. Pacífico, E.R.S. Santos, E.P.M.C. Júnior, L.F.M. Vieira, Fast packet processing with eBPF and XDP: concepts, code, challenges, and applications, ACM Comput. Surv. 53 (1) (May 2020) 1–36, https://doi.org/10.1145/3371038.

[2] R. Sanchez-Iborra, A.F. Skarmeta, TinyML-enabled frugal smart objects: challenges and opportunities, IEEE Circuits Syst. Mag. 20 (3) (2020) 4–18, https://doi.org/10.1109/MCAS.2020.3005467.

[3] A. Boukerche, R.W.L. Coutinho, Design guidelines for machine learning-based cybersecurity in Internet of things, IEEE Netw. 35 (1) (Jan. 2021) 393–399, https://doi.org/10.1109/MNET.011.2000396.

[4] S. Miano, M. Bertrone, F. Risso, M. Tumolo, M.V. Bernal, Creating complex network services with eBPF: experience and lessons learned, in: IEEE 19th International Conference on High Performance Switching and Routing (HPSR), Jun. 2018, pp. 1–8, https://doi.org/10.1109/HPSR.2018.8850758.

[5] T. Høiland-Jørgensen, et al., The eXpress data path, in: Proceedings of the 14th International Conference on Emerging Networking EXperiments and Technologies, Dec. 2018, pp. 54–66, https://doi.org/10.1145/3281411.3281443.

[6] D. Scholz, D. Raumer, P. Emmerich, A. Kurtz, K. Lesiak, G. Carle, Performance implications of packet filtering with Linux eBPF, in: 2018 30th International Teletraffic Congress (ITC 30), Sep. 2018, pp. 209–217, https://doi.org/10.1109/ITC30.2018.00039.

[7] W. Tu, F. Ruffy, M. Budiu, Linux network programming with P4, in: Linux Plumbers 2018, 2018, pp. 1–10.

[8] M. Bachl, J. Fabini, T. Zseby, A flow-based IDS using machine learning in eBPF, http://arxiv.org/abs/2102.09980, Feb. 2021.

[9] C. Liu, Z. Cai, B. Wang, Z. Tang, J. Liu, A protocol-independent container network observability analysis system based on eBPF, in: 2020 IEEE 26th International Conference on Parallel and Distributed Systems (ICPADS), Dec. 2020, pp. 697–702, https://doi.org/10.1109/ICPADS51040.2020.00099.

[10] H. Chang, M. Kodialam, T.V. Lakshman, S. Mukherjee, Microservice fingerprinting and classification using machine learning, in: 2019 IEEE 27th International Conference on Network Protocols (ICNP), Oct. 2019, pp. 1–11, https://doi.org/10.1109/ICNP.2019.8888077.

[11] J. Chen, S.S. Banerjee, Z.T. Kalbarczyk, R.K. Iyer, Machine learning for load balancing in the Linux kernel, in: Proceedings of the 11th ACM SIGOPS Asia-Pacific Workshop on Systems, Aug. 2020, pp. 67–74, https://doi.org/10.1145/3409963.3410492.

[12] M. Wollschlaeger, T. Sauter, J. Jasperneite, The future of industrial communication: automation networks in the era of the Internet of Things and Industry 4.0, IEEE Ind. Electron. Mag. 11 (1) (Mar. 2017) 17–27, https://doi.org/10.1109/MIE.2017.2649104.

[13] D.L. Dutta, S. Bharali, TinyML meets IoT: a comprehensive survey, IEEE Int. Things J. 16 (Dec. 2021) 100461, https://doi.org/10.1016/j.iot.2021.100461.

[14] R. Sanchez-Iborra, LPWAN and embedded machine learning as enablers for the next generation of wearable devices, Sensors 21 (15) (Jul. 2021) 5218, https://doi.org/10.3390/s21155218.

[15] E. Ahmed, I. Yaqoob, A. Gani, M. Imran, M. Guizani, Internet-of-things-based smart environments: state of the art, taxonomy, and open research challenges, IEEE Wirel. Commun. 23 (5) (Oct. 2016) 10–16, https://doi.org/10.1109/MWC.2016.7721736.

[16] A. Roy, A.M.S. Zalzala, A. Kumar, Disruption of things: a model to facilitate adoption of IoT-based innovations by the urban poor, Proc. Eng. 159 (2016) 199–209, https://doi.org/10.1016/j.proeng.2016.08.159.

[17] R. Sanchez-Iborra, J. Sanchez-Gomez, J. Santa, P.J. Fernandez, A. Skarmeta, Integrating LP-WAN communications within the vehicular ecosystem, in: International Symposium on Mobile Internet Security (MobiSec17), 2017, pp. 1–12.

[18] R. Sanchez-Iborra, A. Zoubir, A. Hamdouchi, A. Idri, A. Skarmeta, Intelligent and efficient IoT through the cooperation of TinyML and edge computing, Informatica 34 (1) (Jan. 2023) 147–168, https://doi.org/10.15388/22-INFOR505.

[19] M. Abbasi, A. Shahraki, A. Taherkordi, Deep learning for network traffic monitoring and analysis (NTMA): a survey, Comput. Commun. 170 (Mar. 2021) 19–41, https://doi.org/10.1016/j.comcom.2021.01.021.

[20] A.Y. Nikravesh, S.A. Ajila, C.-H. Lung, W. Ding, Mobile network traffic prediction using MLP, MLPWD, and SVM, in: 2016 IEEE International Congress on Big Data (BigData Congress), Jun. 2016, pp. 402–409, https://doi.org/10.1109/BigDataCongress.2016.63.

CHAPTER

17 A real-time price recognition system using lightweight deep neural networks on mobile devices

Musa Peker, Melek Turan, Hüseyin Özkan, Cevat Balaban, Nadir Kocakır, and Önder Karademir

Özdilek Ev Tekstil San. ve Tic. AŞ, Özveri Ar-Ge Merkezi, Bursa, Turkey

17.1 Introduction

In recent years, the increasing level of automation with AI technology has enabled efficient optimization of many processes and the use of less complex tools for routine tasks [1]. The use of smart technologies in the retail sector is increasing day by day. According to a study by Precedence Research, in 2022, the worldwide market size for artificial intelligence (AI) in retail was valued at USD 8.41 billion [2]. It is expected to reach approximately USD 45.74 billion by 2032, with a compound annual growth rate (CAGR) of 18.45% from 2023 to 2032. So, in the future, innovative retail can be realized entirely with AI technology.

Price tags show important information in the retail industry. People often encounter false information on price tags in stores, which leads to a decrease in customer loyalty and satisfaction [3]. Important information on price tags is usually the product name, product specifications, and price. This information is constantly changing and requires verification. The real price of the product is obtained by reading the barcode on the price tag with a barcode reader. With eye control, the price information on the tag is read and a comparison is made with the real price value obtained from the barcode reader. It is important to automate this process due to frequent changes in price information and errors caused by human control. There are solutions developed to solve this problem. For instance, electronic shelf labels provide an effective solution for validating price tag data. Nevertheless, a comprehensive examination of the associated costs reveals that their implementation can be financially demanding. Upon further analysis, it has been determined that the unit cost of an electronic label typically ranges from $6 to $10 [4]. In the context of large-scale retail environments, such as hypermarkets with a large number of products, the cumulative expenses become substantial. Moreover, the integration of electronic shelf

TinyML for Edge Intelligence in IoT and LPWAN Networks. https://doi.org/10.1016/B978-0-44-322202-3.00022-1

labels necessitates the use of management software, incurring additional expenses in the form of licensing fees and maintenance costs. Additionally, being electronic devices, these labels entail ongoing maintenance expenditures. Thus, while electronic shelf labels offer advantages in terms of data accuracy, their widespread adoption may be limited by the considerable financial investment required for both initial implementation and ongoing operational maintenance. For this reason, paper price tags are still used in the retail industry. In this study, we aimed to automatically read the price information on the price tags by using deep learning technology. The contributions of the developed system are listed below:

- Current optical character recognition (OCR) solutions focus on recognizing all characters in an image, such as text, numbers, and symbols [5]. There are many data on the price tags such as product name, logo, and expiry date. With the existing OCR solutions, unnecessary data are also read while reading the price information, and this causes confusion. To read the price information, it is sufficient to recognize the numbers, the price unit and the decimal point. In this study, a price recognition system focusing only on price information has been developed.
- In obtaining price information, accurately identifying the decimal part is a difficult task and is also the key point that differentiates it from other text and image recognition tasks. In this study, a novel method is proposed for the determination of the decimal part.
- With the proposed method, price information can be read from more than one price tag at the same time. This provides a significant advantage in terms of time.
- It is possible to run deep learning algorithms on mobile devices with fast and high accuracy values. A comprehensive comparative analysis of different algorithms on different smartphone models has been conducted.
- For the price recognition system, a unique dataset consisting of 50,000 images has been created. The dataset consists of price images with different features and difficulty levels (such as lack of lighting, different styles, and different shooting angles). This makes the generalization ability of the developed deep learning model strong.

The remainder of the chapter is organized as follows. Section 17.2 summarizes the literature survey. In Sections 17.3 and 17.4, information about the proposed method is shared. Experimental results and a discussion are provided in Section 17.5. In Section 17.6, a summary of the results obtained and the work planned for the future are shared.

17.2 Related works

In the literature, studies using image processing or deep learning methods for price recognition have been examined. Hussin et al. [6] used the hue, saturation, and value (HSV) model for color filtering. They used various image-processing techniques to obtain fields with product name and price information on price tags. An OCR technique was used to read the information in the obtained fields. Researchers reported that erroneous results were encountered in area detection due to white and black noise. It was also observed that the effectiveness of OCR decreased significantly in two cases. These are images taken from a distance of more than 20 cm and have different size fonts.

Mou et al. [7] focused on finding decimal points, a difficult problem in reading price information. In this context, they proposed a multi-task convolutional neural network (CNN) model. Aliev et al. [1] proposed a practical solution for searching, localizing, and recognizing price labels. This solution is based on the Niblack binarization method and analysis/clustering of connected components. Fomenko et al. [8] developed a solution for detecting price tags and the price information region on a shelf image. They investigated the effect of transfer learning on price information extraction using the fully convolutional network.

Laptev et al. [3] conducted a comparative study using different deep learning-based methods for price tag data analysis. In this context, they conducted a comparative analysis using the MobileNetV2, UNet, YOLOv4-Tiny, and VGG16 algorithms. For text recognition, the EasyOCR API was used. The researchers reported that the most suitable network for segmentation was the YOLOv4-Tiny, with a cross-validation accuracy of 96.92%.

17.3 Machine learning-based techniques for image processing

This section presents an overview of the deep learning models used in this study.

17.3.1 Single Shot MultiBox

The Single Shot MultiBox Detector (SSD) represents an object detection model built upon CNNs. Its primary objective is to efficiently detect objects of varying sizes and aspect ratios within a given input image. To gain a comprehensive understanding of its operational framework, one can refer to Fig. 17.1, illustrating the architecture and workflow of SSD [9,10].

Within the SSD framework, the input image undergoes a partitioning process, forming a grid of varying sizes. Each grid cell takes on the responsibility of detecting objects of different classes and aspect ratios. The detection process involves computing confidence scores for each grid cell, reflecting its alignment with a specific object. An essential clarification lies in the role of the detector, a key component that was not explicitly defined in the original text. The SSD detector employs a combination of convolutional layers, capturing hierarchical features and leading to the computation of class scores for each default bounding box, thereby enhancing the model's accuracy in discerning objects.

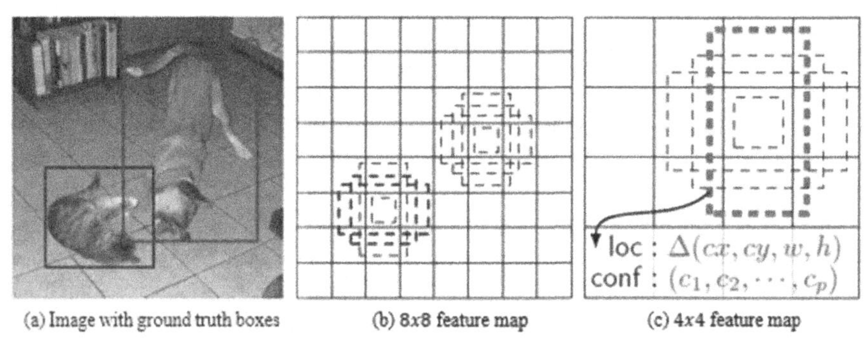

FIGURE 17.1

SSD framework [9].

The non-maximum suppression (NMS) method is a critical component of SSD. NMS acts as a postprocessing technique, effectively filtering out redundant bounding box predictions. Following the initial detection phase, NMS plays a pivotal role in eliminating overlapping bounding boxes, retaining only the most trustworthy and non-overlapping detections. This step is instrumental in refining the final set of accurate object detections.

Additionally, using different grid sizes is a noteworthy strategy employed by SSD to detect objects of varying sizes efficiently. As illustrated in Fig. 17.1, smaller grids are employed when detecting smaller objects such as cats, whereas larger grid sizes are used for larger objects like dogs. This adaptive grid size selection contributes to the efficiency of SSD in object detection.

17.3.2 MobileNets

MobileNet is based on a modern architecture using deeply separable convolutions that enables the development of lightweight deep neural networks with low latency for mobile and embedded devices [11]. In MobileNet, *deep separable convolution* implements a filter to each input channel. *Point-wise convolution* then implements 1×1 convolution to combine the convolution outputs in the depth direction. This model uses two hyperparameters that efficiently make a trade-off between delay and accuracy: width factor (α) and resolution factor (ρ). The α and ρ values are both in the range (0, 1). Decreasing α amounts to decreasing the number of filters, while decreasing ρ amounts to decreasing the image resolution. By adjusting both the α and ρ values, we can get a small network and reduce the computation. This may result in a decrease in accuracy.

In this study, two different versions of MobileNet, MobileNet V1 [11] and MobileNet V2 [12], were used. Fig. 17.2 demonstrates the difference between the MobileNet V1 and MobileNet V2 architectures. Both models have *point-wise convolutional layers* and *depth-wise separable convolutional blocks*. The residual layer

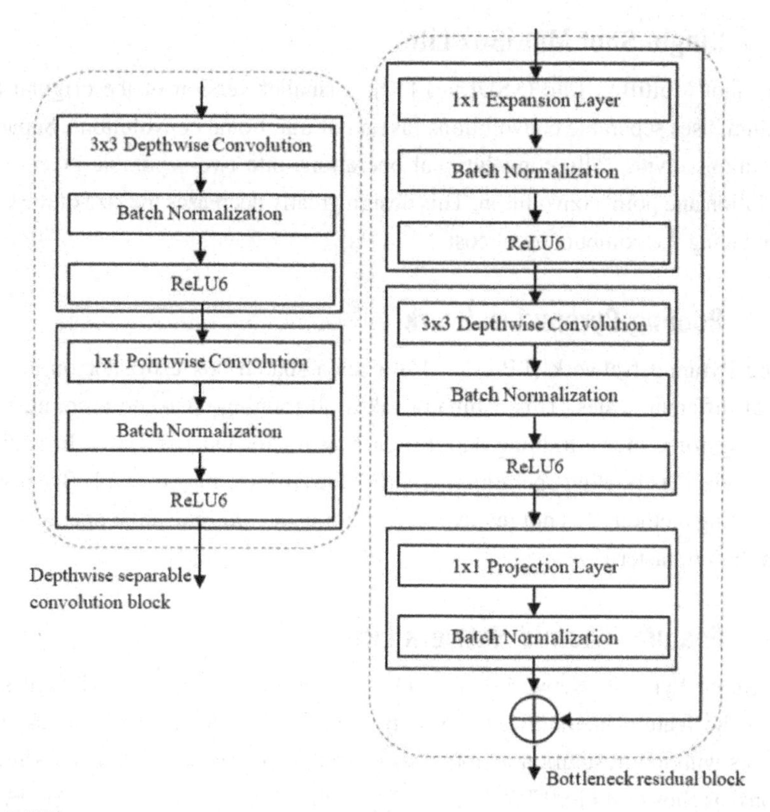

FIGURE 17.2

(a) MobileNet V1 convolutional blocks. (b) MobileNet V2 convolutional blocks [13].

structure of MobileNet V2 is a significant factor that improves the accuracy of the deep convolution layer.

MobileNet V2 substitutes the *depth-wise separable convolution block* with the *bottleneck residual block*, as illustrated in Fig. 17.2. The pivotal alterations in this novel block architecture encompass linear bottlenecks interposed between layers, residual connections linking the bottlenecks, and expansion and projection layers. One of the key attributes of MobileNet V2 lies in its *bottleneck residual block* structure. This architecture regulates information flow between layers through linear bottlenecks, concurrently mitigating gradient loss in deep networks via residual connections between bottlenecks. Expansion layers augment the learning capacity of the network, while projection layers facilitate a transition to lower-dimensional feature maps, thereby reducing computational costs. As evident from this comparison, MobileNet V2 boasts a more intricate structure compared to its predecessor, contributing to heightened accuracy and more efficient learning.

17.3.3 Single Shot MultiBox Lite

Single Shot MultiBox Lite (SSDLite) [12], a smaller version of the original SSD algorithm, uses separable convolutions instead of traditional convolutions. Separable convolutions divide fully convolutional operations into two separable layers, deep convolution and point convolution. This design greatly decreases the size of the CNN, thus reducing the computational cost.

17.3.4 Pooling Pyramid Network

Pooling Pyramid Network (PPN) [14] utilizes a shared box estimator on attribute maps at different scales. This estimator takes all training data into account, rather than using some of the training data for each estimator independently. In addition, PPN utilizes max pooling operations to shrink the attribute map to 1×1. Because the max pooling method does not involve any addition and multiplication operations, the operations are faster.

17.3.5 Feature Pyramid Network Lite

The Feature Pyramid Network (FPN) [15] has strong semantics at all scales and extends the feature hierarchy of CNN. It does this by combining high-resolution attributes with low-resolution features over a top-down trajectory with sideway connections, as shown in Fig. 17.3. The synergy of these mechanisms empowers FPN to effectively capture and preserve meaningful information at various scales, fostering a comprehensive understanding of the visual context. Unlike traditional architectures, these structural modifications do not incur a compromise on representation, memory, or processing speed. Consequently, FPN stands as a testament to the synergy between high- and low-level features, showcasing its prowess in semantic feature extraction across diverse resolutions.

The goal of FPNLite [16] is to get an architecture for faster inference. This architecture is made using the original FPN as the backbone, but by replacing all the convolution layers with deeply separable convolution layers.

17.3.6 Quantized SSD-MobileNet models

In training a deep learning model, 32-bit floating-point weights and activation values are usually utilized. In a quantized model, integers are used instead of floating-point values. This significantly reduces computational complexity. Thanks to this transformation, the trained neural network model is utilized on devices with lower resource requirements [17].

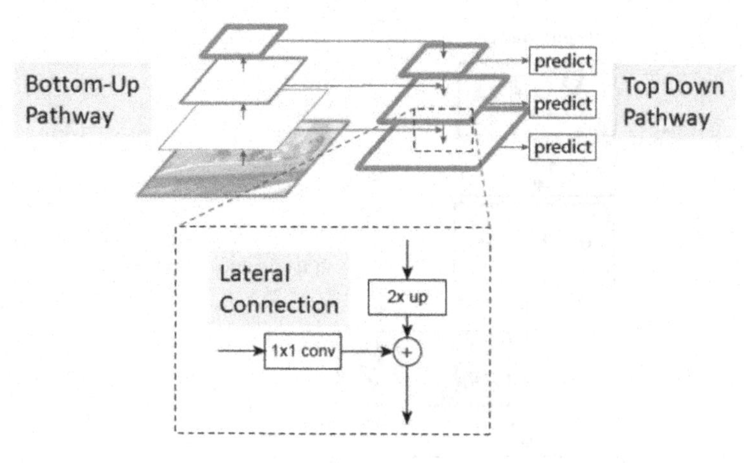

FIGURE 17.3

Feature Pyramid Network (FPN) [15].

17.4 Material and methods

Fig. 17.4 shows the basic steps of the proposed price recognition method. First, the price image presented on the system is presented as input data to the detection algorithm. When this algorithm is applied, the data obtained, such as the coordinates of the class and boundary, are used as input to the recognition algorithm. The recognition algorithm then takes these detected parameters and processes the price image. The developed recognition algorithm identifies and extracts numeric and character components from the image. This critical stage is to ensure the correct detection and localization of relevant information. Finally, the combination of these complex processes leads to the extraction of price information from the initial image. The developed price recognition algorithm plays an important role to correctly parse and interpret the detected numeric and character elements.

In summary, the proposed methodology seamlessly integrates a multi-stage process involving detection and recognition algorithms, effectively extracting price information from submitted images.

17.4.1 Data collection and annotation

The dataset consists of automatically cropped price information fields. This dataset contains 50,000 images obtained from more than 30 stores and markets. The images contain different difficulty levels such as lack of lighting, different size, and different viewpoints. Some sample images of the dataset are presented in Fig. 17.5.

To further strengthen the dataset, 10,000 images obtained using 134 different fonts and 7 different font sizes are included in the dataset. In addition, some of the images

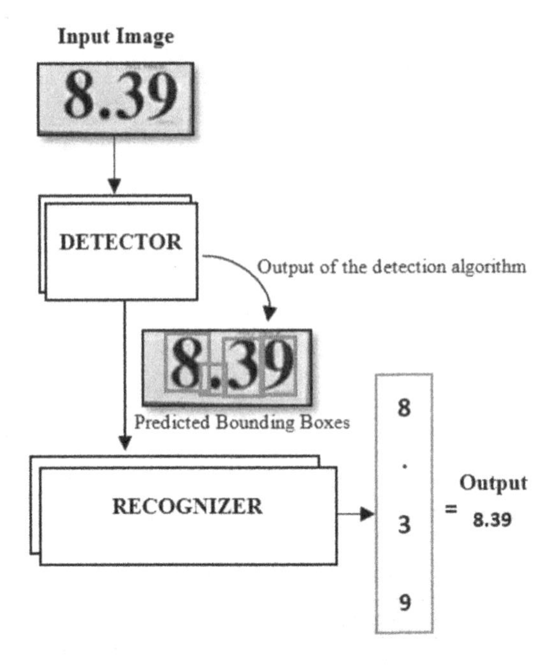

FIGURE 17.4

Schematic overview of the proposed method.

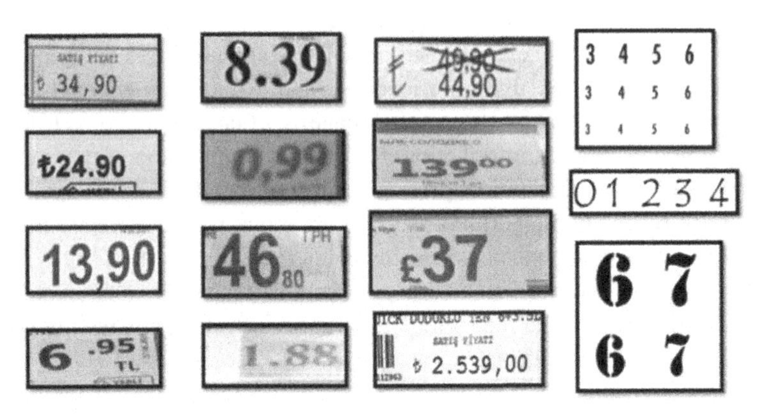

FIGURE 17.5

Some images in the dataset.

taken from different stores and markets were blurred with a 5 × 5 filter and included in the dataset. Of the entire dataset, 90% was used as training data and 10% was used as test data.

Table 17.1 Object detection methods used in this study.

Model	Architecture	Model name	Resolution
Mobilenet v1 ($\alpha = 0.75$)	SSD	SSD_Mobilenet_V1 (0.75_Quantized)	300 × 300
Mobilenet v1 ($\alpha = 1$)	SSD+PPN	SSD_Mobilenet_V1_PPN	300 × 300
Mobilenet v2 ($\alpha = 1$)	SSDLite	SSDLite_Mobilenet_V2	300 × 300
Mobilenet v2 ($\alpha = 1$)	SSD	SSD_Mobilenet_V2	320 × 320
Mobilenet v2 ($\alpha = 1$)	SSD+FPNLite	SSD_Mobilenet_V2_FPNLite	320 × 320
Mobilenet v2 ($\alpha = 1$)	SSD (quantized)	SSD_Mobilenet_V2_Quantized	300 × 300

17.4.2 Detection algorithms

The deep learning models used in this study are hybrid versions of different methods described in Section 17.3. Information on the methods used is presented in Table 17.1.

TensorFlow Lite conversion has been carried out so that these models can work efficiently on edge devices such as mobile phones. TensorFlow Lite enables to quickly run machine learning models with low latency on mobile and embedded devices [18]. In this way, machine learning can be performed on these devices without using an external API or server. The development and deployment processes of this process are presented in Fig. 17.6.

Fig. 17.7 shows the step-by-step creation process of the TFLite file. In this study, transfer learning was used to retrain all used neural network models. To perform this operation, the bounding box data were first serialized to the TFRecord data format. After this operation was performed, the data were stored as a series of binary streams. After the models were retrained, they were transferred to frozen graphs. Frozen graphs reduce the size of a model by removing unnecessary information. Finally, these graphs were converted to TensorFlow Lite format.

17.4.3 Recognition algorithm

The price information on the price tags can be in different formats. While some images have decimal separators, these are not available in some other. In some label types, the decimal part is positioned above, like a superscript. In some label types, it is positioned as a subscript.

In the price tag recognition task, accurately identifying the decimal point that covers a small area of pixels or is skipped directly is a challenging task, and is also the key point that sets it apart from other text and image recognition tasks. All these significantly affect price recognition performance. An algorithm has been developed to detect the decimal part in price images. This algorithm uses the bounding box and class information, which are the outputs of the detection algorithm.

Fig. 17.8 delineates the intricacies of the recognition algorithm employed for price recognition. The process begins with the identification of characters from the price image, along with their corresponding bounding boxes. Subsequently, these characters are sorted from left to right, forming the basis for further analysis.

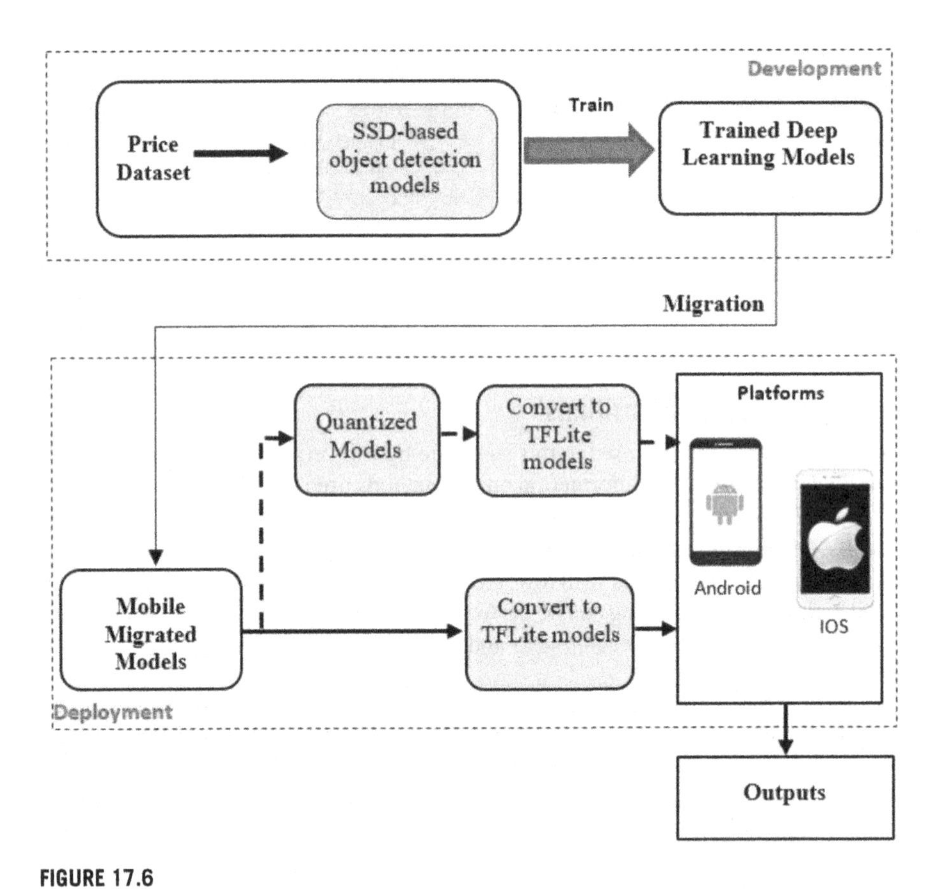

FIGURE 17.6

Development and deployment processes of the detection algorithm.

The algorithm then checks the presence of a decimal separator within the acquired data. If such a separator exists, the algorithm proceeds to determine if multiple separators are present. In the case of multiple separators, only the rightmost one is considered, disregarding others. This meticulous approach ensures accurate identification and isolation of the decimal point. Following this determination, the algorithm enters the price recognition phase and concludes its operation.

In instances where no decimal separator is detected, the algorithm initiates a calculation of the height of all bounding boxes. The box with the highest height value is identified, and a subsequent assessment is made to ascertain whether there are boxes that are 60% smaller than this reference height. If such smaller boxes exist, the characters within them are recognized as belonging to the decimal part, leading to the price recognition stage and the subsequent conclusion of the algorithm. Conversely, if smaller boxes are not detected, this price image has no decimal point and the price information is integer. The price is read and the algorithm terminates. Here, the value

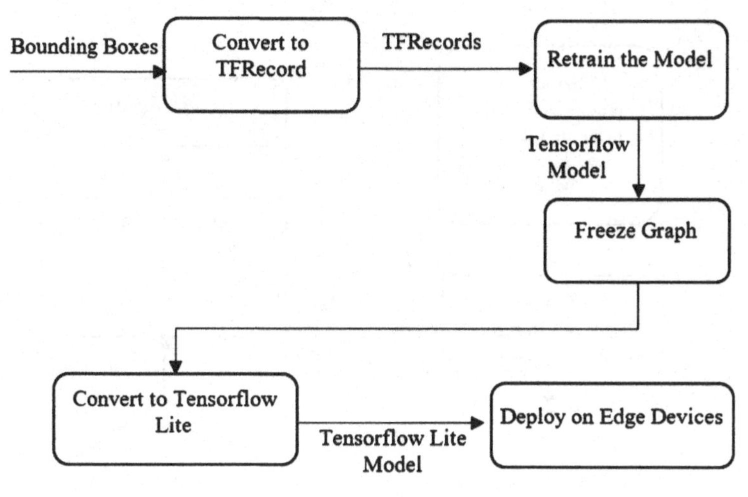

FIGURE 17.7

Training procedure flow [10].

of 60% was determined as a result of the analysis performed on the dataset. It is a parametric value. This value can be changed according to the type of problem.

This algorithmic flow, adept at handling diverse formats of price information, is specifically designed to address challenges inherent in accurately identifying decimal points, which may vary in position or be absent altogether. The proposed algorithm ensures a systematic and reliable approach, setting it apart from conventional text and image recognition tasks. The detailed steps presented in the figure serve as a comprehensive guide to the nuanced process of price recognition.

17.5 Experimental results and discussion

Comprehensive analyses have been conducted to evaluate the performance of detection and recognition algorithms.

17.5.1 Performance evaluation metrics

The results obtained by applying the price recognition algorithm on the test data were analyzed. Accuracy and F1 score were used as performance evaluation metrics. To calculate these metrics, TP (true positives), FP (false positives), TN (true negatives), and FN (false negatives) were utilized:

$$Accuracy = \frac{TP + TN}{TP + FP + FN + TN}, \tag{17.1}$$

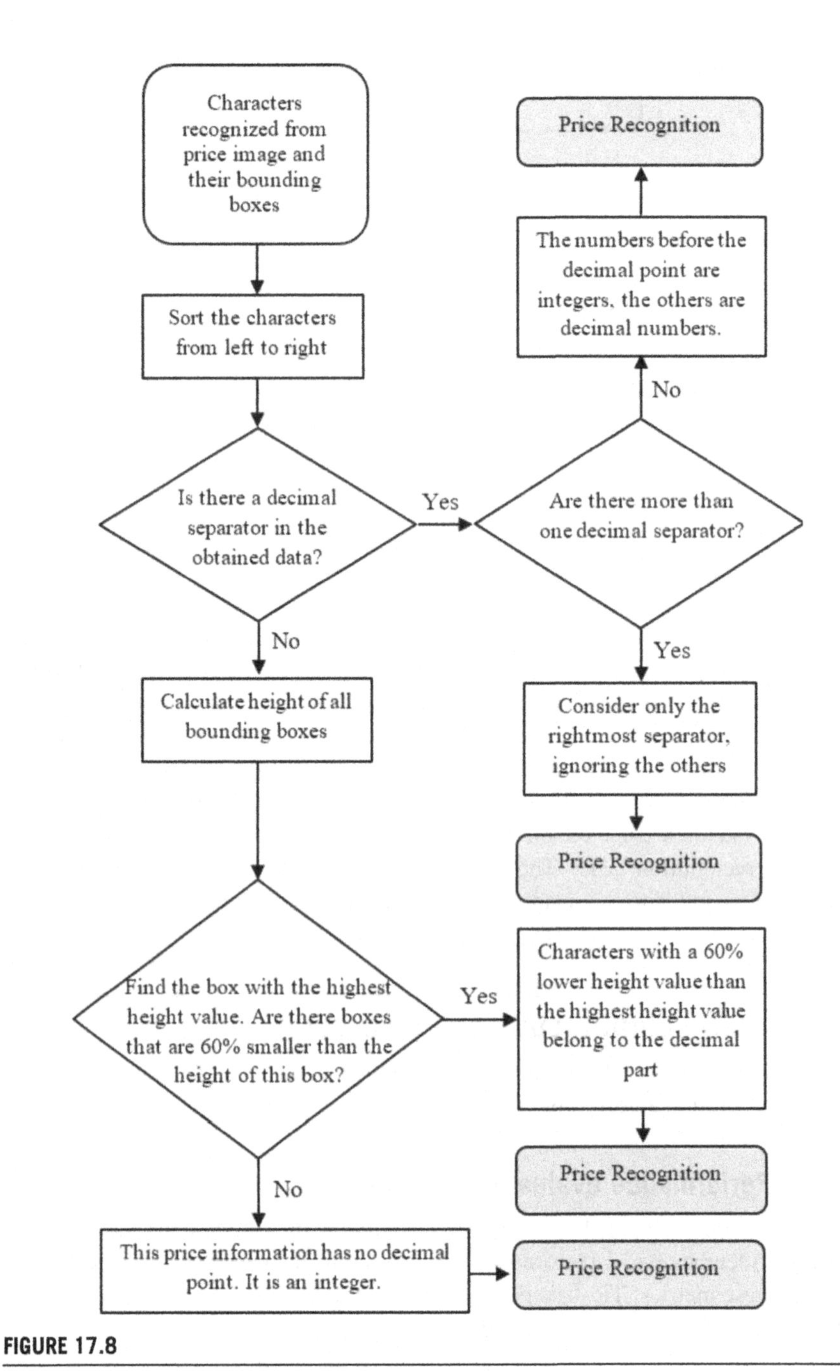

FIGURE 17.8

The flowchart of the price recognition algorithm.

$$Precision = \frac{TP}{TP + FP}, \tag{17.2}$$

$$Recall = \frac{TP}{TP + FN}, \tag{17.3}$$

$$F1 = \frac{2 \times Recall \times Precision}{Recall + Precision}. \tag{17.4}$$

These are key metrics for measuring the performance of a model. Accuracy measures the percentage of correctly estimated observations and is a good measure of quality. The F1 score is computed by the weighted average of precision and recall. Accuracy is preferred when TPs and TNs are more significant while F1 score is preferred when FNs and FPs are very significant.

17.5.2 Experimental studies on detection algorithms

- **Training procedure**

Training and testing of deep learning models were carried out on an Ubuntu system with an Intel Core processor i9, 32 GB RAM, and an Nvidia RTX 3080 Ti GPU with 11 GB RAM. The training parameters are shown in Table 17.2. Hyperparameter values have a significant impact on performance [19]. The models used in this study already had hyperparameter values analyzed to achieve optimal performance for each model. These values are used by default in transfer learning. However, in this study, some values were changed to achieve better performance after the experimental processes. To make these changes, the configuration file for each model was edited.

Batch size is one of the hyperparameters. It is the number of training examples utilized in one iteration. A smaller batch size means less memory and faster training. In this study, the batch size was set to 16. To obtain the ideal number of iterations, tests were performed to observe when the training error curve began to converge. Accordingly, the number of training steps was set to 300,000. Parameters such as momentum, learning rate, and others were not changed, and the values that came with the original model have been preserved.

- **Performance evaluation of detection algorithms**

During the training process, the success of deep learning models on test data was evaluated. The accuracy rates and F1 values obtained on the test data as a result of the training process are shared in Table 17.3. Fig. 17.9 shows the loss values obtained as a result of training for each model. In the figure, looking at the results of the 150,000th step number, it is seen that the lowest loss value is obtained with SSD-Mobilenet-FPNLite. This model achieved an accuracy rate of 96.14% on the test data. SSD-Mobilenet-V2 also achieved good results with an accuracy rate of 94.25%.

The worst result was obtained with the SSD_Mobilenet_V1_0.75_Quantized_300x300 model. In general, models trained with TensorFlow 2 gave better results than models trained with TensorFlow 1. Quantized models were unsuccessful compared to other models. Considering the outputs, the SSD-Mobilenet-FPNLite model

Table 17.2 The training parameters of deep learning models.

Model	Basic learning rate	Number of iterations of training	Batch size	Optimization algorithm	Momentum	Decay	Framework
Model 1*	[0.059, 0.200]	300,000	16	Momentum	0.899	0.97	TensorFlow 1
Model 2*	[0.133, 0.699]	300,000	16	Momentum	0.899	0.97	TensorFlow 1
Model 3*	[0.004]	300,000	16	RMSProp	0.899	0.99	TensorFlow 1
Model 4*	[0.133, 0.800]	300,000	16	Momentum	0.899	0.99	TensorFlow 2
Model 5*	[0.026, 0.079]	300,000	16	Momentum	0.899	0.99	TensorFlow 2
Model 6*	[0.004]	300,000	16	RMSProp	0.899	0.99	TensorFlow 1

**Model 1: SSD_Mobilenet_V1_0.75 Quantized_300 × 300*
**Model 2: SSD_Mobilenet_V1_PPN_300 × 300*
**Model 3: SSDLite_Mobilenet_V2_300 × 300*
**Model 4: SSD_Mobilenet_V2_320 × 320*
**Model 5: SSD_Mobilenet_V2_FPNLite_320 × 320*
**Model 6: SSD_Mobilenet_V2_Quantized_300 × 300*

Table 17.3 Accuracy and F1 scores obtained from the models.

Model	Accuracy (%)	F1 (%)
SSD_Mobilenet_V2	94.25	95.38
SSD_Mobilenet_V1_0.75_Quantized	20.12	26.16
SSD_Mobilenet_V2_Quantized	61.40	69.25
SSD_Mobilenet_V2_FPNLite	96.14	96.85
SSD_Mobilenet_V1_PPN	85.65	90.06
SSDLite_Mobilenet_V2_300x300	50.42	56.50

was preferred for this study. In addition, real-time tests have shown that the SSD-Mobilenet-FPNLite model gives more accurate and stable results.

- **Experimental studies on the recognition algorithm**

As a result of the tests, the most successful detection algorithm was determined as SSD-Mobilenet-FPNLite. The outputs of this algorithm are presented as input data to the recognition algorithm. The successful performance of the recognition algorithm has been tested on nine different phone models with different hardware and features. These smartphones and their features are presented in Table 17.4. Testing was carried out with a total of 1000 price images in different formats in a supermarket in real time. Mean prediction times are presented in Fig. 17.10 and accuracy rates are presented in Fig. 17.11.

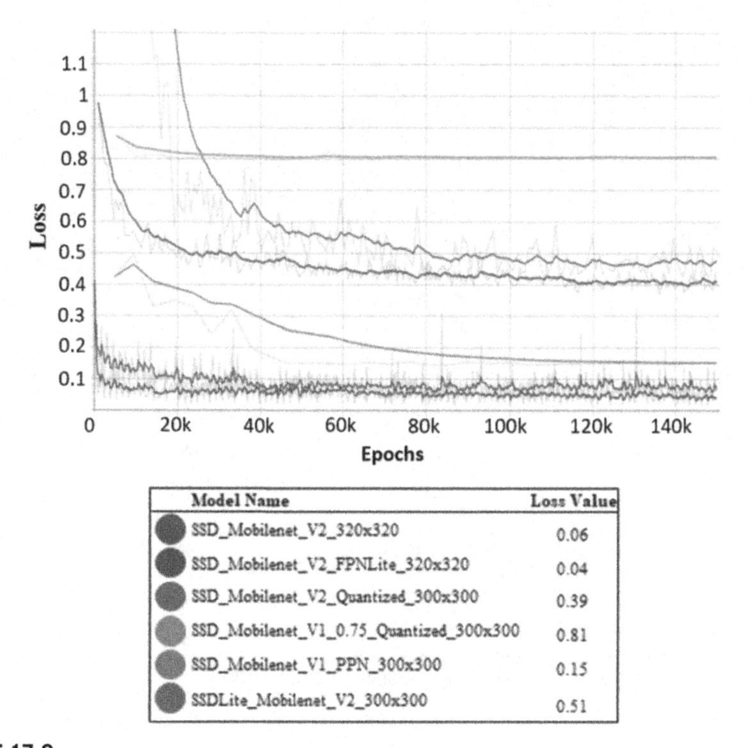

FIGURE 17.9

The training phase of the object detection method.

The mean prediction time and classification accuracy vary according to the hardware capabilities of smartphones. In general, phones with IOS have a lower mean prediction time than phones with Android. Among the phones with Android, the Samsung Galaxy S21 gave the best results. The mean prediction times of the four phones with IOS are close to each other. When we evaluated the phones with Android, we observed that the mean prediction time decreased as the RAM and CPU specs improved. When we examined the phones with IOS, the mean prediction times are close to each other despite the varying amounts of RAM. The reason for this is thought to be the similarity of CPU features.

The most important factor affecting the classification accuracy is the camera feature. The accuracy of the model increases with clearer and higher-quality images. A higher accuracy rate has been achieved with phones with IOS. The camera features of these phones are very close to each other. Among the phones with Android, the lowest accuracy rate was obtained with Samsung Galaxy J7 Core and Huawei P Smart phones with low camera resolution. The highest accuracy rate was obtained with the Samsung Galaxy S21. Camera features of Xiaomi Redmi Note 9 Pro and Huawei P40 lite phones are better than those of Samsung Galaxy S21. However, a higher accuracy

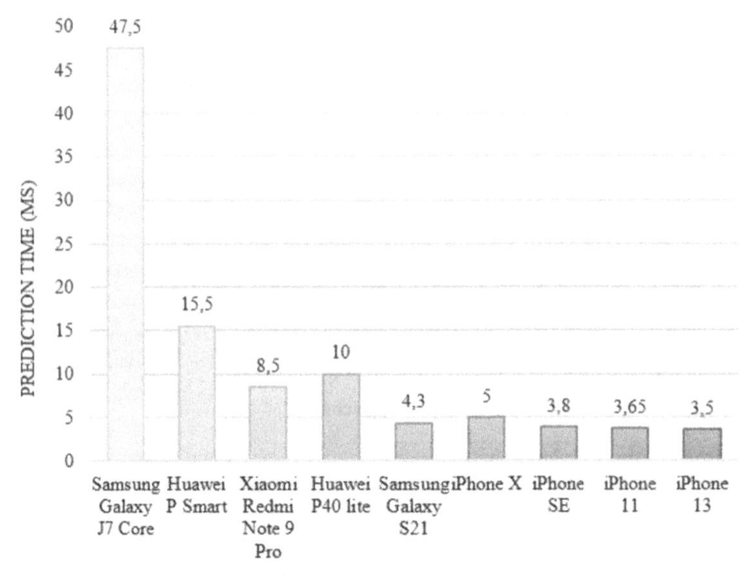

FIGURE 17.10

Mean prediction times of deep learning models on smartphones.

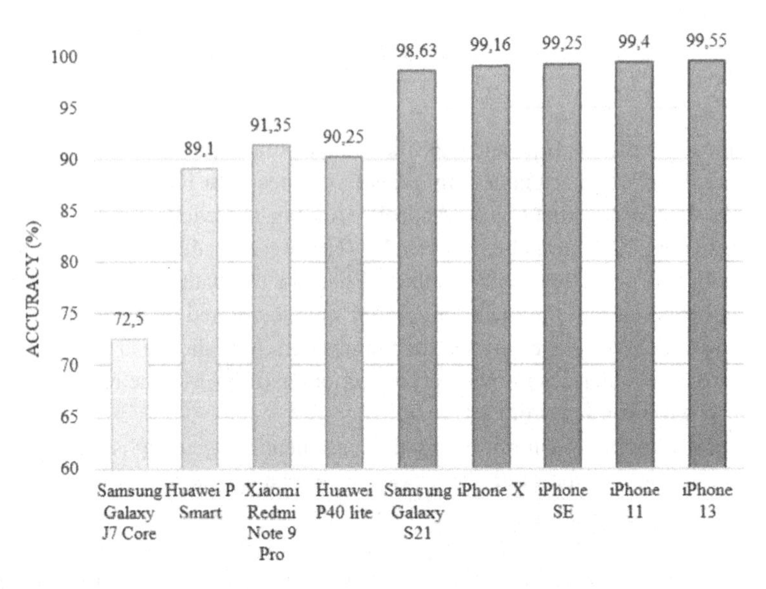

FIGURE 17.11

Accuracy of deep learning models on smartphones.

Table 17.4 Smartphones used in this study and their features.

Model	CPU	GPU	RAM	OS	Camera
Samsung Galaxy S21	Octacore (3× 2.80 GHz Cortex-A78 & 1× 2.9 GHz Cortex-X1 & 4× 2.2 GHz Cortex-A55)	Mali-G78 MP14	8 GB	Android	12 MP, f/1.8, 26 mm (wide), 1/1.76″, 1.8 μm, Dual Pixel PDAF, OIS
Samsung Galaxy J7 Core	Octacore (1.6 GHz Cortex-A53)	Mali-T830 MP1	2 GB	Android	13 MP, f/1.9, AF, 28 mm (wide)
Huawei P Smart	Octacore (4× 1.7 GHz Cortex-A53 & 4× 2.2 GHz Cortex-A73)	Mali-G51 MP4	3 GB	Android	13 MP, f/1.8, PDAF
Xiaomi Redmi Note 9 Pro	Octacore (6× 1.8 GHz Kryo 465 Silver & 2× 2.3 GHz Kryo 465 Gold)	Adreno 618	6 GB	Android	64 MP, f/1.9, 26 mm (wide), 1/1.72″, 0.8 μm, PDAF
Huawei P40 lite	Octacore (6× 1.88 GHz Cortex-A55 & 2× 2.27 GHz Cortex-A76)	Mali-G52 MP6	6 GB	Android	48 MP, f/1.8, 26 mm (wide), 1/2.0″, 0.8 μm, PDAF
iPhone X	Hexacore (2.39 GHz; 2× Monsoon & 4× Mistral)	3-core graphics	3 GB	IOS	12 MP, f/1.8, 28 mm (wide), 1/3″, 1.22 μm, dual pixel PDAF, OIS
iPhone SE	Hexacore (2× 2.65 GHz Lightning & 4× 1.8 GHz Thunder)	4-core graphics	3 GB	IOS	12 MP, f/1.8 (wide), PDAF, OIS
iPhone 11	Hexacore (4× 1.8 GHz Thunder & 2× 2.65 GHz Lightning)	4-core graphics	4 GB	IOS	12 MP, f/1.8, 26 mm (wide), 1/2.55″, 1.4 μm, dual pixel PDAF, OIS
iPhone 13	Hexacore (2× 3.23 GHz Avalanche + 4× 1.82 GHz Blizzard)	4-core graphics	4 GB	IOS	12 MP, f/1.6, 26 mm (wide), 1/1.9″, 1.7 μm, dual pixel PDAF, sensor-shift OIS

rate has been achieved with the Samsung Galaxy S21. This is thought to be due to the dual pixel phase detection autofocus (PDAF) feature of the Samsung Galaxy S21.

The dual pixel PDAF greatly improves both the speed and accuracy of autofocus compared to conventional systems [20]. It achieves this by simultaneously receiving color information while each pixel in the sensor is used for phase detection. This technology provides rapid and accurate autofocus even under challenging conditions such as fast moving subjects or in low-light environments.

17.6 Conclusion

In this study, an algorithm has been developed for price recognition on price tags in real time on smart phones. The developed algorithm uses the outputs of a lightweight object detection method. As a result of experimental studies carried out according to different performance evaluation metrics, it has been observed that the proposed method works very quickly and with a high accuracy value. When the results are examined, it is seen that the phone model used has a significant effect on the performance. Therefore, in order for the proposed method to work successfully, it must be run on a mobile phone with appropriate hardware features.

It is important that the price control process, which is an important issue in terms of customer satisfaction and also a daily retail operation, is carried out quickly and accurately. The results obtained with the proposed solution method in this study are remarkable. In future studies, we will aim to further optimize the deep learning model used in the proposed method. In this way, we plan to reduce the mean prediction time on smartphones that do not have very good hardware features.

References

[1] Michael Aliev, Dmitry Bocharov, Irina Kunina, Dmitry Nikolaev, A low computational approach for price tag recognition, in: The 12th International Conference on Machine Vision, Netherlands, 2020, p. 143314.

[2] Precedence Research, Artificial Intelligence in Retail Market, https://www.precedenceresearch.com/artificial-intelligence-in-retail-market. (Accessed 18 April 2024).

[3] Pavel Laptev, Sergey Litovkin, Sergey Davydenko, Anton Konev, Evgeny Kostyuchenko, Alexander Shelupanov, Neural network-based price tag data analysis, Future Internet 14 (3) (2022) 1–14.

[4] Shelf Labeling, How much do electronic price tags cost?, https://www.solumesl.com/en/insights/how-much-do-electronic-price-tags-cost. (Accessed 5 April 2023).

[5] Niddal H. Imam, Vassilios G. Vassilakis, Dimitris Kolovos, OCR post-correction for detecting adversarial text images, Journal of Information Security and Applications 66 (2022) 103170.

[6] M.N.A. Hussin, A.H. Ahmad, M.A. Razak, Price tag recognition using HSV color space, Journal of Telecommunication, Electronic and Computer Engineering 9 (3–9) (2017) 77–84.

[7] Yong-Qiang Mou, et al., Towards accurate price tag recognition algorithm with multi-task RNN, Acta Automatica Sinica 45 (2020) 1–7.

[8] Vladimir Fomenko, Dmitry Botov, Julius Klenin, Text scene detection with transfer learning in price detection task, in: 7th International Conference on. Analysis of Images, Social Networks, and Texts (AIST), Moscow, Russia, 2018, pp. 173–178.

[9] Wei Liu, Dragomir Anguelov, Dumitru Erhan, Christian Szegedy, Scott Reed, Cheng-Yang Fu, Alexander C. Berg, SSD: single shot multibox detector, in: European Conference on Computer Vision (ECCV), Amsterdam, Netherlands, 2016, pp. 21–37.

[10] André Silva Pinto de Aguiar, et al., Vineyard trunk detection using deep learning—An experimental device benchmark, Computers and Electronics in Agriculture 175 (2020) 105535.

[11] Andrew G. Howard, et al., MobileNets: efficient convolutional neural networks for mobile vision applications, Computing Research Repository (CoRR), arXiv:1704.04861 [abs], 2017, pp. 1–9.

[12] Mark Sandler, et al., MobileNetV2: inverted residuals and linear bottlenecks, in: Computer Vision and Pattern Recognition (CVPR), Salt Lake City, UT, USA, 2018, pp. 4510–4520.

[13] Yassine Bouafia, Larbi Guezouli, Hicham Lakhlef, Human detection in surveillance videos based on fine-tuned MobileNetV2 for effective human classification, Iranian Journal of Science and Technology, Transactions of Electrical Engineering 46 (4) (2022) 971–988.

[14] Pengchong Jin, Vivek Rathod, Xiangxin Zhu, Pooling pyramid network for object detection, Computing Research Repository (CoRR), arXiv:1807.03284, 2018, pp. 1–3.

[15] Tsung-Yi Lin, et al., Feature pyramid networks for object detection, in: Computer Vision and Pattern Recognition (CVPR), Honolulu, HI, USA, 2017, pp. 2117–2125.

[16] Golnaz Ghaisi, et al., Learning scalable feature pyramid architecture for object detection, in: Computer Vision and Pattern Recognition (CVPR), Long Beach, CA, USA, 2019, pp. 7036–7045.

[17] Khadijeh Alibabaei, et al., A review of the challenges of using deep learning algorithms to support decision-making in agricultural activities, Remote Sensing 14 (3) (2022) 638.

[18] Victor Gonzalez-Huitron, et al., Disease detection in tomato leaves via CNN with lightweight architectures implemented in Raspberry Pi 4, Computers and Electronics in Agriculture 181 (2021) 105951.

[19] Musa Peker, Classification of hyperspectral imagery using a fully complex-valued wavelet neural network with deep convolutional features, Expert Systems with Applications 173 (2021) 114708.

[20] Samsung Newsroom, Dual Pixel Pro: improved autofocus to allow you to capture every precious moment, https://news.samsung.com/global/video-dual-pixel-pro-improved-autofocus-to-allow-you-to-capture-every-precious-moment. (Accessed 23 April 2022).

CHAPTER

TinyML network applications for smart cities

18

Zeinab E. Ahmed[a,b], Aisha A. Hashim[a], Rashid A. Saeed[c], and Mamoon M. Saeed[d]

[a]*Department of Electrical and Computer Engineering, International Islamic University Malaysia, Kuala Lumpur, Malaysia*

[b]*Department of Computer Engineering, University of Gezira, Madani, Sudan*

[c]*Department of Computer Engineering, College of Computers and Information Technology, Taif University, Taif, Saudi Arabia*

[d]*Department of Communications and Electronics Engineering, Faculty of Engineering, University of Modern Sciences (UMS), Sana'a, Yemen*

18.1 Introduction

Tiny machine learning (TinyML) is an emerging field that combines embedded systems and machine learning [1]. Its main objective is to implement machine learning algorithms on microcontroller units (MCUs) that have low power consumption and cost. The primary goal of TinyML is to enable real-time data collection, processing, analysis, and interpretation on-site [2]. This is accomplished by utilizing battery-powered MCUs and streaming applications, resulting in advantages such as decreased latency, efficient power usage, and improved data privacy. By conducting these tasks directly on the edge devices, TinyML eliminates the need to transmit data to the cloud, thereby avoiding energy expenses and potential data loss [3]. The survey provides an assessment of the current status of TinyML, focusing on its role in enabling AI on small and low-power edge devices. The study specifically investigates the TensorFlow Lite framework, which is an important platform for developing TinyML systems. Additionally, the survey identifies key factors that contribute to the expansion and growth of TinyML technology [4]. The ability of TinyML to perform computations close to sensors allows for innovative data processing methods in resource-limited environments. Its flexibility, efficiency, and simplicity have the potential to bring significant changes to the Internet of Things (IoT) ecosystem. Various performance metrics, such as the transition to smart IoT devices, optimization of network bandwidth, considerations for security and privacy, latency reduction, energy efficiency, reliability, data filtration, and cost effectiveness, highlight the effectiveness of TinyML as a critical tool in this field [5].

The wide range of applications has fueled the growth of TinyML, contributing to the generation of valuable datasets and economic expansion. The versatility of

TinyML for Edge Intelligence in IoT and LPWAN Networks. https://doi.org/10.1016/B978-0-44-322202-3.00023-3

TinyML extends beyond vision-related tasks to encompass natural language processing, predictive modeling, pattern recognition/classification, and data analysis. This overview provides insights into the diverse application domains of TinyML, highlighting the associated challenges and opportunities. Currently, TinyML finds effective utilization in various fields, including smart objects, industrial control and monitoring, healthcare, surveillance and security, administration and finance, and numerous other aspects of daily life and communities [6]. As cities become smarter and interconnected, the demand for smart devices capable of processing data efficiently is increasing. TinyML emerges as a transformative technology that revolutionizes the usage and interaction with smart devices in smart cities [7]. By combining embedded systems with machine learning, TinyML empowers devices to make intelligent decisions based on data analysis without relying on cloud connections or powerful computers. This scalability reduces costs and enhances efficiency in deploying smart devices. The potential of TinyML extends to various domains within smart cities, such as traffic management, energy efficiency, public safety, and healthcare. Collaboration among disciplines like machine learning, optimization, hardware design, computer architecture, and signal processing are crucial for progress in these areas [8].

The development of smart cities has gained significant attention in recent years, with the aim of improving the quality of life of citizens by using technology to optimize resource utilization, enhance public services, and ensure sustainable development. However, traditional machine learning models require significant computational resources and high power consumption, making them unsuitable for deployment on small, low-power devices commonly found in smart city applications [1]. This is where TinyML comes in, offering a solution that enables machine learning inference on battery-powered devices with ultralow power consumption, low cost, and high efficiency [8]. The motivation for exploring TinyML's potential in enhancing smart cities is evident in the benefits it can provide. By using machine learning models to analyze data from sensors and other sources, smart cities can optimize energy consumption, improve traffic flow, enhance public safety, and provide better services to citizens [4]. However, to achieve these benefits, the machine learning models must be optimized to operate on low-power devices while maintaining accuracy and performance. This chapter explores the potential of TinyML to enhance smart cities by enabling machine learning on low-power IoT devices. The contributions of this chapter include:

- analyzing the challenges of implementing machine learning in smart cities and how TinyML can overcome these challenges,
- identifying potential applications of TinyML in smart cities, such as traffic management, energy management, and waste management, among others,
- discussing the hardware and software requirements for deploying TinyML in smart cities, including the selection of appropriate algorithms and optimization techniques,

- presenting case studies and real-world examples of the deployment of TinyML in smart city applications and their impact on efficiency, cost effectiveness, and sustainability,
- evaluating the benefits and limitations of using TinyML in smart cities and discussing future research directions for the advancement of this field.

The chapter begins with an introduction to the concept of TinyML and its potential to enhance smart cities. The first section explains the motivation and contributions of this chapter, highlighting the importance of utilizing TinyML to improve the efficiency and effectiveness of smart city systems, and then delves into the technical details of TinyML, including its hardware and software requirements, algorithms, and frameworks. The following section explores the current state of smart cities and their limitations, highlighting the need for new technologies like TinyML to address these challenges. The next section discusses TinyML applications in smart cities and provides examples of how TinyML can be used to improve various systems, such as transportation, energy management, and public safety. Finally, the chapter concludes with a summary of the key points and future directions for research and development in TinyML-enhanced smart cities.

18.1.1 TinyML

TinyML is an evolving concept in edge computing that combines embedded systems, machine learning, and privacy considerations to achieve low-power and cost-effective solutions [6]. It enables machine learning inference on battery-powered intelligent devices and is specifically designed for small embedded edge devices with efficient power consumption of a few milliwatts or less [9]. By incorporating power management modules and hardware acceleration, IoT-based edge devices can operate with reduced power requirements. The software used in TinyML is highly compact to conserve power, and its primary goal is to optimize machine learning models for improved accuracy while operating under resource constraints. The initial step in training a TinyML model involves collecting data from real devices. Although current advancements in TinyML primarily focus on on-device inference, on-device learning capabilities are still limited. Fig. 18.1 illustrates the process of creating a TinyML model.

While TinyML shows a lot of promise, there are still some major constraints that are currently hindering its growth [8]. These include:

- Limited processing power: TinyML devices have limited processing power, which makes it difficult to run complex machine learning algorithms.
- Limited memory: Similarly, TinyML devices have limited memory, which makes it difficult to store large datasets and models.
- Energy efficiency: TinyML devices must be highly energy-efficient to operate on batteries for extended periods. This requires careful optimization of both hardware and software [10].

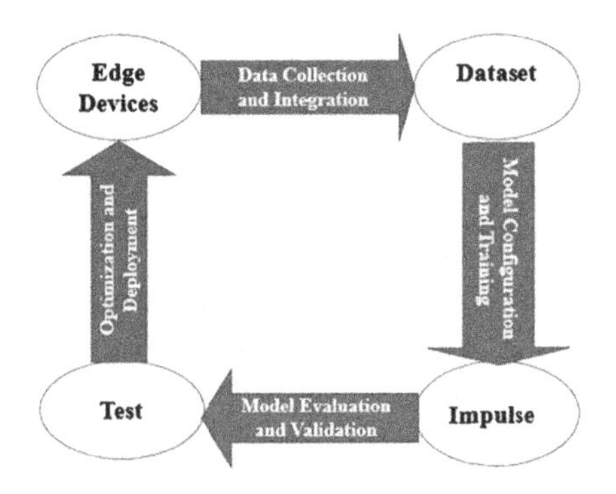

FIGURE 18.1

TinyML Algorithm Building Process with Edge Impulse.

- Security and privacy: As with any connected device, security and privacy are major concerns for TinyML devices. Ensuring that data are transmitted and stored securely is essential for adoption.
- Lack of standards: As TinyML is still an emerging field, there are currently no widely accepted standards for development and deployment. This can lead to fragmentation and compatibility issues.

TinyML and machine learning are both related to the use of algorithms that can learn patterns and make predictions based on data. However, there are some key differences between them:

- Hardware: TinyML is designed to run on small, low-power embedded devices with limited resources such as memory, processing power, and energy. In contrast, machine learning is typically run on high-performance computing systems with more powerful hardware.
- Efficiency: TinyML is optimized for energy efficiency and is capable of running on a few milliwatts of power or less, while machine learning is less concerned with energy efficiency and can consume a lot of power.
- Size: TinyML algorithms are designed to be as compact as possible, so they can fit within the limited memory and storage available on embedded devices. In contrast, machine learning algorithms can be much larger and more complex.
- Deployment: TinyML is primarily used for edge computing, which involves running algorithms on devices that are closer to the data source. In contrast, machine learning is typically run on centralized servers or cloud-based systems.
- Applications: TinyML is well suited for applications that require real-time processing, such as gesture recognition, voice commands, and object detection. Machine learning is more commonly used for applications such as predictive analytics, natural language processing, and image recognition.

TinyML encompasses key elements, including hardware, software, and algorithms, as illustrated in Fig. 18.2. It is specifically tailored to operate effectively on

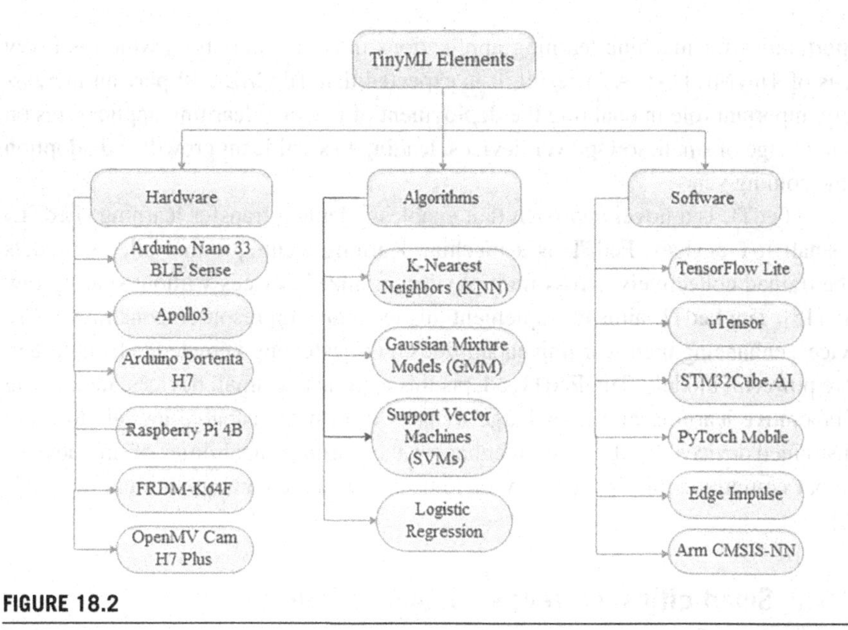

FIGURE 18.2

TinyML Elements.

small embedded edge devices with constraints in terms of memory and processing power. The diagram provides an overview of the essential components that contribute to the efficient functioning of TinyML on these resource-limited devices. To gain further insights into the specific hardware, software, and algorithmic aspects depicted in Fig. 18.2, referring to additional details or accompanying documentation is recommended [11]. The hardware used in TinyML is designed to be small, consumes little power, and includes power management modules and hardware acceleration. Software used in TinyML is highly compact and optimized for power savings. The algorithms used in TinyML can be applied for several tasks, such as clustering, regression, classification, anomaly detection, etc. It is crucial to choose the right algorithm based on specific requirements and available hardware and software resources. TinyML combines these key components to deploy machine learning on low-power devices, opening new possibilities for IoT and other applications.

The future growth and adoption of TinyML are expected to be significant due to several reasons. Firstly, the increasing demand for low-power and low-cost intelligent devices in various applications, including healthcare, agriculture, and smart cities, is driving the need for TinyML [12]. Secondly, the advancements in hardware technology, such as microcontrollers and sensors, are making it possible to develop small, low-power devices with higher processing capabilities [13]. Thirdly, the development of new software and algorithm frameworks specifically designed for TinyML is making it easier for developers to create machine learning applications for small devices. Finally, the increasing availability of data generated by IoT devices is creating new opportunities for machine learning applications in edge computing, which is a key focus of TinyML [14]. As a result, it is expected that TinyML will play an increasingly important role in enabling the deployment of machine learning applications on a wide range of small, low-power devices, leading to significant growth and adoption in the coming years.

TinyFedTL is a novel approach that enables federated transfer learning (FedTL) on small IoT devices. FedTL is a machine learning technique that allows models to be trained collectively across multiple decentralized devices without sharing raw data [15]. TinyFedTL aims to implement this technique on resource-constrained IoT devices, enhancing their learning capabilities by transferring knowledge from larger, more powerful models. TinyFedTL adapts this approach to small devices, facilitating collaborative learning and knowledge transfer from more capable models to these constrained devices. By doing so, it enhances the learning capabilities of tiny devices without compromising data privacy or imposing excessive computational demands [16].

18.1.2 Smart cities: current state and limitations

Smart cities are urban areas that utilize advanced technologies and data to enhance the quality of life for its residents, improve sustainability, and optimize services. These technologies may include sensors, IoT devices, and big data analytics to monitor and manage various aspects of urban life, such as traffic flow, energy usage, and waste management. However, despite the potential benefits, there are still limitations and challenges associated with the current state of smart cities. One of the major challenges is the high cost of implementing smart city technologies, which can be a barrier for many cities, especially smaller ones with limited resources [17]. Additionally, there is a lack of interoperability among the different systems used in smart cities, leading to difficulties in data sharing and analysis. Another issue is the potential privacy and security risks associated with collecting and analyzing sensitive data, which can lead to ethical concerns and public distrust.

Furthermore, there is a need to ensure that the benefits of smart city technologies are accessible and equitable for all members of the community, including those who may not have access to technology or digital literacy skills. This digital divide can further exacerbate existing social and economic inequalities in the community. These limitations highlight the need for new technologies like TinyML that can address these challenges and make smart cities more accessible, efficient, and equitable for all residents. By enabling low-power, low-cost, and privacy-preserving machine learning on edge devices, TinyML can provide more accurate and real-time data analysis, leading to better decision-making and improved urban management [18]. It can also enable the deployment of smart city technologies in areas with limited resources and infrastructure, making these technologies more accessible and inclusive for all residents.

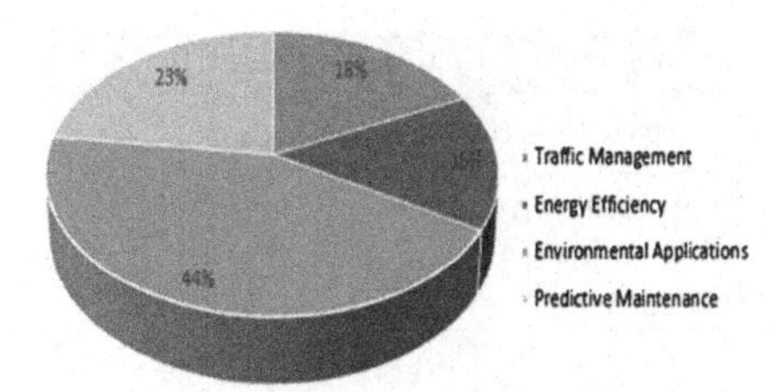

FIGURE 18.3

The Percentage Distribution of Total Articles for Related Work.

18.2 TinyML applications in smart cities

TinyML has the potential to revolutionize various systems in smart cities. It can enable the deployment of machine learning models on small, low-power devices to process data at the edge, reducing the need for sending large amounts of data to centralized servers for processing [19]. This can improve the overall efficiency and speed of various systems. In transportation, TinyML can be used for traffic prediction and management, as well as for improving the safety of autonomous vehicles. For example, TinyML algorithms can analyze data from traffic cameras and sensors to predict traffic congestion and adjust traffic flow in real time. TinyML can also be used for detecting and classifying various traffic participants, such as pedestrians and vehicles, to improve the safety of autonomous vehicles. In energy management, TinyML can be used for energy consumption prediction and optimization [20]. By analyzing data from smart meters, TinyML algorithms can predict energy consumption patterns and adjust energy distribution accordingly. This can lead to more efficient use of energy resources and reduced energy waste. In public safety, TinyML can be used for anomaly detection and predictive maintenance. For example, TinyML algorithms can analyze data from surveillance cameras to detect anomalies, such as people acting suspiciously or unusual traffic patterns. TinyML can also be used for predictive maintenance of infrastructure, such as bridges and buildings, by analyzing data from sensors to detect signs of wear and tear before a failure occurs. Fig. 18.3 illustrates the distribution of articles across different applications such as traffic management, predictive maintenance, energy efficiency, and environmental applications within the related field.

18.2.1 Traffic management

Traffic management in smart cities has been revolutionized by the implementation of TinyML, enabling real-time analysis and decision-making at the edge of the network. TinyML, with its compact machine learning models optimized for low-power devices, facilitates the deployment of intelligent sensors and devices throughout urban landscapes. A previous study [21] proposed a TinyML-based soft sensor to estimate CO_2 emissions in vehicles, which was tested with various configurations including quantization methods [21]. Fig. 18.4 presents a comprehensive overview of the hard-

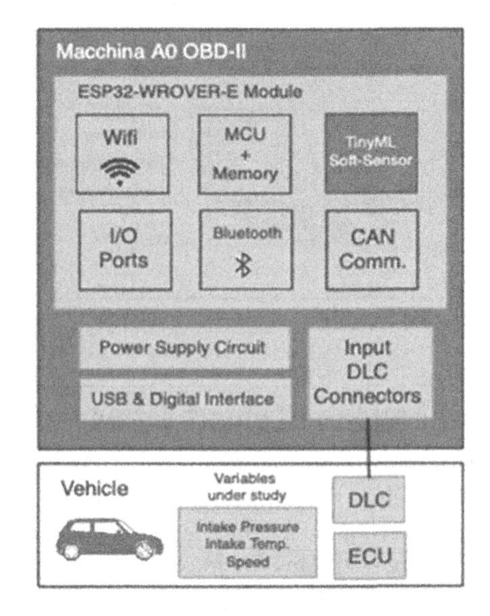

FIGURE 18.4

Hardware Architecture for TinyML Regressor Model Deployment.

ware architecture specifically employed for the deployment of the TinyML regressor model also known as the soft sensor and its integration with the vehicle. Fortunately, CO_2 emission can be derived from other variables according to the following equation:

$$CO_2\,(g/s) = \frac{MAF\,(g/s)}{AFR \cdot Density_{Fuel}\,(g/l)} \cdot CO_2PL, \tag{18.1}$$

assuming petrol is used as the fuel and the mass airflow (MAF) per second and the air/fuel ratio (AFR) are known. Results indicated that quantization methods could enhance compression efficiency and reduce deployed artifact size by an average of 92%. Additionally, model performance was improved and the mean absolute percentage error (MAPE) was reduced by 50%. This behavior was consistent across all experiments and intensified with increasing network complexity.

A battery-powered sensor with a three-axis magnetometer and recurrent neural network (RNN) for vehicle classification was developed in [22]. The sensor operates continuously for multiple years without recharging or maintenance. To extend its lifespan, the authors used optimizations such as reducing the sampling rate and delaying inference after initial detection. These strategies resulted in an impressive 11.8-fold reduction in energy cost per inference, enabling a battery lifetime of up to 2.8 years. This represents a significant improvement over the original baseline model. In [23], a methodology for deploying and improving the reliability of tiny convolutional neural networks (TinyCNNs) in a low-power autonomous mini-vehicle is presented. The authors propose an onboard machine learning predictor that intelligently

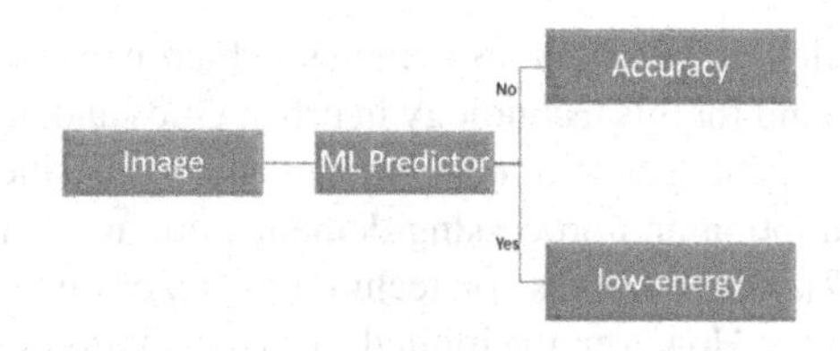

FIGURE 18.5

Adaptive Selection.

selects between a low-energy, accurate TinyCNN and a high-energy, accurate TinyCNN based on the input image, as shown in Fig. 18.5. This optimization maximizes energy efficiency by using the more resource-intensive network only for challenging inputs. A predictive approach continuously evaluates the input image and selects the appropriate network for inference. The system employs a parallel ultralow-power architecture with an energy usage of 3.9 μJ per inference to meet the CNN's latency requirements. An online predictor enables dynamic switching of the CNN during runtime, resulting in a significant 3.2-fold energy reduction with minimal accuracy loss.

Roshan et al. [24] presented a machine learning algorithm that automates traffic scheduling and enhances traffic system efficiency. By utilizing vehicle density, the algorithm operates independently without human intervention. Instead of relying on traditional preset traffic signal timers, which are not suitable for dynamic traffic conditions, the proposed approach employs piezoelectric sensors in each lane to detect vehicles. The two-point time ratio method is used to identify vehicles based on sensor data, categorizing them according to their pick-up speed. Furthermore, a TinyML-based model is introduced to predict green signal timings, offering valuable improvements to the overall traffic management system. Another study [25] addressed the issue of driving under the influence of alcohol and the challenges faced by existing solutions. It highlights the need for an IoT-based system that can measure alcohol levels and prevent impaired individuals from driving. Traditional IoT approaches have limitations such as internet dependency, data privacy concerns, and bandwidth constraints. To overcome these limitations, the paper proposes the use of TinyML, a promising technology that enables real-time intelligence with affordable and compact devices. The paper presents a TinyML-based system that detects alcohol levels and alerts the driver. Experimental results validate the effectiveness of this technology in preventing alcohol-related accidents.

In [26], the state-of-the-art fast accurate stable tiny gated recurrent neural network is introduced, a novel model designed for resource-constrained edge devices. The model focuses on acoustic single- and multi-tone classification and improves performance metrics while reducing model size compared to previous methods. By utilizing fewer parameters, enhancing efficiency, and incorporating a noise reduction algorithm, the model achieves these advancements. It is implemented as an acoustic AI module, suitable for applications like sound identification and localization in autonomous cars. The integration of localization techniques opens up new possibilities for multi-tone classifiers in autonomous vehicles, which is particularly relevant due to the increasing demand for this technology in urban cities and developing countries. With the growth of IoT, data generation and transmission have increased, resulting in higher energy consumption and processing demands due to irrelevant or redundant data transmission [27]. Data compression techniques can be applied to IoT devices to address these challenges. However, the limited storage and processing power of existing devices pose constraints. TinyML, which enables machine learning on low-power devices, offers a solution to these limitations. This article assesses the performance impact of the Tiny Anomaly Compress (TAC) compression algorithm on a microcontroller in the context of vehicles. The study demonstrates that the embedded algorithm has minimal impact on the microcontroller's processing time.

In recent years, advancements in vehicle technology have led to increased instrumentation and the availability of smart devices for embedded processing. This has given rise to the Internet of Intelligent Vehicles, where edge computing plays a crucial role in processing raw data streams near their source. In [28], an unsupervised TinyML approach is proposed for detecting anomalies on roads using a microcontroller and an embedded accelerometer in vehicles. Real experiments yielded promising results, with mean F1 scores of 0.76 for the first driver and 0.78 for the second. These findings indicate the effective performance of the classifier model in the specified scenarios.

18.2.2 Energy efficiency

Smart agriculture researchers are actively introducing new tools, including AI, to improve productivity and sustainability in farming. In [29], the use of TinyML and LoRaWAN in agriculture is investigated to propose an energy-efficient model for fruit detection. Experimental results show an accuracy rate of 90%, and the model is three times more energy-efficient than a cloud-based model with the same objective. In [30], power consumption challenges in unmanned aerial vehicles (UAVs) are addressed by combining edge computing and deep learning. The proposed solution integrates an OpenMV microcontroller into a DJI Tello MAV, enabling machine learning-based inference for navigation and mission objectives. By leveraging TinyML capabilities through OpenMV, including offline inference, low latency, energy efficiency, and data security, this approach enhances UAV decision-making while optimizing energy usage. The effectiveness of the approach is demonstrated through a practical application of onboard mask detection in a crowded environment.

In [31], the benefits of using log-gradient input images in CNN for TinyML computer vision applications are explored. The study demonstrates that log gradients offer advantages such as aggressive quantization of first-layer inputs, potential reductions in CNN resource usage, and robustness to illumination changes. The findings are validated using the PASCAL RAW image dataset and experiments involving neural architecture search and a fixed three-layer network [32]. Training on log gradient images leads to higher filter similarity, facilitating CNN pruning. These combined

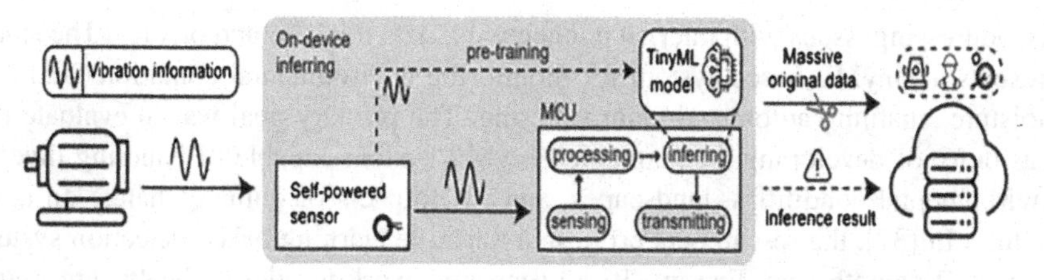

FIGURE 18.6

Architecture of the Ultra-Low-Power and Low-Cost Anomaly Detection System.

benefits enhance the efficiency of TinyML computer vision (CV) applications. In another study [33], a real-time machine learning board suitable for various applications, including environmental sensing, is introduced. The paper also presents an energy-efficient IoT platform (iBUG) with integrated sensors. An algorithm utilizing TensorFlow Lite accurately predicts CO_2 levels based solely on gas resistance measurements.

The paper [34] emphasizes the importance of anomaly detection in IoT technology for Industry 4.0. Traditional monitoring systems rely on energy-intensive servers for sensing and data transmission. To address this, the paper proposes a self-powered sensing system using TinyML for anomaly detection. Instead of commercial IMUs, a lightweight piezoelectric self-powered sensor (SPS) is employed. The system operates on a low-cost embedded system, enabling low-power in situ inference, as shown in Fig. 18.6. Experimental analysis demonstrates that a compressed deep neural network (DNN) achieves 97.6% accuracy using only eight-point normalized SPS data. The TinyML model is deployed on embedded systems for on-device inference and condition-based monitoring. Energy measurements reveal that the self-powered sensing approach can save up to 66.74% of energy compared to IMU-based systems.

In [35], the authors highlight that AI and IoT are revolutionizing industries like agriculture by utilizing real-time data. To optimize resource usage and minimize the environmental impact, artificial IoT (AIoT) is crucial. However, it requires computational resources that can handle AI workloads while ensuring autonomous operation and low energy consumption. This study explores the use of TinyML and edge computing platforms to predict indoor greenhouse temperatures. The study evaluates the computational and energy trade-offs of these platforms and adopts two neural networks for deployment. The results demonstrate that microcontroller-based devices offer an energy-efficient solution for lightweight machine learning tasks. These platforms have the potential to effectively support AIoT applications in agriculture, optimizing resource utilization and reducing environmental impact.

18.2.3 Environmental applications

Tsoukas et al. [36] discuss the challenges of implementing IoT technology in agriculture and emphasize the potential of TinyML as a solution. TinyML provides cost-effective and efficient hardware for local execution of machine learning models, addressing issues with internet connectivity, data transfer, and privacy. The study presents a TinyML-based system for monitoring environmental conditions and soil moisture, enabling automated plant watering. The primary goal was to evaluate the feasibility of developing an adaptable TinyML system capable of handling diverse environmental conditions, landscapes, and the impacts of climate change on agriculture. In [37], the researchers propose a machine learning-based detection system using a Bluetooth Low Energy (BLE) sensor network to address healthcare issues like incontinence, night wandering, and pressure ulcers. The machine learning model, implemented on an edge device with the Edge Impulse platform, provides real-time information to nurses. Data collection utilized a BME400 accelerometer and nRF52 development kit. Two approaches, a neural network with frequency components and a convolutional network, were tested, with the convolutional network showing slightly better performance. Future research aims to improve the model's real-world performance by collecting data from nursing homes, considering individual and bed variability.

In [38], an embedded machine learning pipeline is introduced for creating comprehensive solutions in various applications. The pipeline includes data collection, training a regression algorithm using CNN, fine-tuning the model for specific hardware, generating optimized code, and deploying it on a Sony Spresense setup. Initial testing demonstrates the pipeline's capability for real-time analysis of plant data in a semi-autonomous environment, even on a limited MCU like Spresense. The system achieves impressive performance, accurately predicting leaf area and plant growth with 96.2% accuracy and a root mean square error (RMSE) of 1.86 cm^2. TinyML is gaining recognition as a promising method for creating autonomous and secure devices that can locally collect, analyze, and process data. In [39], the authors present a TinyML-based system designed for detecting hazardous gas leaks. The system is trained to identify abnormalities and promptly notify occupants through BLE technology, sending messages to smartphones or displaying notifications on an integrated screen. The system's performance was evaluated in two test cases, achieving an F1-score of 0.77 in the smoke detection test and 0.70 in the ammonia test.

In [40], the authors focused on using cough sounds for COVID-19 detection. The study includes stages like training, testing, deploying on Arduino, and model evaluation using Edge Impulse. With an impressive 99.5% accuracy, the model can be deployed on the cloud for data classification and applied to body monitoring and disease recognition. The study highlights the significance of cough sounds for COVID-19 detection and proposes potential enhancements. The impulse is integrated into an embedded device's firmware, allowing real-time classification in local and live settings. The paper [41] presents the development of a cough detection system utilizing Edge Impulse Studio and Arduino 33 BLE Sense. The system is designed to distinguish between cough signals and background noise. Edge Impulse Studio is used to train a large dataset comprising cough samples and undesired noise. The implementation involves a TinyML model that leverages efficient machine learning techniques to analyze the resonance of cough signals in real-time. Experimental findings reveal that the proposed system achieves an outstanding recognition accuracy of approximately 97%.

In [42], AERSUT, an Automatic Emotion Recognition System based on TinyML, is introduced. The system includes three stages: input data processing, feature extraction, and emotion recognition. Raw speech data are processed using TinyML, extracting features like MFCC, energy, and zero crossing rate to determine the user's emotional state. Emotions are classified into eight categories, and the most accurate TensorFlow model is employed for emotion detection. Sudharsan et al. [43] presented a novel technique called Over-The-Air (OTA) for deploying TinyML models on IoT devices. OTA enables remote tasks like model updates, firmware reflashing, reconfiguration, and repurposing. The article discusses the challenges of OTA machine learning deployment on IoT devices, including scientific and engineering considerations. Initial testing validates the OTA-TinyML approach using popular models and IoT development boards from the TinyML community. The results demonstrate the successful retrieval, storage, and execution of multiple machine learning models on a single IoT device with limited memory capacity.

In [44], precision agriculture (PA) or smart farming (SF) is examined as a solution for current agricultural challenges, including environmental harm and climate change vulnerability. PA integrates AI and IoT technologies to provide valuable insights for weather forecasts, pest detection, and irrigation management. However, concerns about privacy, latency, security, and energy consumption arise from cloud-based AI processing. To tackle these issues, TinyML enables AI computations on IoT devices, reducing dependence on the cloud. The paper investigates the feasibility of Firmware Update Over-The-Air (FUOTA) for TinyML models using LoRaWAN in agriculture. A prototype TinyML sensor is developed to assess FUOTA, analyzing energy consumption and packet delivery ratios across different network scenarios. In [45], an affordable embedded device is proposed to address communication challenges in medical emergency situations for individuals with speech impairments. The device utilizes a 1D CNN model that analyzes data from an onboard inertial measurement unit (IMU) to classify specific American Sign Language (ASL) words related to medical emergencies. The CNN model is trained offline and deployed on the embedded device for real-time ASL word classification. Pilot tests showed promising results, with an offline accuracy of 91.2% and an average online accuracy of 92% for the optimized 8-bit model.

In [46], the researchers develop wearable hand gesture devices to enhance human–machine interfaces and digital communication. They propose a novel device that combines an electrostatic sensing circuit with a six-degree-of-freedom IMU. This wearable bracelet achieves continuous gesture detection for over 11 hours on a single battery charge. Real-time gesture detection is enabled through a TinyML model implemented on an onboard ARM Cortex-M33 microcontroller, with an accuracy of up to 87%. The integration of the electrostatic sensing bracelet enhances the performance of the IMU gesture recognition system, improving accuracy and overall functionality. In [47], touchless gesture recognition is explored for advancing human–machine interface technology. The researchers introduce a compact and low-power time-of-flight (ToF) sensor by STMicroelectronics, integrated with a TinyML neural network model. The model combines temporal convolutional networks and is optimized for small embedded ARM Cortex-M4 processors with limited RAM. Training and evaluation use a dataset collected with the ToF sensor, achieving over 96% accuracy in recognizing six gestures.

In [48], the focus is on the use of embedded machine learning (TinyML) to control a robotics system with limited computing resources, offering hands-on experience to students in a robotics class for applying AI in robot control. The paper delves into the fundamentals of embedded machine learning, encompassing the training process of a specific CNN and the preparation of the dataset. It examines the employed technology and data, with a particular emphasis on the hardware features utilized to control robot motors through speech recognition using keyword spotting. Validation of the data is addressed, and potential future enhancements for the speech-controlled system are explored. In [49], an embedded experimentation-driven methodology is proposed for developing portable edge AI-based disease prediction kits, specifically focusing on early detection of chronic obstructive pulmonary disease (COPD) using exhalation signatures. The study validates the accuracy of edge AI inference for COPD prediction through simulation experiments, showing comparable results to cloud-based inference. It explores the impact of real-time resource variations, such as the number of breath sensors, on inference accuracy and embedded processor memory requirements. The paper emphasizes the importance of considering software and hardware components together in the co-design of precise edge AI systems, beyond conventional optimization approaches.

In [50], a solution is presented for challenges in traditional monitoring systems, such as in elderly care, child care, and intrusion detection. The proposed approach combines TinyML and single-chip radar technology for on-edge sensing and human activity detection. It addresses issues like latency, data theft, and privacy concerns associated with infrared (IR) and camera-based systems. The solution leverages edge computing in a compact form, ensuring data privacy, and utilizes radar technology to provide enhanced privacy through point cloud information. The system operates effectively in adverse environmental conditions and utilizes the Texas Instruments IWR6843 mmWave radar board for signal processing. A CNN is used for human activity classification, trained on a dataset of four activities across six subjects. The implemented real-time inference engine has a small model size of 1.44 kB and achieves an impressive subject-independent accuracy of 96.43%. The authors of [51] address the challenges of mosquito-borne diseases, specifically focusing on the *Aedes aegypti* mosquito and its role in spreading viruses. The introduction of "MosquIoT" combines traditional ovitraps with IoT and TinyML technologies to detect and quantify *Ae. aegypti* mosquito eggs. By offering cost effectiveness, autonomy, reliability, and ease of implementation, MosquIoT transforms traditional surveillance methods into a proactive and predictive digital approach. It presents a promising solution for understanding mosquito population behavior in urban areas, enhancing mosquito control efforts, and reducing public health risks associated with mosquito-borne diseases.

To ensure resident safety in smart cities, an earthquake early warning system (EEWS) integrates IoT and machine learning technologies for data collection and analysis. In [52], a comprehensive survey on the key elements of a successful EEWS is provided. It explores the role of IoT and categorizes machine learning models into linear and non-linear types. Evaluation metrics for machine learning models, especially in seismology, are examined. The paper introduces a taxonomy showcasing advancements in machine learning and IoT for EEWS and proposes a generic architecture based on IoT and machine learning. It also discusses the application of machine learning in observing earthquake parameters, enhancing the efficiency of an EEWS. Authors propose a solution that applies machine learning techniques on resource-constrained edge devices (TinyML) for healthcare [53]. The system uses MCU-powered edge devices with electrocardiography (ECG) and photoplethysmography (PPG) sensors to predict blood pressure metrics through machine learning inference. Despite limitations in computation, power, and memory, the solution achieves comparable results to server-based methods and meets the BHS standard for grade B (C in extremely constrained devices). The correlation between ECG, PPG features, and blood pressure is analyzed to identify minimal features that satisfy edge constraints and the standard's accuracy. Importantly, all inference tasks are performed on MCU-based edge devices, eliminating the need for cloud infrastructure and enhancing robustness, accessibility, reliability, security, and data privacy.

18.2.4 Predictive maintenance

In [17], a novel method is introduced to detect and notify users about potential lavatory accidents. By combining advanced technologies like the ZAC888DP and ESP32 devices, along with various sensors, data on water, temperature, humidity, light, and motion were collected. The model, based on supervised learning using linear regression with a sigmoid function, was developed using TensorFlow for seamless integration with TinyML. Performance evaluation was conducted using several metrics, including those outlined in Eqs. (18.2), (18.3), and (18.4). The implemented system achieved a high accuracy of 94.11% and a perfect recall score of 100%. Although the precision was lower, at 63%, this was due to the algorithm's more conservative classification approach, set by an aggressive threshold value of 0.23. Accuracy, precision, and recall were calculated as follows:

$$\text{Accuracy} = \frac{TP + TN}{TP + TN + FN + FP}, \tag{18.2}$$

$$\text{Precision} = \frac{TP}{TP + FP}, \tag{18.3}$$

$$\text{Recall} = \frac{TP}{TP + FN}, \tag{18.4}$$

where TP is true positives, TN is true negatives, FP is false positives, and FN is false negatives. In [54], a CNN model is developed to detect falls using sound technology. The goal is to train the model to automatically recognize the sound of an elderly person falling, even in the presence of background noises. The researchers collect their own dataset, convert audio recordings into Mel spectrograms, and extract relevant features for the CNN. The trained model achieves a recognition accuracy of over 95% for identifying falling sounds. They suggest combining sound technology with acceleration and gyroscope sensors for effective fall detection in the elderly population. Liu et al. proposed an efficient approach using TinyML for channel estimation and signal detection [55]. By replacing large dense layers in DNNs with small cascading sublayers, the proposed method has reduced the computation and storage requirements. We introduce a rank-restricted backpropagation algorithm for lightweight training. The authors focus on machine learning-based channel estimation and signal detection in multipath OFDM wireless communication systems. Extensive simulations validate the effectiveness of our approach, with the TinyML model achieving significant acceleration in both training and inference, i.e., approximately 2.5–3 times faster training and 4.5 times faster inference, compared to traditional DNNs. It also enables substantial storage reduction, approximately 4.5 times compared to fully connected DNNs.

In [56], machine learning algorithms are used to forecast greenhouse temperatures with a focus on employing TinyML for low-power environments. The study aims to understand the limitations and complexity achievable in machine learning models on such devices. The findings show that TinyML-enabled devices, like Arduino Nano 33 BLE Sense, can effectively execute lightweight machine learning models while consuming minimal power. The study utilizes a basic multilayer perceptron (MLP) and demonstrates its accuracy in temperature predictions using metrics such as RMSE, MAE, and R^2. The significance lies in the lightweight and straightforward nature of these models, making them notable findings in the study. In [57], the authors emphasize the importance of digital technologies in agriculture for addressing nutritional demands and environmental degradation. Their focus is on conserving natural resources, particularly water, by utilizing microcontrollers, low-power radios, and software tools. They aim to develop a power-efficient system for monitoring water usage and detecting misuse, with minimal network bandwidth requirements. The paper addresses issues such as water wastage from broken pipes and negligence. They propose an experimental system that integrates edge AI, IoT, Edge Computing (EC), and machine learning techniques, enabling automated detection while reducing energy consumption. The paper also discusses implementation details and challenges and presents preliminary evaluation results.

The article [58] presents a solution to address vehicle emissions in urban areas. It utilizes the OBD-II interface in vehicles to indirectly measure and monitor emissions. The proposed solution employs soft sensor techniques to estimate pollution levels by processing engine combustion data. An embedded system retrieves vehicle data through the OBD-II interface and applies different algorithms based on available data. An unsupervised TinyML approach is introduced to remove outliers and enhance the accuracy of the soft sensor without relying on cloud-based servers, as illustrated in Fig. 18.7. Preliminary results demonstrate the feasibility of the solution, achieving

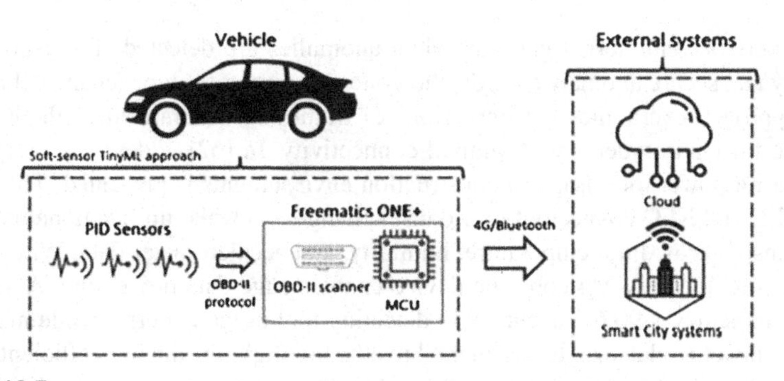

FIGURE 18.7

Architecture of the Proposed Vehicle CO_2 Calculation Soft-Sensor Solution.

a data acquisition frequency of 1 Hz and measuring emissions in grams of CO_2 per kilometer using the Freematics ONE+ board.

In [59], the focus is on sound event detection (SED) pipelines used in smart city applications to identify and categorize events in audio streams. Real-time SED with large models on centralized servers faces scalability and data volume challenges. To overcome this, the study explores performing SED directly on embedded devices, providing advantages in energy consumption, bandwidth, and privacy. The paper investigates the impact of width scaling parameters, α and t0, in the PhiNets architecture for urban sound detection. Experimental analysis on the UrbanSound8K dataset reveals that when training the model from scratch, only the number of parameters affects performance, with α and t0 being interchangeable. However, when applying knowledge distillation from a pretrained teacher model, larger values of t0 result in improved performance. In production, inspecting industrial products is crucial for yield improvement and cost reduction. However, implementing complex deep learning algorithms for quality inspection on low-cost, low-power embedded devices poses challenges due to limited memory and energy resources. In [60], a sensor system with three MCU-based TinyML cameras for automatic artifact and anomaly detection in plastic components is presented. The system includes a top camera for shape defect identification and two side cameras for color anomaly detection. Leveraging TinyML enables local data processing, minimizing data transmission. The study evaluates MobileNetV2 and SqueezeNet, two state-of-the-art CNN architectures. Results demonstrate that with suitable compression techniques, both architectures are suitable for resource-constrained microcontrollers, achieving an impressive 99% classification accuracy.

In industrial settings, monitoring asset status is crucial for preventing failures. The paper [61] suggests an adaptable anomaly detection system for extreme industrial environments, focusing on submersible pumps. The system combines IoT, edge computing, and TinyML Ops methodologies, utilizing an IoT sensing kit with an ESP32 microcontroller and MicroPython firmware near the data source. It collects data, trains an anomaly detection model using the isolation forest algorithm, and sends alerts to an external gateway when anomalies are detected. To ensure transparency and secure anomaly records, the system incorporates blockchain technology. This approach overcomes the limitations of traditional centralized methods and is suitable for environments with limited connectivity. In [62], a device prototype for monitoring power tool usage in construction environments is presented. The device uses BLE and NFC connectivity for data exchange and wake-up functionality, along with sensors providing temperature, humidity, and accelerometer data. With an embedded ARM Cortex-M4 core, the device enables edge data processing. A TinyML algorithm achieves 90.6% accuracy in detecting tool usage patterns, while maintaining a compact model size. Experimental results highlight the device's efficient power consumption during sleep mode and TinyML execution, ensuring extended battery life for prolonged field operation. This research contributes to device degradation analysis and comprehensive utility evaluation, bridging the gap between edge processing and thorough data analysis.

18.3 Analysis and discussion

The studies [21–28] examined in this analysis shed light on the significant advancements and contributions made in the field of TinyML for various vehicle-related applications. These studies collectively contribute to the advancement of TinyML in vehicle applications, paving the way for further innovation and progress in this field. The studies collectively emphasize the potential of TinyML in revolutionizing various aspects of vehicle technology, including emission estimation, vehicle classification, energy-efficient inference, traffic management, alcohol detection, acoustic classification, and data compression, as illustrated in Table 18.1. These advancements have significant implications for enhancing sustainability, safety, and efficiency in the automotive industry and transportation systems as a whole. The findings of these studies pave the way for further research and innovation in the field of TinyML and its application in intelligent vehicles. One study [21] focused on TinyML's role in accurately estimating CO_2 emissions, crucial for environmental monitoring and sustainable transportation. Another study [22] explored the development of maintenance-free vehicle classification systems, benefiting traffic monitoring and infrastructure planning. Other studies delved into employing TinyML, such as TinyCNNs [23], for resource-constrained devices in autonomous vehicles, improving their feasibility. Moreover, TinyML aids in traffic management [24], enhancing flow and safety, while an IoT-based system [25] was used to detect alcohol levels in real time, bolstering road safety. Advancements like an efficient acoustic classification model [26] promise improved performance for AI systems like autonomous cars. Lastly, the TAC algorithm [28] showcases efficiency in data compression on microcontrollers, offering energy-saving solutions while maintaining performance standards.

The studies [29–35] demonstrate the significant potential of TinyML, edge computing, and AI in revolutionizing various aspects of agriculture. These technologies have shown remarkable accuracy improvements and energy efficiency across differ-

Table 18.1 TinyML's revolutionary impact on vehicle technology.

Refs.	Contribution	Limitations	Advantages	Criteria
[21] 2002	This paper demonstrates the feasibility of deploying this resource-efficient soft-sensor on constrained IoIV platforms, optimizing computational needs while maintaining accuracy.	• Scalability Concerns • Model Complexity Management • Data Variability and Generalization	• Resource Efficiency. • Real-time Decision Making • Improved Vehicle Performance	• Performance • Resource Utilization • Adaptability
[22] 2021	The study showcases how TrafficNNode plays a pivotal role in enhancing urban traffic management by enabling real-time vehicle sensing, assisting in congestion monitoring, and optimizing traffic flow within city environments	• Scalability and Coverage • Sensing Accuracy in Complex Environments • Data Processing and Analysis	• Energy Efficiency • Real-time Traffic Monitoring • Deployment Flexibility.	• Sensing Accuracy • Energy Efficiency • Scalability and Deployment
[23] 2021	It introduces innovative strategies aimed at enhancing the reliability and robustness of TinyML models specifically tailored for miniaturized autonomous vehicles.	limitations related to computational constraints and real-time processing in these miniature vehicles could impact the deployment of sophisticated TinyML models.	The paper's main advantage lies in its focus on enhancing the reliability of TinyML models deployed in autonomous minivehicles.	• Performance Optimization • Platform Adaptability
[24] 2021	The study contributes novel algorithms and models leveraging TinyML to optimize traffic flow, reduce congestion, and enhance overall urban mobility.	Challenges could arise in ensuring the scalability and universal applicability of these systems in cities with varying traffic patterns, infrastructure, and socioeconomic factors.	The paper presents several advantages, such as efficient computational processing, optimizing traffic flow, and reducing congestion.	• Traffic Optimization • Resource Efficiency
[25] 2022	The system's foremost contribution lies in enhancing road safety by providing timely alerts to drivers, potentially averting alcohol-related accidents and significantly improving overall road safety standards.	These could encompass challenges related to the system's accuracy and reliability under various environmental conditions and individual physiological variations.	The key advantage of this system is its potential to markedly enhance road safety by detecting alcohol impairment within vehicles through TinyML technology.	• Detection Accuracy • Impact on Road Safety

ent applications, as shown in Table 18.2. For instance, in [29] the use of TinyML and LoRaWAN in fruit detection is highlighted, showcasing the viability of battery-powered sensors for computer vision in smart farming. In [30], an innovative approach is introduced by integrating edge computing and deep learning in UAVs, enabling energy optimization and informed decision-making. The advantages of log gradient input images in TinyML computer vision models, as presented in [31], include improved efficiency through aggressive quantization, reduced resource usage, and robustness to changes in illumination. In [33], the effectiveness of real-time machine learning on edge devices for environmental sensing is emphasized, offering decentralized computation and efficient data processing. In [34], a novel self-powered sensing system with TinyML is presented for anomaly detection, addressing energy consumption challenges in IoT technology while achieving high accuracy. Lastly, the authors of [35] showcase the feasibility of using microcontroller-based devices with TinyML and edge computing for greenhouse temperature prediction, providing an energy-efficient computational solution for AIoT applications in agriculture. Overall, these studies highlight the transformative potential of TinyML, edge computing, and AI in optimizing agricultural processes and minimizing environmental impact.

The studies [36–53] showcase the wide range of applications and benefits of AI, IoT, and TinyML in various domains. These technologies provide solutions to diverse challenges across fields such as precision agriculture, medical communication, gesture recognition, robotics control, disease prediction, human activity detection, mosquito control, disaster management, and healthcare. The research papers demonstrate the feasibility and effectiveness of TinyML in different scenarios. For example, in precision agriculture [38], TinyML and IoT technologies provide valuable insights to farmers while considering privacy and energy consumption. In the medical field, TinyML facilitates real-time classification of ASL words [43], improving communication for individuals with speech impairments during medical emergencies. The integration of electrostatic sensing, IMU, and TinyML enables wearable devices for accurate hand gesture recognition, enhancing human–machine interfaces [46]. Similarly, touchless gesture recognition using low-power sensors and TinyML algorithms demonstrates high accuracy and the feasibility of implementing gesture recognition on small embedded processors [47]. Furthermore, the studies emphasize the potential of TinyML in addressing specific challenges. For instance, the application of TinyML in disease prediction enables early detection of conditions like COPD through exhaled breath signatures [48]. In mosquito control, TinyML combined with IoT technology and ovitraps offers a proactive and predictive approach to monitor and manage mosquito populations [51]. The use of IoT and machine learning in earthquake early warning systems contributes to efficient disaster management in smart cities [52]. Additionally, TinyML enables healthcare applications by predicting blood pressure metrics using ECG and PPG sensors, meeting standard accuracy requirements. Overall, these studies highlight the significance of AI, IoT, and TinyML in addressing real-world challenges, improving efficiency, accessibility, and user experience across different domains. As these technologies continue to advance, they

Table 18.2 Performance advancements and energy efficiency of TinyML.

Refs.	Contribution	Limitations	Advantages	Criteria
[29] 2022	The paper's contributions lie in addressing the energy efficiency challenges prevalent in IoT-based precision agriculture, offering a significant advancement in sustainable fanning practices.	Challenges related to the accuracy of data collection, especially in diverse agricultural landscapes or fluctuating weather conditions, could impact the sensor's reliability.	The primary advantage of the TinyML-based smart sensor lies in its ability to significantly reduce energy consumption within IoT precision agriculture platforms.	• Energy Efficiency • Data Accuracy
[30] 2021	This paper introduces novel methodologies and algorithms tailored for onboard inference, allowing UAVs to conduct data processing on the edge	limitations could include the constraints associated with the computational capacity of TinyML algorithms within the limited resources available on UAVs.	The integration of TinyML algorithms allows for real-time decision-making on the UAV, reducing the dependence on external computational resources and enabling more autonomous functionalities.	• Energy Efficiency. • Environmental Adaptability. • Operational Enhancement
[31] 2023	The proposal aims to maintain the system's resilience to varying input data, showcasing advancements in resource efficiency, critical for resource-constrained tinyML applications.	The adaptability of this method across diverse datasets and complex visual scenarios may necessitate further exploration	This technique extends battery life and operational longevity in resource-constrained environments while maintaining the system's resilience against input variations.	• Energy Efficiency. • Inference Robustness.
[32] 2022	This paper introduces a novel approach to enhance the energy efficiency and robustness of tinyML-based computer vision systems by employing log-gradient input images.	limitations might include the trade-off between image processing complexity and computational efficiency.	By utilizing log-gradient input images, it effectively reduces computational requirements during inference while maintaining model accuracy	• Energy Efficiency • Robustness Assessment

have the potential to revolutionize industries and transform the way we interact with and utilize various systems and devices.

In this analysis, we delve into a set of studies [17] [54–62] that showcase the broad range of applications and advantages of TinyML in the domain of predictive maintenance, as demonstrated in Table 18.3. In [17], the focus is on the prevention of lavatory accidents using innovative solutions and technologies. The authors employ supervised learning, specifically linear regression, and integrate TensorFlow with TinyML. In [54], the researchers strive to develop a CNN model for fall detection using sound technology. The model achieves an impressive recognition accuracy of over 95% when identifying falling sounds. In [55], a fresh approach for efficient channel estimation and signal detection by incorporating TinyML is introduced. The researchers replace large dense layers in DNNs with smaller cascading sublayers to reduce computational and storage requirements. In [56], the use of TinyML-powered machine learning algorithms for greenhouse temperature forecasting is investigated. Utilizing TinyML-enabled devices like Arduino Nano 33 BLE Sense, they execute lightweight machine learning models with minimal power consumption. In [57], the importance of embracing digital technologies in agriculture is emphasized, particularly for water usage monitoring and detecting misuse. They propose an experimental system that integrates edge AI, IoT, edge computing, and machine learning techniques to create an energy-efficient detection mechanism. In [58], a solution for estimating and monitoring vehicle emissions through the OBD-II interface is proposed. The authors introduce an unsupervised TinyML approach to identify and eliminate outliers, thereby improving the accuracy of the soft sensor. The study in [59] delves into SED pipelines for urban sound detection. The researchers investigate the impact of width scaling parameters in the PhiNets architecture, aiming to simplify model design based on memory and computational resources. Lastly, in [60], the researchers present a sensor system equipped with three TinyML cameras to automatically detect anomalies and defects in industrial components [61–65]. These networks achieve high classification accuracy while maintaining real-time performance, underscoring the potential of TinyML for cost-effective quality inspection [66–74].

18.4 Conclusion

The rapid growth of smart cities heavily relies on IoT devices that generate and process large volumes of data. However, traditional machine learning techniques are challenging to implement on resource-constrained IoT devices due to their high computing power requirements. TinyML provides a promising solution by enabling machine learning on small, low-power devices, enabling real-time decision-making and reducing reliance on cloud computing. Integrating TinyML in smart cities has the potential to revolutionize resource utilization, public services, and sustainable development. By enabling machine learning on small devices, TinyML opens up new possibilities for improving the lives of citizens and enhancing city quality. TinyML combines embedded systems with machine learning to achieve low-power, cost-

Table 18.3 TinyML-empowered predictive maintenance.

Refs.	Contribution	Limitations	Advantages	Criteria
[54] 2021	Utilizing TinyML enables real-time analysis of sound data, enhancing the speed and efficiency of fall detection systems.	Sound-based detection might face limitations in environments with high ambient noise levels, potentially impacting the accuracy and reliability of fall detection.	Leveraging TinyML optimizes power usage, making it suitable for deployment in low-power edge devices for continuous monitoring without significant energy consumption.	• Accuracy and Reliability • Scalability • Cost-effectiveness
[55] 2022	Provides a practical demonstration of applying machine learning at the edge (Tiny-ML) to address challenges in wireless communication systems, potentially enhancing performance and reliability.	The effectiveness of TinyML models might rely heavily on the availability and quality of training data, which could limit performance in scenarios with limited or biased datasets.	The paper suggests that using TinyML could enhance the accuracy and efficiency of channel estimation and signal detection.	• Performance Improvement • Resource Efficiency
[56] 2022	Exploring the efficiency of TinyML models for forecasting temperature, which could optimize energy usage in devices deployed for continuous monitoring.	TinyML models might face limitations in handling complex algorithms or large datasets due to hardware or memory constraints in edge devices.	Utilizing TinyML for temperature forecasting at the edge minimizes the need for frequent data transfers to centralized servers, enhancing data privacy and reducing network traffic.	• Robustness and Adaptability • Resource Utilization • Accuracy
[60] 2023	Proposes a potentially cost-effective method for quality control by leveraging TinyML, which could minimize the need for expensive equipment or extensive human intervention.	TinyML-based inspection systems might struggle with adaptability to diverse environmental conditions or changes in product characteristics, affecting their robustness.	The paper suggests that utilizing TinyML could enhance the accuracy of product quality inspection.	• Accuracy • Robustness and Adaptability • Resource Utilization

effective solutions while ensuring privacy. This chapter explores the potential of TinyML in enhancing smart cities by enabling machine learning on small edge devices commonly found in smart city applications. The study assesses the current state of TinyML, with a focus on its role in enabling AI on small IoT devices. It specifically examines the TensorFlow Lite framework, an important platform for developing TinyML models. The chapter discusses the hardware and software requirements for deploying TinyML in smart cities, including algorithm selection and optimization techniques. Real-world case studies are presented to illustrate the impact on efficiency, cost effectiveness, and sustainability.

References

[1] H. Han, J. Siebert, TinyML: a systematic review and synthesis of existing research, in: 4th International Conference on Artificial Intelligence in Information and Communication, ICAIIC 2022 - Proceedings, Institute of Electrical and Electronics Engineers Inc., 2022, pp. 269–274, https://doi.org/10.1109/ICAIIC54071.2022.9722636.

[2] C. Banbury, et al., MLPerf tiny benchmark, http://arxiv.org/abs/2106.07597. (Accessed 15 May 2023), Jun. 2021.

[3] R.F. Mansour, et al., Optimal deep learning based fusion model for biomedical image classification, Expert Systems 39 (3) (2022) e12764.

[4] D.L. Dutta, S. Bharali, TinyML meets IoT: a comprehensive survey, Internet of Things 16 (Dec. 2021) 100461, https://doi.org/10.1016/J.IOT.2021.100461.

[5] N. Schizas, A. Karras, C. Karras, S. Sioutas, TinyML for ultra-low power AI and large scale IoT deployments: a systematic review, Future Internet 14 (12) (Dec. 2022), https://doi.org/10.3390/fi14120363.

[6] V. Tsoukas, E. Boumpa, G. Giannakas, A. Kakarountas, A review of machine learning and TinyML in healthcare, in: 25th Pan-Hellenic Conference on Informatics, ACM, New York, NY, USA, Nov. 2021, pp. 69–73, https://doi.org/10.1145/3503823.3503836.

[7] G. Delnevo, S. Mirri, C. Prandi, P. Manzoni, An evaluation methodology to determine the actual limitations of a TinyML-based solution, Internet of Things 22 (Jul. 2023) 100729, https://doi.org/10.1016/j.iot.2023.100729.

[8] M. Nada, et al., Internet of vehicle's resource management in 5G networks using AI technologies: current status and trends, IET Communications 16 (2022) 400–420, https://doi.org/10.1049/cmu2.12315.

[9] S.B. Lakshman, N.U. Eisty, Software engineering approaches for TinyML based IoT embedded vision, in: Proceedings of the 4th International Workshop on Software Engineering Research and Practice for the IoT, ACM, New York, NY, USA, May 2022, pp. 33–40, https://doi.org/10.1145/3528227.3528569.

[10] S.A.R. Zaidi, A.M. Hayajneh, M. Hafeez, Q.Z. Ahmed, Unlocking edge intelligence through tiny machine learning (TinyML), IEEE Access 10 (2022) 100867–100877, https://doi.org/10.1109/ACCESS.2022.3207200.

[11] A. Osman, U. Abid, L. Gemma, M. Perotto, D. Brunelli, TinyML platforms benchmarking, in: International Conference on Applications in Electronics Pervading Industry, Environment and Society, 2022, pp. 139–148, https://doi.org/10.1007/978-3-030-95498-7_20.

[12] A. Raha, et al., Special session: approximate TinyML systems: full system approximations for extreme energy-efficiency in intelligent edge devices, in: 2021 IEEE 39th International Conference on Computer Design (ICCD), IEEE, Oct. 2021, pp. 13–16, https://doi.org/10.1109/ICCD53106.2021.00015.

[13] A. Moin, M. Challenger, A. Badii, S. Gunnemann, Supporting AI engineering on the IoT edge through model-driven TinyML, in: 2022 IEEE 46th Annual Computers, Software, and Applications Conference (COMPSAC), IEEE, Jun. 2022, pp. 884–893, https://doi.org/10.1109/COMPSAC54236.2022.00140.

[14] Y. Song, et al., BSC: block-based stochastic computing to enable accurate and efficient TinyML, in: 2022 27th Asia and South Pacific Design Automation Conference (ASP-DAC), IEEE, Jan. 2022, pp. 314–319, https://doi.org/10.1109/ASP-DAC52403.2022.9712585.

[15] K. Kopparapu, E. Lin, J.G. Breslin, B. Sudharsan, TinyFedTL: federated transfer learning on ubiquitous tiny IoT devices, in: 2022 IEEE International Conference on Pervasive Computing and Communications Workshops and other Affiliated Events, PerCom Workshops 2022, Institute of Electrical and Electronics Engineers Inc., 2022, pp. 79–81, https://doi.org/10.1109/PerComWorkshops53856.2022.9767250.

[16] K. Kopparapu, E. Lin, TinyFedTL: federated transfer learning on tiny devices, http://arxiv.org/abs/2110.01107, Oct. 2021.

[17] A. Zacharia, et al., An intelligent microprocessor integrating TinyML in smart hotels for rapid accident prevention, in: 7th South-East Europe Design Automation, Computer Engineering, Computer Networks and Social Media Conference, SEEDA-CECNSM 2022, Institute of Electrical and Electronics Engineers Inc., 2022, https://doi.org/10.1109/SEEDA-CECNSM57760.2022.9932982.

[18] O.R. Dirar, et al., Persistent overload control for backlogged machine to machine communications in long term evolution advanced networks, Journal of Telecommunication, Electronic and Computer Engineering (JTEC) 9 (3) (Dec. 2017).

[19] R. Sanchez-Iborra, A.F. Skarmeta, TinyML-enabled frugal smart objects: challenges and opportunities, IEEE Circuits and Systems Magazine 20 (3) (Jul. 2020) 4–18, https://doi.org/10.1109/MCAS.2020.3005467.

[20] S.O. Ooko, M. Muyonga Ogore, J. Nsenga, M. Zennaro, TinyML in Africa: opportunities and challenges, in: 2021 IEEE Globecom Workshops, GC Wkshps 2021 - Proceedings, Institute of Electrical and Electronics Engineers Inc., 2021, https://doi.org/10.1109/GCWkshps52748.2021.9682107.

[21] T. Flores, et al., A TinyML soft-sensor for the internet of intelligent vehicles, in: 2022 IEEE International Workshop on Metrology for Automotive, MetroAutomotive 2022 - Proceedings, Institute of Electrical and Electronics Engineers Inc., 2022, pp. 18–23, https://doi.org/10.1109/MetroAutomotive54295.2022.9855110.

[22] J. Nguyen, R. Grimsley, B. Iannucci, TrafficNNode: low power vehicle sensing platform for smart cities, in: Proceedings - 5th IEEE International Conference on Smart Internet of Things, SmartIoT 2021, Institute of Electrical and Electronics Engineers Inc., 2021, pp. 278–282, https://doi.org/10.1109/SmartIoT52359.2021.00051.

[23] M. de Prado, M. Rusci, A. Capotondi, R. Donze, L. Benini, N. Pazos, Robustifying the deployment of TinyML models for autonomous mini-vehicles, Sensors (Switzerland) 21 (4) (Feb. 2021) 1–16, https://doi.org/10.3390/s21041339.

[24] A. Navaas Roshan, B. Gokulapriyan, C. Siddarth, P. Kokil, Adaptive traffic control with TinyML, in: 2021 International Conference on Wireless Communications, Signal Processing and Networking, WiSPNET 2021, Institute of Electrical and Electronics

Engineers Inc., Mar. 2021, pp. 451–455, https://doi.org/10.1109/WiSPNET51692.2021.9419472.

[25] A. Gkogkidis, V. Tsoukas, A. Kakarountas, A TinyML-based alcohol impairment detection system for vehicle accident prevention, in: 7th South-East Europe Design Automation, Computer Engineering, Computer Networks and Social Media Conference, SEEDA-CECNSM 2022, Institute of Electrical and Electronics Engineers Inc., 2022, https://doi.org/10.1109/SEEDA-CECNSM57760.2022.9932962.

[26] R. Rawat, S. Gupta, S. Mohapatra, S.P. Mishra, S. Rajagopal, Intelligent acoustic module for autonomous vehicles using fast gated recurrent approach, in: 2021 4th International Conference on Recent Developments in Control, Automation & Power Engineering (RDCAPE), IEEE, Oct. 2021, pp. 345–350, https://doi.org/10.1109/RDCAPE52977.2021.9633681.

[27] O.O. Khalifa, et al., An IoT-platform-based deep learning system for human behavior recognition in smart city monitoring using the Berkeley MHAD datasets, Systems 10 (2022) 177, https://doi.org/10.3390/systems10050177.

[28] P. Andrade, et al., An unsupervised TinyML approach applied for pavement anomalies detection under the internet of intelligent vehicles, in: 2021 IEEE International Workshop on Metrology for Industry 4.0 & IoT (MetroInd4.0&IoT), IEEE, Jun. 2021, pp. 642–647, https://doi.org/10.1109/MetroInd4.0IoT51437.2021.9488546.

[29] C. Nicolas, B. Naila, R.C. Amar, TinyML smart sensor for energy saving in internet of things precision agriculture platform, in: International Conference on Ubiquitous and Future Networks, ICUFN, IEEE Computer Society, 2022, pp. 256–259, https://doi.org/10.1109/ICUFN55119.2022.9829675.

[30] W. Raza, A. Osman, F. Ferrini, F. De Natale, Energy-efficient inference on the edge exploiting TinyML capabilities for UAVs, https://doi.org/10.3390/drones, 2021.

[31] Q. Lu, B. Murmann, Enhancing the energy efficiency and robustness of tinyML computer vision using coarsely-quantized log-gradient input images, ACM Transactions on Embedded Computing Systems (Apr. 2023), https://doi.org/10.1145/3591466.

[32] Q. Lu, B. Murmann, Improving the energy efficiency and robustness of tinyML computer vision using log-gradient input images, http://arxiv.org/abs/2203.02571, Mar. 2022.

[33] R.A. Saeed, et al., Machine-to-machine communication, in: Encyclopedia of Information Science and Technology, third edition, IGI Global, July 2014, pp. 6195–6206, https://doi.org/10.4018/978-1-4666-5888-2.

[34] R. Saeed, M. Saeed, Z. Ahmed, A.H. Hashim, Enhancing medical services through machine learning and UAV technology: applications and benefits, in: Applications of Machine Learning in UAV Networks, IGI Global, 2024, pp. 307–343, https://doi.org/10.4018/979-8-3693-0578-2.ch012.

[35] J. Morales-García, A. Bueno-Crespo, R. Martínez-España, J.-L. Posadas, P. Manzoni, J.M. Cecilia, Evaluation of low-power devices for smart greenhouse development, Journal of Supercomputing 79 (9) (Jun. 2023) 10277–10299, https://doi.org/10.1007/s11227-023-05076-8.

[36] V. Tsoukas, A. Gkogkidis, A. Kakarountas, A TinyML-based system for smart agriculture, in: ACM International Conference Proceeding Series, Association for Computing Machinery, Nov. 2022, pp. 207–212, https://doi.org/10.1145/3575879.3575994.

[37] K. Tjonck, C.R. Kancharla, J. Vankeirsbilck, H. Hallez, J. Boydens, B. Pang, Real-time activity tracking using TinyML to support elderly care, in: 2021 XXX International Scientific Conference Electronics (ET), IEEE, Sep. 2021, pp. 1–6, https://doi.org/10.1109/ET52713.2021.9579991.

[38] D. Sheth, B. Sudharsan, J.G. Breslin, M.I. Ali, Poster abstract: embedded ML pipeline for precision agriculture, in: Proceedings - 21st ACM/IEEE International Conference on Information Processing in Sensor Networks, IPSN 2022, Institute of Electrical and Electronics Engineers Inc., 2022, pp. 527–528, https://doi.org/10.1109/IPSN54338.2022.00064.
[39] A. Gkogkidis, V. Tsoukas, S. Papafotikas, E. Boumpa, A. Kakarountas, A TinyML-based system for gas leakage detection, in: 2022 11th International Conference on Modern Circuits and Systems Technologies, MOCAST 2022, Institute of Electrical and Electronics Engineers Inc., 2022, https://doi.org/10.1109/MOCAST54814.2022.9837510.
[40] R. Shankar, K.M. Gautham Mythireyan, N.R. Nalla, M. Venkateshkumar, Cough recognition using TinyML, in: 2022 IEEE Industrial Electronics and Applications Conference, IEACon 2022, Institute of Electrical and Electronics Engineers Inc., 2022, pp. 111–116, https://doi.org/10.1109/IEACon55029.2022.9951763.
[41] A. Rana, Y. Dhiman, R. Anand, Cough detection system using TinyML, in: Proceedings - 2022 International Conference on Computing, Communication and Power Technology, IC3P 2022, Institute of Electrical and Electronics Engineers Inc., 2022, pp. 119–122, https://doi.org/10.1109/IC3P52835.2022.00032.
[42] J. Tharian, R. Nandakrishnan, S. Sajesh, A.V. Arun, C.K. Jayadas, Automatic emotion recognition system using tinyML, in: 2022 International Conference on Futuristic Technologies, INCOFT 2022, Institute of Electrical and Electronics Engineers Inc., 2022, https://doi.org/10.1109/INCOFT55651.2022.10094330.
[43] B. Sudharsan, et al., OTA-TinyML: over the air deployment of TinyML models and execution on IoT devices, IEEE Internet Computing 26 (3) (2022) 69–78, https://doi.org/10.1109/MIC.2021.3133552.
[44] C. Nicolas, B. Naila, R.C. Amar, Energy efficient firmware over the air update for TinyML models in LoRaWAN agricultural networks, in: 2022 32nd International Telecommunication Networks and Applications Conference, ITNAC 2022, Institute of Electrical and Electronics Engineers Inc., 2022, pp. 21–27, https://doi.org/10.1109/ITNAC55475.2022.9998338.
[45] M. Deji Dere, R.O. Dere, A. Adesina, A.R. Yauri, SmartCall: a real-time, sign language medical emergency communicator, in: Proceedings of the 5th International Conference on Information Technology for Education and Development: Changing the Narratives Through Building a Secure Society with Disruptive Technologies, ITED 2022, Institute of Electrical and Electronics Engineers Inc., 2022, https://doi.org/10.1109/ITED56637.2022.10051420.
[46] E. Reinschmidt, C. Vogt, M. Magno, Realtime hand-gesture recognition based on novel charge variation sensor and IMU, in: Proceedings of IEEE Sensors, Institute of Electrical and Electronics Engineers Inc., 2022, https://doi.org/10.1109/SENSORS52175.2022.9967346.
[47] A.A. Fahad, et al., Machine learning techniques in the internet of UAVs for smart cities applications, Journal of Intelligent & Fuzzy Systems 24 (4) (2021) 1–24, https://doi.org/10.3233/JIFS-211009.
[48] J.M. Phillips, J.M. Conrad, Robotic system control using embedded machine learning and speech recognition, in: IEEE 19th International Conference on Smart Communities: Improving Quality of Life Using ICT, IoT and AI, HONET 2022, Institute of Electrical and Electronics Engineers Inc., 2022, pp. 214–218, https://doi.org/10.1109/HONET56683.2022.10019106.
[49] S.O. Ooko, D. Mukanyiligira, J.P. Munyampundu, J. Nsenga, Edge AI-based respiratory disease recognition from exhaled breath signatures, in: 2021 IEEE Jordan International

Joint Conference on Electrical Engineering and Information Technology, JEEIT 2021 - Proceedings, Institute of Electrical and Electronics Engineers Inc., 2021, pp. 89–94, https://doi.org/10.1109/JEEIT53412.2021.9634140.

[50] L.E. Alatabani, et al., Deep learning approaches for IoV applications and services, in: N. Magaia, G. Mastorakis, C. Mavromoustakis, E. Pallis, E.K. Markakis (Eds.), Intelligent Technologies for Internet of Vehicles. Internet of Things (Technology, Communications, and Computing), Springer, Cham, 2021, https://doi.org/10.1007/978-3-030-76493-7_8.

[51] J. Aira, T. Olivares, F.M. Delicado, D. Vezzani, MosquIoT: a system based on IoT and machine learning for the monitoring of Aedes aegypti (Diptera: Culicidae), IEEE Transactions on Instrumentation and Measurement 72 (2023) 1–13, https://doi.org/10.1109/TIM.2023.3265119.

[52] M.S. Abdalzaher, H.A. Elsayed, M.M. Fouda, M.M. Salim, Employing machine learning and IoT for earthquake early warning system in smart cities, Energies (Basel) 16 (1) (Jan. 2023) 495, https://doi.org/10.3390/en16010495.

[53] K. Ahmed, M. Hassan, tinyCare: a tinyML-based low-cost continuous blood pressure estimation on the extreme edge, in: 2022 IEEE 10th International Conference on Healthcare Informatics (ICHI), IEEE, Jun. 2022, pp. 264–275, https://doi.org/10.1109/ICHI54592.2022.00047.

[54] K. Fang, Z. Xu, Y. Li, J. Pan, A fall detection using sound technology based on TinyML, in: Proceedings - 11th International Conference on Information Technology in Medicine and Education, ITME 2021, Institute of Electrical and Electronics Engineers Inc., 2021, pp. 222–225, https://doi.org/10.1109/ITME53901.2021.00053.

[55] H. Liu, Z. Wei, H. Zhang, B. Li, C. Zhao, Tiny machine learning (Tiny-ML) for efficient channel estimation and signal detection, IEEE Transactions on Vehicular Technology 71 (6) (Jun. 2022) 6795–6800, https://doi.org/10.1109/TVT.2022.3163786.

[56] M.F. Alati, G. Fortino, J. Morales, J.M. Cecilia, P. Manzoni, Time series analysis for temperature forecasting using TinyML, in: Proceedings - IEEE Consumer Communications and Networking Conference, CCNC, Institute of Electrical and Electronics Engineers Inc., 2022, pp. 691–694, https://doi.org/10.1109/CCNC49033.2022.9700573.

[57] D. Loukatos, K.A. Lygkoura, C. Maraveas, K.G. Arvanitis, Enriching IoT modules with edge AI functionality to detect water misuse events in a decentralized manner, Sensors 22 (13) (Jul. 2022), https://doi.org/10.3390/s22134874.

[58] Rania Salih Ahmed, et al., Machine learning in cyber-physical systems in Industry 4.0, in: Ashish Kumar Luhach, Atilla Elçi (Eds.), Artificial Intelligence Paradigms for Smart Cyber-Physical Systems, IGI Global, 2021, pp. 20–41, https://doi.org/10.4018/978-1-7998-5101-1.ch002.

[59] A. Brutti, F. Paissan, A. Ancilotto, E. Farella, Optimizing PhiNet architectures for the detection of urban sounds on low-end devices, in: 2022 30th European Signal Processing Conference (EUSIPCO), IEEE, Aug. 2022, pp. 1121–1125, https://doi.org/10.23919/EUSIPCO55093.2022.9909572.

[60] A. Albanese, M. Nardello, G. Fiacco, D. Brunelli, Tiny machine learning for high accuracy product quality inspection, IEEE Sensors Journal 23 (2) (Jan. 2023) 1575–1583, https://doi.org/10.1109/JSEN.2022.3225227.

[61] H. Mona, et al., Artificial intelligence in IoT and its applications, in: Intelligent Wireless Communications, IET Digital Library, pp. 33–58, https://doi.org/10.1049/PBTE094E_ch2, (Telecommunications, 2021), Chap. 2.

[62] O.O. Khalifa, et al., Vehicle detection for vision-based intelligent transportation systems using convolutional neural network algorithm, Journal of Advanced Transportation 2022 (2022) 9189600, https://doi.org/10.1155/2022/9189600.

[63] E.S. Ali, et al., Machine learning technologies on internet of vehicles, in: N. Magaia, G. Mastorakis, C. Mavromoustakis, E. Pallis, E.K. Markakis (Eds.), Intelligent Technologies for Internet of Vehicles. Internet of Things (Technology, Communications, and Computing), Springer, Cham, 2021, https://doi.org/10.1007/978-3-030-76493-7_7.

[64] E. Sayed, et al., Machine learning technologies for secure vehicular communication in internet of vehicles: recent advances and applications, Wiley-Hindawi, Journal of Security and Communication Networks (SCN) 2021 (2021), https://doi.org/10.1155/2021/8868355.

[65] R.H. Aswathy, et al., Optimized tuned deep learning model for chronic kidney disease classification, Computers, Materials & Continua 70 (2) (2022) 2097–2111.

[66] M.A. Sara, et al., Smart IDS and IPS for cyber-physical systems, in: Ashish Kumar Luhach, Atilla Elçi (Eds.), Artificial Intelligence Paradigms for Smart Cyber-Physical Systems, IGI Global, Hershey, PA, 2021, pp. 109–136, https://doi.org/10.4018/978-1-7998-5101-1.ch006.

[67] F.M. Osman, et al., Cyber-physical system for smart grid, in: Ashish Kumar Luhach, Atilla Elçi (Eds.), Artificial Intelligence Paradigms for Smart Cyber-Physical Systems, IGI Global, Hershey, PA, 2021, pp. 301–323, https://doi.org/10.4018/978-1-7998-5101-1.ch014.

[68] S. Ali Ahmed, et al., Algorithms optimization for intelligent IoV applications, in: Jingyuan Zhao, V. Vinoth Kumar (Eds.), Handbook of Research on Innovations and Applications of AI, IoT, and Cognitive Technologies, IGI Global, Hershey, PA, 2021, pp. 1–25, https://doi.org/10.4018/978-1-7998-6870-5.ch001.

[69] B. Hassan, et al., Machine learning for industrial IoT systems, in: Jingyuan Zhao, V. Vinoth Kumar (Eds.), Handbook of Research on Innovations and Applications of AI, IoT, and Cognitive Technologies, IGI Global, Hershey, PA, 2021, pp. 336–358, https://doi.org/10.4018/978-1-7998-6870-5.ch023.

[70] A. Khan, Jian Ping Li, et al., PackerRobo: model-based robot vision self-supervised learning in CART, Alexandria Engineering Journal 61 (12) (2022) 12549–12566, https://doi.org/10.1016/j.aej.2022.05.043.

[71] R.A. Saeed, M. Omri, S. Abdel-Khalek, et al., Optimal path planning for drones based on swarm intelligence algorithm, Neural Computing & Applications (2022), https://doi.org/10.1007/s00521-022-06998-9.

[72] M. Antonini, M. Pincheira, M. Vecchio, F. Antonelli, An adaptable and unsupervised TinyML anomaly detection system for extreme industrial environments, Sensors 23 (4) (Feb. 2023) 2344, https://doi.org/10.3390/s23042344.

[73] M. Giordano, N. Baumann, M. Crabolu, R. Fischer, G. Bellusci, M. Magno, Design and performance evaluation of an ultralow-power smart IoT device with embedded TinyML for asset activity monitoring, IEEE Transactions on Instrumentation and Measurement 71 (2022) 1–11, https://doi.org/10.1109/TIM.2022.3165816.

[74] L.E. Anatabine, et al., Deep and reinforcement learning technologies on internet of vehicle (IoV) applications: current issues and future trends, Journal of Advanced Transportation 2022 (2022) 1947886, https://doi.org/10.1155/2022/1947886.

CHAPTER 19

Emerging application use cases and future directions

Sheetal N. Ghorpade and Sachin Chougule
Rubiscape Private Limited, Pune, India

19.1 Introduction

The aspiration for an improved lifestyle and the increasing set of challenges confronted on the globe impose advancements in technology. "Edge intelligence" (EI) seeks to leverage the collective potential of the Internet of Things (IoT) and machine learning to address distinct challenges on a large scale, precisely at the location where the data are generated. Cameras with local processing abilities and networks are deployed in various intelligent embedded vision frameworks or will be used in edge systems in numerous industrial and domestic applications. At each moment, gigabytes of data are gathered by IoT sensor networks, necessitating real-time processing by their computational systems. This enables the delivery of valuable and actionable insights to individuals and other systems as well [1]. Numerous complex issues that previously relied on human expertise, as well as intricate hardware and software setups for decision-making, are now being automated. This automation is achieved through the implementation of optimized machine learning models on efficient embedded IoT platforms.

In the realm of EI, the primary focus lies on sensory systems, encompassing camera-based setups, audio sensors, and functionalities such as traffic monitoring in smart city environments. Essentially, EI serves as an expansive sensory network, continually observing and interpreting events in the surroundings. In a holistic technological approach, the gathered data can then be transmitted to the cloud for further analysis.

Over the last decade, there has been significant advancement in software engineering for machine learning and IoT embedded systems, addressing an array of obstacles encountered by machine learning developers in leveraging the capabilities of deep learning models [2]. EI excels in scenarios where quick resolutions and instant responses to time-critical data are necessary. Differentiation of edge AI application domains can be based on specific needs such as power sensitivity, size, and load restrictions. Power sensitivity is particularly critical, given that edge devices are generally low-power gadgets utilized in smartphones, wearables, or IoT setups. AI

TinyML for Edge Intelligence in IoT and LPWAN Networks. https://doi.org/10.1016/B978-0-44-322202-3.00024-5

models implemented on these devices must be optimized for efficient power consumption to preserve battery life and extend operational duration.

Edge devices usually feature tiny designs, necessitating AI models to be lightweight and space-efficient. It becomes particularly relevant when incorporating edge devices into drones, robotics, or wearable technology, as the size and weight have a direct impact on how well they perform and how user-friendly they are.

19.2 Overview of TinyML

Software engineering workflow to develop and deploy TinyML models is presented in this section. This procedure can be split into two stages: one conducted in a training environment utilizing CPU/GPU and one executed with lightweight edge hardware for real-world deployment, as illustrated in Fig. 19.1.

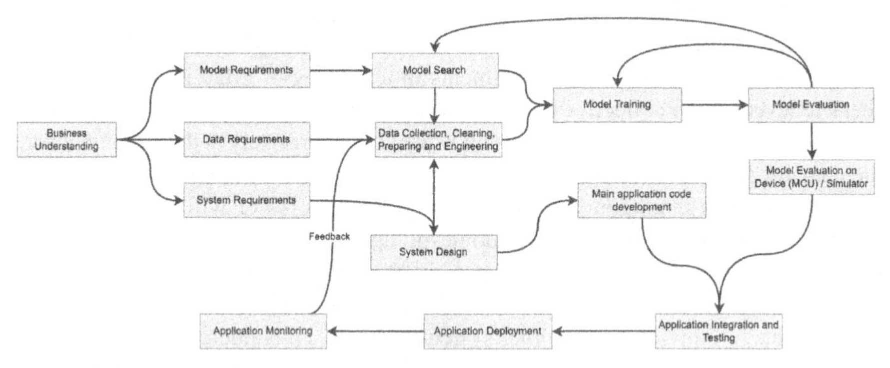

FIGURE 19.1

Development and deployment workflow of TinyML model.

The customer/business requirements are initially divided into three interconnected components:

1. model requirements,
2. data requirements,
3. system requirements.

This methodology facilitates the distribution of responsibility for delivering specific components among distinct entities or groups. Additionally, it grants autonomy to these teams, allowing them to comprehend needs and harness proficiency in various aspects such as system design, model exploration, and establishing pipelines for data collection, cleaning, and labeling.

The process flow will execute in the following manner:

- Once the system design is generated, it provides further insights to steer model exploration and data engineering endeavors. Considering the need for the application to function on a compact microcontroller unit (MCU) resource, the system design sets constraints on the computational and memory complexities of algorithms employed in data engineering for preprocessing, postprocessing, and machine learning. Primary application development can progress in parallel with model development.

- Simultaneous development enables embedded developers to construct and debug heuristic pipelines for sensing and actuation, guided by expected outcomes from the machine learning model. They can also debug this code using calibration functions to simulate outputs akin to those of the final machine learning model. The model development process can proceed beyond the model search and data engineering phases.
- By following established best practices in model training, it is possible to derive a suitable model and evaluate it against a validation dataset. Insights gained from the evaluation phase can then direct model developers toward either refining the model through further training or adjusting its hyperparameters during the model search process.
- Once desired expectations are met by the model validation, the model can advance to compilation steps, which are seamlessly managed by the TensorFlow Lite Micro (TFLM) framework. Specific hardware libraries tailored to the target MCUs can be employed during this phase to harness any custom hardware features available.
- Following successful compilation of the model, evaluation of the compiled model can be conducted either in a simulator within the training environment or in an on-device inference environment, depending on availability. If postcompilation the model performance declines significantly, it should be communicated back to the model compilation step for essential adjustments.
- Once the compiled model showcases performance akin to the original machine learning model, it can be transferred to the application integration team. This team will then incorporate the model into the application code, proceed with testing using a test dataset, and finally deploy it in real-world settings.
- Monitoring gathering real-world performance data from application usage enables valuable insights that can inform continuous integration practices. These insights can help in identifying areas for improvement or the need to incorporate new data from real-world distributions to enhance the system.

19.3 Workflow steps, challenges, and proposed solutions

There is limited knowledge regarding effective, optimal methods for developing, deploying, and maintaining IoT embedded vision products for TinyML applications. There is currently a gap in established agile development practices specifically tailored for TinyML engineering. Developing TinyML applications demands proficiency in embedded systems, computer vision, and machine learning. Due to its early stage, there is a scarcity of experienced engineers with the necessary skills to tackle challenges in TinyML engineering across different applications. Table 19.1 presents the steps involved in developing and implementing TinyML applications and also the associated challenges and proposed solutions.

Table 19.1 Steps, challenges, and solutions in TinyML application development.

Step 1: Business (customer) understanding	
Challenges	• Customer requirements are evolving rapidly. • TinyML engineering demands proficiency across various domains, including AI, machine learning, and embedded systems. SMEs/business users may not have proficiencies to understand the inputs required for these systems.
Solution	• Separating teams responsible for system creation, data handling, and model development facilitates concurrent work [3]. • Optimizing the TinyML engineering process, version control is essential to meet customer needs. This allows for incremental enhancements on specific features demanded by the consumer [4].
Step 2: System requirements	
Challenges	• Deploying in real time environs poses significant challenges. • Ensuring compatibility between hardware systems and the movability of models around various hardware configurations is necessary. • Addressing performance degradation in real-time scenarios is a primary concern for the edge deployment of TinyML systems. • The absence of standardized tools for comparing performance across different algorithms and MCU systems is a notable issue [5].
Solution	• Establishing shared implicit assumptions concerning visual data, such as illumination conditions and shadows. • Managing the exponential increase in hardware device variations by strictly controlling the types of devices and chipsets utilized in TinyML systems. • Making design choices based on the environmental obstacles anticipated by the TinyML application. • Augmenting the training dataset with data is crucial for this purpose [6].
Step 3: Model requirements	
Challenges	• Assessing the impacts of different levels of quantization and precision, along with model peak memory requirements, presents significant challenges. • Documentation regarding the selection of models suitable for TinyML projects is lacking. • Identifying practical models for resource-constrained edge environments is a complex endeavor.
Solution	• TinyML systems must choose target MCU hardware for their projects according to requirements such as model size, inference speed, and power constraints specific to the application [6]. • Constrained neural architecture search (NAS) can help narrow down model architectures that are relevant to the system design choices [7,8].

continued on next page

Table 19.1 (*continued*)

Step 4: System design	
Challenges	• At present, there is no clear-cut framework or standard benchmark to aid developers of TinyML-powered IoT embedded vision applications in determining the most suitable hardware and model designs that align with particular application, market, or customer needs. • Maintaining model development tools for distinct target hardware systems can be laborious. • Model development becomes highly complex while addressing a wide range of applications. • Vision data from videos frequently includes temporally redundant information, resulting in added unessential computing on resource-constrained edge hardware.
Solution	• Ensuring the correct setup of the TinyML TFLM framework infrastructure from the outset will facilitate faster iteration [9]. • Ensuring efficient management of models and maintaining version control are essential for optimizing TinyML development efforts. • Utilizing event-driven methods for visual data processing can reduce the need for repetitive computations [10].
Step 5: Data engineering	
Challenges	• Curated datasets extracted from IoT embedded vision sensors for various applications are lacking. • Model data obtained from various sensors and sources may come in different file formats. • Datasets may not adequately represent real-world conditions found in distributions.
Solution	• Access to data is only available through programmatic interfaces. • Data augmentation involves emulating real-world scenarios by introducing variations like scaling, noise, shadows, color, and lighting conditions.
Step 6: Model training	
Challenges	• Basic joint optimization methods such as NAS, quantization, and pruning might interact in manners that result in less than optimal outcomes [11]. • The absence of emulation tools for TinyML engineering leads to increased time and effort required for model development [12]. • The hyperparameter search process may encounter challenges in reproducing results. • Developing TinyML models with a performance-oriented approach presents challenges. • Automation of TinyML engineering via web-based tools.
Solution	• Using deterministic randomness, such as exposed random seeds, assists in iterating during hyperparameter tuning. • TinyML application developers ought to leverage AutoML tools to automate the training process, especially when confronted with abundant training and validation data. • Startups like Edge Impulse and Qeexo provide AutoML capabilities accessible over the web, which can be directly integrated into edge applications.

continued on next page

Table 19.1 (*continued*)

Step 7: Model evaluation	
Challenges	• TinyML models must demonstrate robustness to handle environmental anomalies and noise effectively. • There is a dearth of uniform TinyML evaluation metrics.
Solution	• Test-driven development should focus on assessing both local and global adversarial robustness [13]. • ML Commons has introduced the ML PerfTiny benchmark, establishing the first industry standard benchmark suite for ultralow-power machine learning [14].
Step 8: Model compilation	
Challenges	• Since MCU designs vary widely, it is advantageous to utilize a framework with general compiler capabilities to ensure design flexibility across different platforms. However, such a generic compiler may not fully exploit the distinct hardware features present in certain MCU families. • Ensuring the security of embedded devices poses a significant challenge in IoT applications [15,16]. • IoT embedded vision devices should minimize communication to conserve power consumption. • IoT embedded devices exhibit significant heterogeneity. While manual coding and code generation can harness tailored optimizations, they may sacrifice flexibility in the process. • Compiler configurations should be fine-tuned to match the hardware system targeted by the chosen system design options. • Implementing device training poses challenges for devices with limited memory footprint [17]. • Assessing performance is only possible after deploying an application onto the edge device system.
Solution	• Compilers should offer admittance to performance metrics for MCUs due to their single-threaded nature. • A hardware-aware compiler can automatically make advantageous trade-offs without requiring deep knowledge of MCU architectures [18,19].
Step 9: Compiled model evaluation	
Challenges	• Debugging of compiled models on edge devices is challenging due to computational limitations. • Limitations in generic frameworks hinder the hardware awareness of the TinyML workflow.
Solution	• TinyML development necessitates the use of device simulators and emulators. • Device simulators can aid in comparing the performance of raw models and different versions of compiled TinyML models.

continued on next page

Table 19.1 (*continued*)

Step 10: Application integration	
Challenges	• Given that edge applications often necessitate time-sensitive decisions, ensuring the reliability of TinyML systems is paramount. It is essential to embed robustness into TinyML models to enable secure fail-over mechanisms in uncertain environments [16]. • Power profiling is intricate because data pathways and preprocessing phases can differ substantially among devices. TinyML-capable devices exhibit vastly different power consumption, complicating the task of maintaining consistent accuracy across them. • Real-world circumstances can swiftly evolve because of the sensing environment changes or the senescence of vision sensors. TinyML addresses a substantial challenge in online learning by offering a solution despite the limited compute resources available. • Optimizing design trade-offs is challenging until the model itself is optimized.
Solution	• Automated tools can assist in exploring trade-offs among performance, precision, and power [20].
Step 11: Application testing	
Challenges	• Debugging application issues becomes more challenging as the system encompasses heuristic code alongside trained TinyML models. • TinyML-powered applications are streamlined, lacking robust tools for data measurement and visualization. This presents supplementary hurdles for debugging throughout model development and deployment because of memory limitations [17]. • Testing TinyML models can be wrecked by biases in the tests.
Solution	• Examining bug taxonomy for IoT applications can help preempt critical issues in application code [21,22]. • Having devoted quality assessment squads can facilitate fair debugging by mitigating developer bias.
Step 12: Application deployment	
Challenges	• Models installed on TinyML systems should support upgrades with minimal intervention or system downtime. • Deploying TinyML models in practical settings continues to be difficult despite successful model validation during training. • Ensuring that applications can run seamlessly across different devices and vendors presents a significant obstacle in TinyML engineering. • Rectifying bugs in embedded application deployment presents significant challenges.
Solution	• Achieving reproducibility relies on the distinct identification and accessibility of algorithms, parameters, and data/labels in real-world settings [6]. • Having over-the-air updates is crucial for seamlessly deploying bug fixes and upgrades for both firmware and TinyML models [23].

continued on next page

Table 19.1 *(continued)*

Step 13: Application monitoring	
Challenges	• Enhancing performance in real-world settings poses challenges due to constraints on edge hardware. • Measuring the performance of TinyML applications in real-time scenarios is difficult.
Solution	• Incorporation of labeling and retraining into continuous integration processes, informed by feedback from surveillance, can enhance application performance by adapting to new data distributions confronted in real-time scenarios. • Utilizing device level surveillance tools can enhance debuggability in real-world deployments.

19.4 Applications of TinyML and future directions

The remarkable expansion of the TinyML movement is driven by the diverse applications it enables. These applications not only create crucial datasets but also contribute to the growth that fuels ongoing research, fostering the need for more efficient TinyML systems and continuous enhancements in the overall design process.

In recent years, TinyML applications have expanded their scope from primarily focusing on vision to include natural language processing, predictive modeling, pattern recognition/classification, and data analysis. This section offers a summary of the application areas, along with a broad assessment of the challenges and prospects involved. Presently, TinyML demonstrates its efficacy across diverse applications such as smart devices from sensors to urban infrastructure, industrial supervision and control, healthcare, security and surveillance, administrative tasks, and financial operations, among others. These applications have tangible effects on our everyday lives and communities.

Over the past two years, TinyML applications have evolved beyond their initial emphasis on vision, now encompassing natural language processing, predictive modeling, pattern recognition/classification, and data analysis. This section provides an overview of the application domains, accompanied by a high-level examination of the associated challenges and opportunities. Currently, TinyML proves effective in various applications, including smart objects ranging from sensors to cities, industrial control and monitoring, healthcare, surveillance and security, administration, and finance, among others, impacting our daily lives and communities [24–29].

19.4.1 Smart vehicles

An intelligent vehicle is defined as a vehicle with computing capabilities, storage, and communication facilities that enable learning from its environment and making conclusions consequently. Sensors and multi-interface cards are used for equipping vehicles inside and outside. The increasing prevalence of smart vehicles endowed with onboard wireless devices and sensors like radar and lidar has led to a focus on efficient management and transportation applications. The goal is to improve traffic flow by reducing travel time and preventing jamming.

Smart vehicles possess a range of novel features, including information exchange and location info. These functionalities assist specialized applications, such as security communication and warnings. Vehicles inside a vehicular edge computing (VEC) system typically have onboard wireless devices, particularly onboard units (OBUs). During disaster alarm schemes, sensors play a crucial role in verifying if airbags were deployed during an accident.

19.4.1.1 Smart vehicle services

Smart vehicles offer a diverse range of services. Some key services, such as assisted driving, autonomous vehicles, platooning, and parking solutions, are discussed below.

19.4.1.1.1 Assisted driving

In contemporary times, vehicles such as cars, buses, and trains are designed with the capability to convey valuable information, including details about accidents, road closures, and traffic congestion. This is achieved through the integration of sensors, actuators, and processors, enhancing safety and navigation for these vehicles. The data on traffic patterns, made available by these intelligent features, can prove advantageous for all types of organizations [30]. Intelligent vehicles are categorized into five layers by the National Highway Traffic Safety Administration [31].

19.4.1.1.2 Autonomous vehicles

As smart vehicles progress towards autonomous driving, establishing robust connectivity amongst smart vehicles becomes imperative. Vehicular networks, on the rise due to this evolution, play a pivotal role in shaping intelligent transportation systems and smart cities. These systems are anticipated to support a spectrum of advanced applications, ranging from road safety and enhanced traffic efficiency to automated driving and seamless admittance to Internet facilities [32,33].

The global acceptance of automated vehicles has sparked a transformation in the automobile sector. Nevertheless, challenges such as invulnerability, fidelity, and secrecy persist in realizing completely automated vehicle editions. Notably, the susceptibility of automated vehicles to security threats is a concern; a single attack on the software of an autonomous vehicle could lead to multiple mishaps. Additionally, interconnected systems on the Internet face risks of unauthorized access, presenting unknown threats. Vehicle design addresses safety-critical issues by enabling the vehicle to anticipate and respond to potential dangers while continuously monitoring road conditions throughout the journey. The assumption in the design is that the driver provides the destination or navigation. However, it may not be in regulation throughout the excursion, emphasizing the role of automated vehicular systems in ensuring safe operations [34]. While automated vehicular systems differ from connected vehicular technology, they share some similarities.

19.4.1.1.3 Smart parking

In metropolitan regions, the number of vehicles parked in parking lots is substantial, distributed across various locations such as street parking and outer parking. Unlike moving vehicles, parked vehicles remain stationary for extended periods. Although they do not transport info from location to location, parked vehicles equipped with wireless communication devices and rechargeable batteries as part of smart street vehicles (SSVs) serve as communication infrastructures with unique characteristics. This allows parked SSVs to transmit information among each other and also link to neighboring moving SSVs, functioning as static backbones to enhance communication amongst vehicles. The number of parked vehicles in a parking slot and their duration of stay are critical factors influencing their role as communication infrastructures [35]. Collaboration among parked SSVs, particularly in parking lots, enables the execution of heavy computation tasks under favorable communication conditions. Individual vehicles, constrained by limited resources, may struggle to meet substantial computation demands. Parked SSVs address this challenge by providing powerful and underutilized computation resources, efficiently accomplishing allocated tasks in less time. This environment can be likened to tiny data centers capable of handling intricate tasks that require significant computational capability.

19.4.1.2 Smart vehicle applications

The emergence of VEC and the utilization of smart vehicles as infrastructures have paved the way for a multitude of associated vehicular applications. These applications span various domains, including driving safety, augmented reality (AR), infotainment services, and video streaming. Particularly in scenarios wherever higher computational processing is essential, VEC networks plays a crucial role in accelerating computing processes, in this manner curtailing delays. For instance, in the event of an accident, quick computations are needed to formulate solutions such as rescheduling traffic lights and efficiently dissipating a large traffic backlog. Meeting such demands places an exceptional requirement on computational resources [36]. In this context, applications are categorized into two groups: safety and non-safety. VEC proves to be supportive of both types of applications, as discussed below.

19.4.1.2.1 Safety applications

It emphasizes enhancing security by minimizing the likelihood of accidents. These applications monitor the driving environment and alert drivers to potentially harmful situations to prevent accidents. One such application involves the use of a global camera sensor mounted at a traffic monitoring signal, capable of detecting movement in its region by recognizing number plates within the detection field. This sensor records the location and vehicle number, sending this information to the local edge server. A smart local camera sensor (LCS) positioned at the front of the vehicle observes the driver's activities. The LCS issues warning messages to the driver for such activities, aiding in accident prevention. Repetitive warnings at appropriate times help drivers avoid hazardous situations and ensure their safety. The LCS is equipped to generate these warning messages and, later broadcasting a specific number, informs the edge server about any interrupting activities involving the vehicle. This report involves activity evidence and vehicle identification [37–41].

Context-aware systems are also employed, utilizing information related to the user to adapt operations based on environmental conditions. Context-aware applications adjust their operations according to the user's context, sensing information specific to the environment. These applications involve context acquisition, processing, and acting [42]. Leveraging contextual knowledge allows the generation of concise, context-aware information, reducing the radio resource requirements for transmission. Users can extract coveted content from the context using suitable decoders and big data analytics techniques such as natural language processing [43,44].

19.4.1.2.2 Non-safety applications

Applications of VEC extend beyond safety services to include the development of non-safety applications, such as multimedia services like video streaming, AR, and infotainment. The surge in streaming applications has notably contributed to a significant portion of network traffic, particularly in IoT communication, where video streaming plays a pivotal role [45]. This is particularly evident in smartphone applications like video crowdsourcing [46]. The Internet of Vehicles (IoV) supports various applications, including intelligent transportation systems and mobile multimedia. In IoV, users connect their mobile devices to the internet to access multimedia content from remote servers. However, maintaining quality of service (QoS) becomes challenging, given factors like jitter, buffering, throughput, and transmission delays in video streaming applications, exacerbated by the high mobility of vehicles in IoV. A proposed solution in [47] introduces distributed reliable real-time streaming in vehicular cloud-fog networks. A utility function is utilized to improve QoS and fairness in resource reservation among mobile devices, considering content provision for streaming and the number of tokens for content reservation from service providers, edge, and cloud. Mobile devices in the network query their probable location, the amount of data for streaming, and required tokens for content provisioning, facilitating effective reservation of streaming content from computing service providers and enhancing streaming utility reliability.

Addressing the parking lot monitoring issue, in [48] an edge computing-based scheme is proposed where each vehicle uploads street contents collected by the camera for video analytics. This enables ParkMaster to estimate precise locations and track parked vehicles using information from the vehicle's camera, GPS, and inertial sensors.

AR is an evolving multimedia application that seamlessly integrates real scenes into virtual scenes, overlaying virtual content onto the real environment to enhance traditional image information. AR can improve traffic awareness for vehicles or pedestrians near drivers, with the head-up display (HUD) reducing distractions and enhancing driving safety. An exploration of the HUD-based navigation system with AR-based content is detailed in [49], illustrating its potential for safety and convenience services [50]. A novel application, walk navigation, utilizes a camera and GPS for a car navigation system with AR technology, providing real-time navigation without compromising safety. The device's camera output is analyzed by an edge computing application to overlay viewed objects with AR content. Given the intricate storage and processing demands of AR, VEC is considered the optimal solution to meet the specific requirements of AR applications in a vehicular network, including mobility, location awareness, and low latency.

19.4.2 Anomaly detection

Anomaly detection, or outlier detection, refers to the use of methods to recognize patterns or data points that differ from the anticipated behavior. These methods have practical applications across IoT smart systems, such as smart cities, monitoring systems, and power management. Typically, sensors play a crucial role as input devices in these applications, generating significant amounts of data transmitted to cloud servers for analysis, decision-making, and storage [51].

The literature presents a range of methods for anomaly detection in data, spanning geometrical, statistical, and machine learning approaches [52]. There has been a notable trend toward adopting data-driven techniques, particularly machine learning algorithms and deep learning, to detect abnormal situations [53]. This approach involves building models from extensive sensor data to distinguish between normal and abnormal instances. Cloud-based architectures are commonly utilized in various anomaly detection applications, such as condition monitoring [54,55]. For instance, in condition monitoring, cloud computing is leveraged to monitor machine health using vibration signals. Vibration data are collected by edge devices and transmitted to the cloud for storage. Machine learning models in the cloud then identify abnormal machine behavior. However, processing data in the cloud server does come with some drawbacks, including latency and security concerns.

TinyML has the potential to revolutionize anomaly detection by utilizing machine learning directly at the edge. This strategy allows for real-time monitoring and detection of unusual patterns or behaviors without relying on a connection to a more robust computing device.

19.4.2.1 Condition monitoring

An implementation of a TinyML model on the STMicroelectronics STM32H743Z12 aims to detect anomalies in rotating machinery. This MCU device captures vibration signals using an accelerometer, processes them to extract features, and stores the data locally on the MCU. Subsequently, it trains an autoencoder machine learning model.

In another application, a system designed for anomaly detection in submersible pumps at wastewater management plants utilizes a retrofitting kit equipped with an MCU [56]. This kit integrates temperature and vibration sensors, an ESP32DEVKIT MCU, and power line communication equipment installed in the terminal chamber of the pump. The MCU collects data from the sensors, extracts features through processing, and utilizes this information to locally train an isolation forest model for anomaly detection in the data stream. After the training phase, the MCU transitions to the inference model, wherein it continues to process and learn to update the model, even in this mode [57].

19.4.2.2 Predictive maintenance

A TinyML model deployed on the ESP-WROOM-32 MCU device is utilized for anomaly detection in thermal images [58]. The device communicates data to the server using the Message Queuing Telemetry Transport (MQTT) protocol solely when it detects an anomaly on a machine. Operating on a convolutional neural network (CNN) model developed with Keras and converted using the tinymlgen library from Eloquent Arduino, the MCU device executes the detection process.

For water distribution systems, an online learning anomaly detection model known as Deep Echo State Network (DeepESN), based on a recurrent neural network (RNN), has been designed to evaluate the reliability of such systems and is suitable for deployment on MCUs [59].

A block-based Binary Shallow Echo State Network (BBS-ESN) model serves as a deeply quantized anomaly detector for oil leaks in wind turbines [60]. This TinyML model utilizes DeepESN, binarized images, and 1-bit quantization of weights and activation functions. The performance evaluation is conducted using the STMicroelectronics NUCLEO-H743ZI device.

19.4.2.3 Internet of intelligent vehicles

The MCU device is integrated into a vehicle to detect road anomalies such as potholes, bumps, and obstacles. The Arduino Nano 33 IoT is utilized for gathering accelerometer data and running an unsupervised TinyML algorithm called TEDA. Real-world experiments are carried out on asphalt pavement to validate the system's performance [61].

19.4.3 Healthcare

The evolution of TinyML has inevitably made its mark on the healthcare sector, a crucial component of societies worldwide. Given that the human body emits signals from various organs, deploying sensors to gather these data becomes instrumental in addressing healthcare challenges. Utilizing TinyML can effectively address these issues. Previous studies [62–64] suggest that integrating these edge devices could revolutionize healthcare, elevating the well-being of individuals.

TinyML presents a precise and dependable solution for healthcare monitoring, leading to advancements in various health products. Moreover, this technology holds the promise of introducing innovative methods of medical monitoring, diagnostics, and therapy, thereby enhancing the quality of care, and ultimately improving patient outcomes. By enabling real-time data processing in compact devices, TinyML opens up new opportunities for healthcare professionals to monitor and treat patients with greater effectiveness and efficiency [65]. Biological signals originating from the human body are captured by specialized sensors and transmitted to the core of embedded systems, which incorporate machine learning or deep learning algorithms interacting with cloud environments. This collaboration allows the device to identify deficiencies, monitor, predict, or provide assistance to the patient. In this section, we explore promising TinyML-based solutions in the healthcare sector, highlighting some of its emerging applications.

19.4.3.1 Cardiovascular disease

Cardiovascular diseases stand as a leading cause of global mortality. The mortality rate due to heart disease continues to rise, despite significant technological advancements in recent decades. It remains a leading cause of death, particularly among the elderly, with identified major risk factors including high blood pressure, high cholesterol, smoking, and diabetes. Early and accurate prediction of an individual's susceptibility to heart disease holds the potential to not only reduce mortality but also considerably reduce the number of cases.

Forecasting the onset of heart diseases poses a complex challenge. Nevertheless, the incorporation of TinyML into healthcare systems has emerged as a noteworthy approach for monitoring patients' well-being and identifying anomalies. The prevalent sensors employed include those for electrocardiography (ECG), blood pressure, heart rate, and body temperature. ECG signals depict the heart's electrical activity during rest, offering insights into heart rate, rhythm, and aiding in the diagnosis of conditions such as heart enlargement resulting from high blood pressure, elevated heart rate, dysrhythmia, or heart attacks.

A healthcare system based on the IoT is developed to evaluate heart diseases by utilizing a modified deep CNN (MDCNN) [66]. This system incorporates a smartwatch and a heart monitoring device to collect blood pressure and ECG information, which is then transmitted to the server via LoRa. Utilizing these transmitted data, the MDCNN model categorizes patients into normal or abnormal heart conditions, with alerts sent to doctors in the event of abnormalities.

Methods for heart diagnosis through ECG analysis and classification within an embedded IoT system are seamlessly integrated with wearable sensors, enabling continuous 24/7 monitoring. The algorithm was subsequently coded and implemented on Intel's Galileo IoT board. ECG signals were read and digitized at a sampling rate of 360 samples per second, with analysis performed every 3000 samples [67].

Acute coronary syndrome is a severe medical condition caused by an imbalance in the body's metabolic demands. It is characterized by chest pain that may radiate to the neck and left arm. Healthcare providers typically advise a range of laboratory tests and ECG based on each patient's specific condition.

The evolution of TinyML systems continues, promising additional technological advancements. With increased data availability, refined research methods, and advanced techniques, future estimates are poised to benefit from improved accuracy. Alongside ongoing research in ECG analysis, another challenge lies in the creation of economically viable wearable devices that can be worn consistently. The integration of modern wearable solutions holds the potential to elevate the entire process, contributing to the development of a more sophisticated healthcare system, ultimately enhancing decision-making capabilities.

A smart healthcare system has been developed to assist physicians in diagnosing cardiac conditions through intelligent analysis of ECG signals. This system is designed to automatically identify and categorize various cardiovascular diseases, utilizing a 1D CNN. Data acquisition and processing are carried out using MCU devices, and TinyML offers a reliable edge solution for the system.

Exploring the potential of TinyML in healthcare, researchers have introduced a solution named "TinyCare" to monitor high-level blood pressure [68]. This cloud-independent TinyML approach relies solely on patient-derived data. This systematic approach commences with data preprocessing based on physiological signals, followed by feature extraction. Multiple machine learning algorithms are deployed on three edge devices: Arduino Uno, ESP32 Wrover Board, and AdaFruit PyBadge.

19.4.3.2 Neurological disorders

Epilepsy, a prevalent neurological disorder worldwide, impacts individuals of all age groups. Electroencephalography (EEG) serves as a fundamental tool for diagnosing this disorder. It enables doctors to examine the brain's electrical activity, with disruptions potentially indicating epilepsy. Monitoring patients with neurological disorders like epilepsy and Alzheimer's through EEG-based systems stands as a pivotal contribution of smart healthcare systems. While EEG reports are traditionally subject to visual inspection by diagnosticians, this approach has limitations and a significant error margin, depending on the diagnostician's expertise.

EEG, a diagnostic method utilizing electrodes on the scalp to detect brain electrical signals, is widely employed for this purpose. However, another brain signal diagram, magnetoencephalography (MEG), holds promise as a diagnostic tool due to its less attenuated signals compared to EEG.

An AI-based framework designed for EEG data classification aims to categorize a patient's EEG signals as either pathological or normal while providing guidance for subsequent actions [69]. The model involves embedding various sensors in the patient's vicinity to capture data such as body temperature, heartbeat, blood pressure, voice, facial expressions, body movement, and EEG. The engine is described as being founded on AI and deep learning algorithms. Upon analyzing the data, the engine determines whether further analysis is necessary. If required, the data are directed to a classifier model within the cloud system, which employs CNN models such as VGG 16 and AlexNet for detection and classification. The outcomes are then communicated back to the cognitive engine (cloud), which initiates additional actions if the patient's signals are deemed pathological. Both the patient and relevant healthcare professionals receive alerts and suggestions.

The utilization of preictal EEG data for predicting and mitigating seizures in patients was showcased. A collection of data from individuals with epilepsy was utilized to construct a model, which was subsequently deployed on a microcontroller through TinyML. The microcontroller demonstrated effective performance in predicting epileptic seizures [70].

Neurological applications of AI are increasingly crucial, particularly in the context of Parkinson's disease, a progressive and chronic movement disorder characterized by impaired movement, balance issues, limb stiffness, tremors, and speech disorders. Annually, approximately 60,000 Parkinson's cases are diagnosed in the

US, while the number rises to a million in India. While there is no definitive cure, early detection can substantially aid in patient management. Parkinson's disease continues to be a central focus of active research, with both pharmaceutical companies and researchers involved in drug development and monitoring systems [71].

The data used to tackle the challenge of improving Parkinson's classification utilizing nodes consist of voice recordings obtained from digital home virtual assistant devices, specifically the Lee Silverman Voice Therapy (LSVT) voice data [72]. Integrating cognitive technology with the IoT holds significant promise for enhancing the healthcare system, particularly in remote monitoring. The latest technology in this regard involves monitoring and diagnosing epileptic seizures through deep learning methods within a cognitive healthcare IoT (CHIoT) framework, providing a more comprehensive approach for healthcare applications [73].

CHIoT facilitates a system that integrates physiological and psychological aspects to manage medical emergencies with prompt responses. This healthcare framework utilizes wearable sensors to capture signals, primarily scalp EEG, and employs a cognitive engine based on deep CNNs for feature extraction, followed by a stacked autoencoder. Initial monitoring occurs at the edge by observing the patient's gestures, movements, facial expressions, and EEG signals to determine their state.

TinyML presents a plethora of embedded solutions and optimization techniques within the healthcare wearable device domain [74]. A noteworthy application involves its use in detecting and mitigating freezing of gait (FoG) in Parkinson's patients. A machine learning model deployed on an ATMega2560 microcontroller effectively manages power consumption and processing speed [75]. Expanding its application in healthcare, TinyML demonstrates its value in identifying cardiac arrhythmia, characterized by irregular heartbeat stemming from abnormal electrical activity in the heart. Utilizing low-power microcontrollers, the CMSIS-NN library is employed to deploy a CNN.

An ultralow-power wearable device designed for detecting epileptic seizures employs a random forest machine learning model as a classification algorithm on an STM32L476 ARM Cortex-M4 microcontroller [76]. This setup enables up to 40.87 hours of monitoring on a single battery charge, equivalent to a full day of operation, with a reduced number of false alarms [77].

Furthermore, TinyML is introduced as a medical embedded system for diagnosing focal liver lesions. A deep neural network model trained on liver images obtained through ultrasound imaging is implemented using computer-aided diagnosis. This comprehensive process is condensed into a small memory footprint that operates on an edge platform [78,79].

19.4.3.3 Identification of deadly mosquitoes

Healthcare applications also encompass the detection of harmful species that can precipitate diseases and infectious illnesses. Researchers have employed CNNs for classifying audios captured through smartphones and introduced TinyML technology to enhance the process [80]. Hazardous mosquitoes are detected using a model sensor that translates wing beats into audios. These audios are then classified by a tiny embedded system, leveraging both machine learning and the Edge Impulse platform. A prototype based on Arduino Nano BLE 33 Sense, equipped with a trained machine learning model, is capable of classifying collected audio data of wing beats to identify mosquito species. The model achieved an accuracy of 88.3%, demonstrating promising results, particularly considering the low-cost and low-energy consumption aspects of the system [81].

19.4.3.4 Assistive systems

Assistive systems function as rehabilitative frameworks or offer support in daily activities for individuals with disabilities. This category holds significant importance within TinyML frameworks, as many of these systems depend on accurate assessments of a disabled person's ability to perform various tasks. This may encompass tasks such as limb movement [82], where assistive systems can facilitate physical rehabilitation, or ensuring proper hearing through devices like cochlear implants. The following section aims to explore key applications in assistive systems based on EEG and surface electromyography (sEMG) signals, highlighting the role of machine learning in processing these signals.

19.4.3.4.1 EEG-based

A crucial application of brain–computer interface (BCI) devices is to empower physically disabled individuals to overcome the limitations of their disabilities and lead a more conventional life [83]. This is achieved by providing them with BCI devices that respond to their natural impulses, eliminating the need for physical movement of their body parts to perform actions. The utilization of EEG signals derived from brain neurons serves as the control mechanism for these systems. The ongoing advancements in BCI technology pave the way for the development of next-generation prosthetic implants. These prosthetic implants, along with other assistive systems driven by BCI, hold the potential to bring about life-changing transformations for individuals with disabilities.

EEG signals are analyzed to capture ocular information, specifically eye blinking, and classify it for use in control applications [84]. These applications could include a variety of functions, from home automation to navigation, allowing individuals to overcome physical limitations and interact with their environment. The EMOTIV EPOC headset was used to capture EEG signals corresponding to eye closure.

BCI has been integrated with a comprehensive deep learning framework for various applications, such as developing typing systems based on EEG signal patterns and controlling robots for domestic assistance. These systems are particularly relevant for individuals with speech and/or visual impairments [85].

19.4.3.4.2 sEMG-based

An alternative to BCI through EEG sensors involves utilizing sEMG signals, which consist of both EMG and neural electrical signals on the skin surface of superficial muscles. These signals can be detected by sensors and linked to actuators to execute the actions intended by the patient. Gait phase detection is explored to facilitate the natural movement of lower limb auxiliary robots, employing a kernel linear discriminant analysis (LDA)-based nonlinear fusion model. Modeling the complex stance and swing motion of lower limbs with realistic variations poses a challenge [86].

An intelligent wearable armband equipped with sensors to capture sEMG signals for hand gesture identification was developed by the authors of [87]. The recorded information is transmitted to a robotic arm, enabling it to replicate these movements and assist in stroke rehabilitation. The data collected by these sensors undergo processing and classification through principal component analysis. For the classification process, six distinct sEMG features are employed.

19.4.3.4.3 Neural speech enhancement for hearing aids

Noise suppression and hearing aids represent crucial medical solutions for individuals with impaired hearing, helping to alleviate challenges associated with listening, particularly in noisy environments. While several models have been developed using RNNs [88], the authors of [89] expanded on these models by incorporating TinyML. Leveraging the hardware foundation of a hearing aid, the authors utilized RNNs and pruning techniques to create a model aimed at enhancing speech perception. Despite the limitations imposed by the edge device, which required training large neural networks with substantial datasets due to storage constraints, the authors achieved a moderately satisfactory rating for their model.

19.4.4 Smart farming

The global population is rapidly increasing, leading to a growing demand for agricultural products to sustain future needs. However, challenges such as population growth, climate change, soil erosion, and the degradation of natural resources are placing significant pressure on food production. This underscores the necessity for novel farming methods to address the escalating demand. Consequently, agriculturists, researchers, and agricultural sectors are adopting cutting-edge advancements and approaches, including IoT, UAV, AI, big data, cloud and fog computing, and edge computing. These advancements are transforming conventional agriculture into ecological, clever, competent, and environmentally friendly practices, often referred to as smart agriculture or smart farming [90].

The transformation is significantly influenced by IoT, where farms can now actively monitor crop and soil health, identify potential diseases affecting plants, and utilize drones to track plantation growth, among other capabilities. The electronics industry has witnessed remarkable progress, resulting in the accessibility of superior and economical parts like MCUs, microcontroller boards, sensors, and wireless transceivers. The latest MCU models not only excel in typical sensing and control jobs, they also enable intricate operations, including the execution of machine learning models. Additionally, advancements in contemporary radio technology enable long-range transmissions with reduced energy expenditure.

The smart farming IoT network utilizes IoT devices to collect data on various aspects such as soil conditions, crops, greenhouses, irrigation, and weather, employing cameras or sensors. This data are then transmitted to the cloud via wireless sensor networks (WSNs) providing farmers with the ability to monitor crop health and detect potential diseases in the plants. Through the analysis and interpretation of these data, farmers gain insights into the relationships between different agricultural factors, such as soil characteristics and climate variables, enabling informed decision-making and efficient planning [91]. In this context, machine learning plays a pivotal role in modeling the intricate patterns within the data, forming the foundation of decision support systems.

In smart farming, various architectures are employed based on specific applications, with common ones including two-layer (physical–edge), three-layer (physical–edge–cloud), and four-layer (physical–edge–fog–cloud) structures. Many smart farming applications adopt the three-layer architecture [92].

The physical layer comprises sensors, actuators, and drones that gather data from the soil, animals, greenhouse, and weather. These collected data are transmitted to edge nodes through WSNs.

The edge layer consists of computing devices with low or medium computing resources, which interpret and analyze data received from the physical layer.

The cloud layer serves as the central hub and data storage in the smart farming architecture, possessing high storage capacity to store data generated by various sensors. Additionally, it aggregates and extracts insights to provide machine learning models for decision-making purposes.

Advanced machine learning algorithms require significant computing resources to operate optimally, leading to the adoption of cloud computing. Cloud computing offers the capacity to handle large machine learning models containing millions of parameters, as well as high-speed processors and ample memory necessary for effective model execution. However, this presents challenges in certain regions [93], such as Africa, where internet bandwidth may be insufficient for quick transmission to the cloud. Insufficient internet connectivity leads to notable issues, including considerable latency, data loss, and reliability concerns. Furthermore, heavy dependence on the cloud increases vulnerability to cyberattacks and reduces data security [94,95].

To overcome the challenges linked with cloud computing, alternative approaches have emerged, notably involving bringing computational processes closer to end devices through fog and edge computing. Recently, there has been growing interest in integrating TinyML into smart farming. TinyML enables sensor devices to perform machine learning tasks, such as monitoring crop health, identifying plant diseases, and forecasting optimal crops, without depending on the cloud. TinyML offers several advantages over cloud, fog, and edge computing in terms of privacy, security, latency, and energy efficiency [92].

TinyML shows great promise in the realm of smart farming, especially in regions like Africa where embedded systems and AI are currently underutilized [94]. An example of this potential is the Plant Village initiative [96], an open-source project managed by Penn State University. Within this initiative, the Nuru app has been developed to aid farmers in identifying and combatting plant diseases. By utilizing machine learning via TensorFlow Lite (TFLite) on mobile phones, the app provides real-time solutions even without internet access, which is crucial for farmers in remote areas. The future evolution of this system is set to capitalize on TinyML and TFLM for deploying sensors across remote farms, thereby enhancing tracking and analysis capabilities [97].

TinyML has the potential to make a significant impact on environmental problem-solving. It enables the deployment of intelligent devices capable of monitoring and collecting data on various environmental factors, such as air quality, water quality, and meteorological conditions. These devices are engineered to withstand noise and changes in data distribution. Moreover, they can operate effectively in challenging environments, such as extreme weather conditions or areas with limited or no access to power. The widespread deployment of TinyML devices facilitates the availability of real-time environmental data, enabling prompt identification and resolution of issues. Smart farming initiatives encompass the following.

19.4.4.1 Crop management

Crop management involves employing various techniques to cultivate and sustain crops efficiently and sustainably. These techniques encompass appropriate irrigation and fertilization, pest and disease management, and optimizing crop yields. TinyML can be leveraged to collect data on factors such as soil moisture and temperature and analyze these data to make predictions and offer insights that enhance crop yields and efficiency. Recently, numerous TinyML-based innovations have been proposed by researchers to assist farmers and scientists in this field. Below are a few examples presented in this section.

An embedded machine learning pipeline enables users to develop an embedded machine learning model applicable to various plants in labs, greenhouses, farms, or gardens [98]. The process begins with data collection, with suggested best practices for gathering plant data. Subsequently, a CNN is trained to estimate the leaf area index and predict the plant growth stage. After training, the machine learning model is compressed and converted to TFLite format for deployment on MCU devices. For testing purposes, the authors utilized the Sony Spresense setup as the target device.

A TinyML solution for detecting drought stress in soybeans comprises a Raspberry Pi Zero W and a Sony IMX219 camera module [99]. The Raspberry device runs a CNN model on the captured image to identify crop drought stress and then transmits predictions to a web platform. The CNN model is converted to TFLite format for deployment on resource-constrained devices.

A low-power and real-time image detector for grape leaf esca diseases was developed based on a compressed CNN model [100]. Several compression techniques, including CP decomposition, Tucker decomposition, and tSVD, were evaluated to determine the method with the optimal compression factor and accuracy. CP decomposition was chosen and applied to the CNN model. After training and validation, the model was compressed using posttraining quantization with TFLite to produce an 8-bit model. The compressed model was deployed on an OpenMV Cam STM32H7, mounted on an agricultural vehicle moving at a constant speed through the cultivation field.

19.4.4.2 Plant disease

To identify plant diseases in crops, recent research has shifted towards image-processing-based deep learning systems, deviating from the traditional RNA analysis approach for timely disease identification, thereby preventing significant economic losses for farmers. The prevalent image-based deep learning framework employs a multi-layer CNN architecture that autonomously learns features from labeled training image data in the initial stage. These learned features are subsequently input to an artificial neural network in the second stage, utilizing neural weights, bias functions, and non-linear activation functions to achieve high classification accuracy [101].

19.4.4.3 Agricultural advisory system

Multiple chat bot-driven systems have been designed to address farmers' queries, providing them with tailored responses [102]. The system utilizes online web resources, such as documents, as training data and employs natural language processing to construct a deep learning-based RNN framework. Farmers can pose a variety of questions, spanning from inquiries about crop types grown in specific geographical regions to inquiries about the usage of pesticides and fertilizers.

19.4.4.4 Smart irrigation

An agro-environmental management system integrates moisture sensors and real-time video analysis of soil images [103]. This system utilizes the VGG-19 model to classify image types and determine the required irrigation water quantity, considering the specific crop type.

In another approach, a cost-effective device constructed on the MCU ESP32-CAM is introduced [104]. This device employs a camera to gather data from digital water meters for monitoring central pivot irrigation systems. The device incorporates a TinyML model for image processing to read the water meter, followed by transmission to a server via LoRaWAN.

19.4.4.5 Smart greenhouse

A greenhouse is a specially designed structure created to provide a controlled environment for plant growth. This controlled environment includes variables such as temperature, humidity, light, and nutrient levels, which can be adjusted to optimize crop development. Greenhouses enable the year-round cultivation of various crops, including fruits, vegetables, and flowers, regardless of external weather conditions. In greenhouses, TinyML can be utilized to enhance crop growth by implementing machine learning on low-power devices for data collection and predictive analysis.

One widely accepted approach involves a multi-label TinyML model utilizing a multilayer perceptron (MLP) architecture for micro-climate management in a strawberry agricultural greenhouse [105]. Another noteworthy initiative is a system where IoT devices and edge nodes collaborate to monitor plant status in a greenhouse and make decisions regarding sprinkler operation. Each IoT device employs its embedded TinyML model based on a decision tree to evaluate plant needs, transmitting this information to the edge node. The edge node aggregates decisions from all sensors associated with a specific sprinkler and utilizes a machine learning model to determine the final action, such as "no action," "irrigation," or "fertigation" [106].

19.4.5 Smart environments

19.4.5.1 Water quality monitoring

A system which integrates machine learning and TinyML-based devices for monitoring and assessing water quality collects sensor data on various water quality parameters like temperature, pH, and chemical concentrations and transmits the information to a Raspberry Pi connected to a data center [107]. Subsequently, a machine learning model analyzes the data, and the results are sent to the cloud for further examination. This system helps in water conservation and also creates the ability to monitor multiple factors influencing water quality.

19.4.5.2 Vehicular emission

The greenhouse effect, a significant contributor to environmental concerns, largely originates from automobile emissions, which contain significant amounts of CO_2. To address this issue, TinyML is utilized alongside the onboard diagnostics system, a feature present in most automobiles, to monitor CO_2 emissions per liter [108]. The collected data are then transmitted to Typicality and Eccentricity Data Analytics (TEDA), an unsupervised anomaly detection algorithm. This approach quantifies the quantity of CO_2 produced in grams per second, storing the data on a microSD card for cloud processing by external systems via Bluetooth or 4G.

19.4.5.3 Temperature prediction

Temperature forecasting through TinyML is a crucial application in environmental monitoring, offering insights for decisions such as optimal energy utilization in buildings, strategic timing for crop planting, and appropriate crop selection [109].

19.4.5.4 Water disease detection and prediction

Edge AI is proposed to protect water supplies from hazardous contaminants like cholera. The existing alkaline peptone water (APW) method and edge computing are used to establish an experimental setup for monitoring the physicochemical parameters of water to prevent cholera outbreaks. Due to the lack of wireless connectivity in remote areas, an offline model is employed. The device was eventually integrated with on-tap inference, enabling the monitoring of water safety metrics.

19.4.5.5 Underwater imaging

Conventional underwater imaging systems require active power sources, which are typically unavailable in many underwater environments. These systems are essential for research purposes, including studying marine species, climate change, marine geology, aquaculture farms, particulate organic carbon transport, and maritime archaeology. Recent research has investigated the use of fully submerged, battery-free cameras and acoustic backscatter for on-site wireless underwater imaging. However, the limited bandwidth of the underwater acoustic channel results in inefficiencies in both image acquisition and communication [110].

19.4.5.6 Earthquake detection

Researchers are focusing on earthquake detection as a crucial initial step in earthquake early warning (EEW) systems. The emphasis is on providing real-time alerts, and robust EEW systems must demonstrate high detection accuracy, low detection latency, and a high sensor density. Traditional EEW systems rely on fixed sensor networks or, more recently, networks of mobile phones equipped with microelectromechanical system (MEMS) accelerometers.

A novel approach is proposed for global-scale earthquake detection and warning based on IoT edge devices with TinyML capabilities and always-on, always-connected stationary MEMS accelerometers. This approach explores and evaluates deep learning machine learning algorithms for earthquake detection using a limited-resource Arduino Cortex-M4 microcontroller (256 kB of RAM) [111].

19.4.5.7 Wildlife conservation

Sensor technology plays a crucial role in wildlife conservation endeavors, especially in safeguarding endangered species. TinyML-based solutions include animal sensors for monitoring movement and bioacoustics sensors to collect data on wildlife and the surrounding environment. These methods contribute to enhancing comprehension of biodiversity and the ecosystem. However, the limitations and challenges include low latency and limited capacity, a lack of substantial data volume, and the expansive nature of habitats [112].

Another application concerns the protection of sea turtles, which are threatened by factors such as uncontrolled fishing and sea pollution. The TinyML framework was utilized to bolster conservation efforts by implementing advanced real-time vision-based TinyML capabilities. The study aimed to leverage this technology for the real-time tracking of sea turtles, offering immediate data on their position, activity, and population to improve the efficiency and effectiveness of conservation operations. This approach can be adopted not only for sea turtle protection but also for other conservation tasks, such as combating illegal hunting, logging, and ensuring animal welfare.

Utilizing TinyML can play a significant role in monitoring essential biological activities across various species, providing valuable insights into their behavior, movement, and ecosystems. Ultrasonic sensors have been utilized to monitor bat activity. These sensors capture sound waves and generate spectrograms, which are then analyzed by TinyML-based hardware to classify the signal and extract information about the bat's location, its genus, and the time of recording. Such an approach offers valuable data on biodiversity loss and other critical ecological aspects [113].

References

[1] W. Shi, J. Cao, Q. Zhang, Y. Li, L. Xu, Edge computing: vision and challenges, IEEE Int. Things J. 3 (5) (2016) 637–646.

[2] R. Oshana, M. Kraeling, Software Engineering for Embedded Systems: Methods, Practical Techniques, and Applications, Newnes, 2019.

[3] S. Amershi, A. Begel, C. Bird, R. DeLine, H. Gall, E. Kamar, N. Nagappan, B. Nushi, T. Zimmermann, Software engineering for machine learning: a case study, https://www.microsoft.com/en-us/research/uploads/prod/2019/03/amershi-icse-2019_Software_Engineering_for_Machine_Learning.pdf.

[4] G. Giray, A software engineering perspective on engineering machine learning systems: state of the art and challenges, J. Syst. Softw. 180 (2021) 111031.

[5] C.R. Banbury, V.J. Reddi, M. Lam, W. Fu, A. Fazel, J. Holleman, X. Huang, R. Hurtado, D. Kanter, A. Lokhmotov, D. Patterson, D. Pau, J. sun Seo, J. Sieracki, U. Thakker, M. Verhelst, P. Yadav, Benchmarking TinyML systems: challenges and direction, https://doi.org/10.48550/arXiv.2003.04821, 2021.

[6] D. Doria, I. Ernst, B. Kadlec, CVPR18: Tutorial: Software Engineering in Computer Vision Systems, ComputerVisionFoundation, Jun 2018.

[7] R.P. Adams, M. Mattina, P.N. Whatmough, SpArSe: Sparse architecture search for CNNs on resource-constrained microcontrollers, https://doi.org/10.48550/arXiv.1905.12107, 2019.

[8] F. Paissan, A. Ancilotto, E. Farella, PhiNets: a scalable backbone for lowpower AI at the edge, https://doi.org/10.1145/3510832, 2021.

[9] R. David, J. Duke, A. Jain, V.J. Reddi, N. Jeffries, J. Li, N. Kreeger, I. Nappier, M. Natraj, S. Regev, R. Rhodes, T. Wang, P. Warden, TensorFlow Lite Micro: Embedded machine learning on TinyML systems, https://doi.org/10.48550/arXiv.2010.08678, 2021.

[10] Sironi, TinyML talks: Machine learning for event cameras, tinyML, https://www.tinyml.org/event/tinyml-talks-machine-learning-for-event-cameras/, Oct 2021.

[11] M. Shafique, T. Theocharides, V. Reddy, B. Murmann, TinyML: current progress, research challenges, and future roadmap, in: 2021 58th ACM/IEEE Design Automation Conference, DAC 2021, Proceedings - Design Automation Conference, Institute of Electrical and Electronics Engineers Inc., Dec. 2021, pp. 1303–1306, https://doi.org/10.1109/DAC18074.2021.9586232.

[12] M. Gielda, Running TF Lite on microcontrollers without hardware in Renode, https://cms.tinyml.org/wp-content/uploads/talks2020/, Feb 2022.

[13] Lomuscio, TinyML talks: Verification of ML-based AI systems and its applicability in edge ML, tinyML, https://cms.tinyml.org/wp-content/uploads/talks2021/tinyML_Talks_Alessio_Lomuscio_211005.pdf, Oct 2021.

[14] Banbury, V.J. Reddi, P. Torelli, J. Holleman, N. Jeffries, C. Kiraly, P. Montino, D. Kanter, S. Ahmed, D. Pau, U. Thakker, A. Torrini, P. Warden, J. Cordaro, G.D. Guglielmo, J. Duarte, S. Gibellini, V. Parekh, H. Tran, N. Tran, N. Wenxu, X. Xuesong, MLPerf Tiny Benchmark, https://dl.acm.org/doi/10.1145/3528227.3528569, 2021.

[15] M. Loukides, TinyML: the challenges and opportunities of low-power ML applications, https://www.oreilly.com/radar/tinyml-the-challenges-and-opportunities-of-low-power-ml-applications/, Oct 2019.

[16] M. Shafique, M. Naseer, T. Theocharides, C. Kyrkou, O. Mutlu, L. Orosa, J. Choi, Robust machine learning systems: challenges, current trends, perspectives, and the road ahead, IEEE Des. Test 37 (2) (2020) 30–57.

[17] H. Cai, C. Gan, L. Zhu, S. Han, Tiny transfer learning: towards memory-efficient on-device learning, CoRR, arXiv:2007.11622v1 [abs], 2020.
[18] L. Lai, N. Suda, V. Chandra, CMSIS-NN: Efficient neural network kernels for Arm Cortex-M CPUs, https://doi.org/10.48550/arXiv.1801.06601, 2018.
[19] D. Xu, T. Li, Y. Li, X. Su, S. Tarkoma, T. Jiang, J. Crowcroft, P. Hui, Edge, intelligence: Architectures, challenges, and applications, https://doi.org/10.48550/arXiv.2003.12172, 2020.
[20] C. Banbury, C. Zhou, I. Fedorov, R.M. Navarro, U. Thakker, D. Gope, V.J. Reddi, M. Mattina, P.N. Whatmough, MicroNets: Neural network architectures for deploying TinyML applications on commodity microcontrollers, https://doi.org/10.48550/arXiv.2010.11267, 2021.
[21] A. Makhshari, A. Mesbah, IoT bugs and development challenges, in: 2021 IEEE/ACM 43rd International Conference on Software Engineering (ICSE), 2021, pp. 460–472.
[22] M.J. Islam, G. Nguyen, R. Pan, H. Rajan, A comprehensive study on deep learning bug characteristics, https://doi.org/10.1145/3338906.3338955, 2019.
[23] V. Mehta, Challenges of edge AI inference, https://semiengineering.com/author/vinay-mehta/, Jul 2021.
[24] R. Sanchez-Iborra, A.F. Skarmeta, TinyML-enabled frugal smart objects: challenges and opportunities, IEEE Circuits Syst. Mag. 20 (2020) 4–18.
[25] Z. Wu, M. Jiang, H. Li, X. Zhang, Mapping the knowledge domain of smart city development to urban sustainability: a scientometric study, J. Urban Technol. 28 (2021) 29–53.
[26] I. Fedorov, M. Stamenovic, C. Jensen, L.C. Yang, A. Mandell, Y. Gan, M. Mattina, P.N. Whatmough, TinyLSTMs: efficient neural speech enhancement for hearing aids, in: Proceedings of the Interspeech 2020, Shanghai, China, 25–29 October, 2020.
[27] D. Rossi, F. Conti, A. Marongiu, A. Pullini, I. Loi, M. Gautschi, G. Tagliavini, A. Capotondi, P. Flatresse, L. Benini, PULP: a parallel ultra low power platform for next generation IoT applications, in: Proceedings of the 2015 IEEE Hot Chips 27 Symposium (HCS), Cupertino, CA, USA, 22–25 August, 2015, pp. 1–39.
[28] M. Monfort Grau, TinyML: From Basic to Advanced Applications, Bachelor's Thesis, Universitat Politècnica de Catalunya, Barcelona, Spain, 2021.
[29] U.S. Shanthamallu, A. Spanias, Machine and deep learning applications, in: Machine and Deep Learning Algorithms and Applications, Springer, Berlin/Heidelberg Germany, 2022, pp. 59–72.
[30] L. Atzori, A. Iera, G. Morabito, The internet of things: a survey, Comput. Netw. 54 (2010) 2787–2805.
[31] X. Cheng, C. Chen, W. Zhang, Y. Yang, 5G-enabled cooperative intelligent vehicular (5GenCIV) framework: when Benz meets Marconi, IEEE Intell. Syst. 32 (2017) 53–59.
[32] L. Liang, H. Peng, G.Y. Li, X. Shen, Vehicular communications: a network layer perspective, IEEE Trans. Veh. Technol. 66 (2017) 10647–10659.
[33] H. Ye, G.Y. Li, B.-H.F. Juang, Deep reinforcement learning based resource allocation for V2V communications, IEEE Trans. Veh. Technol. 68 (2019) 3163–3173.
[34] A Look at the Future of 5G. Available online: https://spectrum.ieee.org/computing/software/a-look-at-the-future-of-5g/. (Accessed 28 November 2023).
[35] X. Hou, Y. Li, D. Wu, S. Chen, Vehicular fog computing: a viewpoint of vehicles as the infrastructures, IEEE Trans. Veh. Technol. 65 (2016) 3860–3873.
[36] M. Eltoweissy, S. Olariu, M. Younis, Towards autonomous vehicular clouds, in: Proceedings of the Springer Conference on International Conference on Ad Hoc Networks, Edmonton, AB, Canada, 20–22 August 2010, pp. 1–16.

[37] K. Hong, D. Lillethun, U. Ramachandran, B. Ottenw, B. Koldehofe, Mobile fog: a programming model for largescale applications on the internet of things, in: Proceedings of the ACM SIGCOMMWorkshop on Mobile Cloud Computing, Hong Kong, China, 16 August 2013, pp. 15–20.

[38] Sachin B. Chougule, Bharat S. Chaudhari, Sheetal N. Ghorpade, Marco Zennaro, Exploring computing paradigms for electric vehicles: from cloud to edge intelligence, challenges and future directions, World Electr. Veh. J. 15 (2) (2024) 39, https://doi.org/10.3390/wevj15020039.

[39] S. Roy, R. Bose, D. Sarddar, A fog-based DSS model for driving rule violation monitoring framework on the internet of things, Int. J. Adv. Sci. Technol. 82 (2015) 23–32.

[40] G. Vashitz, D. Shinar, Y. Blum, In-vehicle information systems to improve traffic safety in road tunnels, Transp. Res., Part F Traffic Psychol. Behav. 11 (2008) 61–74.

[41] S.J. Miah, R. Ahamed, A cloud-based DSS model for driver safety and monitoring on Australian roads, Int. J. Emerg. Sci. 1 (2011) 634.

[42] H. Vahdat-Nejad, A. Ramazani, T. Mohammadi, W. Mansoor, A survey on context-aware vehicular network applications, Veh. Commun. 3 (2016) 43–57.

[43] M. Baldauf, S. Dustdar, F. Rosenberg, A survey on context aware systems, Int. J. Ad Hoc Ubiquitous Comput. 2 (2007) 263–277.

[44] T.E. Bogale, X. Wang, L.B. Le, Machine intelligence techniques for next-generation context-aware wireless networks, Comput. Sci. Inf. Theory (2018), https://api.semanticscholar.org/CorpusID:34702660.

[45] Q. He, J. Liu, C. Wang, B. Li, Coping with heterogeneous video contributors and viewers in crowdsourced live streaming: a cloud-based approach, IEEE Trans. Multimed. 18 (2016) 916–928.

[46] G. Zhuo, Q. Jia, L. Guo, M. Li, P. Li, Privacy-preserving verifiable set operation in big data for cloud-assisted mobile crowdsourcing, IEEE Int. Things J. 4 (2017) 572–582.

[47] C. Huang, K. Xu, Reliable real time streaming in vehicular cloud-fog computing networks, in: Proceedings of the IEEE Conference on Communications in China, Chengdu, China, 27–29 July 2016, pp. 1–6.

[48] G. Grassi, P. Bahl, K. Jamieson, G. Pau, Park Master: An in vehicle, edge-based video analytics service for detecting open parking spaces in urban environments, in: Proceedings of the ACM/IEEE Symposium on Edge Computing, San Jose, CA, USA, 18–21 April 2017, p. 16.

[49] S.N. Ghorpade, M. Zennaro, B.S. Chaudhari, GWO model for optimal localization of IoT-enabled sensor nodes in smart parking systems, IEEE Trans. Intell. Transp. Syst. 22 (2021) 1217–1224.

[50] H.S. Park, M.W. Park, K.H. Won, K.-H. Kim, S.K. Jung, In-vehicle AR-HUD system to provide driving-safety information, ETRI J. 35 (2013) 1038–1047.

[51] J. Manokaran, G. Vairavel, Smart anomaly detection using data driven techniques in IoT edge: a survey, in: Proc. 3rd Int. Conf. Commun., Comput. Electron. Syst., Springer, Cham, Switzerland, 2022, pp. 685–702.

[52] A. Chatterjee, B.S. Ahmed, IoT anomaly detection methods and applications: a survey, IEEE Int. Things J. 19 (Aug. 2022) 100568.

[53] B. Nassif, M.A. Talib, Q. Nasir, F.M. Dakalbab, Machine learning for anomaly detection: a systematic review, IEEE Access 9 (2021) 78658–78700.

[54] A. Xenakis, A. Karageorgos, E. Lallas, A.E. Chis, H. González-Vélez, Towards distributed IoT/Cloud based fault detection and maintenance in industrial automation, Proc. Comput. Sci. 151 (Jan. 2019) 683–690.

[55] A. Mostafavi, A. Sadighi, A novel online machine learning approach for real-time condition monitoring of rotating machines, in: Proc. 9th RSI Int. Conf. Robot. Mechatronics (ICRoM), Nov. 2021, pp. 267–273.
[56] M. Antonini, M. Pincheira, M. Vecchio, F. Antonelli, A TinyML approach to non-repudiable anomaly detection in extreme industrial environments, in: Proc. IEEE Int. Workshop Metrology Industry 4.0 & IoT (MetroInd4.0&IoT), Jun. 2022, pp. 397–402.
[57] M. Lord, A. Kaplan, Mechanical anomaly detection on an embedded microcontroller, in: Proc. Int. Conf. Comput. Sci. Comput. Intell. (CSCI), Dec. 2021, pp. 562–568.
[58] V.M. Oliveira, A.H. Moreira, Edge AI system using a thermal camera for industrial anomaly detection, in: International Summit Smart City 360o, Springer, Cham, Switzerland, 2022, pp. 172–187.
[59] D. Pau, A. Khiari, D. Denaro, Online learning on tiny microcontrollers for anomaly detection in water distribution systems, in: Proc. IEEE 11th Int. Conf. Consum. Electron. (ICCE-Berlin), Nov. 2021, pp. 1–6.
[60] M. Cardoni, D.P. Pau, L. Falaschetti, C. Turchetti, M. Lattuada, Online learning of oil leak anomalies in wind turbines with block-based binary reservoir, Electronics 10 (22) (Nov. 2021) 2836.
[61] P. Andrade, I. Silva, G. Signoretti, M. Silva, J. Dias, L. Marques, D.G. Costa, An unsupervised TinyML approach applied for pavement anomalies detection under the Internet of intelligent vehicles, in: Proc. IEEE Int. Workshop Metrology for Industry 4.0 & IoT (MetroInd4.0&IoT), Jun. 2021, pp. 642–647.
[62] V. Tsoukas, E. Boumpa, G. Giannakas, A. Kakarountas, A review of machine learning and TinyML in healthcare, in: Proc. 25th Pan-Hellenic Conf. Informat., Nov. 2021, pp. 1–7.
[63] P.P. Ray, A review on TinyML: state-of-the-art and prospects, J. King Saud Univ, Comput. Inf. Sci. 34 (4) (Apr. 2022) 1595–1623.
[64] M.S. Diab, E. Rodriguez-Villegas, Embedded machine learning using microcontrollers in wearable and ambulatory systems for health and care applications: a review, IEEE Access 10 (2022) 98450–98474.
[65] Sheetal N. Ghorpade, Marco Zennaro, Bharat S. Chaudhari, IoT-based hybrid optimized fuzzy threshold ELM model for localization of elderly persons, Expert Syst. Appl. (ISSN 0957-4174) 184 (2021) 115500, https://doi.org/10.1016/j.eswa.2021.115500.
[66] M.A. Khan, An IoT framework for heart disease prediction based on MDCNN classier, IEEE Access 8 (2020) 34717–34727.
[67] D. Azariadi, V. Tsoutsouras, S. Xydis, D. Soudris, ECG signal analysis and arrhythmia detection on IoT wearable medical devices, in: Proc. 5th Int. Conf. Mod. Circuits Syst. Technol. (MOCAST), May 2016, p. 14.
[68] K. Ahmed, M. Hassan, TinyCare: a TinyML-based low-cost continuous blood pressure estimation on the extreme edge, in: Proc. IEEE 10th Int. Conf. Healthcare Informat. (ICHI), Jun. 2022, pp. 264–275.
[69] S.U. Amin, M.S. Hossain, G. Muhammad, M. Alhussein, M.A. Rahman, Cognitive smart healthcare for pathology detection and monitoring, IEEE Access 7 (2019) 10745–10753, https://ieeexplore.ieee.org/document/8605339/.
[70] M.O.E. Aoueileyine, Tiny machine learning for IoT and eHealth applications: epileptic seizure prediction use case, in: S. Motahhir, B. Bossoufi (Eds.), Digital Technologies and Applications. ICDTA 2023, in: Lecture Notes in Networks and Systems, vol. 669, Springer, Cham, 2023, https://doi.org/10.1007/978-3-031-29860-8_25.

[71] Y. Bodyanskiy, I. Perova, O. Vynokurova, I. Izonin, Adaptive wavelet diagnostic neuro-fuzzy network for biomedical tasks, in: Proc. 14th Int. Conf. Adv. Trends Radioelecrtron., Telecommun. Comput. Eng. (TCSET), Feb. 2018, pp. 711–715.

[72] S. Panda, G. Panda, Intelligent classification of IoT traffic in healthcare using machine learning techniques, in: Proc. 6th Int. Conf. Control, Automat. Robot. (ICCAR), Apr. 2020, pp. 581–585.

[73] M. Alhussein, G. Muhammad, M.S. Hossain, S.U. Amin, Cognitive IoT-cloud integration for smart healthcare: case study for epileptic seizure detection and monitoring, Mob. Netw. Appl. 23 (6) (Dec. 2018) 1624–1635, http://link.springer.com/10.1007/s11036-018-1113-0.

[74] F. Sabry, T. Eltaras, W. Labda, K. Alzoubi, Q. Malluhi, Machine learning for healthcare wearable devices: the big picture, J. Healthcare Eng. 2022 (Apr. 2022) 1–25.

[75] H. Gokul, P. Suresh, B.H. Vignesh, R.P. Kumaar, V. Vijayaraghavan, Gait recovery system for Parkinson's disease using machine learning on embedded platforms, in: Proc. IEEE Int. Syst. Conf. (SysCon), Aug. 2020, pp. 1–8.

[76] A. Faraone, R. Delgado-Gonzalo, Convolutional-recurrent neural networks on low-power wearable platforms for cardiac arrhythmia detection, in: Proc. 2nd IEEE Int. Conf. Artif. Intell. Circuits Syst. (AICAS), Aug. 2020, pp. 153–157.

[77] R. Zanetti, A. Aminifar, D. Atienza, Robust epileptic seizure detection on wearable systems with reduced false-alarm rate, in: Proc. 42nd Annu. Int. Conf. IEEE Eng. Med. Biol. Soc. (EMBC), Jul. 2020, pp. 4248–4251.

[78] D. Caleanu, C.L. Sîrbu, G. Simion, Deep neural architectures for contrast enhanced ultrasound (CEUS) focal liver lesions automated diagnosis, in: Proc. Int. Symp. Electron. Telecommun. (ISETC), 2021, pp. 1–9.

[79] T.M. Ingolfsson, A. Cossettini, X. Wang, E. Tabanelli, G. Tagliavini, P. Ryvlin, L. Benini, S. Benatti, Towards long-term non-invasive monitoring for epilepsy via wearable EEG devices, in: Proc. IEEE Biomed. Circuits Syst. Conf. (BioCAS), Oct. 2021, pp. 1–4.

[80] M.S. Fernandes, W. Cordeiro, M. Recamonde-Mendoza, Detecting Aedes aegypti mosquitoes through audio classification with convolutional neural networks, Comput. Biol. Med. 129 (Feb. 2021) 104152.

[81] K. Trivedi, H. Shroff, Identification of deadliest mosquitoes using wing beats sound classification on tiny embedded system using machine learning and edge impulse platform, in: Proc. ITU Kaleidoscope, Connecting Phys. Virtual Worlds (ITUK), 2021, pp. 1–6.

[82] U. Tripathi, R.S. J, V. Chamola, A. Jolfaei, A. Chintanpalli, Advancing remote healthcare using humanoid and affective systems, IEEE Sens. J. 22 (18) (2022) 17606–17614, https://doi.org/10.1109/JSEN.2021.3049247.

[83] V. Chamola, A. Vineet, A. Nayyar, E. Hossain, Brain-computer interface-based humanoid control: a review, Sensors 20 (13) (Jun. 2020) 3620, https://www.ncbi.nlm.nih.gov/pmc/articles/PMC7374399/.

[84] S.S. Poorna, K. Anuraj, S. Renjith, P. Vipul, G.J. Nair, EEG based control using spectral features, in: Proc. 2nd Int. Conf. I-SMAC (IoT Social, Mobile, Anal. Cloud), Aug. 2018, pp. 788–794.

[85] X. Zhang, L. Yao, S. Zhang, S.S. Kanhere, Q.Z. Sheng, Y. Liu, Internet of Things meets brain-computer interface: a unified deep learning framework for enabling human-thing cognitive interactivity, CoRR, arXiv:1805.00789 [abs], Oct. 2018, pp. 1–8.

[86] F. Peng, W. Peng, C. Zhang, D. Zhong, IoT assisted kernel linear discriminant analysis based gait phase detection algorithm for walking with cognitive tasks, IEEE Access 7 (2019) 68240–68249, https://ieeexplore.ieee.org/document/8708273/.

[87] G. Yang, J. Deng, G. Pang, H. Zhang, J. Li, B. Deng, Z. Pang, J. Xu, M. Jiang, P. Liljeberg, H. Xie, H. Yang, An IoT-enabled stroke rehabilitation system based on smart wearable armband and machine learning, IEEE J. Transl. Eng. Health Med. 6 (2018) 1–10.

[88] D. Takeuchi, K. Yatabe, Y. Koizumi, Y. Oikawa, N. Harada, Realtime speech enhancement using equilibriated RNN, in: Proc. IEEE Int. Conf. Acoust., Speech Signal Process. (ICASSP), May 2020, pp. 851–855.

[89] I. Fedorov, M. Stamenovic, C. Jensen, L.-C. Yang, A. Mandell, Y. Gan, M. Mattina, P.N. Whatmough, TinyLSTMS: efficient neural speech enhancement for hearing aids, Tech. Rep., Cornell Univ., Ithaca, NY, USA, 2020.

[90] A. Mitra, S.L.T. Vangipuram, A.K. Bapatla, V.K.V.V. Bathalapalli, S.P. Mohanty, E. Kougianos, C. Ray, Everything you wanted to know about smart agriculture, arXiv: 2201.04754, 2022.

[91] S. Condran, M. Bewong, M.Z. Islam, L. Maphosa, L. Zheng, Machine learning in precision agriculture: a survey on trends, applications and evaluations over two decades, IEEE Access 10 (2022) 73786–73803.

[92] S.N. Ghorpade, M. Zennaro, B.S. Chaudhari, Towards green computing: intelligent bio-inspired agent for IoT-enabled wireless sensor networks, Int. J. Sens. Netw. 35 (2) (Feb 2021) 121–131, https://doi.org/10.1504/IJSNET.2021.113632.

[93] G. Singh, A. Singh, G. Kaur, Role of artificial intelligence and the Internet of Things in agriculture, in: Artificial Intelligence to Solve Pervasive Internet of Things Issues, Elsevier, Amsterdam, The Netherlands, 2021, pp. 317–330.

[94] S.O. Ooko, M.M. Ogore, J. Nsenga, M. Zennaro, TinyML in Africa: opportunities and challenges, in: Proc. IEEE Globecom Workshops (GC Wkshps), Dec. 2021, pp. 1–6.

[95] V.K. Quy, N.V. Hau, D.V. Anh, N.M. Quy, N.T. Ban, S. Lanza, G. Randazzo, A. Muzirafuti, IoT-enabled smart agriculture: architecture, applications, and challenges, Appl. Sci. 12 (7) (Mar. 2022) 3396.

[96] Plant Village, https://plantvillage.psu.edu/. (Accessed 5 February 2023).

[97] N. Schizas, A. Karras, C. Karras, S. Sioutas, TinyML for ultra-low power AI and large scale IoT deployments: a systematic review, Future Internet 14 (12) (Dec. 2022) 363.

[98] D. Sheth, B. Sudharsan, J.G. Breslin, M.I. Ali, Embedded ML pipeline for precision agriculture, in: Proc. 21st ACM/IEEE Int. Conf. Inf. Process. Sensor Netw. (IPSN), May 2022, pp. 527–528.

[99] P. Ramos-Giraldo, S.C. Reberg-Horton, S. Mirsky, E. Lobaton, A.M. Locke, E. Henriquez, A. Zuniga, A. Minin, Low-cost smart camera system for water stress detection in crops, in: Proc. IEEE SENSORS, Dec. 2020, pp. 1–4.

[100] L. Falaschetti, L. Manoni, R.C.F. Rivera, D. Pau, G. Romanazzi, O. Silvestroni, V. Tomaselli, C. Turchetti, A low-cost, low-power and real-time image detector for grape leaf esca disease based on a compressed CNN, IEEE J. Emerg. Sel. Top. Circuits Syst. 11 (3) (Sep. 2021) 468–481.

[101] N. Gobalakrishnan, K. Pradeep, C.J. Raman, L.J. Ali, M.P. Gopinath, A systematic review on image processing and machine learning techniques for detecting plant diseases, in: Proc. Int. Conf. Commun. Signal Process. (ICCSP), Chennai, India, Jul. 2020, pp. 0465–0468, https://doi.org/10.1109/ICCSP48568.2020.9182046.

[102] P.Y. Niranjan, V.S. Rajpurohit, R. Malgi, A survey on chat-bot system for agriculture domain, in: Proc. 1st Int. Conf. Adv. Inf. Technol. (ICAIT), Chikmagalur, India, Jul. 2019, pp. 99–103, https://doi.org/10.1109/ICAIT47043.2019.8987429.

[103] S.W. Mohammed, N.R. Soora, N. Polala, S. Saman, Smart water resource management by analyzing the soil structure and moisture using deep learning, in: J. Choudrie, P. Mahalle, T. Perumal, A. Joshi (Eds.), IoT with Smart Systems, Springer Nature, Singapore, 2023, pp. 709–719.

[104] D.M. Matilla, A.L. Murciego, D.M. Jiménez-Bravo, A.S. Mendes, V.R. Leithardt, Low-cost edge computing devices and novel user interfaces for monitoring pivot irrigation systems based on Internet of Things and LoRaWAN technologies, Biosyst. Eng. 223 (Nov. 2022) 14–29.

[105] I. Ihoume, R. Tadili, N. Arbaoui, M. Benchrifa, A. Idrissi, M. Daoudi, Developing a multi-label TinyML machine learning model for an active and optimized greenhouse microclimate control from multivariate sensed data, Artif. Intell. Agric. 6 (Jan. 2022) 129–137.

[106] R. Sanchez-Iborra, A. Zoubir, A. Hamdouchi, A. Idri, A. Skarmeta, Intelligent and efficient IoT through the cooperation of TinyML and edge computing, Informatica 34 (1) (2023) 147–168.

[107] D. Loukatos, K.-A. Lygkoura, C. Maraveas, K.G. Arvanitis, Enriching IoT modules with edge AI functionality to detect water misuse events in a decentralized manner, Sensors 22 (13) (Jun. 2022) 4874.

[108] P. Andrade, I. Silva, M. Silva, T. Flores, J. Cassiano, D.G. Costa, A TinyML soft-sensor approach for low-cost detection and monitoring of vehicular emissions, Sensors 22 (10) (May 2022) 3838.

[109] M.F. Alati, G. Fortino, J. Morales, J.M. Cecilia, P. Manzoni, Time series analysis for temperature forecasting using TinyML, in: Proc. IEEE 19th Annu. Consum. Commun. Netw. Conf. (CCNC), Jan. 2022, pp. 691–694.

[110] N. Naeem, T. Boroushaki, W. Chen, Efficient Ultra Low Power Underwater Imaging.

[111] Timothy Clements, Earthquake detection with TinyML, Seismological Research Letters 94 (4) (2023) 2030–2039, https://doi.org/10.1785/0220220322.

[112] D. Tuia, B. Kellenberger, S. Beery, B.R. Costelloe, S. Zuffi, B. Risse, A. Mathis, M.W. Mathis, F. van Langevelde, T. Burghardt, Perspectives in machine learning for wildlife conservation, Nat. Commun. 13 (1) (2022) 792.

[113] S. Gallacher, D. Wilson, A. Fairbrass, D. Turmukhambetov, M. Firman, S. Kreitmayer, O.M. Aodha, G. Brostow, K. Jones, Shazam for bats: Internet of Things for continuous real-time biodiversity monitoring, IET Smart Cities 3 (3) (Sep. 2021) 171–183.

Index

D

I

J

K

L

N

O

Q

R

T

U

V

W

Z

Printed in the United States
by Baker & Taylor Publisher Services